BUDGETING, FINANCIAL MANAGEMENT, AND ACQUISITION REFORM IN THE U.S. DEPARTMENT OF DEFENSE

A volume in
Research in Public Management

Series Editors:
Lawrence R. Jones and Nancy C. Roberts
Naval Postgraduate School

Research in Public Management

Lawrence R. Jones and Nancy C. Roberts, Series Editors

*Communicable Crises: Prevention, Response, and
Recovery in the Global* (2007)
edited by Deborah E. Gibbons

*From Bureaucracy to Hyperarchy in Netcentric and
Quick Learning Organizations: Exploring Future
Public Management Practice* (2007)
by Lawrence R. Jones and Fred Thompson

The Legacy of June Pallot: Public Sector Financial Management (2006)
edited by Susan Newberry

*How People Harness Their Collective Wisdom to
Create the Future* (2006)
by Alexander N. Christakis with Kenneth C. Bausch

*International Public Financial Management Reform
Progress, Contradictions, and Challenges* (2005)
edited by James Guthrie, Christopher Humphrey,
L. R. Jones, and Olov Olson

*Managing the Electronic Government:
From Vision to Practice* (2004)
by Kuno Schedler, Lukas Summermatter, and
Bernhard Schmidt

Budgeting and Financial Management for National Defense (2004)
by Jerry L. McCaffery and L. R. Jones

Budgeting and Financial Management in the Federal Government (2001)
By Jerry L. McCaffery and Lawrence R. Jones

BUDGETING, FINANCIAL MANAGEMENT, AND ACQUISITION REFORM IN THE U.S. DEPARTMENT OF DEFENSE

by

L. R. Jones and **Jerry L. McCaffery**
Naval Postgraduate School

Information Age Publishing, Inc.
Charlotte, North Carolina • www.infoagepub.com

Library of Congress Cataloging-in-Publication Data

Jones, Lawrence R.
  Budgeting, financial management, and acquisition reform in the U.S. Department of Defense /
by Lawrence R. Jones and Jerry L. McCaffery.
     p. cm. -- (Research in public management)
  Includes bibliographical references.
  ISBN 978-1-59311-870-9 (pbk.) -- ISBN 978-1-59311-871-6 (hardcover) 1. United States.
Dept. of Defense--Appropriations and expenditures. 2. Budget--United States. I. McCaffery,
Jerry. II. Title.
  UA23.J65 2008
  355.6'220973--dc22

                                                                                    2007050760

ISBN 13: 978-1-59311-870-9 (paperback)
            978-1-59311-871-6 (hardcover)

Printed in the United States of America

DEDICATION

We dedicate this book to our wives for putting up with us, and without us, as we worked many long hours in research and writing this book.

CONTENTS

ACKNOWLEDGMENTS

The authors want to acknowledge the contributions made to the research conducted for this book by our colleagues and graduate students. Colleagues who were helpful in various ways include Phillip Candreva, John Dillard, Paul Dissing, Richard Doyle, Clinton Miles, John Mutty, Paul Posner, John Raines, Rene Rendon, and Corey Yoder. Students whose research contributed to this volume include Joshua Bigham, Richard Buell, Sean Donohue, Lina Downing, David Duma, Kory Fierstine, Paul Godek, Thomas Goudreau, Douglas Harold, Bernard Knox, Mark Kozar, Steven McKinney, Ernie Phillips, James Reed, Brian Sandidge, John Skarin and Brian Taylor, and Aaron Traver.

It should be noted that much of the information in this book was derived from ongoing dialogue with a number of Department of Defense officials to whom we have guaranteed anonymity. For this reason their names are not mentioned by specific reference in this book.

We are grateful for research funding provided through sponsorship of the Admiral George F. A. Wagner Chair in Public Management from the PEO C4I Command, and for support from the Naval Air Forces Command, without which this book could not have been written. We also wish to thank James B. Greene, RADM, USN, (Ret.), the Acquisition Chair, Graduate School of Business and Public Policy, Naval Postgraduate School, Professor Keith Snider, and the Acquisition Research Program at the Naval Postgraduate School for support for work presented in this book.

L. R. Jones
Jerry L. McCaffery
Monterey, California

LIST OF TABLES, FIGURES, AND EXHIBITS

Tables

Figures

CHAPTER 1

BUDGETING IN
THE FEDERAL GOVERNMENT

INTRODUCTION: BUDGETS AS MULTIPURPOSE INSTRUMENTS

A myth of the colonial period was that Americans could defend them-selves by keeping a rifle in the closet and grabbing it and marching off to do battle in times of crisis. Unfortunately, defense is more complicated than that now; indeed it was more complicated than that then. During the Revolutionary War, General Washington's struggles to form a standing army supported by a logistics and supply organization and to get funding for both from the revolutionary congress are well known. Defense requires planning and resourcing in advance. Reacting at the instant of crisis is too late. Moreover, production of defense goods has long lead times and involves decisions that have consequences for decades. This includes selecting and training personnel as well as designing, buying, and field-ing a vast array of ground weapons, ships, aircraft and other supporting items. Also, the decision to buy a major defense asset sets in motion a chain of decisions likely to endure for decades. For example, buying an aircraft carrier for 5 billion dollars presupposes the purchase of support ships, force defense ships, aircraft, and training of pilots and crews that will cost an additional $50 billion or more. And, it is not unusual for an aircraft carrier and other ships in a carrier battle group to have a service life of 40 years or more. Such decisions are resourced through the budget

Budgeting, Financial Management, and Acquisition Reform in
The U.S. Department Of Defense, pp. 1–33
Copyright © 2008 by Information Age Publishing

process, a planned yet somewhat disorderly system for deciding how to allocate scarce resources in a manner that culminates with congressional and presidential approval. In this chapter we examine the concept and practice of budgeting, provide an overview of the policymaking process, and analyze the most significant features of the federal government resource management system.

DEFINING BUDGETING AND THE BUDGET

In sending his proposal to create an executive budget system to Congress in 1912, President Taft said, "The Constitutional purpose of a budget is to make government responsive to public opinion and responsible for its acts" (Burkhead, 1959, p. 19). Taft noted that a budget served a number of purposes, from a document for congressional action, to an instrument of control and management by the President, to a basis for the administration of departments and agencies. Thus the multiple purposes of the budget have long been known, but no one has been more eloquent in describing their complexity than Aaron Wildavsky. In his classic 1964 book *The Politics of the Budgetary Process*, Wildavsky explained that a budget is:

1. "Concerned with the **translation** of financial resources into human purposes."
2. A mechanism for making choices among alternative expenditures ... **a plan**; and if detailed information is provided in the plan, it becomes a **work plan** for those who administer it.
3. An instrument to attempt to achieve **efficiency** if emphasis is placed on obtaining desired objective at least cost.
4. A **contract** over what funds shall be supplied and for what purposes:

   – Between Congress and the President
   – Between Congress and the Departments and Agencies
   – Between Departments and Agencies and their subunits

These "contracts" have both legal and social aspects. Those who give money expect results; those who are due to receive money expect to have the funds delivered on time to execute their programs effectively. Both superiors and subordinates have rights and expectations under such contracts, and mutual obligations are present.

5. A set of both **expectations** and **aspirations are contained in proposed budgets** from submitting agencies. Agencies expect to get money, but they may aspire to much more than they are given. The budget process regularly allows them to ask for what they aspire. What they are given in dollars reveals the preferences of others about their agency's budget. This is important information for the next budget cycle.

6. A **precedent:** something that has been funded before is highly likely to be funded again (this is defined as budgetary incrementalism).

7. A tool to **coordinate and control**: to coordinate diverse activities so they complement each other, to control and discipline subordinate units, e.g., by limiting spending to what was budgeted or by providing money to or taking it from pet political projects.

8. A **call to clientele** to mobilize support for the agency, when programs appear to be underfunded or losing ground to other programs.

9. "A **representation** in monetary terms of governmental activity." (Wildavsky, 1964, pp. 1-4)

In the American context under the Constitutional separation of powers among the executive, congressional, and judicial branches of government, the budget process begins in the executive branch of government, where the budget plan is developed, and proceeds into the legislative branch where it is reviewed, reformulated, (sometimes even rejected in total), amended and enacted. The process usually concludes in the executive branch where the President, as chief executive of the executive branch (and commander-in-chief of the armed forces), signs the appropriation bill or bills into law. At the federal level, the chief executive may veto bills he does not like, but eventually he must sign some sort of compromise bill. In contrast, most state governors have an "item veto" authority that allows them to change bills in different ways depending upon the precise nature of the veto power, for example, in California the Governor can use the line-item veto to cut money from an appropriation for a program approved by the legislature, but he cannot add money through the veto. Other governors may veto the modifying language as well, thus changing the nature and effect of a provision entirely.

Budgetary power is a shared power, employed under a system of checks and balances made possible by the separation of powers, although the Constitution is clear that the "power of the purse" rests with the Congress. Article I, Section 9, clause 7 of the United States Constitution requires that "No money shall be drawn from the treasury, but in consequence of appropriations made by law; and a regular statement and account of receipts and expenditures of all public money shall be published from time to time." This clause flows from the basic "power of the purse" granted in Article I Section 8, authorizing Congress to "pay the debts and provide for the common defense and general welfare of the United States." The first Secretary of the Treasury, Alexander Hamilton, said:

> The House of Representatives cannot only refuse, but they alone can propose, the supplies requisite for the support of the government. They, in a word, hold the purse.... This power over the purse may, in fact be regarded as the most complete and effectual weapon with which any constitution can arm the immediate representatives of the people. (McCaffery & Jones, 2001, p. 53)

Under the Articles of Confederation, Congress did try to organize itself to run the government, but this was largely a failure. U.S. history from 1790 to 1921 reflects the gradual growth of executive competence and power over the budget, capped by the Budget and Accounting Act of 1921, which directed the President to submit an annual budget and equipped him with a staff office, the Bureau of the Budget, for assistance. Subsequently the President submitted a collective budget each year for the executive branch, but Congress still held the power of the purse and feels free to put its own imprint on that budget. Congress also feels a Constitutional obligation to exercise precise and detailed oversight of how programs are administered as part of its budgetary responsibility. It does this through authorization of programs and program review in the appropriation process.

In aggregate dollar terms, Congress may not make great changes in the President's budget from year to year. However, while it often appears that Congress is a "marginal modifier," of the President's budget, Congress may take a position that is very different from the President with respect to agencies and programs, notwithstanding that the final dollar numbers are not very far apart. This is particularly true when control of the Congress and the presidency is divided. When one party holds the executive branch and the other party holds one or both houses in the legislative branch, the budget process can be both heated and extended. In the 1980s, several presidential budgets were termed "dead on arrival" when submitted to Congress, because Congress did not even consider them as a base for spending negotiation. This was basically true also for the fiscal year (FY) 2008 budget. Such a lack of confidence in a parliamentary system of government might have led to a vote of no confidence and a general election to choose a new governing party. Instead, in the United States it led to late appropriation bills, summit meetings between leaders of the executive and legislative branches, and, to some extent, policy gridlock, sometimes followed by budget process reform efforts.

Once the President has signed the appropriation bills, it is the function of the executive branch agencies to execute the budget as enacted, and not as submitted. This is not as simple as it seems. Changing conditions may lead the executive branch to try to defer spending or rescind (cut) programs; sometimes emergency supplemental appropriations are sought to fund emergencies, for example, natural disasters or military action. A continuing traffic exists in reprogramming (moving money within appropriations) and transfers (moving money between appropriations), some of which agencies can do on their own and some that Congress must approve. In short, budgeting seems to be a never-ending activity. At any point during the budgetary year, federal budgeteers must deal with three budgets: the current budget under execution, the budget for the next year under review in Congress, and the budget for the following year under preparation by agencies in the executive branch. To thin out these cycles and leave more time for analysis, some reformers have suggested that the federal government pursue a biennial budget process where budgeting is done every 2 years. This form of budgeting was found in 19 American states in the 1990s, but it has few serious advocates in Congress.

At its heart, the budget process is a planning process. It is about what should happen in the future. For nondefense agencies, this planning process may involve agency estimation of the amount of services to be provided in the next year, and to whom services will be supplied. For income security and welfare programs, planning may involve estimation of what it will take to provide a decent standard of living for the poor. For defense agencies, it may involve estimation of the consequences of U.S. foreign policy commitments and defense resource planning in terms of threat response capacity and the personnel and support resources necessary for threat management and deterrence. While numbers and quantification give the budget document the aura of precision, it is still a plan; this is most clearly evident in budget execution, where agencies struggle to spend the budget they have received in an environment inevitably changed from the one for which the budget was developed. Consequently, all budget systems provide some capacity to modify the enacted budget during budget execution, for example, fund transfers and reprogramming, emergency bills and supplemental additions of new funding.

Finally, it is important to recognize that budgeting is not done within government in a vacuum. Both in formulation and execution, various stakeholders outside of government attempt to influence budget decisions and outcomes. In the defense environment, these stakeholders range from corporations that do defense business, to state and local governments where the corporations are located, hire people, make purchases, and pay taxes, to employees and employee unions, to lobbyists and legislators who represent these and more general interests and to those who would either like to share in the defense spending pie or diminish it in order to have it spent on other policy areas. Each policy area seems to have major players. In defense, the major players exist in the DoD and the military departments, in Congress in the defense authorization committees and in the appropriations committees and subcommittees. Program advocates focus on these critical players. They articulate demands and show support for those demands. Voters, citizens, lobbyists, political action committees all help articulate demands and gather support. The various outlets of the news media are very important in seizing on issues and helping the public understand what is at stake, even if they sometimes prefer the relatively unimportant but titillating, to the important but obscure and complex.

Most observers of the American system assume that nothing is written in stone, thus information about current policy outcomes may be immediately fed back into the policymaking mechanism to help correct flaws in current policy. This is not as easy to do, as it is to say. Moreover, some things do seem to be written in stone; for example, subsidy programs and entitlements are very difficult to change. Tax laws, especially when they increase taxes, usually only get the necessary support if they are written to occur at a point in time far enough in the future that a majority of the potential taxpayers feel that they can arrange their affairs so the tax will not affect them. Thus, flawed policies do endure and are hard to change, particularly when a small, but intensely vocal group favors the current arrangement. Some focused arrangements have developed historically where the interests of a particular company or clientele, an executive branch agency and a congressional committee or committees combine to make and sustain policy favorable to those specific parties, perhaps at the expense of the public. This is true to some extent in all policy areas. These relationships are commonly referred to as "iron triangles" to denote their power.

AN OVERVIEW OF THE FEDERAL BUDGET PROCESS CYCLE

Government budgeting is a process that matches resources and needs in an organized and repetitive way so that collective choices are properly funded. The product of this process is the budget document—an itemized and programmatic estimate of expected income and operating expenses for a given unit of government over a set time period. Budgeting is the process of arriving at such a plan and executing it. Once a FY has begun, the budget becomes a plan for tracking and managing the collection of taxes, fees, and other revenues, and for distributing and disbursing these revenues to attain the goals specified in the budget. A variety of financial management functions are performed throughout the FY, in conjunction with taxing and spending in attempt to coordinate the financial activities of government, and to ensure accountability, safety, legality, and propriety in the raising and expending of public monies. At the end of the year, the budget process produces reports that allow for comparison of the achievements of government relative to the commitments made when the budget was enacted. In democratic systems, these commitments represent the will of the people as expressed during the politics of the budget process.

Budgeting and financial management are not performed in a vacuum. They are part of a public policy cycle in which (a) public service demands and preferences are articulated, (b) public policy is developed to respond to these demands and preferences by elected officials, (c) resources are generated and allocated to various public and private purposes, (d) programs and implementation strategies and tactics are developed and executed, (e) spending is incurred in the delivery of services and benefits, (f) the outcomes of policies and programs are reported and analyzed. Citizens consume this information and respond to the manner in which services are delivered and the amounts of services supplied, and again articulate their service demands and preferences to their representatives in government. In democratic political systems, it is assumed that the role of government in large part is to meet the demands and preferences of citizens with resources afforded relative to the condition of the national economy, and to do so in a manner that promotes social equity, economic efficiency, and social and economic stability.[1] Figure 1.1 depicts a simple graphic to help explain the public policy process.

This graphic helps us understand that budgeting is not done within government in a vacuum. Both in formulation and execution various stakeholders outside of government attempt to influence budget decisions and outcomes. This environment is depicted in the "stakeholder space" graphic in which the Boeing Corporation (other nongovernmental organizations or public interest lobby groups, etc., could be substituted in this diagram to reflect their input) is shown as an example of how private firms play a role in the budget process. What this diagram shows is that before policy can be made, demands and support must be articulated. A "good idea" that gathers little or no support has no chance of making it through the policymaking apparatus whereas an "average" idea that has intense support from relatively small groups may well become policy. Voters, citizens, lobbyists, political action committees all help articulate demands and gather support. The various outlets of the news media are very important in seizing on issues and helping the public understand

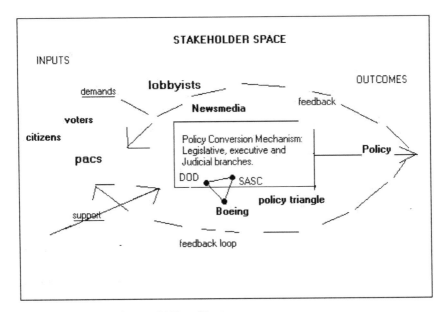

Source.   McCaffery and Jones, 2001, p. 22.

Figure 1.1.   A stakeholder space model of public policy.

what is at stake, even if they sometimes prefer the relatively unimportant but titillating to the important but obscure and complex.

The budget process is often portrayed as a process where disaggregated groups of experts make policy, experts on the appropriations subcommittees and their staffs, experts on the authorization committees and their staffs, experts in the Office of Management and Budget, and experts in the agencies. Their decisions are later tested and ratified—or rejected—by others in the budget process. This is another way of describing the "iron triangle" mechanism. The anchoring of the triangle in public space to represent interest groups, lobbies, citizens, and corporations only indicates that they too can be experts in a policy area and bring to lawmakers information about impacts or consequences that may not be available to or provided by participants within the administration or by the policy experts in Congress. This is the positive case for iron triangles. On the negative side is the worry that the iron triangle mechanism will result in unfair or inefficient or ineffective decisions. For example, in early 2007, the Pentagon's acquisition chief, Kenneth Krieg was asked "How do you break the so-called 'iron triangle' of the Pentagon, the Congress and the defense industry? Does it force you to make unwise decisions?" Krieg responded:

I worry about it greatly. I worry about getting the best product I can for the war fighter at the best price in a changing strategic environment, which means you need to be able to move in different directions. And that is really hard when people are anchored in a different world view. I have a requirement for a Lamborghini; I drive a 104,000-mile Town and Country minivan. The reality is we make trades all the time between what we want and what we buy. (Bennett & Muradian, 2007, p. 30)

Some might think the iron triangle mechanism might make Krieg's job easier, but the reality is that the military departments have their own allies at different points in iron triangle space and their influence in any particular case might be strong enough to overturn a "DoD" perspective or recommendation. The Pentagon rarely speaks with a single voice. For example, earmarks buying more aircraft or ships than requested or improving bases where no improvement was requested in the annual budget submittal are examples of iron triangle mechanisms subverting the main DoD budget submission.

An earmark to do just this was offered in the summer of 2007 in the House of Representatives. Seven house members inserted a $2.42 billion earmark into the defense appropriation bill, despite the fact that the Air Force had not requested any such aircraft (or at least they did not make it through the Office of the Secretary of Defense cuts). This was the second consecutive year that 10 unbudgeted C-17s made their way into the defense bill. Not only did the DoD not ask for the aircraft in the FY 2008 bill, it set aside half a billion dollars to stop the program. Officially the Air Force took the same view, but admitted that if it were given the aircraft it knew how to use them. Boeing had been lobbying for the earmark; it gave more than $72,000 in campaign contributions to the seven house members who sponsored it. Despite DoD's decision to close the C-17 line, Boeing said it was prepared to risk millions to keep the production line open "because the company was getting indications from Capitol Hill that orders for 30 C-17s would be coming." Danielle Brian, executive director of the Project on Government Oversight, observed that the congressional sponsors are "not ashamed of the earmarks, they're proud of them." She said that "it's part of the fellowship between the service, the contractor and their patrons in the Congress, and they work very hard not to leave anyone hanging out to dry." Brian also said that while such a large earmark might help protect thousands of jobs at Boeing and its subcontractors, it could easily pull money away from more important weapons needs (White, 2007, p. A01.) Not only is this an example of the iron triangle at work, but it is also an illustration of the different forces (and voices) within the Pentagon. We discuss this later under 'turf wars.'

To a limited extent, the struggle over the deficit in the 1990s, caps on discretionary expenditures and pay-go provisions may have changed the dimensions of this struggle so that players within each triangle are forced to limit their aspirations to taking money away from other players within the triangle, rather than preying upon the treasury at large, or another policy area. This is only, however, to a limited extent and the growing use of earmarks in the mid-2000s only illustrates a different aspect of the iron triangle game, which is how it can be driven by self-concerned interests in Congress. Also, a policy area that is blessed by the President or majority interests in Congress can be assured of an increasing budget. This is all part of the politics of the budgetary process and serves to remind that real benefits are distributed through the budget process. Citizens who are not aware of this will end up paying for benefits enjoyed by others as they wonder where their taxes went or how long it will take their children to pay down the national debt.

In a formal sense, public policy consists of a set of goals or objectives, strategies and tactics to achieve these goals, a commensurate commitment of resources, an implementation plan based on strategy and limited by resource availability, and a means for measuring, reporting and evaluating the extent to which goals have been met and other outcomes achieved. If these basic components are not present, then policy may be viewed as not properly formulated. Unfortunately, policy analysis often determines that important components are omitted in government policy development and application. Analysis of policy outcomes may reveal that the costs of service development and delivery by government are lower than the benefits achieved. However, in some cases, benefits are exceeded by costs, that is, net negative benefits have resulted from government action. Where this occurs, policy and service delivery may be questioned relative to their effectiveness. In democratic systems of governance, it is intended that such questions be addressed regularly in the public budget process.

Within stakeholder space as described above, the budget process may be understood to operate in two main phases, formulation and execution. The model of the budget process shown in Table 1.1 is useful for delineating the major phases and activities of the budgetary process.

The American governmental system is an open system, thus almost all of the stages of the budget process above must be understood to be open to the impacts of stakeholders and would-be stakeholders who seek to maximize their own goals and aspirations. Some are much more powerful than others, of course, and not all stages are equally open to all participants, but it would be incorrect to view any model of the budget process as a closed model immune from what is happening in the broader society. The audit stage is perhaps the most technical and most closed, but this stage is also open to public impact as a result of the way the other executive branch participants react to an audit or in terms of the way Congress uses audit findings as a basis for future public policy decisions.

Thus we offer our model of the stages of the budget process as a starting point for interpretation of what the process looks like when extracted from its environment.

Also, while Table 1.1 is linear and based on a single year evolution, readers must understand that at least 3 budget years are in play at any one time. As agency and department leaders prepare estimates for 1 year, for example, FY 2005 in the summer of 2003 (IA in Table 1.1), they also execute the current budget (IIA to C in Table 1.1) and testify in Congress on the FY 2004 budget (IB in Table 1.1). The budget under execution (FY 2003) has various reporting dates and deadlines, quarterly plans, and reports to both OMB and Congress as funds are allotted and obligated, and transfers and reprogrammings are made. The budget under scrutiny in Congress is the focus of hearings in a number committee and subcommittee venues, including the budget, authorization, and appropriations committees in each chamber. To some extent, what members of these committees say about budget proposals under review must be considered by the agency as it begins preparation of the budget for the following year. Hearings usually begin in February and last through May, and should be understood as part of the negotiation process noted above. As appropriation bills progress through Congress, departments and agencies begin to prepare their initial obligation plans for OMB in mid-August, with the final plan due by October 1 or within 10 days of passage of their appropriation bill. Agencies within the department have to prepare financial and staffing plans for the upcoming FY for the department budget office The budget preparation process for the following FY (BY+1: 2005)

Table 1.1. Phases of the Budget Cycle

I. Budget Formulation
    A. Preparation of estimates (executive branch)
    B. Negotiation
        1. Executive: departments, agencies, budget offices
        2. Legislative: budget, authorizing, and appropriating committees analysis and hearings, amendment and voting
C. Enactment
        3. Legislative: debate, amendment, conference committees, regular and continuing appropriations voting
        4. Executive: lobbying, signing or vetoing appropriations
II. Budget Execution (executive branch)
    A. Apportionment of appropriations to departments and agencies, Allotment within agencies and spending
    B. Monitoring and control of spending
    C. End year accounting and reconciliation to appropriations
    D. Financial audit, management audit, program evaluation, and policy analysis

---

Source: McCaffery and Jones, 2004, as modified from McCaffery and Jones, 2001, p. 3.

runs concurrently with the execution of the current budget (CY: 2003) and testimony on the next budget (BY, 2004).

More elaborate time horizons exist within defense. For example, the defense PPBE system is focused around a multiple year array of budget data in the future year defense plan or Future Years Defense Program FYDP). This display shows the current year, the budget year, and 5 future years. The FYDP allows defense planners to examine resource-planning profiles over an extended time horizon. Any budget changes that have future year consequences must be carefully tracked and changed as necessary over the course of the FYDP. DoD budget insiders remark that the first 2 years of the FYDP are of "budget quality," but the "out-years," the last 2 or 3 years, are not of budget quality. Even so, they serve as target planning estimates.

The President's Budget

Budget preparation begins with a guidance letter from OMB in late winter, is followed up with a spring preview session where the department and OMB may work certain issues at dispute, and concludes with the routines of budget preparation as dictated in OMB circular A-11, the budget preparation circular. In 1999, A-11 was issued on July 12 and contained over 580 pages of explanation, definitions, and instructions. In 1999 and 2000, efforts were made to rewrite A-11 in plain language to make it more accessible to the departments and agencies that must use it. This has been largely successful, but the budget process remains a stunningly complex process, where confusion abounds and the necessity for negotiation is obvious. Various amendments and clarifications are added to A-11 during the year as is necessary.

The executive budget process timetable has five discernable stages:

1. **April-June**: Agencies begin development of their budget requests based on prior year programs, problems and issues, and new initiatives. The President, assisted by OMB, reviews the current budget and makes decisions to guide policy for budget decisions. These may be conveyed to agencies at the spring preview or through budget instructions.

2. **July-August:** OMB issues policy directions to agencies and provides guidance for the agencies formal budget decisions. Agencies usually prepare and submit budgets for agency budget review. This may be done at several levels in large agencies and involve various hearings and negotiation stages. Ultimately, issues will be decided by the head of the agency and a final budget will be prepared.

3. **Early Fall:** Agencies will submit their budget requests to OMB. In defense the military departments go through a formal budget preparation process in the summer with hearings and cuts and appeals, arrive at a final budget and then submit it to DoD in the early fall where it is diligently reviewed. Both at the military department level and the DoD level, most issues are negotiated and solved by analysts, but some major issues can only be settled at the secretary of defense level. The defense budget is then submitted to OMB.

4. **November-December:** OMB and the President review and make decisions about agency requests and "passback" their decisions to the agencies. Agencies then revise their budget submission for inclusion in the President's budget. Agencies may appeal analyst decisions to the OMB director or even to the President when they cannot negotiate a satisfactory solution with line budget analysts. Only major issues go to the President. When the major agencies, for example, DoD, appeal issues to the President, they are expected to bring suggested solutions with them to the table, for example, they can not solve a problem by using money from another agency's budget.

5. **January-early February**: OMB and the President continue to make decisions on the budget. The President is required by law to submit the budget to Congress by the first Monday in February. At some point during this period, the decision process must give way to the needs of the government printing office; hence, the budget database is locked and the budget is printed and subsequently presented to Congress, where it undergoes a review and revision process that may last from February to November or later, notwithstanding that the new FY begins on October 1.

Each new presidential regime requires a transition year with basically ad hoc procedures. For example, in 2000, departments prepared their current services budget (with very little or no policy change) until an early amendment to A-11 was issued by Mitchell Daniels, President George W. Bush's OMB director. On February 14, 2001, Daniels ordered the departments to carry on with the Government Performance and Results Act initiative and to surface the Bush Administration priorities. This transmittal letter said, in part:

Most agencies submitted an initial version of their FY 2002 performance plan to OMB last fall. The performance goals in this initial plan were set using a current services funding level, and did not anticipate policy and initiative decisions by the new Administration. You should immediately begin making all necessary changes to your FY 2002 performance goals to reflect both the agency's top-line allowance and any applicable policy and initiative decisions. The top-line allowance will need to be translated into goal target levels for individual programs and operations. Your FY 2002 performance plans and budget materials should reflect the focus on bringing about a better alignment of performance information and budget resources...These plans should be sent to OMB at least two weeks prior to being sent to Congress to ensure that the President's decisions, policies, and initiatives are appropriately reflected. (Daniels, 2001)

We should note that this letter sent in February 2001 was intended to have an impact on the appropriation bills that would be passed in Congress in the summer of 2001 to fund departments beginning in October 2001 (FY 2002). This shows the desire of the new administration to get its priorities included in the budget as soon as possible.

Normally, for the United States federal government, each year the programs and spending that departments initially propose to begin the budget cycle are prepared and reviewed in close detail first by agency and then department budget staff. These budgets are based upon instructions from the executive so departments will have some notion of how much they may ask for in total, thus the spring preview by OMB and the midsession reviews. The instructions include policy guidance and directions about the form and format of the budget. The federal budget process did not always operate in this top-down manner, but it has done so since the early 1980s. After this review, budgets are sent to the President's OMB where hearings are held, decisions made, and passed back to the departments and appeals are heard. December is spent preparing the multiple products that comprise the President's budget, updating, and locking-up electronic databases, and preparing congressional justifications. Even in normal years, this process lasts until the end of January.

By the first Monday in February, the President is required by law to submit to Congress his proposed budget for the next FY. As part of his submittal, he delivers a myriad of exhibits, tables, graphics, and thousands of pages of text that show where revenues come from and on which programs he proposes they be spent. For example, for FY 2002, the federal budget was composed of four volumes, *The Budget of the U.S. Government: FY 2002*; *Analytical Perspectives*; *Historical Tables*; and *A Citizen's Guide to the Budget*. These are presented to Congress and widely disseminated in the media because they are supposed to make the gargantuan sums of money collected and distributed by the federal budget comprehensible to the average citizen. Unfortunately, given the complexity of the data and the general absence of knowledge and interest of the citizenry in budgeting and what the government does with money, it is doubtful that this objective is achieved.

The Congressional Budget Process

It is important to understand that Congress never appropriates the President's budget proposal exactly as it is proposed. This is because the Constitution provides the power to enact taxes and budgets to the Congress, and not the President. An old and true budget aphorism is, "The President proposes and the Congress disposes." Congress spends 8 or more months of effort each year scrutinizing the President's proposal in great detail in numerous committees and subcommittees, debating alternatives and amendments of their own origination, asking questions of witnesses they call, and listening to testimony from the President's administration and a variety of other advocates and interest group lobbyists in hearings on the budget. Congress then writes separate authorization and appropriation bills that may include substantial changes to what the President proposed, votes to approve these huge bills (typically, there are thirteen separate appropriation bills) and sends them to the President for his signature or veto.

In addition to appropriation bills, Congress may also have to pass other legislation to complete the budget plan, for example, bills that change existing tax laws, enact new ones, or modify benefit structures for entitlement programs, for example, social security.

Legislative products that affect the budget include:

- **The Concurrent Resolution on the Budget**: The budget resolution sets aggregate spending and taxing totals and estimates the resulting deficit or surplus. It also sets spending totals by functional area, for example, defense, transportation, and so on. The budget resolution is a plan; once it is adopted, Congress tries to stick to it through "scorekeeping" mechanisms enforced by points of order. The appropriation committees take the amounts allotted to them by the budget resolution and divide them up among the subcommittees that produce appropriation bills. As these bills progress through Congress, members of the Budget Committees assisted by the Congressional Budget Office keep score to ensure that the functional total does not exceed the amount assigned to it in the budget resolution. The rules of Congress call for the budget resolution to be reported out of the Senate Budget Committee by April 1 and passed both chambers by April 15, although this rarely happens. Budget resolutions usually set targets for the budget year and a number of future years, usually 3 or 4, but this has been extended to as many as 7 years to coincide with some key electoral strategy. Without doing too much violence to the concept, it may be said that the budget resolution is Congress's plan for spending and taxing, and just as much a budget as is the President's budget, although at this stage the President's budget is much richer in details. At the end of the process, the budget resolution and the appropriation bills comprise a comparable level of detail, and indeed, govern in detail how agencies will operate, at least for the discretionary parts of the budget.

• **Reconciliation Bills:** A reconciliation instruction may be added to the budget resolution to affect tax or mandatory spending changes. When this is done, it results in a reconciliation bill drafted by various committees at the direction of the Budget Committee and submitted to Congress by the Budget Committees. Almost all of the major budget and tax changes of the last two decades have come as a result of reconciliation bills, starting with the Reagan tax cut of 1981. Usually reconciliation bills do not affect defense, but remain focused on tax and entitlement matters.

• **Appropriation Bills**: Discretionary spending for the federal government is provided by annual appropriation bills, including the defense appropriation bill. The budget resolution comes first. Once it has been passed, serious work may begin on the appropriation bills. The leaders of this process are the appropriation subcommittees for each bill, for example, defense, transportation, agriculture, and so on. These subcommittees hold hearings, question witnesses from the agency or department as well as independent experts and lobbyists or other interested parties (e.g., defense contractors, Government Accountability Office), review the departmental request, listen to committee staff, and react to the chair's mark (suggested list of changes). When the chairman provides a mark, the committee generally supports that mark; after all, the chairman holds his position because his party controls a majority of votes on the committee. The bill must then stand for full committee scrutiny and then pass on to floor deliberation, debate, and amendment. When each chamber has passed its version of the bill, a conference committee is appointed to resolve differences between the two versions. As a guideline, appropriation bills are supposed to be out of the House by the end of June and enacted before the new FY begins. In general, the

House often meets the end of June test, but appropriation bills are seldom passed by October 1.

- **Continuing Resolution Appropriation (CRA)**: When no new appropriation has been passed and the FY is about to begin, Congress passes a CRA to cover the gap. The CRA provides agencies with Budget Authority to operate in the interim. The amount of money provided may be the current rate or an amount set in a bill passed by one chamber or one committee in one chamber. It is usually set at the current operating rate. For FY 2006, the initial CRA was clear to stipulate the FY 2005 rate, or the lower of the bills passed the House or Senate, or the lowest of the FY 2005 rate or the bills passed the House or Senate (For example, if the CRA were for 30 days the computation would be 30/365ths of the FY 2005 appropriation). The intention is clearly to fund at the lowest possible level. This means that no new personnel can be hired, no new programs started, no new equipment purchased, and so on. The purpose of a CRA is meant to be quite restrictive, with no or minimal new activities. This is seen in how the DoD was treated by the September 30, 2005 CRA. This specified for DoD

> no new production of items not funded in FY 2005 (the preceding year fiscal year), no increase in production rates sustained with FY 2005 funds or the initiation, resumption, or continuation of any project or activity ... for which appropriations ... were not available during fiscal year 2005.

The CRA also advised that no multiyear procurement programs could be entered into. The CRA did give the secretary of defense the authority to initiate projects or activities required for "force protection" purposes, using funds from the Iraq Freedom Fund, following notification of the congressional defense committees, normally the House and Senate Armed Services Committees and the appropriations subcommittees on defense of each chamber. In allowing purchases for force protection needs, this CRA did allow for "new" items/projects not in the current budget base, if the secretary of defense deemed it necessary, so the blanket statement can not be made that new programs are never permitted. In general CRAs are not meant to be controversial. When it votes a CRA, Congress picks the appropriate period for it, a morning, a day, a week, a month, or whatever it decides is necessary. The time chosen indicates roughly how long congressional leadership thinks it will take to come to a compromise and pass the remaining appropriation bill or bills. As individual bills pass, each ensuing CRA may cover fewer and fewer appropriations, until finally all appropriations have been provided. In some years, compromise is very difficult and an omnibus continuing resolution appropriation may be passed to include all remaining appropriation bills for the remainder of the FY. This has been the rule, rather than the exception during the last decade.

- **Authorization Bills**: Authorization bills create programs. They establish the department and its mission and any changes to it. Defense has an annual authorization cycle, but other policy areas may have different cycles for the authorization process, from 3 to 5 years to permanent authorization. In defense, annual authorizing bills may set limits on what appropriators may appropriate for the program created in the authorizing bill, but appropriators do not have to follow authorization dictates. An authorization bill does not make money available; only the appropriation bill does this. The defense authorizing committees see themselves as helping inform the appropriators on major issues, thus they try to keep the authorization bill ahead of or even with the appropriation bill in the congressional cycle. This does not always work; sometimes the authorizers get involved with treaties, test ban limits, when to commit American troops in foreign lands, and other controversial issues, with the result that the authorization bill is passed after the appropriation bill. When this happens, it is good to remember that it is the appropriation bill that provides the money. Recently the authorization bill has led in pay, personal benefits, and other quality of life issues for uniformed personnel in defense and is always of interest to defense contractors and weapon system suppliers who hope to get additional systems authorized for procurement.

In the summer of 2006, the House and the Senate approved a 3-year buy of 60 more F-22 Raptors, in their versions of the authorization bills. Usually aircraft are bought annually; what the two authorization bills did was provide for buys of 20 a year for 3 years. Multiple year buys are doable, but there are tests the weapons system must meet, related to cost savings, a stable weapon system, and a stable mission. According to GAO, the F-22 did not meet four of the six requirements to qualify for multiple year purchase. In fact the Government Accountability Office said that the multiyear buy appeared to drive costs up, not down. What made this debate particularly interesting was that the Senate approved its version of the bill 70 to 28, over the resistance of the Chairman of the Senate Armed Services Committee, Senator Warner (R-VA) and that of the ranking minority member Senator Levin (D-MI) and Senator McCain (R-AZ) the chair of the Airland subcommittee of the Senate Armed Services Committee. Senator McCain then held a hearing in late July during the conference committee time period to question the multiyear buy, a hearing at which Senator Warner appeared and said that he fully supported McCain's actions. The multiyear buy was inserted into the authorization bill during Senate floor debate on the authorization bill, lead by Senator Saxby Chambliss (R) from Georgia where the F-22s are built. The Armed Services Committee opposed the purchase, but it should be noted that not only was Senator Chambliss a member of the Senate Armed Services Committee, he was also a member of McCain's Airland subcommittee. Some observers felt that this was a triumph of constituent interest over party and committee discipline. They also noted that the F-22 buy had been capped by Secretary Rumsfeld at 183 and that the multiyear buy would extend the program into 2011, into a new administration and a new secretary of defense, thus perhaps creating an opportunity for the Air Force to renegotiate its goal of 381 F-22s (*Defense News*, July 17, 2006, "Hurdles Remain for Multiyear Raptor Buy"). The point here is that this giving guidance process is not as simple as it looks. In this case the contractor, Lockheed Martin, Senator Chambliss from Georgia, and the Air Force appeared to have formed a coalition that trumped both committee and party discipline and a settled decision by the Secretary of Defense.

- **Supplemental Appropriation Bills**: Supplemental appropriations occur when emergency needs dictate, for natural disasters and for defense needs, such as the $48 billion supplemental passed after September 11, 2001. In defense, these bills generally supply funding to replenish accounts drawn down in response to mission tasking generated by the President that was not foreseen in the annual budget such as evacuation of American citizens or of embassy personnel or by providing aid and comfort to victims of earthquakes, floods, and other natural disasters in foreign countries. Supplementals are meant to be largely noncontroversial; they allow for quick response to an unpredictable emergent need, the money used out of current funds and then reimbursed later, but still within the current FY. However, the "war supplementals" passed after 2003 are somewhat different. In 2006, despite supplementals amounting to more than $100 billion for the war on terrorism, in July the Army was in the uncomfortable position of freezing travel and hiring, and laying off temporary employees, at some bases, while other bases were running at full capacity, these inequities seemingly due to the way the supplementals were accounted for. This will be discussed later.

While no two legislative sessions are identical, benchmarks do exist. The following are suggested key dates for monitoring the legislative budget and appropriation processes:

1. First Monday in February: President sends budget to Congress
2. April 1: Senate Budget Committee reports out the budget resolution
3. April 15: Conference committee report on the budget resolution passes both chambers.
4. June 30: All appropriation bills passed by the House.
5. October 1: All appropriation bills passed.
6. Anytime: The Defense Authorization Bill precedes the defense appropriation bill.
7. By mid-August: Defense supplemental is passed (if any). If passed later than this, the supplemental may be caught up in the end of FY politics.

The final steps to coordinate bills in Congress involve appointing a conference committee of leaders from each chamber to meet and reconcile the provisions that are different in each bill. When a bill is passed in the House or Senate, the different constituencies will result in different provisions in a bill, thus it is the job of the conference committee to iron out the differences and get a unified version that will be supported by both chambers. The conference committee only exits for the time period it takes to meet and hammer out a compromise that will stand in both chambers. If it is a defense appropriation bill, the conference committee will include the appropriations subcommittee chairman and ranking member (senior member from the other party on the committee) other members of the defense appropriations subcommittee from each chamber, and these will be supplemented by key party leaders from the authorization committees, and perhaps the Budget Committees or the party leadership. Both parties are represented. Interestingly, the conference committee is not necessarily bound to what is in either bill before them; if a compromise solution that will enable passage of the bill rests on an idea or proposal not in either bill, the conference committee can include it in the report. Conference committees are the focus of intense lobbying efforts. By law, when the House and Senate versions of the DoD appropriation bill differ, DoD may submit an appeal to the conference committee favoring its position. This may or may not be successful. For example, in 2000, the House had cut $48 million from the $305 million request for DD-21 destroyer class ships and the conference committee allowed $292 million. Other interests also lobby the conference committee. In another example from 2000, the President's budget asked for four C-130 cargo planes, the House and Senate gave five and the conference committee allowed six (Congressional Quarterly Almanac, 2000, pp. 2-51). The conference committee report does have to gain a majority vote of each chamber in an all or nothing vote process. Conference committees are very powerful, but they are disciplined by the full membership when it votes on the conference committee report.

The budget and appropriations process is described in the graphic below. In general, the Budget Committees should finish their work before the appropriation committees. Both the appropriation committees and authorization committees may send views and estimates letters to the Budget Committees to help them decide how much to set aside for the programs under their jurisdiction. The budget resolution conference report includes 302a allocations for the appropriations committees. When these are passed on to the appropriation subcommittee for separate bills (defense, agriculture) they become 302b allocation targets. The Senate Appropriations Committee issued the information (Figure 1.2) in a press release on July 19, 2003. It contains the 302b allocations for the appropriations bills measured against what was enacted the previous year and what the President requested for the current year.

When reconciliation is called for in the budget resolution, the conference report also contains reconciliation instructions for the authorizing committees advising each committee of how much it is expected to save in its programs to meet the reconciliation changes. While the budget resolution is not signed by the President and does not become law, the reconciliation bill does become law and must be signed by the President first.

It is a fact of life that because Congress rarely passes budgets before the beginning of the FY, departments, and agencies often begin each FY under a CRA.

Budget Execution

After the appropriation bill has been passed and signed by the President, the process for providing spending authority to departments and agencies begins. OMB apportions money to the departments who in turn allot money to their subunits. Each agency head then uses allotments to delegate to subordinates the authority to incur a specific amount of obligations. These allotments may be further subdivided into allocations for lower administrative levels. Following these allotments and allocations, obligations can be incurred (e.g., a contract let) and outlays paid when the work or service is completed or the equipment delivered.

The apportionment, allotment, and allocation processes are the actual planning for when funds will be spent, by quarter or month and by administrative level. This process also requires departments to resubmit their budgets to OMB for approval, indicating how actual appropriations, rather than the proposals included in the President's budget, will be spent. Department requests must be approved jointly by OMB and the treasury before money is approved for expenditure and made available for obligation in department and agency accounts maintained by the treasury. In effect, the apportionment/allotment process represents a separate minibudget cycle within the executive, although its major focus is on when dollars will be spent within the FY and to a lesser extent, what the mix of consumables will be within the categories approved in the appropriations bill.

Once the appropriations have been allotted, departments allocate their budgets to their subunit agency budget staff, which then prepare and issue spending authority and guidance to the various program components where spending obligations are incurred, services delivered, and resources consumed. It is important to recognize that some agencies are very large and have huge budgets. In the DoD, the Office of the Secretary of Defense and the DoD comptroller receive and allocate the budget for national defense appropriated by Congress. Among the agencies to which funding authority is provided are the departments of the Army, Navy, and Air Force, each of which may spend more than $75 billion annually.

In allocating the budget, the central budget staffs of departments and large agencies do not completely free the programmatic side of their enterprises to spend as they wish, despite the desire for such flexibility on the part of those who spend. Rather, spending is accompanied by constant monitoring and control by central budget office staff as well as by budget, accounting, and audit staff internal to the program units. Spending is monitored in terms of actual rates versus those projected, and by other variables, including legality and purpose of expenditure (of utmost importance, spending has to conform to the appropriation and other attendant control language), schedule and timing, location, measures of production and volume, and other variables. In essence, monitoring looks for variances between planned and actual spending that then have to be accommodated through management and control.

Where funding is not available in places most needed, reprogramming or transfers (defined subsequently) are requested. Where funding is exhausted due to exceptional circumstances (e.g., natural disasters), supplemental funding requests are sent to and approved by Congress (more on this later). Moving money to the highest priority and executing the full

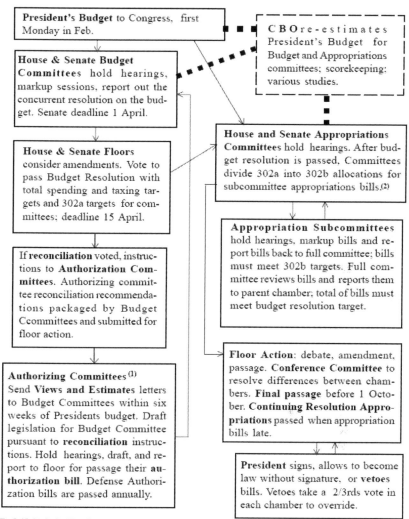

Exhibit 2.2: Budget and Appropriation Process

(1) **House Authorizing Committees:** Agriculture; Banking & Financial Services; Housing & Urban Affairs; Commerce, Economic & Educational Opportunities; Government Reform & Oversight; House Oversight; International Relations; Judiciary; National Security; Resources; Science; Select Intelligence; Small Business; Transportation & Infrastructure; Veterans' Affairs; Ways & Means. **Senate Authorizing Committees:** Agriculture, Nutrition & Forestry; Armed Services; Banking, Housing & Urban Affairs; Commerce, Science & Transportation; Energy & Natural Resources; Environment & Public Works; Finance; Foreign Relations; Government Affairs; Indian Affairs; Judiciary; Labor & Human Resources; Rules & Administration; Select Intelligence; Small Business; Special Aging; Veterans' Affairs.

(2) **House & Senate Appropriations Bills:** Agriculture & Rural Development; Commerce, Justice, State; Judiciary; District of Columbia; Energy & Water Development; Foreign Operations; Interior; Labor, HHS, Education; Legislative Branch; Military Construction; Defense; Transportation; Treasury and General Govt.; VA, HUD, and Independent Agencies.

Source: McCaffery and Jones, 2004. p. 36.

Figure 1.2.   Congressional budget and appropriation process.

amounts appropriated are the key tasks of budget execution. Execution is often taken for granted, but it is a very important phase of the budget cycle. Execution is where services are either provided or not, are provided efficiently or not, and where needs are met or neglected.

While the act of budgeting is a planning process, budget execution is a management process. In budget execution, agencies obligate or commit funds in pursuit of accomplishing their program goals. Following plans made in the budget preparation cycle, employees or contractors are engaged, materials and supplies purchased, contracts let, and capital equipment purchased. The basic assumption is that the budget will be executed as it was built once it has been approved. In many jurisdictions, this concept has legal backing; the summary numbers that appear in the budget documents stand for all the little numbers behind them. Executing the budget simply means going back to the detailed spreadsheets and following the numbers. Many also assume that this execution phase is a relatively simple task compared to preparing and passing the budget, since all it involves is returning to the budget building documents and executing the plans described in them, as modified by the final version of the appropriations bill. The reality of administrative life is somewhat different.

A substantial portion of budget execution is driven by the necessity of rescuing careful plans from unforeseen events and emergencies and unknowable contingencies. At the close of the FY, in the aggregate and on average, budget execution may appear to have been a matter of uninteresting routine dominated by financial control procedures, but it is unlikely to have appeared so uneventful to the department budget officer and his staff or the manager charged with carrying out the program. The usual occurrence is for most of the budget year to unroll as planned, but for execution of a small percentage of the budget to consume a major investment of managerial and leadership effort. All jurisdictions have set rules and procedures to help guide agencies through the budget execution process. The tensions in this process involve issues of control, flexibility, and the proper use of public funds.

The final phase of the budget process involves audit and evaluation. In this phase the disbursement of public money is scrutinized to assure officials and the public, first, that funds were used in accordance with legislative intent and that no monies were spent illegally or for personal gain and, second, that public agencies are carrying out programs and activities in the most efficient and effective manner pursuant to legal and institutional constraints.

The first type of audit is called a financial audit. It concentrates on reviews of financial documents to ensure that products and services are delivered as agreed on, that payment is accurate and prompt, that no money is siphoned off for personal use, and that all transactions follow the legal codes and restrictions of the jurisdiction. Such audits are generally carried out or supervised by agents external to the entity undergoing audit. These agents include the Government Accountability Office and agency inspectors general at the federal level as well as internal auditing agencies (for example, the Navy Audit Service), elected or appointed state auditors at the state level and by private sector accounting firms at all levels of government. Some accounting occurs between levels of government as the federal government audits use of federal monies in state and local programs and as state governments audit certain local fiscal practices.

Financial audits evaluate honesty and correctness in handling money. As the role of government expanded after World War II, more effort was spent to measure the efficiency and effectiveness of government. This led to experiments with performance budgeting at different levels of government and ultimately to performance auditing. Here the auditor attempts to ensure that the agency is conducting programs in a manner consistent with the laws and regulations that authorize the program and to ensure that the agency has taken judicious action in resource deployment to attain programmatic ends. Basically, the auditor is attempting to judge efficiency in resource use and effectiveness in program delivery. Findings from such audits help policy makers enhance program outcomes while minimizing the resources required to operate the programs. Often the same agencies responsible for financial audits also do program audits, but since the audit focus differs, personnel with different skills are required. The reports often become part of the budget process, especially during the legislative stage, and may lead to changes in the laws and rules that guide the program and to the managerial practices of the agency that administers it.

Audit findings may also be reviewed in special hearings by oversight committees, outside the budget process, who want to ensure that agencies behave responsibly. Generally, these audits are published and available to the public. Both financial audits and performance audits help ensure that a jurisdiction is getting the most for its tax dollars. In so doing, they help maintain the vital element of trust in government and in those who disburse its monies, and create and administer its programs. In general, the same entities who audit for propriety may also be called upon to audit for performance, including the Government Accountability Office at the federal level and legislative audit agencies at the state level. OMB also has a management review staff and at state and local levels, the budget agencies often require that budget analysts combine managerial analysis work with budget analysis.

After all of this activity concludes at the end of the FY and accounts are closed-out to prevent further obligation, budget accounts are audited internally and externally by agency auditors, Inspectors General, the Government Accountability Office and other audit agents. In some cases, private firms conduct parts of or all of these audits. After or in conjunction with auditing, programs are evaluated to see to what extent they have met the objectives and commitments promised to Congress and the executive, and policies are analyzed for their value. All of the information thus developed becomes grist for the mill for preparation of future budgets, as the cycle operates continuously.

A final note is required regarding the revenue side of budgets. In general, it is only at the top level of the budget staff of Congress, the Budget Committees and Congressional Budget Office and the executive OMB for the President, where revenue and expenditure estimates for the upcoming year are totaled based upon what Congress passes in appropriations. Simultaneously, as the President and Congress propose and enact the spending plan for the upcoming year, expenditure budgeting management and control of the current year budget is the pervasive activity that occurs throughout the executive branch departments and agencies as organizational subunits compete for their share of funds. Ordinarily, departments and agencies do not concern themselves with revenues other than fees and charges they are allowed to retain in current or future year budgets, and others they collect incidental to their own operation, for example, national park admission fees or federal license fees.

For the most part, departments and agencies compete for their share of general revenues accumulated from the tax mechanism. For example, the DoD and the Department of Health and Human Services do not concern themselves much with general government tax policy and revenue generation. Instead, they rely on Congress and the President to tell them if their needs can be accommodated within revenues approved for the current and the following budget years.

BUDGET FORMATS

The budget of the federal government, indeed of all governments, has many forms. It can be rendered simply as a single line with a few descriptors (e.g., agency name) and funding proposed or available for spending, or as complexly as is imaginable, containing all kinds of programmatic and even performance—related information and associated funding. Essentially, the only constant in budgets is that funding (dollars in the United States) is shown in nominal values, that is, not discounted over time.

All of the different budget formats have their particular uses, as is the case for the myriad of ways in which expenditure plans and programs are displayed. For example, the display by function for defense includes expenditures in other departments that is defense-related, for example, defense related expenditures in the Department of Energy, civil defense programs in FEMA (Federal Emergency Management Agency), Selective Service programs related to the draft, and defense-related activities of some other agencies including the coast guard and the FBI (Federal Bureau of Investigation). The display by appropriation for defense for FY 2002 is what the appropriations committee has given to its defense sub-committee, pursuant to targets given it by the Budget Committee. The total for 2002 in the departmental display is the amount the President's budget recommended for defense for FY 2002. These numbers are different because they measure something different (function versus agency) or the same thing at a different point in the budget process (President's budget recommendation versus defense appropriation bill, see Table 1.2).

That the format of the budget depends on its intended use comes as no surprise; what is surprising is that the budget is used in so many different venues and contexts. To understand why this is so requires considerable familiarity with the decision process cycle employed to determine and manage resource allocation.

Why are so many formats—probably too many—used in federal budgeting? Technically, the fundamental task in budgeting is twofold; it lies in predicting the future and evaluating the past. Visions about the future are complicated by differing visions of what the future should be, as well as arguments about what it will be given current understandings of causal relationships. Since government is a coercive arrangement and can extract money and time from its citizens and deliver goods and services unequally, say for example more to those who are more successful in lobbying and legislating, budgeting tends to be a conservative process where players examine, in excruciating detail, the evidence on decisions that have been made and information about decisions that are about to be made. Consequently, the great problem in budgeting is information overload. Reformers have attempted to address this problem in a variety of ways, with the executive budget movement, the creation of central budget bureaus and legislative staff agencies, public hearings, and published documents describing the budget. Much of this effort involves

**Table 1.2.   Three Views of Defense Budget:
By Function, Agency, and Appropriation Bill**

	2002 Estimates
National defense function	$319 billion
Defense as an agency	$310 billion
Defense Appropriation Bill (House 302b)	$300 billion

a rationalistic sense that a better budget process results from better information.

Reformers have also attempted to solve the information overload problem by creating better budget systems. Historically, the basic budgeting system has been an object of expenditure system, which features objects of expenditures, arrayed in lines of items in spreadsheet format with their names and proposed expenditures, hence the name line item budget. The items are usually personnel and supporting expenses, like travel, telephone, and offices supplies. In its early form, this type of budget format was closely identified with the accounting system, thus making it relatively simple to execute and audit. In small jurisdictions, with few functions and a limited and homogeneous role for government, this type of budget is still simple to construct, review, execute, and audit.

The format of a line item budget presentation includes the line items, what each category cost the previous year, its authorized level in the current year, and the amount requested for funding in the budget year. If this is a biennial presentation, 2 budget years will be shown. Usually, these documents will also show the percentage of change from the current year to the budget year, thus enabling reviewers to scan the presentation and select objects for review, for example, that which increased the most. This type of budget emphasizes control and fiscal accountability, rather than management or planning (Schick, 1966). Usually the budget document will have introductory narrative explaining the purposes of the unit; this usually changes minimally from year to year and is almost universally referred to as "boilerplate." Budget officers feel that once they have their narrative right, it is a waste of effort and time to do anything other than fine-tune it with this year's emphases. Most, but not all, line item systems also will have some explanation of changes, usually very brief, and very carefully written. Writers know that readers will not read long justifications, but they also know that the justification will focus how the budget request is perceived. In many budget shops, the less experienced budget analysts write the boilerplate and the experts write justification paragraphs.

At the agency level, the line item display usually carries with it a list of all personnel employed in the unit and their salary cost. This total is transferred to the personal services line on the budget display. Supporting expense categories will be displayed by item, for example, postage, travel, office supplies, and so on. A list of minor capital outlay items (desks, chairs,) will be constructed; this may or may not accompany the budget, but the agency budget analyst will use the exhibit to monitor the agency purchases during the budget year. It is easy to see how this type of budget lends itself to control purposes, both for accounting and for managerial control. It is also easy to see that this type of budget would not work well for large and complex organizations, thus DoD uses its own system.

TYPES OF BUDGETS

Budgets may be divided into three basic types: operating, capital, and cash. The operating budget contains funds to be spent on short-term consumables including the personnel payroll and the goods and services that keep government agencies in business on a day-to-day basis. The capital budget plans for and purchases long-term assets such as land for parks and buildings. The cash budget is used to manage the cash flow of a government or government agency.

The Operating Budget

The operating budget includes salaries and benefits for employees, funding for utilities, money to contract for trash management, maintenance and supplies, space rental and lease payments, pens, pencils, copy machine paper and so on. Typically, employee salaries and benefits are the highest cost items in the operating budget, representing anywhere from 60% to 90% of spending. Personnel are usually listed by position by type and grade and perhaps seniority, with a certain percentages added on to cover the cost of fringe benefits including vacation, sick leave, health, life and disability insurance, and retirement plan costs.

The operating budget funds the annual operational needs of the jurisdiction, ranging from road maintenance, to park and recreation supervision, tax collection, education, welfare, public safety, and defense. The time period for spending the funds in the operating budget is almost always 1 year at state and local levels. At the federal level, the time period of obligational availability varies depending on the type of appropriation.

The operating budget usually includes minor capital outlay amounts for desks and office machines and the like, below a specified item cost threshold, for example, $10,000. Consumable items range from computer parts, to telephone installations, to pencils. In line-item budgets, the line includes the name of the item, the budgeted amount for the current year and a requested amount for the next year. Many jurisdictions also include what was spent on that item the previous year. Revenue sources may be broken into equally excruciating detail, not only general taxes, for example, sales, property, and income, but also revenue from fees, charges and miscellaneous sources, including parking fines, dog and bicycle licenses, and garage sale permits. When the line-items are reorganized into programs, the budget emphasizes program accomplishment and is called a program budget. A performance budget changes the mix so that consumables are attributed to activities accomplished and a cost may then be attributed to each activity output, such as the cost to construct a mile of road or the cost to collect a ton of refuse. The same budget may be displayed in different ways; the defense budget is commonly displayed by appropriation for congressional purposes and force structure and mission within the DoD.

The Capital Budget

The capital budget contains funding for purchase of long-lived physical assets such as buildings, bridges, land for parks, and high cost equipment such as aircraft, ships, and so forth. Many, but not all, state government and other jurisdictions in the United States have separate capital and operating budgets. The federal government does not have a separate capital budget, but it has distinct capital accounts, mostly used to pay for acquisition of military hardware in the defense department. The capital budget is usually used for long-term investment type functions, including buying parklands and constructing buildings, bridges, and roads, where consumption spans over a period of years or decades. Thus, the capital budget is appropriated and consumed on a multiyear basis. In state and local governments, capital projects are usually paid for through issuance of bonds whose principal provides the money for the capital budget projects. These bonds normally are paid back over a time period that approximates the consumption of the asset, for example, 30 years. In contrast, the U.S. federal government appropriates money for capital consumption out of the annual operating budget, and obligates and outlays it over multiple year terms as the project is executed. This is how the interstate highway system was funded and built in the 1950s and 60s. On the defense side, weapons systems such as aircraft carriers and certain aircraft have enjoyed lifespans measured in decades, thanks to outstanding initial design and continuous modification processes. Both initial purchase and subsequent modifications were funded by annual appropriation bills.

It also should be noted that capital and operating budgets must be linked to provide services. Where the capital budget pays for construction of an office building, it is the operating budget that pays for the furniture and equipment to make the building ready for use, and for daily activities of maintenance in and around the building. This is true both in the defense and nondefense sectors. In defense, a new aircraft will have to have pilots, a pilot training program, weapons, hangar space, and spare parts funded out of the annual operating budget. On the nondefense side, the capital budget will prepare a project that will be taken over and operated out of the operating budget, including maintenance personnel, paint and other maintenance and other items, such as snow removal for highways. Good budget analysis always means checks are made to ensure that capital projects scheduled to come on line are fully funded on the operating side.

Cash budgets are used to manage the cash flow demands of operation on a daily, weekly, monthly, and annual basis. The objective of cash budgeting is to insure the liquidity and solvency of government and its agencies. Liquidity may be defined as having the cash and readily convertible assets that are used to pay the government's bills, that is, its short-term liabilities. In contrast, solvency refers to the ability of government to sustain operations over the long-term, that is, greater than 1 year. Typically, the cash budget is managed in close coordination with the investment functions of the Treasury Department or other cash controlling entity, for example, the comptroller or department of finance. The task for proper cash management is to have enough cash available to cover current liabilities, but not too much, so that the opportunity to invest and earn interest from surplus cash is wasted. In the private sector, what is termed the "quick ratio" is used as a measure of cash flow management. A ratio of 1.1 or 1.2 cash on hand to 1.00 liabilities is appropriate. Ratios lower than this imperil the ability to meet payment obligations; a higher ratio wastes investment opportunity.

PERSPECTIVES ON BUDGET THEORY

Scholars in political science and public administration tend to examine budgeting in terms of who gets what, why, and under what conditions. Those in public administration typically are concerned with how the budget is proposed, decided upon, how services are produced, and how budget information is presented and analyzed. They take great interest in the internal workings of government and governance, a propensity shared by political science from which public administration was born. Matters of policy and program effectiveness are of primary concern from this perspective. Public administration attempts to improve the welfare of citizens through attempts to make governance more responsive and government more effective. Often, their prescription for improving effectiveness includes employing more staff and spending more money to achieve unequivocally reasonable and desirable objectives.

Criticism of the public administration perspective on budgeting is that it is too focused on process, overly concerned with function versus output and outcomes, and obsessed with the details of congressional, presidential, and agency decision-making and service production, to little consequence. However, it may be pointed out that many of the reforms that have made governance more responsive and governments more effective in the past 100 years or so are wholly or in part attributable to reformist efforts led by scholars in public administration.

Economists tend to examine budgeting from the perspectives of equity (who pays and who benefits, that is, the distributive consequences of tax and allocation decisions), allocative efficiency, and stability. Our interpretation of this perspective, generally speaking, is that it assumes that the role of government is to assess and determine the validity of arguments made by various claimants for shares of the distribution of public money, that is, to define equity in practice. Equity in the context of budgeting means what is "fair" given that how we define "fairness" is always a debatable question in a democracy, and that some degree of competition for resources is inevitable in any socioeconomic system, because demand always exceeds supply. How we define what is fair varies over time and is always a primary consideration in the numerous forums of government decision making.

To some, distributional equity refers to policies that transfer income from one set of citizens to another through tax policy, spending, and other government actions. Others vehemently disagree with this perspective and generally do not acquiesce to the redistributive role of government. Clearly, the definition of what is fair in a democracy is virtually always up for grabs—a work in progress—ever changing relative to the political will of the people and their elected representatives. We are quite aware that fair is sometimes, but certainly not always, defined as equal. But, given variances in need, income, wealth and other variables, distribution of spending based on an "equal share for all" rule would, in many instances, not be judged as fair.

To economists, efficiency means how the decisions of the government affect the productivity of the private sector and the economy as a whole. Efficiency in this context is not concerned with whether the internal operations of government operate in a managerially efficient manner. Rather, efficiency is determined in essence by economic decisions about what should be produced, how and by whom. The means of production and what is produced are determined through the interaction of the public and private sectors of the economy, and in choices over which goods and services should be provided by government and which are better provided by the private sector. To some, the government is regarded as a net drag on the private economy, causing the sacrifice of efficiency in pursuit of equity objectives. From this view, the best government is that which governs least. To others, the essence of the role of government is to supply equity, a good that cannot and will never be supplied purely as a function of the pursuit of efficiency in the private sector. The mixed capitalist-socialist form of economy that prevails in the U.S. represents a compromise between these polar perspectives. As such, the trade-offs that are forced constantly between equity and efficiency are the very stuff that makes the government budget process intensely competitive.

From an economic perspective, stability refers to policies pursued through the budget to stabilize the economy, in conjunction with the fiscal and monetary policies of the government and the actions and productivity of the private sector. Stability is measured in terms of prices, employment, growth of the gross domestic product and other indices. Just as trade-offs are made between equity and efficiency, inevitably potential trade-offs may be considered between each of these and stability under some circumstances. Balancing these exchanges is in part a result of decisions made in the ongoing cycle of government budgeting and financial management. We subscribe to the perspectives of both political science—public administration and economics. Our hybrid view attempts to draw in an interdisciplinary manner on both perspectives, that is to say that neither view is wrong—but that each emphasizes different aspects and ways of understanding budgeting and spending outcomes. Now that we understand some of the basics of budgeting, we need to understand something about the history of federal government budgeting and financial management. Then, in chapter 3 we consider the intricacies of defense budgeting.

NOTE

1. Different schools of thought exist on how individuals and groups gain and exert power in society. In this book we pursue an institutional model embedded in a pluralist model of democracy. This is our way of saying that groups and individuals have access to the institutions of governmental power and use that access to govern through the distribution of budget (and tax) policy. We believe there is no set of dominant groups that always controls American society and that it is not controlled by the elite few who control and profit from the efforts of the many. We believe laws, institutions, and individual behavior matter and that individuals are well-advised to gather into like-interested groups to affect political action. There are other schools of thought on these matters, from those who would argue that a focus on institutions is too narrow, that groups control everything and the individual is powerless, or that elites rule behind the scenes and what is seen publicly is a charade. There are also schools of thought about the policy process, that it is rational or incremental or that game theory or public choice theory provide better explanations for how decisions get made or can be explained. In these debates, we tend to be midlevel rational focusing on the science of muddling through, the virtues of incremental behavior and the necessity to work the numbers. We suggest that this is typical for budgeteers. For those interested in the larger debates about society and power, we suggest starting with the work of Thomas Dye, S. M. Lipsett, or the classical political theorists, for example, Plato, Aristotle, Machiavelli, Hobbes, and so forth.

CHAPTER 2

HISTORY AND DEVELOPMENT OF FEDERAL GOVERNMENT BUDGETING

Executive and Legislative Branch Competition

Defense spending has been at the center of budgetary history before this country was born. Taxing for defense of the colonies (against the French and along the borderlands) led to resentment of the King and charges of taxation without representation. Staggering debts were accrued during the Revolutionary War and the existence of the new nation was threatened if it could not develop a stable currency and pay off its war debt. Debate occurred over what the new President's budgetary power should be and Congress first tried to run the country and provide for its defense without an executive branch. Later, with the beginnings of a budget system, arguments transpired over whether lump sum or line item appropriation was better, over the idea of control and how much was appropriate, particularly for important functions like the Army and Navy. Presidents did not have a budget office until 1921 and different Presidents took different views of the budget power. This notwithstanding, crises continued to occur and with them came stress for the fiscal system,

Budgeting, Financial Management, and Acquisition Reform in
The U.S. Department Of Defense, pp. 35–73
Copyright © 2008 by Information Age Publishing
35

especially after World War I and World War II that resulted in budget reforms and responsibilities. It is clear that the Constitution gives the President great latitude to act as commander in chief when the nation is attacked or threatened, but Congress also has been given the power of the purse as a check on the power of the President. By and large this sharing of power has resulted in satisfactory outcomes as the President was able to act swiftly in the nation's defense (e.g., President Truman's commitment of troops to Korea in 1950) and then explain and justify his actions later to Congress. There have been times however when this sharing of power has been tested, in the latter years of the Vietnam conflict and in 2006 and 2007 when Congress sought to direct how the President should bring the war in Iraq to a successful conclusion by including detailed guidance in the supplemental appropriation bills, especially in 2007. In this chapter we review some of this history to explain how we got where we are presently.

Elsewhere we have suggested that a review of the history of budgeting in the United States reveals debate over two prominent questions: how the budgetary power should be divided between Congress and the President, and how the budget be employed as a tool to better govern and manage.(McCaffery & Jones, 2001, chapter 2) The question of how power is shared between Congress and the executive branch is particularly important for defense, since confusion arises over the correct division of power between the President who must function as commander in chief and Congress which has been given the power of the purse by the Constitution. The struggle for power (or over how best to discharge their lawful responsibilities) between the executive and the legislative branches has been a recurrent theme in the American budget process since the founding of the nation. Shortly after the Revolutionary War, Congress appeared to have taken the initiative in competition over the proper role of the two branches in debating whether the executive was legally empowered under the Constitution to make budget estimates, and whether the secretary of treasury could or should submit a budget framework to Congress (Burkhead, 1959, p. 3). Members of Congress were opposed to having the secretary of treasury even submit plans to Congress for the following fiscal year (FY) (Browne, 1949, p. 12). In contrast, beginning with the Budget and Accounting Act of 1921, by the late 1960s the steady accretion of power in the executive branch, and within it in the Bureau of the Budget, provided the chief executive significant leverage in the budget process. presidential use of impoundment power and the politicization of the Bureau of the Budget essentially consolidated gains for the presidency over Congress, and over the executive branch as well.

However, just as nature abhors a vacuum, the American federal system abhors an imbalance of power. The Congressional Budget Impoundment and Control Act of 1974 reestablished Congress's role in the budget process. In 1980, Congress used the provisions of this Act to rewrite the President's budget. While President Ronald Reagan used the reconciliation instruction from the 1974 Act to seize a great victory in Congress in 1981, Congress again rewrote the President's budget in 1982. The years since 1974 have witnessed a reversal of form of most of the twentieth century practices that had seen Congress increasingly relegated to the role of making marginal and incremental changes in the President's budget. Since 1976 Congress has had its own budget plan to contrast to the President's budget request. Notwithstanding this, any enlightened observer would have to cede the weight of power to the executive branch and within it the President's budget office (the Office of Management and Budget). Such an observer would also have to conclude that congressional procedures have not led to a timely or orderly appropriations process and that no 2 years are alike. Allen Schick (1990) has termed this "improvisational budgeting" (pp. 159-196). Indeed in recent years, Congress sometimes has not been able to agree on a budget plan. However, even given all this turbulence, there is no doubt but that Congress has legitimate power over the purse and intends to exercise it. In retrospect, the Republican Congresses of 1995 and 1997 seized much of the initiative from the President and drove much of the budget planning for a balanced budget with 7- and 5-year budget plans.

In 1789, arguments over the balance of power centered on the extent to which the President should prepare a budget; and although this is no longer debatable, the limits to the advantage that preparation gives the executive branch still is arguable. This theme is still as current as it was in 1789. Within this recurrent theme is an idea central to American democracy: power is balanced at the national level between the executive and legislative branches. When a serious imbalance occurs, corrective action ensues to restore and ensure the balance so that when one side is a leader the other remains a powerful modifier.

During the first century of the nation's existence, simple forms seemed sufficient for simple functions, a premise that held true through the opening decades of the twentieth century. Then, as the functions and responsibilities of government expanded, changes were made in budget technology and technique. This seems to be a linear and expanding process, with more reforms attempted in the last 50 years than in the first 175 of the American experience. Nonetheless, those in charge at the earliest stages of budgeting in this country recognized the need for different budget forms. As early as 1800, civilian agency budgets were presented in carefully detailed object-of-expenditure form, while military expenditures tended to be appropriated as lump sums not unlike specific program categories. Early debate on budget development focused on flexibility and program accomplishments rather than on strict agency accountability.

Early American budgetary patterns were both part of and separate from their predominantly English colonial heritage. They were part of that heritage in that the American colonies inherited the full line of English historical experience with a limited monarchy and expanded legislative powers. This historical legacy may be dated to 1215, when a group of dissident nobles forced the king of England to accede to and sign the Magna Carta. Of the 69 articles in this document, the most important is that which stated, "No scutage on revenue shall be imposed in the kingdom unless by the Common Council of the Realm" (Caiden & Wildavsky, 1974, p. 25) The Common Council preceded Parliament, and the statement that revenue could be raised only with the consent of a legislative assembly remained constant. This is often hailed as a beginning of popular government, but it is useful to note that this was basically a sharing of power between the king and the most powerful nobles in the realm, the two top tiers in an elite-dominated society where status was conferred mainly by birth. Nonetheless, by the end of the thirteenth century, the principle was established that the Crown had available only those sources of revenue previously authorized by Parliament.

In England, the Magna Carta was only the beginning of a long process of movement toward popular government, a process completed in the twentieth century when the House of Lords lost the power to reject money bills. By the middle of the fourteenth century, the House of Commons was established and its leaders realized that a further check upon the power of the king would result from legislative control over appropriations. At first, revenue acts were phrased broadly, and once the money authorized was available the king could spend it as he wished. However, Parliament began to insert appropriation language in the Acts of Supply and other similar legislation, stating that the money be used for a particular purposes. Moreover, rules were made for the proper disposal of money, and penalties were imposed for noncompliance (Thomson, 1938, p. 206). Consequently, by the middle of the fourteenth century, fiscal practices included a check on the Crown's right to tax and spend; bills from Parliament carried notice of intent designating what money was to be used for, rules for disbursement of money, and penalties when rules were not followed.

Refinement of this system would take centuries, and its progress was not linear. Some kings were more skillful, personable, or powerful than others, and Parliament's role manifested steady evolution only in the most general terms. Wildavsky (1975) suggests that if a benchmark is needed, formal budgeting can be dated from the reforms of William Pitt the Younger (p. 272). As Chancellor of the English Exchequer from 1783 to 1801, Pitt faced a heavy debt burden as a result of the American Revolution. In response to this, Pitt consolidated a maze of customs and excise duties into one general fund from which all creditors would be paid, reduced fraud in revenue collection by introducing new auditing measures, and instituted double-entry bookkeeping procedures (where each transaction is entered twice, as a credit to one account and a debit to another). Moreover, Pitt established a sinking fund schedule for amortization of debt, requiring that all new loans made by government impose an additional 1% levy as a term of repayment (Rose, 1911a, 1911b, 1912). Pitt raised some taxes and lowered others to reduce the allure of smuggling. The legacy Pitt left was a model that encompassed a royal executive with varying degrees of strength and a legislative body attempting to exert financial control over the Crown by requiring parliamentary approval of sources of revenue and expenditure. Approval was provided through appropriations legislation. In this way, administrative officials ultimately were held accountable to Parliament (Browne, 1949, p. 15).

The history of the American colonies has been described as a replication of the struggle between Parliament and the Crown, with the colonies, like Parliament, gradually winning a more independent position (Labaree, 1958, p. 35), even before the Revolutionary War. For example, the colonies turned the power of the purse against the English royal governors. Colonial legislatures voted the salaries of governors and their agents, appropriating them in annual authorizations rather than for longer periods. Indeed, one colonial governor's salary was set semiannually. In theory, the royal governors had extensive fiscal powers, but in fact these powers were often exercised by colonial legislative assemblies. These included raising taxes, appropriating revenues, and granting salaries to the royal governors and their officers. Caiden and Wildavsky (1974) remark that the colonists were thoroughly in the English tradition of denying supply (budget dollars) to the colonial governors to force compliance with the will of colonial legislatures. Not only were salaries voted annually, but taxes were also often reenacted annually. "Royal Governors were allowed no permanent sources of revenue that might make them 'uppity.' " (p. 25). Appropriations were specified by object and amount and appropriation language was used to specify exactly what funds could and could not be used for, for example, "no other purpose or use whatsoever" (p. 26). Royal governors even were prevented from using surplus funds or unexpected balances; these were required to be returned to the treasury. Various mechanisms were used to impose further restrictions on the power of the royal governors. In some colonies, independent treasurers were elected to manage funds. Several colonies required legislative approval prior to disbursement of funds; in emergencies this might necessitate a special appropriation. Some colonial legislatures appointed special commissioners accountable not to the royal governor but to the legislature as a further check on the power of the governor. Caiden and Wildavsky (1974) conclude that "power, not money was the issue" (p. 32). Thus, from the earliest days, budget decisions in the American colonies focused around the issue of the correct balance of power between the colonial legislatures and royal governors, a discussion that continues to occur every year at the national level, but rarely at the state level.

Generally, neither expenditures nor taxes were heavy during the colonial period; England did not extract much revenue from the colonies except in periods of war (Browne, 1949, p. 16). What was troublesome to the colonists was that the Crown could and did impose duties and excises intended to regulate trade and navigation without the colonists' approval, hence the revolutionary complaint of "taxation without representation." Up to the Revolutionary War, the colonies followed the British pattern of a gradually developing budget power. Decisions about taxation were paramount, and the exercise of budget power was basically sought as a check upon royal power. Development of the instruments of taxation, appropriations, and accounting were all evidenced in this pattern, but a formal budget system did not yet exist.

The colonies departed from English tradition when they gained independence after the Revolutionary War. Like Pitt, the founders of this new nation faced a heavy debt burden; unlike Pitt their primary concern seemed to be focused on creating a country that could operate without an executive branch in a decentralized format almost dependent upon the voluntary contributions of the individual colonies. Under the Articles of Confederation, a fiscal system was created in which there was no executive branch. Power was vested in various legislative arrangements. As a result of fear of central government inherited from their experience with the British, the powers of the first Congress established under the Articles of Confederation were very weak. In fact, this fear was evident in the manner in which powers were delegated to both the legislative and executive branches as the Constitution was drafted. Also the colonists were averse to a system of national taxation. Taxation imposed a special hardship on the colonies because hard coinage was scarce and bills or letters of credit were used irregularly. Consequently, the colonies were chronically short of cash and coinage schemes abounded. During the Revolutionary War, borrowing and promising to pay either with bills of credit or by coining paper money became endemic as the colonists pursued the war and made expenditures without tax revenues. Washington's continuing struggle to adequately equip his armies is well known, with the winter at Valley Forge standing for all time as a symbol of heroic efforts to contend with a new nation's ineffectual and rudimentary governmental systems.

The costs of war were great. Thomas Jefferson estimated the cost at $140 million from 1775 to 1783. By contrast, the federal government operating budget in 1784 was $457,000. Bills of credit were issued both by the states and by Congress from 1775 to 1779. Bills of credit rapidly depreciated. In 1790 Congress was forced to admit that a dollar of paper money was worth less than two cents and passed a resolution to redeem bills of credit at one fortieth their face value (Dewey, 1968, p. 36). Paper currency did not become legal tender again until after the Civil War.

The Articles of Confederation provided that revenues were to be raised from a direct tax on property in proportion to the value of all land within each state, according to a method stipulated by Congress. These limitations upon congressional taxing power left it dependent upon the states. Congress was not disposed to provide effective fiscal leadership to states, in part because Congress was debating issues related to its own budgetary procedures and its leverage vis a vis the states regarding fiscal power. Congress was attempting to act as both the executive and legislative branch in a system where the preponderance of power was held by the individual states. Not only was this a departure from the English tradition but, it was a model of government that would be short-lived in this country.

Constitutional government began, then, with a long history of British practices further shaped by both the inefficiencies of the Confederation and the cost of the Revolutionary War. If American institutions were influenced by an antiexecutive trend, they were also affected by the chaotic nature of legislative government under the Articles of Confederation. This period was marked by extraordinary negligence, wastefulness, disorder, and corruption, as Congress in its committees prepared all revenue and appropriation estimates, legislated them, and then attempted to exercise exacting control over accounts (Bolles, 1896/1969, p. 358). As Vincent J. Browne (1949) observes:

> Until the framing of the Constitution, the future of the States was almost as much imperiled by financial indiscretions as it had been previously jeopardized by the forces of George III. (p. 17)

Legislative dominance began to give way when the Continental Congress created the post of Superintendent of Finance in early 1781. Robert Morris, the first Superintendent of Finance, was charged with oversight of the public debt, expenditures, revenues, and accounts to the end that he would, "report plans for improving and regulating the finances, and for establishing order and economy in the expenditure of the public money," (Powell, 1939, p. 33) as well as perform oversight of budget execution, purchasing and receiving, and collecting delinquent accounts owed the United States.

The enabling legislation has been referred to as "a bit radical for the times," because of the vast authority it delegated to one man (Browne, 1949, pp. 21-22). Morris's pressure for revenue collection seemed to have angered some members of Congress. Consequently, in 1784, a treasury board or committee was established. However, the benefits derived from a single executive, albeit not the President, equipped with broad powers was evident and this pattern would reappear. As a practical matter, the whole period of Confederation was a time of experimentation within the context of antimonarchical rule. The events of this period seem somewhat confusing, but there existed no model financial system to follow. Let it be remembered that William Pitt was the contemporary of Morris and the founding fathers. Pitt did not take office until after 2 years after Morris had been appointed, and the system Pitt created operated in a system where law and tradition still gave the balance of power to the Crown. Pitt was the king's minister. The Americans were busy negotiating the mechanics of representative government, influenced by the Confederation model of strong legislative assemblies.

However, by the time of the constitutional convention, it was clear that this experiment in representative democracy operated through the

Table 2.1.   Constitution Sets Roles for Legislative— Executive in Defense

Article I, The Legislative Branch, Section 8: Powers of Congress

The Congress shall have Power To lay and collect Taxes, Duties, Imposts and Excises, to pay the Debts and provide for the common Defence and general Welfare of the United States;
To raise and support Armies, but no Appropriation of Money to that Use shall be for a longer Term than two Years;
To provide and maintain a Navy;
To make Rules for the Government and Regulation of the land and naval Forces;

Article I, The Legislative Branch, Section 9: Limits on Congress

No Money shall be drawn from the Treasury, but in Consequence of Appropriations made by Law; and a regular Statement and Account of the Receipts and Expenditures of all public Money shall be published from time to time.

Article II, The Executive Branch, Section 2: Civilian Power over Military

The President shall be Commander in Chief of the Army and Navy of the United States, and of the Militia of the several States, when called into the actual Service of the United States.

Source: U.S. Constitution (see www.usconstitution.net)

legislature and by committees within it was not a practical solution to administering government, whatever its virtues in representing the people, thus the founding fathers created the presidency (despite their fear of kings) and laid out the design for the taxing and spending power. The Constitution provided the right to tax to Congress and set forth four qualifications on spending power:

1. No money shall be drawn from the Treasury but in consequence of appropriation.

2. A regular statement and account of all receipts and expenditures must be rendered from time to time.

3. No appropriations to support the Army shall run for longer than 2 years.

4. All expenditures shall be made for the general welfare (U.S. Const. art. I, §8).

The first two points contained in U.S. Const. art. I §8 are the corner-stones of the budget process. On the revenue side, all money bills were directed to originate in the House because of its proportional and direct representation of the people. The role of the Senate was debated, with the compromise that the Senate could concur with the House, or it could propose amendments to revenue bills. Fiscal power would be developed within an environment where the Congress was expected to be supreme at the federal level, and the states were expected to be jealous guardians of their powers. Under the impact of the American Revolution, planning had been nonexistent, management had been a legislative responsibility, and control for propriety was honored more in its absence than in its presence. However, the period 1789 to 1800 marked the beginning of the movement toward executive management and perhaps could be called the first stage of U.S. budget reform. The talents of Alexander Hamilton strongly influenced development in this period (Caldwell, 1944; Fesler, 1982, pp. 71-89; Seiko, 1940, p. 45).

Hamilton was a man of great talent and achievement. He learned applied finance at the age of 11 while a clerk in a countinghouse on the island of St. Croix in the West Indies. He learned quickly and was promoted to bookkeeper and then to manager. Before Hamilton was 21, he had impressed friends with his abilities to the extent that they sponsored him in a course of studies, first at a preparatory school and then at the predecessor to Columbia University in New York. Here he quickly gained a reputation as an adroit protagonist for the cause of the American colonies (Caldwell, 1982, pp. 71-89; Miller, 1959). In 1776, he had won George Washington's eye with his conspicuous bravery as an artillery captain at the Battle of Trenton. Washington used him as a staff officer until 1781 when Hamilton, chafing under the limitations of staff routine, seized upon a trivial quarrel to break with Washington and leave his position. Washington seemed to have understood his impetuous subordinate well. He gave Hamilton command of a battalion that attacked a British strongpoint at the siege of Yorktown in October of 1781, a siege that ultimately became the decisive battle of the Revolutionary War.

During the 1780s, Hamilton practiced law in New York City and was active in congressional politics, arguing for a strong central government. Hamilton believed that English government, as then constituted under George III, should be the American model. He proposed a President elected for life, who would exercise an absolute veto over the legislature. The central government would appoint the state governors, who would have an absolute power over state legislation. The judiciary would be composed of a supreme court whose justices would have life tenure. The legislature would consist of a Senate, elected for life, and a lower house, elected for 3 years. In this system the states would have virtually no power.

Hamilton's ideas seem to have had little influence upon the constitutional convention. However, when opponents attacked the document brought forth by the convention. Hamilton, with James Madison and John Jay, authored *The Federalist Papers,* a collection of 85 essays that were widely read and helped mold contemporary opinion; they became one of the classic works in American political literature. This was the man Washington appointed as secretary of the treasury in September of 1789.

Hamilton fused his own goals for a strong central government with the new nation's fiscal needs. His first efforts were directed toward establishing the credit of the new government. His first two reports on public credit urged funding the national debt at full value, the assumption by the federal government of all debts incurred by the states during the Revolutionary War, and a system of taxation to pay for the debts assumed (Hamilton, 1790; 1791).[1] Strong opposition arose to these proposals, but Hamilton's position prevailed after he made a bargain with Thomas Jefferson, who delivered southern votes in return for Hamilton's support for locating the future nation's capital on the banks of the Potomac near Virginia.

Hamilton's third report to Congress proposed a national bank, modeled after the Bank of England. Through this proposal, Hamilton saw a chance to knit the concerns of the wealthy and mercantilist classes to the financial dealings of the central government. This was a very controversial proposal, not only because it was a national bank, but because banks of any sort were almost unknown in colonial America until the 1781 when the Confederation Congress set up a bank of North America, unlike

England where there were "dozens and upon dozens of private and county banks scattered all over" (Wood, 2006, p. 133). Despite heated opposition, Congress passed and Washington signed the bill creating what became the Federal Reserve Bank into law, establishing this national bank based, in part, on Hamilton's argument that the Constitution was a source of both enumerated and implied powers, an interpretation he used to expand the powers of the Constitution in later years. Hamilton's fourth report to Congress was perhaps the most philosophic and visionary. Influenced by Adam Smith's *The Wealth of Nations* (1776), Hamilton broke new ground by arguing that it was in the interest of the federal government to aid the growth of infant industries through various protective laws and that, to aid the general welfare, the federal government was obliged to encourage manufacturing through tax and tariff policy. Hamilton's contemporaries seem to have rejected the latter view; Congress, at least, would have nothing to do with it. Nonetheless, in little more than 2 years Hamilton submitted four major reports to Congress, gaining acceptance of three that funded the national debt at full value, established the nation's credit at home and abroad by creating a banking system and a stable currency, and developed a stable tax system based on excise taxes to fund steady recovery from the debt and to provide for future appropriations. Indeed, Hamilton opposed the popular cry for war with England in the mid-1790s, at a time when France and England were at war, and England was seizing American ships in the Caribbean. He believed that commerce with England and the import duties it provided were crucial to the well-being of the new United States.

Hamilton's essays, published in New York newspapers in 1795, helped avoid war with England, thus helping to save his revenue system. Also, Hamilton was an admirer of the English system with its strong central government, fiscal systems, and professionalized standing Army; in the latter respect, he advocated a strong standing Army for the United States to allow it to subdue any "refractory state" and "to deal independently and equally with the warring powers of Europe" (Wood, 2006, p. 130). While neither of these conditions would come to pass in his lifetime, Hamilton was recalled to service by George Washington as second in command of the Army under Washington in 1798 when it seemed that France might invade the United States.

The elements of the new nation's monetary and fiscal policy were bitterly contested issues, and groups coalesced around various positions. Hamilton became the leader of one faction, the Federalists, and because Washington supported most of Hamilton's program, in effect he became a Federalist. The two most prominent individuals in opposition were James Madison in the House of Representatives and Thomas Jefferson in the Cabinet. Madison and Jefferson were the Republican leaders. Hamilton and Jefferson feuded for several years beginning in 1791, as each tried to drive the other from the Cabinet. Finally, tired, stung by criticism of his operation of the treasury department, and needing to repair his personal fortunes, Hamilton announced his intentions to resign his post as secretary of the treasury at the end of 1794. Hamilton did not, however, retreat to obscurity. He still held presidential ambitions that were narrowly frustrated; he was appointed to high military command, and he remained within the inner circle of the nation's political elite before departing center stage, killed in a pistol duel with Aaron Burr in 1804. Hamilton had made both great accomplishments and bitter enemies. Moreover, all of his suggested reforms became settled and accepted policy, though he was perhaps 50 years too early on using tariffs to protect infant industries.

Hamilton's role in establishing a system for debt management, securing the currency, and providing a stable revenue base make him perhaps the founding father of the American budgeting system. Without faith in the soundness of the nation's currency and credit system, and a productive revenue base, it is difficult to make any budget system work. The federal taxing power alone was a dramatic change from the system envisioned under the Articles of Confederation, which approximated a contributory position by the separate states, hectored by the central government. Only a sure and certain revenue base, providing predictable revenue collections, allows the creation and maintenance of the modern nation-state. It was Hamilton's genius to direct the United States to that pathway.

To Congress, Hamilton represented a transitional figure. Before his appointment, the House of Representatives had a tax committee, the committee on ways and means, established in the summer of 1789. But this committee fell into disuse when a secretary of the treasury was appointed. In fact, from 1789 to 1795 when Hamilton resigned as Secretary of Treasury, Congress discharged its committee on ways and means and stated that it would rely on Hamilton for its financial knowledge (Wood, 2006, p. 129). At this juncture in history, Congress viewed the Treasury Department as a legislative agency and the secretary of treasury as its officer (Browne, 1949, p. 34). The first appropriations bill for an operating budget came about because the House ordered the secretary of treasury to "report to this House an estimate of the sums requisite to be appropriated during the present year; and for satisfying such warrants as have been drawn by the late Board of the Treasury and which may not heretofore have been paid" (1 Annals of Congress: 929).

When the articles of the Constitution were being debated, Hamilton wrote:

> The House of Representatives cannot only refuse, but they alone can propose, the supplies requisite for the support of the government. They, in a word, hold the purse.... This power over the purse may, in fact, be regarded as the most complete and effectual weapon with which any constitution can arm the immediate representatives of the people for obtaining a redress for every grievance and for carrying into effect every just and salutary measure. (Hamilton, as cited in Miller, 1959)

Whatever the flaws of the act creating a department of the treasury, it seems clear that its intent was to make the Congress alone responsible for the budget process. That there was little room for executive leadership is demonstrated in the fact that the act mentions the President only in connection with the appointment and removal of officers. Furthermore, while the act was being debated, opinion was divided over the wording of the duties of the secretary of the treasury with respect to whether he was to digest and *report* revenue and spending plans or whether he was to digest and *prepare* plans. Those congressmen hostile to strong executive power believed that giving the secretary the power to digest and report plans would take the fiscal policy initiative away from the House. The secretary would report only what he had already done; this would deprive the House of its ability to exercise a prior restraint on the actions of the secretary. The word *report* was deleted from the legislation and *prepare* was inserted and carried by the majority (Browne, 1949, p. 31).

Some observers have mistaken Hamilton's approach to appropriation as having established an executive budget system. The traditional model of an executive budget system would encompass a presidential review of departmental documents, revision of estimates, and a unified submission by the President or his agent of those estimates to Congress for approval. Hamilton, as Secretary of the Treasury, did not wish the system to function in this manner, and personally he acted as an agent of Congress. The development of an executive budget system occurred in more gradual process, with a steady line of evolution leading to the authorization of a formal presidential budget—but not until 1921.

The first appropriation act of Congress was brief and general:

> That there be appropriated for the service of the present year, *to be* paid out of the monies which arise, either from the requisitions heretofore made upon the several states, or from the duties on impost and tonnage, the following sums. *viz.* A sum not exceeding two hundred and sixteen thousand dollars for defraying the expenses of the *civil* list, under the late and present government: a sum not exceeding one hundred and thirty-seven thousand dollars for defraying the expenses of the department of war; a sum not exceeding one hundred and ninety-six thousand dollars for discharging the

warrants issued by the late board of Treasury and remaining unsatisfied; and a sum not exceeding ninety-six thousand dollars for paying pensions to invalids. (I Statutes at Large, U.S. Congress, Ch. XXIII, Sept. 29, 1789: 95)

Although salaries are the largest single item in this list (civil list), mandated expenditures—bills and pensions—comprised 45% of the budget, defense 21% and entitlements 14%. De facto uncontrollability was high. True to modern practice, this appropriation bill was not the only money bill passed by Congress. Between the summer of 1789 and May of 1792, numerous bills were passed to provide for a variety of expenses, including defense, Indian treaties, debt reduction, and establishment of the federal mint[2] (3 Annals: 1,259).

The first three bills were written as lump sum general appropriations, for the civil list, the department of war, invalid pensions, the expenses of Congress, and contingent charges upon government. Appropriating by lump sum caused resentment among some congressmen. One wrote of the appropriations bill of 1790 in his diary:

The appropriations were all in gross, and to the amount upward of half a million. I could not get a copy of it. I wished to have seen the particulars specified, but such a hurry I never saw before.... Here is a general appropriation of above half a million dollars—the particulars not mentioned—the estimates on which it is founded may be mislaid or changed; in fact it is giving the Secretary the money for him to account for as he pleases. (Wilmerding, 1943, p. 21)

Notwithstanding their general nature, appropriation bills were linked to estimates of expense as specified in other bills. Expenditures for salaries were generally governed by laws enumerating the salary and number of the officers stipulated; for example, five associate Supreme Court justices at a salary no more than $3,500 per year. Estimates for the military were assumed to control the appropriations voted for the military. Therefore, even though the appropriations were voted in gross, the calculations adding up to the total were assumed to control the total.

By 1792, Congress was appropriating money in gross but stipulating what the money was to be used for, with "that is to say" clauses: for example, $329,653.56 for the civil list, with a "that is to say" clause followed by specific sums attached to an enumeration of the corresponding general items (Fisher, 1975, p. 61; Wilmerding, 1943, p. 23).[3] Congress was now planning in detail and the executive branch accepted that detail, although knowing full well that the dictates of administering might make it imperative to depart from the detailed plans expressed in the appropriation acts. Budgeting by lump sum was not a characteristic of the routine of American government except in case of emergency appropriations

(Browne, 1949; Wilmerding, 1943). Congress increasingly specified the itemization of appropriation bills, in part as a strategy to control Secretary of the Treasury Hamilton, who was viewed by some as a member of the executive branch. In 1790, the House had 65 members, and most of its business could be carried out as a committee of the whole, but by 1795 it was clear that the Treasury Department could not serve the needs of Congress as well as it could serve the needs of the executive. Therefore, Congress reinstituted the committee on ways and means, initially as a select or special committee, and by 1802 as a standing committee. Also, during this period, Woolcott, Hamilton's successor at Treasury Department, was embroiled in an increasingly bitter argument with Congress over the transfer of appropriations. Although Congress could appropriate in very specific terms, it could not stop the administration from transferring from one account to another when the situation seemed to warrant such transfers. The war and navy departments seemed particularly able to transfer funds, thereby dissolving the discipline of detailed itemization. (This is another issue with modern analogs; see our discussion of transfers and reprogramming in our chapter on budget execution.)

In 1801, when the Federalists were defeated and the Republicans took office, Thomas Jefferson spoke to the need for increased itemization of expenditure in appropriations. Nonetheless, the transfer of appropriations was an accepted practice in the administration, albeit an illegal one (Wilmerding, 1943, p. 48). Jefferson himself made the Louisiana Purchase after a liberal interpretation of executive authority to issue stock when government revenues were insufficient to cover necessary expenditures. Thus Congress's insistence on itemization led to deficiencies in accounts. Later, budget practices also became part of developing party politics. The Federalist Party believed in a strong executive, thus they preferred lump sum appropriations for activities that would give administrators as much flexibility as possible in managing programs. Meanwhile, the Republican party favored specific line item appropriations that limited department heads to doing specifically what Congress intended. However, Congress found out it could not control everything and, beginning with appropriations for the Army and the Navy, line item controls were relaxed and other controls not invoked, for example, penalties for unauthorized transfers.

The tension between delegating power to the executive and retaining appropriate congressional control of the power of the purse has remained an issue to current times. Generally, when relationships between the executive and legislative branch deteriorate, because of divided party control or because an executive agency becomes either too aggressive in its budget practices or not aggressive enough, Congress changes the rules to ensure that its intent is preserved, for example, by changing reprogramming within appropriation thresholds (ceilings), requiring advance notification of Congress for reprogramming changes from one category within an appropriation account to another, or by attaching a legislative interest note to an item to ensure that money is spent in a special way, or that a program is executed within the FY.

By 1800, the initial pattern had been set. Appropriation bills were passed and linked to specific estimates for specific purposes, military and civilian expenditures were treated somewhat differently, and it was recognized that transfer of funds between categories to meet contingencies unforeseen at the time of appropriation was necessary, if technically illegal. The House held major control of the purse. By 1802, Congress had developed a standing committee to deal with revenues and appropriations, while the secretary of treasury had become more and more the President's agent in shaping appropriations bills, collecting revenues, and debt management. Both the legislature and the executive were elected by and responsible to the people. Rudimentary and disconnected as it seems from modern perspectives, no other country had such a budget system. The themes that surfaced during these years still appear: establishing a correct balance between executive and legislative roles, lump sum versus line item, correct use of funds during emergencies, even to the need to vote large sums on the basis of limited information. Still, this was democracy in action, and we reiterate that no other nation had such a budget system. Great Britain also had a system, but ultimate power resided in a king, not the people.

By the early 1800s, the transfer of power to Thomas Jefferson and the Republicans marked the end of the period of creation of the Republic, the end of the process of separating from England and the setting up of a new government. Much remained to be done, but many of the basic mechanisms of governance, and of monetary and fiscal policy, were now in place. In fiscal affairs the federal government had established its powers to tax and to budget, as well as to issue notes of credit when revenues did not match expenditures. Budgeting was in the main a legislative power. The Treasury Department was originally conceived as Congress's assistant. Budgeting power in the Congress was held in the House, which was small enough so that it could operate as a committee of the whole. As the Treasury increasingly served the President, Treasury's power in Congress declined, and Congress chose its own internal review and enactment body, the committee on ways and means, and this committee would gain great power. During the course of the nineteenth century, appropriation bills would be sent to other committees as the ways and means committee work load became heavier, or as political factors dictated. After the Civil War, a committee on appropriations was created.

When it used retrenchment powers to reach into substantive legislation under the jurisdiction of other committees, Congress reacted against this expansion of the appropriation committees' powers and diminished the power of the appropriations committees. It seems that whenever a committee role became too important, Congress changed its procedures to move power away from it. Thus, legislative procedure changed, but budgeting maintained eminence as a vital legislative process.

There was an executive branch component to this process, but different Presidents chose different profiles. Some were quite involved, others not. Fisher suggests that during the nineteenth century a number of Presidents revised departmental estimates before they were sent to Congress, including John Quincy Adams, Martin Van Buren, John Tyler, James K. Polk, James Buchanan, Ulysses S. Grant, and Grover Cleveland, and were assisted in this task by a number of secretaries of the treasury. Some ascribe an even larger role to the executive in this period[4] (Fisher, 1975, pp. 269-270; Smithies, 1955, p. 53, White, 1951, pp. 68-69.)

The first decade of the twentieth century was pivotal in terms of the balance of budgetary power. Government revenues based on customs and excise taxes were insufficient for the task of achieving the nation's "manifest destiny." Although the budget had been in a surplus position from the conclusion of the Civil War, after 1893, the economy and the budget ran into trouble and some policymakers worried that an antiquated revenue system prevented government from meeting new needs. The Spanish American War and the expense incurred in building the Panama Canal created budget deficits. Moreover, customs revenues began to decline. The federal budget was in a deficit position for 11 of the 17 years from 1894 to 1911, including 5 of the 7 years from 1904 through 1910. In addition to these debts arising from emergencies, some also felt that the revenue system was not up to the task of funding America's new and expanding world role. Passage of the 16th Amendment by Congress in 1909, ratified by the states in 1913, authorizing a federal income tax was a major milestone event in U.S. fiscal policy, and was intended in part to remedy this dilemma.

The debate over strengthening presidential spending power was essentially completed by 1912, with issuance of the report of the Taft Commission on Economy and Efficiency. Taft submitted this report to Congress, along with a plan for a national budget system, but his party did not control the House during that session of Congress and the two branches of government could not agree on a new budget process. The commission's position was succinctly stated:

> the budget is the only effective means whereby the Executive may be made
> responsible for getting before the country definite, well-considered,

comprehensive programs with respect to which the legislature must also assume responsibility either for action or inaction. (H. Doc. 854, 62-2, 1912, p. 138; Taft, 1912, pp. 62-6)[5]

Budget reform was further delayed at the national level by the First World War, but reform continued apace at the local and state levels. Indeed, some observers have suggested that budget reform during this period in the American context began at the local level. Reform efforts resulted from indignation over corruption, graft, and mismanagement prevalent in local governments, exposed by journalists and good government movements, and supported by the Progressive party. Budget reform complemented other innovations including establishment of city manager and commission government forms, and the initiative, referendum, recall, and short-ballot electoral procedures. Budget reform in this period may be considered a local affair that eventually carried over to the federal government (Burkhead, 1959, p. 15; see also Schick, 1966, pp. 243-258).[6] The fiscal stress caused by the American commitments to World War I, and President Woodrow Wilson's own interest in budget and administrative reform also precipitated the adoption of the executive budget process.

In his 1917 annual message to Congress, President Wilson stressed his party's platform on budget reform. Although reform seems to have been possible in any of these years, Wilson chose to wait until the end of World War 1. While he waited, the nation incurred a large deficit. In the three years from 1917 through 1919, federal debt grew from $1.2 billion to $25.5 billion. This gave urgency to the case for budget reform. After the peace treaty had been signed, Wilson argued that budget reform would give him a better grasp of the continuing level of defense spending, the effect of the disposal of surplus military property, and the impact of demobilization upon the economy (Fisher, 1975, p. 33).

In 1918 and 1919, a series of bills intended to reform the distribution of budget power were passed, and in 1921 Congress passed the Budget and Accounting Act (Burkhead, 1959, pp. 26-28).[7] This bill created the Bureau of the Budget (BoB), to be located in the Department of the Treasury with a director appointed by and responsible to the President. The BoB was given the authority to, "assemble, correlate, revise, reduce, or increase" departmental budget estimates (42 Stat. 20, 1921). The intention of the writers of this law was to avoid unnecessary friction between the President and his cabinet officers over budget matters by locating the budget review power within the BoB in Treasury. This was intended to avoid setting the BoB against the more powerful cabinet officers. Also, placing the BoB in the Department of the Treasury facilitated the coordination of expenditures and revenues (Fisher, 1975,

p. 34). Later it was moved to the executive office of the President. Under any interpretation, establishment of the BoB in 1921 and the crucial tasking of the President to prepare and submit a budget to Congress shifted power to the executive.

However, in passing reform legislation that increased the power of the executive, Congress also took something back by creating the Government Accountability Office (GAO) to audit and account for expenditures, led by a comptroller general of the U.S. responsible to Congress and appointed for a 15-year term. Any perusal of reports and testimony generated by GAO indicate how important this office has become in providing information for Congress to use in reviewing budgets and making financial management decisions. At the time, creation of the GAO was overshadowed by the attention directed at the BoB.

The halcyon days of the BoB existed from 1939 through the end of the 1940s. During this time, the BoB built and held a reputation for unsurpassed excellence as a neutral, analytic power operating as a staff instrument for the executive. The reputation for excellence gained during these years of depression and war would mantle the BoB into the late 1960s, but then its function changed to match the politics of the time. The BoB was renamed the Office of Management and Budget (OMB) in 1969. Further, OMB would become tainted somewhat by the politics of Watergate but more by the aggressiveness of Richard Nixon in using and abusing his presidential impoundment authority by refusing to spend money appropriated to executive agencies by Congress. OMB would be accused of exerting too much power, of resistance to change in a world interested in policy analysis instead of budget examination, and failure as an intergovernmental program manager for the multiplicity of programs resulting from President Lyndon Johnson's quest for the Great Society (Fisher, 1975, p. 58; see also Davis & Ripley, 1967, pp. 749-769). However, during this entire period, for better or worse, the BoB functioned increasingly as the instrument of executive budget and policy making power.

One way to conceptualize budgetary control of the type wielded by the BoB and all central executive budget control agencies in government is to envision it as a tool that operationalizes fiscal values. These fiscal values are basically economizing values. As identified by Appleby (1980), they include fiscal sense and fiscal coordination: "Fiscal sense and fiscal coordination are certainly values. The budgeting organization is designed to give representation in institutional interaction and decision-making to this set of values" (p. 134) Appleby argued that the budget function is inherently and preponderantly negative because it is against program expenditure and expansion. He explained that this is proper because program agencies and pressure groups are so extensive that there is no danger the values they represent will be overlooked or smothered by budgeteers.

Appleby (1980) conceded that a budget control agency cannot always be negative, for there are ways to save money by spending money, and the controllers have to be on the lookout for these occasions. In the main, however, budget control agencies will be the aggressors, pushed to cut, trim and squeeze spending. Spending agencies, on the other hand, will temper their requests by their judgment of what is wise and practical, and what policy makers and the budget bureau will accept. The executive budget bureau is at the center of this struggle, and yet it is removed from direct contact with many if not most of the political pressures of the politics of budgeting due to its isolation within the executive and the fact that it works for only one political party (the President's) at a time. Consequently, the budget bureau could and should act as a counterweight to ensure that economizing fiscal values are entered into the decision-making calculus.

Wildavsky (1964) characterized the budget process as a competition between the "spenders" and the "cutters," with BoB and agency budget control offices as the primary cutters in the executive branch of government, and the appropriations committees as the cutters in Congress. However, when the appropriations committees play this role, it is often to cut one program so as to add funding to another.

With respect to spenders, program agencies are expected to advocate for their programs and constituencies both within and outside of government. Members of Congress and the President, as elected officials, generally are expected to play the role of spending advocate most of the time. Otherwise, how would they get reelected? In a democracy, what do people send their elected officials to Washington, D.C. to do? The answer in large part is to solve or resolve problems, and to do so requires Congress to spend—from the perspective of the clients of governments and many stakeholders in the economy. Consequently, according to Wildavsky (1964), the spenders vastly outnumber the cutters and this creates a pro-spending bias in government. An understanding of the roles, duties, and expectations related to the players in the budget process is critical to comprehension of budgetary competition for power. It is also important in attempting to understand the proper functioning of budget control agencies.

Neutral competence was the keystone of the philosophy of the BoB. This was typified by the folklore the BoB perpetuated about itself. As Berman (1979) noted, "BoB officials often told the story that if an army from Mars marched on the Capitol, everyone in Washington would flee to the hills, except the Budget bureau staff, who would stay behind and prepare for an orderly transition in government" (p. 29).

The BoB was reorganized in 1969-1970 and became OMB in part to add political acumen by layering political appointees over the career staff (Reorganization Plan No. 2, 1970). After 1970, the BoB's representation of fiscal values would be filtered through nets of political values before they reached the President, a change that may have improved the advice the BoB could give the President but changed the character of its neutral competence. Gone was the pure budget technician, lost in part to the era of policy analysis where the ability to detail the policy consequences of alternative budget decisions was the task at hand.

What happened to merely cutting the budget through close examination and intimate knowledge of the program? A reorientation of the role of BoB to become OMB probably was a necessary change. Schick (1970) observes that the BoB as a simple representative of fiscal values could serve every President with, "fidelity, but it could effectively serve only a caretaker President. It could not be quick or responsive enough for an activist President who wants to keep tight hold over program initiatives." (p. 532) As the functions and responsibilities of the presidency changed, so did the role of the BoB.

Highly respected budgetary scholar Jesse Burkhead (1959) judged the institution of the executive budgetary system in the United States to be a revolutionary change. Burkhead argued: "The installation of a budget system is implicit recognition that a government has positive responsibilities to perform and that it intends to perform them" (pp. 28-29). To do this would require reorganizing administrative authority in the executive branch, said Burkhead, and an increase in publicly organized economic power relative to privately organized economic power. Thus, the institution of executive budgetary systems in the United States clashed with customary doctrine about public versus private economic responsibility, but more importantly it was fundamentally at odds with the basic organizing precepts of the founding fathers. The budget system after 1921 and particularly in the post-World War II years through 1970 was an integrating system that allowed positive movement toward goals by relatively small groups of participants within the political system. It had to work this way, or it could not be an efficient system.

However, this kind of organizational efficiency appears to run counter to the Constitutional doctrines of separation of powers and checks and balances. Consequently, Burkhead (1959) suggested that not only would the practices of government have to be altered before budget systems could be installed and operated, but their development and installation alone were "revolutionary" in the context of American society. Burkhead concluded that although budget systems need not be synonymous with an increase in governmental activities (budget systems can be used for retrenchment), their installation is synonymous with a clarification of responsibility in government.

In reaction to a number of what it saw as abuses of executive power in the early 1970s, Congress reasserted its power with the passage of the Congressional Budget and Impoundment Control Act of 1974[8] (Schick, 1980). This Act sought to correct certain abuses of presidential impoundment powers, but more importantly, it also sought to reorganize the congressional budget power to give Congress a better chance at full partnership in budgeting for the modern welfare state. If the Full Employment Act of 1946 gave the President responsibility for managing the economy, the Congressional Budget and Impoundment Act of 1974 extended the same opportunity to Congress. In addition, Congress equipped itself with more analytic power by creating the Congressional Budget Office, comprised of a neutral staff imbued with a sense of high calling and professionalism similar to that found in the BoB of the 1940s but in a somewhat more complex fiscal world.

The 1974 Congressional Budget Impoundment and Control Act centralized the planning function of the budget in the House and Senate Budget Committees. These two committees have the responsibility to develop a target resolution in the spring of each year containing detailed appropriation, spending and other targets (e.g., lending) to guide the work of the appropriations committees and subcommittees. The target resolution shows the overall situation, total spending, including the level of spending by function, taxing projections, and the level of surplus or debt forecasted. Then, in September, the Budget Committees were to shepherd a second resolution through Congress that matched the early planning target to the final appropriation bills. Through the reconciliation process, these committees may also ask Congress to tell its appropriation and taxing committees what and where reductions are appropriate in order to reconcile the final bills against the target resolution. In June of 1981, Congress attached reconciliation instructions to the first resolution and in effect dictated what would be done later that year in appropriations committee work.

The reconciliation instruction of June 1981 marked a turning point in the American budgetary process. For the first time in the history of the United States, the Congress set budget targets for taxing, spending, and debt, and thereby had a sense of what the national budget ought to be before it started enacting appropriation bills. At no other time since 1789 was this done. After 190 years of titular vesting of the power of the purse in Congress, Congress organized itself to pursue a budget prospectively, rather than adding up the total appropriations and expenditures and calling it a budget.

The 1981 budget was essentially an executive budget due to the fact that the Republican Party controlled Congress to a much greater degree after the 1980 elections and tended to support the budget proposals of recently elected President Ronald Reagan, but still it was endorsed in Congress only after a bitter struggle. However, Congress used its newly developed budget power to develop congressional budgets that were different from the proposed executive budget of 1980, the final Carter budget. Congress asserted itself again in 1982, in deliberations over the second Reagan budget (Peckman, 1983, p. 19). As we note in analysis of the separation of powers, Congress has exercised its budget power both in support of and against the executive. Although the power to prepare and submit budgets remains with the executive, and a formidable power it is, Congress has evolved into a powerful and systematic modifier of budgets.

From 1945 to 1970, congressional scrutiny of budgets was characterized by students of budgeting as one of incremental review and marginal adjustments by appropriation committees to whom the other members of Congress deferred. Incremental behavior was rational, according to Aaron Wildavsky (1979), because in reviewing that in which he or she was most interested, members allowed individual self-interest to protect the public good. Fenno documented the success of final adoption of appropriations committees' recommendations as being 87 per cent. Sharkansky observed that congressional behaviors could be summarized as the concept of contained specialization-elite status, specialized expertise, deference to the acknowledged experts, and conflict management (Fenno, 1966; Sharkansky, 1969).

During this era of stability of review, enormous changes were taking place within society, and the composition of the budget reflected it. Social service and especially entitlement expenditures increased dramatically. Although each bill was intensely scrutinized, no one in Congress knew what all the appropriations bills would total in terms of actual spending (outlays) until the end of the FY. As macroeconomic management became more important to the nation (Jones & Wildavsky, 1995), Congress had no apparent forum of its own to make and enforce economic policy through the budget. Thus, the budget process became less and less useful to the realities of managing a modern welfare state.

Seizing on the Nixon abuses of impoundment power to reorganize the budget process and to make itself a full partner in the process once more was an outcome not totally anticipated by Congress. Some thought that Congress had merely changed the FY in an attempt to give itself more time to process appropriation bills to offset its difficulty in passing them on time. Others saw the 1974 Budget Act as a weighting of the budget power in favor of Congress at the expense of the executive. Although the Act did improve congressional potential, it need not be said that it usurps executive prerogatives. If it did it would have been ruled unconstitutional by the U.S. Supreme Court. In fact, there is more than enough budget power for both branches to share in attempt to satisfy constituents. Congress did become a powerful critic, however, because as a result of passage of the 1974 Budget Act it has the power to redevelop an executive budget when the President's original submission does not match congressional interpretations of the needs of a particular year, and it has the assistance of the Congressional Budget Office to help members exert their will and power in negotiations with the President and OMB.

SEPARATION OF POWERS; CHECKS AND BALANCES

The United States is lawfully set up with three branches of government, legislative, executive, and judicial, with each having certain powers and each equipped with powers to check the other. A short excursion onto the Internet onto such sites as Wikipedia or the state department Web site, or the CIA Fact Book shows how different other countries are from the United States in formal government processes. In particular, the countries that provide the most students to the Naval Postgraduate School are parliamentary democracies where the President is titular head of state, but his main functions are ceremonial and whoever controls Parliament exercises the power of the legislative branch. In Table 2.2 we show some of these countries.

These countries are all parliamentary systems with a unicameral legislature and a long electoral cycle. Except for Lithuania, all of these systems result in one-party dominance in the Parliament even if there are two or

Table 2.2.   Country Characteristics

| Characteristic | Country | | | | |
	Turkey	Greece	Taiwan	Singapore	Lithuania
Number of parties	7	2	1	1	9
Houses in parliament	1	1	1	1	1
One party dominant	yes	yes	yes	yes	no
Percent of seats held by leading party	66	55	69	98	21
Head of government	Prime minister	Prime minister	Prime minister	Prime minister	Prime minister
President's role	ceremonial	ceremonial	ceremonial	ceremonial	ceremonial
Electoral cycle	5 years	4 years	4 years	5 years	4 years

more parties competing at elections. For example, in the last election in Greece the winning party held 45% of the popular vote and the second party had 40%; in number of seats won, the winning party held 55% of the seats and the second party 39%. Thus when the vice President casts the tie-breaking vote in the U.S. Senate, he is doing something which is not going to happen in these countries. First there is no Senate. Second, an officer from the executive branch is casting the deciding vote. Third, the second house in this issue is in effect forcing the other chamber to compromise with whatever the first house wants, whereas in all these other systems the lower house is the only house and in most cases it is ruled by one party. Fourth, the party which holds the presidency gets to vote twice in the instance cited above; the vice President votes in the Senate and then the President may sign or veto the bill as he wishes. Admittedly this does not happen often, but it does happen and it is legally permissible and has to be understood to be a part of the American government mechanism. The countries cited above basically are designed to promote stability (long electoral terms, one party dominance) and efficiency (Prime minister runs country, holds majority in parliament, picks cabinet of ministers to administer departments, dominant party wins all votes, no negotiation with other branch of government, other chamber, other party or parties.) Both the strengths and weaknesses of the separation of powers systems have long been recognized. In fact some believe that the U.S. system is qualitatively better than that of other countries because it results in a unique political structure with an unusually large number of interest groups, because it gives groups more places to try to influence, and creates more potential group activity. Opponents of separation of powers indicate that it also slows down the process of governing, promotes executive dictatorship and unaccountability, and tends to marginalize the legislature. This is somewhat of a modern view given the tendency of the press to build up the power of the presidency, while the founding fathers were quite clear that the legislative power was the most important one.

Much of the U.S. structure was designed and has evolved to limit the power of the other branches of government, while in many other democratic countries power is organized to govern efficiently following the doctrines of the party which has won the most recent election. Also, elections tend to be at least 4 years apart and some are longer, thus guaranteeing a longer time period for the dominant party to implement its programs. Turkey for example has a 5-year electoral cycle, unlike the United States where the House gets elected every 2 years, along with one-third of the Senate. Thus in the United States, all the House members and one-third of the Senate are almost always trying to make a track record for reelection. Here our foreign neighbors have emphasized the value of stability of longer terms, rather than responsiveness to what may be of concern to the public at the moment. This could allow for a more statesman-like stance by legislators in these countries and less churn. In the United States it is necessary to distinguish between what candidates say when they run for election as opposed to what they say and do when they have to govern. To international observers, it may seem like there is a lot of wasted motion in the U.S. political apparatus, because someone is always running for election.

Another very important difference is that most of parliamentary systems are heavily invested in one chamber; in the five countries cited above, there is only one chamber, thus there is no fundamental appreciation for the dynamics between two chambers, on different electoral cycles and representing different constituencies, even when they are controlled by the same political party. In the United States, both the House and the Senate have legitimate roles and political power and their perspectives differ, even when they are controlled by the same party. This is different for many other countries, where even in two-chamber systems like Canada and Great Britain, only one house governs. Thus in addition to negotiation between executive and legislative branches over appropriation bills, there is an area of negotiation within the legislative branch that has to be understood.

Thus, it is not only the separation of powers that international observers have to understand about the U. S.; they also have to understand checks and balances. For example, the powers and responsibilities of the different branches of government intentionally were designed to overlap. Congressional authority to enact laws can be checked by a presidential veto, but that veto can be overridden by a two-thirds majority in both houses. The President serves as commander in chief, but only Congress has the authority to declare war and raise funds to support and equip the armed forces. The President has the power to appoint ambassadors, federal judges, and other high government officials (e.g., director of OMB, secretary of defense), but all appointments must be affirmed by the Senate, and sometimes these appointments are denied, if rarely. Also, the Supreme Court has the power to overturn both presidential and legislative acts. This checking and balancing was designed to ensure that no one branch of government would grow too powerful, and in the early days, that the national government would not grow too powerful at the expense of the states.

In the world of budgeting specifically this means such things as the President proposing budgets, but Congress passing appropriations bills which the President can veto, but Congress can override. The President can impound, delay, and defer budgeted funds, but only within certain limits and is subjected to Congressional review. The President may veto bills, but he does not have an item veto, since that would make him a legislator. The President appoints judges, diplomats, top level civil servants and top level military officers, but Congress has to confirm them. The President can commit troops into immediate military action, but only Congress has the power to declare war and fund the military. The vice President chairs the Senate in a role which is largely ceremonial, but he may cast tie-breaking votes which can be very important. The House can impeach the President and the Senate then tries him, thus the legislative branch has control over the executive for gross dereliction of duty. The House is also given the responsibility of choosing the President if there is no majority in the electoral college, which is in itself another check and balance in the electoral system. Finally the courts constitute a check on the actions of both Congress and the President through the power of judicial review, which may happen long after an issue has lost the public's attention. This checking and balancing is a large part of what makes the budget process so complicated. Later we discuss in more detail the ways Congress modifies, controls and generally "improves" presidential actions.

CONGRESS AS ARBITER OF CIVIL-MILITARY RELATIONS

Another way to examine the roles of Congress and the executive branch is to assess their roles as arbiters of civilian and military relationships. Historically, the challenge of civil-military relations (CMR) in democratic societies has revolved around the dilemma of raising a military strong enough to deter and defeat a state's enemies while at same time controlled sufficiently so as not to threaten its own government. As Peter Feaver (1996) put it, the "challenge is to reconcile a military strong enough to do anything the civilians ask them to with a military subordinate enough to do only what civilian authorities authorize them to do" (p. 149). Over the past half-century, the dialogue on CMR has been dominated by the views of Samuel Huntington and Morris Janowitz, but important new perspectives have emerged.

Huntington (1957) proposed that maximizing the professionalism of the military is the key to assuring a subordinated military. In *The Soldier and the State*, he argues that a professional officer corps is the key to maintaining the balance between military superiority and civilian control. An autonomous, professional military will subordinate itself to legitimate civilian control. Janowitz (1960), in *The Professional Soldier*, also focused on the officer corps, but saw them as more politicized than Huntington. Janowitz likened Cold War missions to a constabulary and saw threats to CMR from the centralization of power in the Department of Defense. He viewed the budget process as an important tool of civilian control, but not one effectively used. He argued for greater oversight by civilian authorities and the establishment of not simply a professional military, but a professional military ethic.

The views of Huntington (1957) and Janowitz (1960) were foundational to the management of the military during the Cold War; for example, they were influential in the design of officer training programs. However with the end of the Cold War came a rethinking of these ideas. Roles and missions—one of the control mechanisms advocated by Janowitz—were changing. Emerging democracies have created a fertile subject for research (Cottey, 2002, pp. 31-56). Newer conceptualizations have also recognized that mature democracies have not suffered military coups. Thus, if the extreme end of the CMR continuum is so unlikely to occur, then what is the relevant range of the continuum (Feaver, 1996)? The answer is not necessarily found in the sociological literature on the military such as we have seen in the Janowitz tradition (Burk, 1993, pp. 167-185). "Their findings ... are of great sociological import but seem less relevant to political scientists concerned with the exercise of power between institutions" (Feaver, 1996, p. 157).

Feaver (1996) argues for a more nuanced view of CMR that incorporates "interest-based and external control mechanisms" and "changing patterns of civilian control" (p. 167). In developing a set of benchmarks for the development of future theory, he suggests that the new problematic is "...about the delegation of responsibility from the notional civilian to the notional military. It is about increasing or decreasing the scope of delegation and monitoring the military's behavior in the context of such delegation. And it is about the military response to delegation, desire for more delegation, and even occasional usurpation of more authority than civilians intended." (Feaver, 1996, pp. 168-169)

Burk (2002) presents a compelling criticism of traditional (i.e., Huntington and Janowitz) CMR theory and summarizes recent trends in CMR theory-building. While not offering a theory of his own, he argues for a model based on protecting and sustaining democratic values in the context of the post-Cold War geopolitical situation and the realities of the blurring of the military and civilian spheres in the United States. He suggests in the conclusion that the answer perhaps lies in a federalist-like division of responsibilities across levels or units of government (pp. 7-9). In their review of emerging democracies in Eastern Europe, Cottey, Edmunds, and Forster (2002) came to similar conclusions: the control of the military is not necessarily an executive function, it is more democratic. They specifically address the role of the legislature or parliament in policy setting, oversight and resource allocation decisions.

Effective parliamentary oversight of the armed forces and defense policy, however, depends on both the formal constitutional or legally defined powers of the legislature and the capacity of the legislature to exercise those powers in an effective and meaningful way in practice. (p. 44)

Since Huzar's classic *The Purse and the Sword*, there have been numerous empirical studies on the power of the purse as a tool to control the military. Most studies focused on the Cold War years (Kanter, 1983), the relationship of military spending to economic activity (Goldsmith, 2003, pp. 551-573; Kollias, 2004, pp. 553-569), or the management of the defense establishment (Gansler, 1989; McCaffrey & Jones, 2004; Thompson & Jones, 1994). There is also a substantial literature on congressional control of the purse that explains why Congress exercises control through the budget and the mechanisms they use to do so.

Since the birth of the nation Congress has chosen to delegate authority to the military departments in cases where it appeared to be required, that is, in times of war or imminent armed conflict. However, in the 1960s the debate in Congress over policy and funding for the war in Vietnam divided members into those willing to delegate increased authority to President Lyndon Johnson and those who sought to restrict the use of money to fight the war under any condition. This debate parallels the dialogue of the past few years in Congress over funding the war on terror and military operations in Afghanistan and, particularly, in Iraq. While most members of Congress support a War on Terrorism, many have reservations about how and where this war should be fought. The defense budget is a tool for showing support and addressing reservations.

The degree of resource decision flexibility and delegation of authority from Congress to the Department of Defense has been examined by observers of congressional defense budgeting and management for decades (Augustine, 1982; Fox & Field, 1988; Kanter, 1983; Jones & Bixler, 1992). The advantages of increased delegation of resource management authority by Congress have long been asserted by DOD leadership. For example, DOD Comptroller Robert Anthony developed an extensive plan for reorganizing defense accounting and budgeting under Project Prime in the mid-1960s, but Congress rejected the proposal (Jones & Thompson, 1999). Defense Secretary Frank Carlucci asked Congress for increased resource and managerial powers at the end of the Reagan administration in 1988. Defense Secretary Dick Cheney proposed six acquisition programs in the 1991-1994 time frame for execution without congressional oversight as a test of the DOD ability to operate efficiently independent of external micromanagement. Despite considerable congressional lip service to the effect that these proposals would increase program management and budget execution efficiency and better "bang for the buck" in defense, Congress supported neither Carlucci's nor Cheney's request.

Most recently, in April 2003, Defense Secretary Donald Rumsfeld offered a set of legislative proposals under the umbrella title, *Defense Transformation for the 21st Century Act*. Included among the proposals were requests for greater autonomy in budget execution. Congress largely ignored the budgetary proposals and devoted most of their attention to the more public proposal (creation of the National Security Personnel System). Given this longstanding absence of trust of DOD management and budgetary judgment it is significant when Congress deviates from traditional patterns of control.

Those traditional patterns of control have been well described in the literature. Research has explained the incentives for and the means by which Congress oversees the Defense Department through the budget and authorization processes. In summary, Congress has traditionally exercised control as constitutional prerogative (Fox, 1990; Owens, 1990, pp. 131-146; Lindsay, 1990, pp. 7-33), to shape defense policy (Jones & Bixler, 1992, pp. 293-302; Mayer, 1993, pp. 293-302), in response to media publicity of defense mismanagement (Fox, 1988; Lindsay, 1990; Jones & Bixler, 1993), and due to partisan congressional-executive branch competition and occasional mistrust (Blechman, 1990; Mayer, 1993; Thompson & Jones, 1994). Since the defense budget has historically accounted for about half of all federal discretionary spending, it is a favorite target for those legislators seeking to influence federal spending even for nondefense matters (Fox, 1988; Halperin & Lomasney, 1999, pp. 85-106; Wildavsky & Caiden, 2001). A final reason is the advocacy or protection of constituent interests, for example, military installations, labor, or defense contractors (Hartung, 1999; Lindsay, 1990; Jones & Bixler, 1999, pp. 29-84).

Table 2.2 drawn from the work of Candreva and Jones (2005) displays these incentives and tools as they were applied in the Defense Emergency Response Fund (DERF). From September 2001 until the balance of funds were transferred to the Iraqi Freedom Fund in October 2003, the DERF was used by Congress to fund the response to the terrorist attacks. It was also a case of delegated Budget Authority to the Department of Defense (Candreva & Jones, 2005). DERF was an existing account designed to provide flexibility in times of crisis, to provide immediate obligation authority when a need arose, but before the specifics were known. The global War on Terrorism was just that sort of scenario. The unconventional warfare made difficult predicting how much funding was required in the usual form of appropriations. Unknown were the specifics of how much for each service, in which account, for reserves or active forces, to buy what, to be used when and the other variables of budget execution.

The administration argued that those decisions belonged to the combatant commander based on military effectiveness and national security, not the constraint of a comptroller's best guess. The administration argued for and requested flexible authority.

DERF was initially created in the Fiscal Year 1990 Department of Defense Appropriation Act (PL 101-65) and was provided $100 million of obligation authority without FY limitation to its use. These funds were to be used for reimbursement of other appropriations when expended by the Defense Department in support of state and local governments. As originally conceived, DERF was not a mechanism for funding Defense Department military activities overseas; instead it was designed to facilitate the use of the department to address domestic problems (e.g., fighting forest fires). Overseas military actions were traditionally funded through existing operations accounts if they were sufficient or through emergency supplemental appropriations as we discuss elsewhere (McCaffrey & Godek, 2003, pp. 53-72). As can be seen from the exhibit, Congress has many tools to hand, but the use of these tools while increasing Congressional control of defense, decreases the flexibility of the executive branch to deal with the crisis at hand.

It is not only important to examine why Congress tends to control defense, it is useful to examine the manner in which that oversight occurs. The literature provides us with a framework of tools including restrictions on the use of funding, tools for gathering information, and accounting requirements. Regarding the use of funding, members of Congress may make line-item adjustments to the budget (Blechman, 1990; Halperin & Lomasney, 1999; Lindsay, 1990; Mayer, 1993; Thompson & Jones, 1994); earmark funds for specific purposes (Owens, 1990; Mayer, 1993); place restrictions on the reprogramming and transfer of funds between accounts (Fox, 1988; Jones & Bixler, 1993; Wildavsky & Caiden, 2001); and restrict funds pending executive compliance with provisions of law or committee reports (Owens, 1990). They gather information formally in congressional hearings and informally between congressional and DoD staff outside of hearings or through the use of reviews, audits, and investigations by committee staffs, GAO, and Congressional Research Service (CRS) (Owens, 1990). Finally, Congress may place requirements on program execution or may specify reports on a myriad of topics (Owens, 1990; Jones & Bixler, 1993).

Strong incentives are present for Congress to actively manage defense policy and budgets, and congressional rules and procedures provide many means by which to control DOD through authorization, appropriation and oversight as we discuss elsewhere. The fundamental tension between the constitutional roles of the executive and legislative branch

Table 2.3. Why and How Congress Exercises Authority Over the Defense Department

	Fox	Owens	Lindsay	Blechman	Jones & Bixler	Mayer	Thompson & Jones	Halperin & Lomasney	Hartung	Wildavsky & Caiden
	1988	1990	1990	1990	1992	1993	1994	1999	1999	2001
Why Does Congress Control Defense?										
Legitimate exercise of constitutional power	X	X	X			X	X			
Policy influence						X				
Response to media publicity of DoD mismanagement	X		X		X					
Partisan politics, competitiveness, mistrust	X			X	X	X	X			
Direct federal spending (Defense budget is half of discretionary spending)	X	X			X			X		X
Advocate or protect constituent interests	X	X	X	X	X			X	X	X
How Does Congress Control Defense?										
Line item adjustments to the budget	X	X	X	X	X	X	X	X		X
Earmarked funds		X				X				
Reprogramming and transfer restrictions	X	X			X					X
Restricting access to funds pending compliance		X								
Formal and informal information gathering	X	X	X	X	X	X	X			X
Reviews/audits/investigations by committee staffs, GAO, CRS, and so forth			X		X					
Structural requirements placed on programs		X			X	X	X			X
Reporting requirements	X	X			X					

Source: P. J. Candreva and L. R. Jones, 2005.

over control and flexibility is longstanding and has no definitive answer other than to seek a balance appropriate for the time/situation.

CONCLUSIONS

Budgeting began in this country as a legislative enterprise, and specifically as a check against executive power. The people exercised the power of the purse through elected representatives. Effective representation of demands was emphasized over the needs of executive efficiency. When this was seen as not efficient enough, we created an executive branch and have gradually released power to it. By the 1960s the executive power had taken the lion's share of power. Since 1974, the legislative branch has gradually organized itself to be a more effective partner in the resource allocation process. Thus, our experience with a prominent executive budget power is relatively limited. In periods of extreme crisis, Congress has tended to cede power to the executive and reclaim it after the crisis has passed. Although there is a good deal of conflict in the budget process, there is also a good deal of reconciliation and adaptation.

The United States was a country born poor. There were no crown jewels, no colonies to exploit. As a consequence of the Revolutionary War, the currency was a wreck and the colonists owed a substantial debt. Alexander Hamilton's efforts have long been recognized for paying down the debt and stabilizing the currency, in effect, putting the country on a sound, but not rich, fiscal footing. This uncertain fiscal history may have been the start of the balanced budget ethic which dominates so much of American budgetary discussion and practice. A country born poor, or just out of debt, must live within its means[9] (See Table 1.1). How this has turned out is interesting. Aaron Wildavsky compared American National budgets to their European counterparts. He found that U.S. budgets have been consistently balanced and that per capita revenue and expenditure ratios consistently lower than that of our European counterparts. What could explain these differences? Wildavsky (1964) suggests it comes about as a social legacy rising out of the revolutionary period.

Wildavsky (1964) observes that the winning side Revolutionary America was composed of three groups. The first, the *social hierarchs*, wanted to replace the king with a native variety better suited to colonial conditions. The second were *emerging market men* who wanted to control their own commerce. The third were *egalitarian Republicans*—a legacy of the continental Republicans. They stressed small, egalitarian, and voluntary association. What allowed these three groups to coexist, create, and operate a successful government was agreement on a balanced budget at low levels of expenditure, except in wartime. Wildavsky suggests that there was no formal declaration of this agreement, nor was it done on a single day. However, this agreement lasted for 150 years until a new understanding was forged in the 1960s. The social hierarchs would have preferred a stronger and more splendid central government, and the higher taxes and spending that went with it. But the emerging market men would have had to pay, thus they preferred a smaller government, except where taxing and spending provided direct aid. Together this coalition led to spending on internal improvements, like railroads, canals, and harbors. However, it was the egalitarian Republicans who did not believe that government spending was good for the common man, meaning the small property-holder and/or skilled artisan. They suggested that unless the scope of government spending was limited, groups of these common men would withdraw their consent to the union. Limiting, not expanding, government was their aim.

Out of these three strains came the impetus for a balanced budget at low levels of taxing and spending. Egalitarian Republicans were able to place limits on central government. Market men won the opportunity to seek economic growth with government subsidy, but the extent of this was limited by an unwillingness to raise revenues. The social hierarchs obtained a larger role for collective concerns provided they were able to convince the others to raise revenue. Wildavsky (1964) suggests that no group got all it wanted, but all got something. The Jacksonian belief that equality of opportunity would lead to equality of outcome helped cement this outcome observed Wildavsky. The result has been an ethic of a balanced budget at comparatively low levels of taxing and spending. Moreover, what is balanced in this equation, according to Wildavsky, is not only the budget, but also the social orders and their supporting viewpoints that helped found the country. This impetus towards balance continues to surround the decisions we make about taxing and spending.

It is clear that at some time during the mid-twentieth century the budget power assumed burdens that made it different in kind from anything that had gone before. The coming together of responsibilities for social welfare (with social security, medicare, and the great society programs of the 1960s), macroeconomic management (beginning with the Full Employment Act of 1946), and defense budgeting predicated on winning the Cold War (1948 to 1989) precipitated the modification that strengthened executive power in the process. The effect of these factors may be seen in the budget trend lines of the early 1970s when the rising trend line for human service expenditures crosses over the declining trend line for defense expenditures in the federal budget. Experts said that at that point the federal government went from a "doer" to a "checkwriter." Figure 2.1 is a mapping of this crossover with defense outlays and payments for individuals as a share of federal budget outlays.

Budget Shares

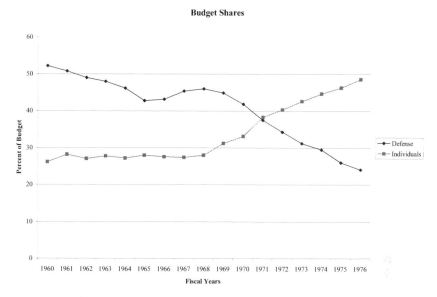

Source: Derived from OMB 2007. Historical Tables, Budget of the United States, FY 2008; Table 6.1 Composition of Outlays 1940-2012 for payments for individuals and Table 3.1: Outlays by Function and Subfunction for defense outlays.

Figure 2.1.   From doer to checkwriter.

The turning from doer to checkwriter may also be seen as a turning from provision of goods and services (like provision of defense) to a guarantor of social welfare through such mechanisms as social security, medicare, and medicaid where a large and growing share of the budget is dedicated to providing support to the aged, the infirm, and the poor to provide an adequate standard of living. By 2006, payments for individuals had reached a 60% share of the federal budget while defense was just under 20%. Most of the remainder was taken up by interest payments (15%). It would not be incorrect to say that the federal government has two main functions: provision of defense and writing checks for individuals. Table 2.3 indicates that this story is more complex.

First, since 1940, the United States has spent a substantial share of its wealth on human resources; this can be seen in the almost 44% share in 1940 and the 4.3% GDP (gross domestic product) share. Second, while its share increased, to 63% by 2006 and almost tripled to 12.8% of GDP, the way it was distributed changed; that is, payments for individuals increased from 17.8% of budget share to 59.9% and from less that 2% of GDP to more than 12%. As the vast majority of those payments go directly to individuals, they give individuals a sense of entitlement and ownership, thus making these patterns hard to change. Defense, on the other hand, is not linear; it responds to the threat as we have demonstrated elsewhere. Here we see it holding about the same budget share as payments for individuals in 1940 and increasing slightly to 2006, while payments for individuals more than tripled. The table shows that in the last 60 years there has been a definite change in American society and while few would argue that it is not for the better, there are questions about its sustainability. Concerning defense, the relevant question for policy makers is not its share of the budget or percent of GDP, but the extent and nature of the threat and if what is being spent is enough.[10]

Robert Hormats (2007) traces U.S. fiscal history and war financing through the Revolutionary Period and to the present. He observes,

> looking back over this nation's more than two hundred years, one central, constant theme emerges: sound national finances have proved to be indispensable to the country's military strength. Without the former, it is difficult over an extended period of time to sustain the latter. (Hormats, 2007, p. xiii)

Hormats notes that during the Revolutionary War Congress had no taxing power and was frequently unable to provide the money General

Table 2.4.   Historical Trends: Defense and Payments to Individuals: 1940-2006

1940	Defense Budget	Human Resources	Payments for Individuals
% Share	17.5	43.7	17.8
% GDP	1.7	4.3	1.7
2006			
% Share	19.7	63	59.9
% GDP	4.0	12.8	12.2

Source: Historical tables, U.S. budget FY 2008.
Table 3.1 for defense and human resources;
Table 6.1 for payments for individuals.

Note: Payments for individuals includes some grants made to state and local governments which are then redirected to individuals. Probably the most familiar source of payments to individuals is social security, but there are others. Human resource spending is the inclusive budget category that includes such spending as that on education, training, and veterans' benefits. In the Table, % share means share of the federal budget and % GDP means share of GDP.

Washington needed, how printing money quickly devalued the worth of the continental dollar, and how in the end loans from France and the Netherlands saved the day.

Hormats (2007) observes that since then military necessity often resulted in new and expanded fiscal powers, like the income tax during the Civil War. Hormats also observes that it is not enough to have well-trained troops, good generals and a good strategy, that the country also needs a sound financial strategy and skillful leaders at the Treasury, in the White House, and in Congress to ensure that sufficient money is available to meet military expenses. He concludes that under the duress of war, the nation's policymakers have produced "dramatic innovations in the nation's tax and borrowing policies, innovations that lasted long beyond the conflict during which they were introduced" (p. xiv). He believes that America's wars have been fiscally as well as politically transforming events, leading to changes that would not have been envisioned or accepted in quieter times, and the result has been that in the twentieth century America was able to generate, "colossal amounts of tax revenues, conduct massive bond drives, and produce great volumes of weapons" (p. xv), which resulted in victory in two world wars and the cold war. However, he warns that leadership is not only about raising huge amounts of money, but it is also about uniting the country behind the war effort and raising money without weakening the economy. This takes skillful political as well as fiscal leadership. Hormats worries that currently the United States is living in a post-9/11 world with a pre-9/11 fiscal policy where a, "heavily debt-laden, over obligated, revenue-squeezed government, highly dependent on foreign capital creates major security vulnerabilities." This is a useful warning to heed. Defense spending as a percent of GDP still remains at a generational low, the lowest since the 1960s, but there has been an increase in foreign holding of public debt. Should a terrorist attack result in an upset in these markets, the result could be a dramatic reduction in capital inflows, a spike in interest rates, and a collapse in value of the dollar, a more profound effect than any terrorist could hope to achieve on the battlefield.

In his historical review, Hormats (2007) lauds three Secretaries of Treasury, Hamilton for his handling of the post-Revolutionary war crisis; Salmon P. Chase who guided the Union through its fiscal and monetary efforts in the War Between the States, and William Gibbs McAdoo who helped the U.S. withstand the global economic storm set off by World War I. Hormats also has kind words to say about President Dwight D. Eisenhower who steered a middle course between a hungry military-industrial complex and a sometimes irresponsible tax-cutting Congress. In his farewell address, mostly known for its warning about the military-industrial complex, President Eisenhower also warned, "We cannot mortgage the material assets of our grandchildren without asking the loss also of their political and spiritual heritage. We want democracy to survive for all generations, not become the insolvent phantom of tomorrow" (p. 203).

Critics including Hormats (2007) worry that the United States is at another turning point in its history where it may be risking its future by imprudent fiscal behavior. This notwithstanding, the critics' main message is that the United States has found appropriate fiscal strategies and competent leaders to implement them in times of danger and war. We may note, however, that the legacies of two Presidents, Johnson and George W. Bush, have been severely marred by the inability to either win or withdraw from publicly contentious and expensive wars. Had the Union not prevailed over the Confederacy in 1865 perhaps Lincoln would have been similarly criticized.

Moreover, the long sweep of history teaches us that techniques of budgeting are not as important as the purposes for which the money is spent and the ideas that drive those purposes and the men who pursue them. While it is in a constant state of change, the budget process itself is a resilient and flexible procedure able to accommodate to changing conditions. By 2007, most observers had agreed that using supplementals to fund ongoing operations in Iraq was a flawed process, yet in separating out and highlighting those costs, the supplementals served a purpose. Part of the genius of the American character resides in its impulse to find a better way, but not to depart radically from tried and tested methods. Social engineers, rationalists, and autocrats might have done better, but experience with the budget process leaves us with a demonstration that representative government works.

NOTES

1. Hamilton's program was outlined in four separate reports: Reports on the Public Credit of January 14, 1790 and December 13, 1790; The Report on a National Bank, December 14, 1790; The Report on Manufactures, submitted to Congress on December 5, 1791.

2. There were 15 appropriations bills passed between the founding of the Federal government and May 8, 1792. A listing of these may be found at 3 Annals 1258-1259. From 1789 to 1792, the United States had a surplus of $21,762 on revenues of $11,017,460. (3 Annals: 1,259) During this period, general appropriations grew steadily from $639,000 in 1789 to $1,059,222 in 1792, and ranged from a high of $2,849,194 for payment of interest on the national debt in 1790 for 1792 to a low of $548 for sundry objects in 1790. Protection of the frontier was a growing expense: $643,500 in 1792, not included in the general appropriation. In 1791, Congress appropriated $10,000 for a lighthouse; in 1792, $2,553 was appropriated for a grammar school. Of the $11 million raised to the end of 1792, over $6.3

million was applied to the debt, either interest or principal. It is clear that these were still transitional years for the federal budget.

3. Fisher (1975) notes that lump-sum appropriations are especially noticeable during periods of war and national depression, when the crisis is great and requirements uncertain. At these times, the Congress tends to delegate power (p. 61).

4. This position was supported by both academics and public administrators. Notable among the former were Arthur Smithies (1955) in, *The Budgetary Process in the United States* (p. 53), and Leonard D. White in his four-volume history of the federal government, notably, *The Jeffersonians*, (1951, pp. 68-69); *The Jacksonians*, (1954, pp. 77-78); and *The Republican Era*, (1958, p. 97). The bureaucrats who argued this included two budget bureau directors, Maurice Stans and Percival Brundage; see Fisher (1975: 269-270). Fisher describes the growth of executive budget power as a steady accretion manifest in numerous statutes, financial panics, wars, "a splintering of congressional controls," and demands from the private sector for economy and efficiency.

5. This commission report was submitted to Congress on June 27, 1912. See also Taft's message, Economy and Efficiency in the Government Service, H. Doc 458 62-2, January 27, 1912.

6. Burkhead (1959) observes that the interest of the business community in reform was the crucial element in this mixture. Business expected lower taxes (p. 15). To understand other variables at work, see Allen Schick, "The Road to PPB: The Stages of Budget Reform," *Public Administration Review*, 26 (December) 1966, pp. 243-258.

7. Burkhead suggests the primary motive of Congress in passing the Budget Act was to reduce taxes, not to improve executive leadership.

8. One of the most comprehensive examinations of the Congressional Budget and Impoundment Control Act is Allen Schick, *Congress and Money: Budgeting, Spending and Taxing* (1980). For a description of reconciliation, see Schick, *Reconciliation and the Congressional Budget Process* (1982).

9. From 1789 to 1849, accounts for the U.S. government show a surplus of $70 million. Notwithstanding any individual year deficits, they practiced what they preached. See Table 1.1 U.S. Budget FY 2008 summary of receipts and expenditures.

10. As the first Secretary of the Treasury in 1790, Alexander Hamilton explained that the huge debt run up during the Revolutionary War was the "price of liberty" and had to be repaid.

CHAPTER 3

BUDGETING FOR NATIONAL DEFENSE

INTRODUCTION: DEFENSE BUDGETING IS DIFFERENT

Defense budgeting presents unique challenges not faced elsewhere in the federal budget process, or in budgeting for state and local governments. The research and writing of budget guru Aaron Wildavsky shaped how students and scholars viewed budgeting for more than three decades, but most of his work explicitly ignored defense budgeting. The original *Politics of the Budgetary Process* in which Wildavsky's (1964) theory of budgeting as incremental behavior was unveiled (pp. 13-16) had no chapter on budgeting for national defense. Wildavsky is not alone in this omission. Most of the literature on public sector budgeting ignores defense budgeting, despite the fact that spending in this area typically represents a large part of the discretionary budget of the federal government. For example, far more is written and published in analysis of the roughly $50 billion the budget allocates annually to social welfare programs versus the $400 billion spent on national defense. Recognizing the deficiency, Wildavsky (1988) added an insightful chapter on defense budgeting to his classic 1964 work when he revised it as *The New Politics of the Budgetary Process* (pp. 348-395). Wildavsky recognized that for the reasons we explore in this chapter, budgeting for national defense might be differentiated from budgeting elsewhere in the federal government.

Budgeting, Financial Management, and Acquisition Reform in
The U.S. Department Of Defense, pp. 75–93
Copyright © 2008 by Information Age Publishing
All rights of reproduction in any form reserved.

Total Spending Matters

Defense and nondefense budgeting differ in important ways. The defense budget is an instrument of foreign policy and other nations react to changes in funding levels and priorities. U.S. defense budgets also respond to spending in nations that pose threats to our interests and those that we rely on as allies. When other nations change their defense allocations, for example, as a percentage of gross domestic product (GDP), U.S. decision makers must determine how such changes should affect our defense spending. Will we have enough trained personnel, ships, and aircraft, spare parts, maintenance capacity, and all the other capabilities that establish and sustain military force readiness to counter changes in levels of threat? Money buys capability. More capability means an increased ability to deter threat or to inflict damage to others who then must counter that increased ability with increases of their own, or find strategic alliances to negate the threat. Other nations do not monitor the total amount the U.S. spends on education or health, or if they do, it does not have the same salience. Defense budgeting is about deterring or countering threats that exist, or that may emerge in the future. Shifts in funding are early warnings that the threat scenario is changing and responses to it must also change. Thus, Wildavsky (1988) observed that one difference between defense and nondefense budgeting is that total spending means something to other nations and the amount of change from 1 year to the next is carefully watched for hints about future behavior.

Recognition of this fact was perhaps the most significant insight of the Presidency of Ronald Reagan. Reagan and the Congress appropriated funding levels for national defense that literally drove the Soviet Union to bankruptcy and contributed to the end of the Cold War. In this respect, U.S. spending on defense during the entire Cold War era represented willingness on the part of American people and their leaders to spend what was necessary to counter what was perceived as an ever-expanding threat from the USSR. After the end of the Cold War, Presidents Bush and Clinton and Congress struggled with the question of how much to spend on defense. The terrorist attack of September 11, 2001 forced the United States into a new posture to counter a type of threat never faced before. The result is increased defense budgets and spending plans for the first decade of the twenty-first century and perhaps longer. President George W. Bush articulated the resolve of the United States of America to do, "whatever is necessary for as long as it takes," to prevail in the War on Terrorism. To a great extent, the American people have supported the view of spending whatever is necessary to counter a threat that is far less easily identified and attacked than any other the nation has faced in it history.

The Defense Budget Provides Opportunity to Reward Constituents

National defense comprised about 21% of discretionary spending in the FY 2008 budget request.

Because so much of total federal spending has been placed into entitlement accounts (e.g., Social Security, Medicare, Medicaid) that are permanently authorized rather than appropriated annually, the defense budget is one of the few discretionary pots of money available where a Congressman can seek to get money to fund local projects. Moreover, since defense appropriations tend to be veto proof (Reagan vetoed one because the out-year funding provided by Congress was too low), Congress is often tempted to attach unrelated items to defense appropriation bills. For example, the fiscal year (FY) 1994 federal budget carried a sum for breast cancer research, as well as dollars for various museums and memorials. The Senate version of the supplemental appropriation to cover the cost of the war in Iraq in 2003 included funding for fisheries management in Lake Champlain, subsidies for catfish farmers, research facilities in Iowa and the South Pole, repair of a

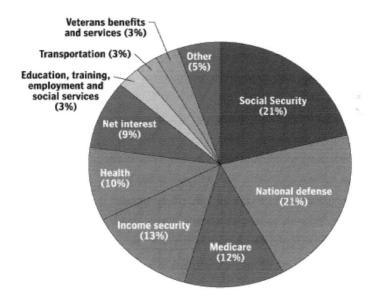

Source: Office of Management and Budget, Graphic: Retrieved February 5, 2007, from http://www.washingtonpost.com

Figure 3.1.   Federal budget spending categories FY 2008.

dam in Vermont, money to designate Alaskan salmon as "organic" and an increase in postage allowances for Senators to communicate with their constituents. The defense appropriation bill is much larger than any other appropriation bill and presents Congress with a continuing temptation to add unrelated items to it. Thus, the defense appropriation sometimes is referred to as a "Christmas tree" bill because it provides an opportunity for everyone to get something.

A Different Relationship With OMB

The Office of Management and Budget (OMB), the President's budget office, has a different relationship with the Department of Defense (DoD) than it has with other federal departments and agencies. OMB generally does not play the same adversarial role toward defense that it plays in examining the budgets for agriculture or education and other domestic programs. Rather, OMB and the Office of the Secretary of Defense (OSD) team up in review of the defense budget, a process done in-house in the Pentagon. Other domestic departments submit their budgets for review to OMB. When OMB cuts them, these agencies may appeal the actions to the President; the pattern is reversed with defense. If OMB wants to cut the defense budget, this first must be negotiated with officials of the OSD. Then, on large spending issues, OMB must appeal decisions not approved by the secretary of defense directly to the President, a tradition that Wildavsky and Caiden (1997, p. 234) date to the Kennedy administration.

Appeals to the President may not support the OMB position versus defense. This depends on the priorities of the President. For example, David Stockman, former Director of OMB under President Reagan, related (Stockman, 1986) that he would tell the President defense spending increases could not be accommodated without increasing the annual budget deficit. Invariably, Stockman was argued down by Defense Secretary Caspar Weinburger in front of the President, and Stockman would be sent back to OMB headquarters, grumbling to himself, after having been told to "rework the numbers" so that defense increases would be "deficit neutral" in the President's budget. The point is that presidential priorities for defense and national security dominate the relationship between OMB and the DoD. With national defense, it is clear that policy drives the budget—not the other way around. This cannot be said for the rest of the domestic portion of the budget. The prominence of policy is highly evident in this decade in post-September 11, 2001 era where defense budgets have increased to fight the war on terrorism.

The types of decisions made solely by OMB with respect to other agencies are made in concert with the Department of Defense at the Office of Secretary of Defense level. The partnership that exists between OMB and DoD is different from that which exits between OMB and other cabinet level agencies. OMB staff work at the Pentagon and are involved not only in budget review, but also in development and review of program structure in the Program Objective Memorandum phase of the Planning, Programming, and Budgeting Execution System (PPBES) process where resources planners decide what program structure needs to be maintained, improved, created or de-emphasized to meet changes in the threat. Tyszkiewicz and Daggett (1998) state, "The defense budget is unique in the extent to which OMB is directly involved throughout the budgeting process" (p. 28). To some extent, this results from the fact that defense is different from other federal departments. Its appropriation is much larger than that of other federal departments; it employs many more people than the others do; and its input mix of people and capital equipment is very different as it provides trained people and equips them to defend the country against various threat scenarios decades into the future. Its focus is also somewhat different as it prepares to meet what might happen as opposed to reacting to what did happen last year. Most importantly, the defense budget is simply too large for OMB to review on its own. It is not staffed to perform this tremendously time-consuming task.

No Ceilings for Potential Defense Spending

Another unique aspect of the defense budget is that the process for defining requirements begins with no ceiling in the PPBES used by DoD to define the threat, and to plan and budget to counter it. However, annually ceilings are introduced in the budget cycle, initially by the President and DoD, and then by Congress. Historically, this ceiling has been depicted in the President's budget in billions of dollars and as a percentage of GDP, a percent of the federal budget, and a real dollar (inflation adjusted) percentage amount over a 5-year period. OMB represents the President's priorities as it helps negotiate overall defense spending in the defense budget process prior to presentation of the budget to Congress.

Cycles Versus Straight Line Growth

Defense spending has exhibited a cyclical long-term pattern different from that of other budget accounts. For example, human resource spending has soared since the 1950s in a relatively straight line, but defense has experienced a feast or famine profile, consisting of roughly 7 lean years and 7 rich years. This pattern is revealed in the build up and down for the Korean and Vietnam wars in the period 1950 through 1980, the increase during the Reagan years, and the decline during the Presidencies of George H. W. Bush and Bill Clinton. With the fall of the Berlin wall and the end of the Cold War, defense budgets declined by about a third in dollars and personnel to the mid-1990s and stayed roughly flat to almost to the end of the decade. Increases in appropriations began to flow into the personnel accounts to keep salaries and benefits equal with inflation and to solve retention problems within the defense community in 2000. Then, the onset of the War on Terrorism initiated a new spike upward in of defense spending.

Absence of Consensus on How Much to Spend for Defense

Economists have depicted budgeting for national defense as a competition between funding "guns or butter." There are few milestones to indicate what is sufficient spending for defense. During the 1960s and again in the 1980s the media often portrayed defense spending as taking from the "poor" (human resource spending) to give to the "wasteful" (defense). One trend identified by Wildavsky (1988) is the disappearance of consensus over national defense policy and budget goals. Defense was budgeted under bipartisan consent from the beginning of World War II to the early 1960s. However, what was determined by consensus in the 1950s became subject for conflict in the 1960s. During this period, considerable consensus existed in other areas of budget policy. However, constrained resources, an increasing national debt, the end of the Cold War, an aging population, and increasing health care costs have driven great fissures into the previous state of budgetary consensus, so that policy that was consensual has now become embroiled in the politics of dissensus. The result of this dissensus is that budgets have tended to be bitterly contested and generally passed late each year. Moreover, part of the price of passage involved spending on projects and programs that benefited certain areas or regions more than other areas. In the 1990s, for example, numerous projects were forced on DoD to maintain local employment. This was sometimes called "maintaining the defense industrial base." In other instances, it was simply called "pork." Observers have also suggested that DoD responds to political pressures by placing programs in as many states and districts as possible to strengthen its political base in Congress (Smith, 1988). The terrorist attack on the United States on September 11, 2001 and the subsequent War on Terrorism produced a new consensus on the need for increased defense spending. Defense spending rose significantly in 2001 and thereafter and large supplemental bills dominated by defense and homeland security interests were also quickly passed.

Increased Participation in Defense Budgeting and Policymaking

Lindsay (1987), Wildavsky (1988), Jones and Bixler (1992) and others have suggested that since the 1960s budgeting for defense has gone from an "insiders" game to an "outsiders" game. In the 1950s, it was possible for just a few members of Congress serving on appropriation subcommittees to dictate decision making. This was true for nondefense budgets as well. However, since the end of the 1960s there has been a loss of power by the powerful committee chairmen who rose to power based on seniority and safe electoral districts and could bestow rewards and punish almost with impunity. To some extent, stability in budget decision making was purchased with the coin of secretive, elitist decision making.

The seniority system for selecting committee chairs and members remains important, but committee positions may also be gained through caucus elections. Powerful committee chairmen can be upset and disciplined through this election process. The traditional power of political parties to enforce voting blocks has been eroded by the formation of shifting coalitions of legislators seeking to unite with others to vote for specific interest that cross party line boundaries. Thus, the power of committee and subcommittee chairs has been reduced and can be moderated. Moreover, when the Democrat or Republican caucus picks a committee member by vote, the vote of the newest freshman is equal to that of the most senior member. Furthermore, as opportunities for funding projects for local constituencies in other parts of the budget have declined, more effort has been exerted to gain access to and influence over the defense budget.

In the 1950s, the defense budget was composed and reviewed in approximately 6 committees. By the mid-1960s, the number had expanded to 10 (Wildavsky, 1988). Since then there has been a proliferation of committees involved in defense policymaking and budgeting. Now, defense policy and budget hearings are held in approximately 28 different committees and subcommittees. This includes the full appropriations committees and five subcommittees (military construction, defense, energy, and water, HUD (Department of Housing and Urban Development), and the independent agencies, and commerce, justice, and state), the Budget Committees and the committees on armed services, commerce, energy, government affairs, select intelligence, small business, and veterans affairs or their analogs in each chamber. The

General Accountability Office estimated that in the period from 1982 to 1986, 1,306 DoD witnesses testified before 84 committees and subcommittees, presenting 11,246 pages of testimony in 1420 hours (Wildavsky & Caiden, 1997, p. 243). Jones and Bixler (1992) found that the size of defense authorization bills increased from one page in 1963 to 371 pages by 1991 (p. 49). Appropriation bills increased in the same period from 18 pages to 59. Similar increases were found in the length of committee reports accompanying the budget. They also found that the number of days the House of Representatives devoted to debate on defense authorization bills rose by a factor of 10 from 1961 to 1986, and the number of proposed amendments increased from 1 in 1961 to 140 in 1986 (pp. 68-69). The number of directives related to the budget made by Congress to DoD increased from 100 in 1970 to 1084 in 1991 (p. 78). DoD made fewer than 10 presentations on the budget to Congress annually in the 1950s, but by the early 2000s, this number was over 100.

It should also be remembered that DoD has a sizeable "black" or secret budget, once inadvertently leaked to the press to be in excess of $30 billion. Wildavsky and Caiden (1997, p. 234) assert that the black budget has increased from about $5.5 billion in fiscal 1981 to $28 billion in 1994. In 2007, it was estimated that DoD was asking for $31.9 billion for classified acquisition programs, about 18% of the acquisition total of $176.8 billion, as compiled by Steven Kosiak of the Center for Strategic and Budgetary Assessments. About $14 billion was for procurement and about $17 billion was for RDT&E (research, development, training, and engineering). Estimates are that about half of the testimony on defense is classified secret and takes place behind closed doors. Thus, even if the public does not have access to this testimony, members of Congress do. Still, experts worry that "black" programs typically get more restrictions on their funding streams, yet less oversight by lawmakers and Pentagon leaders and that this leads to performance problems and undue cost growth. Of the services, Air Force gets the largest share of the black budget, perhaps 80%, because it handles so much of DoD's command, control, communications, and intelligence tasks, including satellite and space launches (Bennet, 2007, p. 22).

More committees mean more places where decisions about defense are made, more opportunities for outsiders to hold congressional committee seats from which influence can be leveraged, and more necessity to coordinate final decisions on the floor where each member has one vote. These committees demand lots of testimony from DoD, for budget making and for oversight. In addition, turf wars between committees, for example the authorization and appropriation committees, must be negotiated among a larger number of participants and sometimes the dialogue spills over to the floors of the House and Senate where outsiders have voice in voting on bills such as the defense authorization, appropriation, and military construction bills and amendments to them.

In summary, the number of players in the defense policy and budget arena has expanded from a handful of members to a cast of hundreds, all supported by personal and committee staff. Thus, budgeting for defense has gone from an insider game to an outsider game, where considerable dissensus exists about defense policy priorities. While the terrorist attacks of 2001 seemed to provide a renewed public appreciation for national defense, anti-war demonstrations took place even as Operation Iraqi Freedom went forward and American and coalition troops were under fire. How much should be spent for defense, and on what, remains a highly debatable and political question.

The Power of the Budget Base and Threat Assessment

Each year the defense budget debate focuses more on the changes to the base—on new plans, programs, and activities (PPAs). Here is where benefits may be sought for constituents. Thus, the dialogue in most congressional committee venues is on the increment of change; the base will be questioned in most cases only to the extent that proposed changes expose its strengths or weaknesses. In the main, this process is incremental, historical, and reactive. The defense budget is large and bewilderingly complex, and so is the process that produces it. Size and complexity make complete review of the annual defense budget impossible.

The main drivers in the defense budget are the policy priorities of the President combined with assessments made in the Planning, Programming and Budgeting System (PPBS) process about the nature of future threat, the direction and demands placed on U.S. foreign policy, the nature of existing defense alliances worldwide, and the disposition and behavior of allies and other nations and their military capabilities. From these assessments, a scenario is created, much like a large mural, describing the world of the near term future. Not all the details are filled-in, nor are all the details clearly seen. This scenario exists some 5 years into the future and the iterative, cyclical nature of the defense PPBS process keeps this scenario alive and moves it forward year by year. The scenario depicts threats to U.S. interests at home and abroad. Each annual budget fleshes out details, but only for the most important parts of the picture at that moment. There is a constant evaluation of the scenario viewed from different levels within the DoD and other parts of the government concerned with national security. A continuous budget process attempts to clarify the near term aspects of the scenario.

The strength of PPBES is asserted to be the supply of long-term stability to defense planning and budgeting. Conflicting political priorities, complicated congressional budget procedures and annual delays in passing defense appropriations make it more difficult for DoD to link the multiple budget years it administers at any one time with proposed funding for the FY sent to Congress by the President. This situation has perplexed a succession of defense leaders. Secretary of Defense Donald Rumsfeld viewed the PPBES cycle as too long and repetitive, and in the period 2001-2003 took steps to shorten it to provide decisions in a more timely manner.

In sum, the threat drives the budget, but antecedent to this is foreign policy and threat definition. Good foreign policy can reduce threats or cause them to disappear, through treaties and alliances, but diplomats know that diplomacy without the threat of force and the ability to project power is useless. DoD supplies the threat of force, but foreign policy defines where force may be needed and risk assessments dictate how much force should be purchased through the budget.

When defense leaders deploy the armed forces, they are using the tools and personnel that their predecessors built. They literally fight with a force and force structure someone else built. Their challenge is to build a force that will meet the demands that will be placed on their successors some 5 to 10 years in the future. In terms of what the budget buys, the military must fight with yesterday's force as they create tomorrow's force. Tomorrow in this sense may be 5 years away, or it may be just over the horizon where a potential opponent introduces new technology, for example, North Korean nuclear capability. In budgeting then, the crucial questions for DoD leaders are "What is the threat?" followed by "How has it changed from last year?" and then "Given this change, how do we need to change force structure and how much will this cost?" The further imperative is that only so many dollars may be allocated to defense. Defense leaders always feel that whatever is allocated is not enough, thus they continually fight to let no dollar go to waste. DoD always budgets in an environment of scarcity. Other departments do not have quite the same conceptual burden of reading the future and interpreting the past that DoD annually faces in the budget process.

Notwithstanding these complexities, some simplifying routines do exist in defense budgeting. The threat scenario is fine-tuned from 1 year to the next. Fundamentally, this is an incremental process. Indeed, in defense some patterns are even more stable than for nondefense programs, for example, treaty commitments are long-lasting, binding agreements that shape the defense configuration. Acquisition of defense hardware, weapons, aircraft, ships, tanks, and similar purchases involve long-term commitments. Once a decision has been made to purchase an aircraft carrier, a type of aircraft, or an advanced submarine, the posture of defense capability has been set for the next 2 to 3 decades or longer. For example, in 2007 the average age of the Air Force KC-135 Stratotanker was 48 years, this for a plane in daily use around the globe.

What happens when a fundamental change takes place, for example, at the end of the Cold War? At this point, the entire threat and response capability must be reevaluated, and this takes time that runs beyond the annual budget cycle. It is arguable that long-term planning systems such as PPBES do not handle revolutionary change very well. A major portion of defense workload disappeared with the end of the Cold War. The 1991 Gulf war hastened the transition from Cold War thinking and planning. Still, it has taken more than a decade and the emergence of the new threat of terrorism to transform U.S. defense forces planning away from a Cold War mentality. Moreover, even in the mid-2000s, some of the defense budget is Cold War influenced and many of the assets in the military force structure are those designed to counter the Cold War threat. Replacing these aging assets has become a major budgetary problem for DOD.

The defense resource allocation process is inextricably entwined with foreign policy, which itself is constantly evolving. New threat and policy response can redefine the workload demand placed on the military forces. By choosing to deal with the threat in a variety of ways, the demand place on defense capability is varied. Recent examples are decisions to go to war in Iraq, but to negotiate the threat posed by North Korea. The point is that the defense resource allocation process may have huge burdens thrust upon it or taken from it in ways that are not characteristic for other parts of the federal government. The PPBES and DoD as an organization have problems in accommodating revolutionary change, not only in deciding how to meet the new threat, but because its capital asset base (trained personnel, aircraft, ships, weapon systems) once purchased and deployed take time to redirect and rebuild to meet emergent threats.

Differences Related to Federal Budget Process Dynamics

The federal budget process contributes to the difficulty of enacting and changing the direction of defense policy and the budget. That the budget process has no end is a lament heard in the Pentagon and Congress about the resource allocation process. The PPBES process is an iterative process, so it purposely recreates and adjusts the threat and response scenario on an annual basis. Additionally, because much of the defense budget is for acquisition of military hardware this means that programs are almost always at risk for elimination, for modification, delay, "stretch-outs" (buying weapons over a longer period of time than intended), and other

changes deemed politically attractive in Congress. For acquisition program managers, their programs are always under scrutiny and decision
closure is hard to get—what is agreed to in 1 year may be undone or
changed later. This is primarily a defense phenomenon due to the high
percentage of the nondefense federal budget that is mandated. Thus,
defense budget players know that each budget process has an end game
in which decisions are finally taken to produce a President's budget or to
get an appropriation bill passed. However, they are also aware that as
soon as decisions have been made, the next budget cycle will crowd in
with its deadlines, crises, problem programs and issues that seemed to
have been settled but must be visited anew. Hence, the adage, "There is
no ninth inning" for the defense budget is accurate.

For many domestic programs, the annual budget process has fairly stable rhythms. In addition, for many domestic agencies funding increments
since the early 1980s have been small or in budgets that have been in
steady state. In contrast, the defense budget is not really an annual budget, although much of it is annually authorized and the majority of it is
appropriated and obligated annually. Rather, it may be viewed as a stream
of decisions and resources whose intent is to sustain a force to safeguard
the future by being ready for use if called upon.

Since the early 1960s, defense funding has gone through cycles of
increase, major decline, significant growth, another major decline, and
most recently, an increase. These changes in funding levels resulted from
the Vietnam war spending boost, post-Vietnam war cuts that left what has
been referred to as a "hollow force," the Reagan defense build-up, the
Cold War "peace dividend" reduction, and new growth to fight the war on
terrorism. In the late 1980s, the Senate Armed Services Committee
required a biennial defense budget intended to provide stability and to
extend the decision time horizons, but the appropriating committees
never adopted it. To a considerable extent, by virtue of the design of the
PPBES, defense resource planners have operated beyond a 2 year horizon
since the 1960s, notwithstanding that most defense dollars are provided
in annual appropriation bills.

Short Versus Long Spending Accounts

In the defense budget, PPAs are created and provide Budget Authority
(BA) through passage of appropriations legislation that is eventually
signed into law by the President. BA is a promise to pay over the life of a
program but not all BA to fund acquisition and other programs is provided in any one FY. Figure 3.2 indicates the amount of new BA recommended for 2008 as well as how much of FY 2008's outlays would result

from BA created in prior years. Thus in total, the U.S. intended to spend $2902 billion in 2008 and added another $628 billion to the stock of BA to be spent in future years, totaling $1474 billion that would be spent in future years, even if no further appropriation action were taken in the future. Thus, although the U. S. is correctly described as an annual budget system, its budget process regularly makes legally appropriated future year BA commitments, whose cumulative size ($1,474b) is half the size (50.7%) of the total of current year outlays.

Total Obligational Authority

In DoD, BA for the current year is merged with BA created in past years, but available in the current year. This is called Total Obligational Authority (TOA) and consists of all funds available for obligation in a single year. At the end of the FY, unexpended funding will be rolled

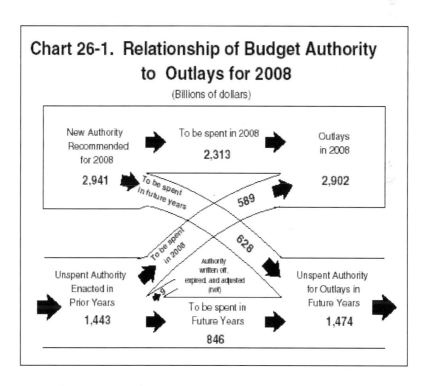

Source: Budget of the United States FY 2008. *Analytical Perspectives,* p. 402.

Figure 3.2.   Budget Authority for the current and future years.

forward as long as the authority for it has not expired. Figure 3.3 shows the relationship of TOA, outlays, and BA provided for defense. The years after 2000 are estimates. TOA may exceed outlays and BA when the stock of unspent authority from previous years falls in any particular year, for example, 1991. BA may exceed TOA for any particular year, because BA is created for this and future years, thus BA is more than TOA in 1990, less in 1991 and about the same as TOA in 1998. Outlays are made as goods and services are delivered to execute obligations. Outlays can be more or less than TOA and BA. From Figure 3.3 it appears that obligations created in 1991 were satisfied and paid for in 1992 through 1995, thus making outlays greater than either TOA or BA. The one relationship not shown on this exhibit is obligations; as we will discuss later, obligations should never be more than TOA. TOA is the stock of authority DOD has to use to create obligations in any 1 year.

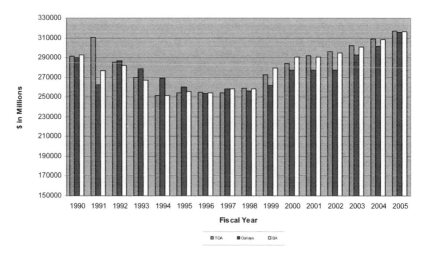

Source: Duma, 2001: extracted from National Defense Budget Estimates for Fiscal Year 2001 Budget, the Green Book, pp. 62-145.

TOA equals Budget Authority created for the current year and Budget Authority created in previous years to be spent in the current year. (See Figure 3.2) TOA tells DOD administrators how much they have to spend in the current year.

BA is the amount of Budget Authority created by Congress in the current year; some of this will be for the current year and some will be for future years.

Outlays consist of dollars actually spent in a particular year; past years are actual amounts while future years are estimates based on historical outlay to TOA relationships.

Figure 3.3.   Defense Total Obligational Authority, outlays, and Budget Authority.

Obligation and Outlay Rates

Defense budget accounts have different outlay or "spend out" rates, for example, military personnel (MILPERS) spends out at about 96% in a single FY, operations and maintenance (O&M) accounts spend out about 82% in their first year and have 5 additional years of expenditure availability, although by the end of the third year 99.8% of the funds were outlayed. Acquisition spending for major programs may be spread over a 5- to 10-year time horizon. For example, aircraft procurement has a 3-year obligational window and 5 years beyond that for outlay. Aircraft procurement Navy (APN) accounts obligate 78% of their funds in the budget year and outlay 16%; ship construction obligates 63% in the first year and outlays 7%. Military construction appropriations may take 20 years to be spent. Thus, the promise is on the books, but as successive budgets are reviewed and approved in Congress, changes in programs may be made and BA may be adjusted accordingly. This is a normal part of the turbulence in the defense budget process.

Managing Multiple Versus Single Years

Defense budget managers have multiple year budgets to monitor and control in execution, and execution issues and problems differ by type of appropriation account. A manpower analyst in the Pentagon may work only with current FY accounts, for example, FY 2004, using money from the MILPERS account. On the other hand, a Navy fleet comptroller will work with current FY accounts and accounts for the two previous FYs to fund (O&M where money to pay for fuel, ship repairs, steaming and flying hours is expended. On the other hand, an acquisition program manager managing the construction of a weapons system may have three past years and 5 to 7 future years to manage to complete a purchase from APN or ship building and construction Navy accounts. In contrast, nondefense federal budgeteers basically operate on a single year time horizon as BA or its equivalent is created, obligated, and outlayed during the FY.

Managing Outlays Versus Obligations

Attempts to control the federal deficit using outlay limits in deficit scorekeeping from 1986 through the late 1990s further complicated defense budgeting by making outlays more important than BA. Prior to this period, defense budget analysts were not as concerned about outlays; BA was what counted. During the late 1980s, outlays became critical targets in attempting to manage the annual operating budget deficit. This presented difficulty in managing accounts with greater than annual spending patterns. Currently this would amount to about 40% of the defense budget.

When spending reductions have to be made because of outlay targets, it also means that the faster spending accounts are cut first (e.g., MILPERS, O&M), because a dollar in BA for them is closer to a dollar in outlay in the budget year. For defense budgeting, time displaced merit in an outlay driven context to an extent not applicable in the nondefense arena. For example, merit might dictate that a ship purchase be cut from the budget, but its first year spend out rate (outlay) may be only 5%. Thus, to meet the outlay savings target might take a cut of ten ships; hence, the cut would be made in personnel, because 98% of outlay in this account is realized immediately. Whenever quick cuts are sought in defense, the temptation is to put at risk those accounts that have a high annual spend-out ratio, irrespective of what this does to total force capability. Because it appears that the federal budget will be in deficit for most of the coming decade, the outlay scoring game may once again become more important for defense budgeting in the future than during the brief 4-year period when annual federal budgets were in surplus.

Maintaining Balance in Managing Budget Accounts

Balance is an important concept within defense due to systems criticality. When one budget category, program, or element is adjusted, typically several others have to be readjusted so that all the elements of a defense program arrive at a predetermined time in the future, for example, aircraft, pilots, trainers, repair and maintenance manuals, weapons systems, avionics, spare parts. Constant attention has to be paid to balance. Nondefense systems generally do not have the same systems criticality and do not need to continuously scan budget lines for balance.

The procurement/acquisition and research, development, test, and evaluation budget accounts make up approximately 40% of the defense budget and are heavily scrutinized by Congress. Balance in managing these accounts is a critical for defending them from reduction by Congress in annual budget review. In addition, within DoD, balance has to be maintained between the military construction account provided as a separate appropriation bill (MILCON) and the main DoD appropriation legislation; a change in one may well affect a program in another. The task of budget analysts is to make the corresponding changes to keep the program in balance.

Overview: Policy and Budget Negotiation With Congress

The stakes are high when DOD budget growth or decline is negotiated with Congress because defense spending is such a sizeable component of the federal budget. Successful lobbying by constituent interests produces a significant payoff in terms of jobs, income, and preservation of a stable defense industrial base. Even when the climate exists to substantially change the defense budget, for example to reduce it and harvest a "peace dividend" in the early 1990s or to increase it substantially to respond to the terrorist attacks of 2001, the demands of electoral politics may be expected to compete with the national interest in providing for a common defense.

Congressional review and enactment of the defense program and budget requires action by six core committees and several others, assisted by their subcommittees. In addition, floor, conference committee, and reconciliation votes are required to pass enabling legislation authored by authorization (Armed Services) and appropriation committees. Given the decentralization and, at times, disorganization of congressional resource decision making, it is little wonder that the DoD, like many other federal departments and agencies, engages in some degree of strategic representation in justifying its program and budget. Both Congress and the DoD attempt to satisfy constituent demands in the budget. After all, pork barrel politics is nothing more or less than democracy in action, albeit at the loss of military effectiveness.

The defense plan and budget are prepared by the OSD, the military departments, and uniformed services on a programmatic basis with requests to Congress divided into 11 program elements. This program structure is cross-walked into appropriation, function, subfunction, object of expenditure, and other budget formats by the DoD for presentation to and review by Congress. Although DoD prepares the program budget, Congress does not review or enact the budget on the DoD programmatic base. Instead, separate authorization and appropriation processes are employed for policy, program, and budget decision making by the six major committees that negotiate and enact the defense budget (House and Senate budget, armed services, and appropriations). This is also the case for the rest of the federal budget. Congress reviews, negotiates, and executes much of the DoD budget proposal on a project and object-of-expenditure basis. Also similar to budget review for other federal departments and agencies, the most detailed analysis of the defense budget outside the DoD occurs in Congress at the subcommittee level.

Congress requires the DoD to submit its budget in a variety of forms and at a highly disaggregated level of detail. Appropriation committees, for example, receive voluminous computerized R-1 and P-1 program exhibits that show proposed spending on a line-item basis for every research and development or procurement program in the defense budget. DoD reports required by Congress indicate the item, quantity of purchase, and cost proposed by the DoD for each procurement by each service branch, in thousands or millions of dollars, and also contain information on previous purchases and product suppliers.

With this level of detail, subcommittee members and staff can attempt to surgically manipulate the DoD budget request to satisfy national security needs as well as the various constituent interests represented effectively by lobbyists in the highly decentralized Congressional decision process. So-called "add-ons" and "plus-ups" that provide funds for programs not requested by the DoD or that increase spending over DoD proposals are a normal element of congressional budgeting. However, in spite of the degree of budgetary power and influence wielded by congressional subcommittees, decisions reached in subcommittee are not final. Many opportunities exist to add or cut programs and money in full committee or on the floor of either house of Congress, in conference-committee negotiation, through reconciliation legislation, in ad hoc budget "summit" conferences, as well as through reprogramming and transfers in oversight of the budget once enacted. Review of supplemental budget requests from the DoD provides additional opportunity for congressional program, budget, and policy direction.

As we have noted, there is *never* a final defense budget—the budget is constantly up for negotiation. Despite the appearance of programmatic vitality or morbidity in the short-term perspective of 1-year budget nego-tiation, programs that seem dead rise again as advocates find new win-dows of opportunity. Conversely, programs that appear to be blessed with everlasting life may be put at risk of reduction or elimination at almost any time in the 9-to-11 month decision cycle that is typical for annual congressional action on the budget, and in the total budget cycle includ-ing preparation, negotiation, enactment, and execution that lasts approx-imately 2½ years. If audit and evaluation phases of the process are included, the total cycle takes 3, 4 or more years, for example, the length of a single presidential administration and double the length of the 2-year terms of members of the House of Representatives.

In resource negotiation with Congress, the DoD advocates its position assertively in a myriad of public and not-so-public forums. Since the 1970s, the number of formal committee and subcommittee hearings held on the defense budget has increased substantially. Despite DoD claims that it is "micromanaged" to incredible excess by Congress, DoD legislative strategy is developed and executed to play in the congressional budgetary system as it is rather than the way many critics believe that it ought to be. This is done through political issue positioning and a considerable degree of strategic representation that is a necessary response to congressional tendencies to micromanage, as with other federal departments and agencies. The details of congressional budgeting for national defense are provided in chapter 6.

CONCLUSIONS

The purpose of this chapter is to provide a basic understanding of some of the characteristics that make defense different from other players in the federal government budget process and cycle. This chapter has reviewed how budgeting for national defense differs from budgeting for other entities in the public sector in the United States, and virtually anywhere else in the world, including how Congress makes decisions on defense spending. In the chapters that follow, we provide detailed analyses of the various components of defense budgeting and resource management. To understand defense budgeting and financial management thoroughly, we analyze how national defense fiscal policy is developed and executed. Then, we delve into the operation of the PPBES process, the intricacies of congressional defense budgeting and supplemental appropriation, DoD budget execution, the roles played by participants in the defense budget process in the Pentagon and the field, financial management issues that confront defense resource managers and decision makers, budgeting for acquisition of weapons, and recent initiatives to transform defense budgeting and the DoD. To these topics we now turn.

CHAPTER 4

NATIONAL DEFENSE
SPENDING AND
THE ECONOMY

INTRODUCTION

The influence of U.S. national defense spending on defense spending in
allied and other nations has been noted in chapter 1. However, the
impact of defense spending on the U.S. economy is so significant that it
deserves special examination and analysis. There are a number of ways of
measuring the level and significance of U.S. defense spending, many of
which we explore in this chapter. We do so with the caveat that no single
measure is adequate on its own to indicate why defense spending is so
important to the U.S. economy or how it best can be measured both in
terms of impact and in identification of historical patterns and trends. In
this chapter we provide a number of methods to view and analyze U.S.
national defense spending related to a number of variables.

EVALUATING DEFENSE SPENDING: PROCESS COMPLEXITIES

Testifying before a House subcommittee on government reform on June
4, 2002, defense official Franklin Spinney (2002) compared the defense
budget plan to "a boiling programmatic soup in which 'low-balled' cost

Budgeting, Financial Management, and Acquisition Reform in
The U.S. Department Of Defense, pp. 95–136
Copyright © 2008 by Information Age Publishing

estimates breed like metastasizing cancer cells," where biased numbers "hide the future consequences of current policy decisions, permitting too many programs to get stuffed into the 'out-years' of the long-range budget plan." Spinney added that this set the stage for "unaffordable budget bow waves, repeating cycles of cost growth and procurement stretch-outs, decreasing rates of modernization and older weapons, shrinking forces and continual pressure to bail out the self-destructing modernization programs by robbing the readiness accounts" (Grossman, 2002, p. 6).

It is obvious that these are serious matters, but the defense budget process is so complex and the terminology so complicated that decision makers sometimes have trouble deciding not only what is at stake, but what the words mean. Certainly, someone unfamiliar with defense at all would have trouble interpreting the paragraph above. In this chapter, we provide a guide for those who would understand the dialogue and rhythms of defense budgeting. We examine basic terms and concepts that surround the defense resource allocation process to better understand the sources of complexity in defense financial management and what drives it. With a capital and operating budget mixed together, over various planning horizons and with different levels of review in different organizational locales, this process is very complex. We hope to organize some of these complexities to promote understanding as we examine the various lenses through which the defense budget is viewed.

A point to recognize at the outset is that the complexity of this process has many sources. For example, in the hearing cited, Spinney (2002) observed that the problems he identified started in the Pentagon, but added, "We get a lot of help from Congress." Moreover, Department of Defense (DoD) personnel turnover cycles also lent complications. For most military officers the standard tour of duty in the Pentagon is 3 years or less. Spinney observed that people come through the system so rapidly that no military corporate memory exists and, as people come and go in the short term, they do not understand the big picture until they leave. Moreover, historical accretion of single-purpose systems that served unique objectives to solve one set of problems but are never discarded have created a substantial problem in system integration. The result is a serious increase in complexity.

For example, the congressional hearing at which Spinney testified opened with the display of a chart depicting the more than 1,000 accounting systems used in the Pentagon (Grossman, 2002). Spinney indicated that many of these systems collected data in unique ways and were unable to communicate with other databases—12 years after the passage of the Chief Financial Officers Act (CFOA) that called for system integration and financial audits. Everyone is clear that this is a serious problem. Rep. Dennis Kucinich (2002) (D-OH), noted that the Pentagon still insisted it would not be able to obtain a clean audit opinion for 8 to 10 years and suggested that DoD leaders could not make informed decisions on current and future defense spending if they could not understand past expenditures (Grossman, 2002).

Everyone is also clear that large sums of money are being lost due to these inefficiencies. Former Secretary of Defense Donald Rumsfeld stated that the Pentagon could save up to $18 billion annually once proper financial management is achieved; other estimates are as large as $30 billion. Defense is a large and complex business where the stakes are high for the security of the country. In this area, a billion dollars misspent might be the critical difference in avoiding disaster.

THE BUDGET STRUCTURE FOR THE DEFENSE DEPARTMENT

Within the DoD, budgetary accounts are assembled in force packages termed program elements within the structure of the Planning, Programming, and Budgeting Execution System (PPBES). A program element is an aggregation of weapons, personnel, and support equipment. These force packages are then grouped into major force programs. Grouping accounts in force packages is helpful to defense planners as they consider the multiplicity of missions that must be carried out by both the military forces and those who support their missions. Major force programs are arrayed in the Future Years Defense Program (FYDP) database. This critical database summarizes resources by category, including Total Obligational Authority (TOA)—total dollars that may be spent, personnel (both military and civilian), and force program. This information is arrayed by fiscal year (FY) for a six FY period and is updated 3 times a year. These updates correspond with major events in the budget cycle—submission of the President's budget to Congress in February, at the close of the planning cycle in May-June, and at the close of the federal government FY on September 30.

Currently the FYDP presents eleven Major Force Programs:

1. Strategic Forces*
2. General Purpose Forces*
3. Command, Control, Communications, Intelligence and Space*
4. Mobility Forces*
5. Guard and Reserve Forces*
6. Research and Development
7. Central Supply and Maintenance
8. Training, Medical and Other General Personnel Activities

9.   Administration and Associated Activities

10.   Support of Other Nations

11.   Special Operations Forces*

* Combat Forces Program

The purpose of the FYDP within the context of PPBES is to present a two-dimensional display or "crosswalk" allowing for explanation of the major force programs in terms of the resources that fund them. The FYDP includes six combat force programs and five support programs. A program is a collection of program elements that comprise a combat or a support mission and contains the resources needed to achieve the plan. The budget is divided into 11 major force programs, composed of thousands of program elements (Hleba, 2002, p. 5).

The FYDP resource display shows for each major force program the dollars appropriated by Congress, including personnel—military end strength and civilian full-time equivalent work years and forces, for example, combat units or other program items such as missiles and aircraft. The display covers the prior year, the current year, 2 budget years and four "out-years." The force structure is displayed for 7 additional years beyond the second budget year.

In 1988, the Senate Armed Services Committee asked DoD to submit a biennial budget. This is one of the reasons why budget year plus one (BY+1) and two are used in this display. DoD still submits its budget to congressional authorization committees that approve defense programs in biennial format, but not much use of this budget is made in Congress outside of these committees (House and Senate Armed Services Committees). In most planning displays, the first two budget years are said to be "of budget quality" meaning the numbers are as accurate as possible, but that the out-years are "not of budget quality," meaning that these are projections and subject to change as the future unrolls. This is a reasonable way to look at numbers in planning horizons that everyone understands are not carved in stone.

In Table 4.1, PY is the FY that closed the previous September 30; CY is the current FY under execution; BY1 is the budget year or budget year one and BY2 is the second budget year, as required by the authorization committees. Then, BY3 through BY6 are the additional years displayed, BY5 and six are commonly called the out-years. All years are fiscal years.

Table 4.1.   Future Year Defense Plan Display Formula

FYDP display formula = PY + CY + (BY1 + BY2) + BY3 + BY4 + BY5 + BY6

Force structure displays show seven years beyond BY2, but without dollars. Following the FYDP notation in Table 4.1, until September 30 2008: 2007 (PY) + 2008 (CY) + 2009(BY1) + 2010 (BY2) + 2011 (BY3) + 2012 (BY4) + 2012 (BY5) + 2013 (BY6). Many of the exhibits in this book that describe defense dollar trends are based on FYDPs. It is useful to remember that a FYDP is a forecast where all the years except PY may be adjusted (CY may be adjusted by supplementals). Also FYDP's where the last year (BY6) is an even numbered year will include two presidential elections and four congressional elections in years still open to change; hence FYDP numbers can change, and quickly. For example, in Table 4.2, we show defense spending by major force program from 2000 to 2008. There is little likelihood that the FYDP planners could have foreseen in 2000 or up to 9/11/2001 September 11, 2001 what would happen to the defense budget between 2000 and 2006, namely a 57% overall change, with an 86% growth in general purposes force programs and a 93% growth in special operations programs as a result of the War on Terrorism and operations in Afghanistan and Iraq. Also it is good to remember that while the executive branch produces the FYDP and it is a statement of administration policy about defense, congressional off-year elections can have important consequences for what the President plans to do, as for example did the off-year elections dominated by the Republicans in 1994 and the Democrats in 2006. Thus it is good to view the out-year predictions of the FYDP with some degree of skepticism. On October 1 of each year, the FYDP rolls forward into the new FY, CY becomes PY and a new year is added (BY6). With all its faults as a forecast, the FYDP is a critical vehicle for tracking resources over the near-term future within DoD, the executive branch, and for Congress. With the long lead times necessary to recruit and train personnel and to develop and deploy weapons systems, a resource allocation system with an annual focus would be biased toward immediate concerns and short term solutions and due to these factors, inefficient, ineffective, and overly expensive.

The force program packages are largely self-explanatory, notwithstanding a few complications. First, the assumption that the strategic forces program contains all the nuclear forces is not accurate; funding to develop strategic nuclear weapons appears in the research and development force package and funding for theater or tactical nuclear weapons is located in the general purpose force package. Moreover, funding for nuclear warhead development appears in the Department of Energy budget rather than in the DoD budget. A second consideration concerns support force programs. In Table 4.2, major force programs 7, 8, and 9—central supply and maintenance; training, medical, and other general personnel activities; and administration and associated activities—provide support for the combat force packages. Reformers sometimes argue that these costs should be

Table 4.2. DoD TOA by Major Force Program (Total Obligational Authority in Millions of Current Year Dollars)

Program	1	2	3	4	5	6	7	8	9	10	11		
FY	Strategic Forces	General Purpose	C3, I & Space	Mobility Forces	Guard & Reserve	Research & Dev.	Support & Maint	Training Med., Pers	Admin	For Other Nations	Special Ops.	Undistrib.	Totals
2000	8,784	124,490	41,105	14,704	30,363	32,814	24,982	65,621	10,512	1,243	4,512	149	359,280
2001	8,322	134,592	43,866	13,210	31,039	34,889	22,700	68,865	10,970	959	3,622	3	373,038
2002	9,813	143,382	46,834	16,686	32,005	40,927	24,634	72,515	26,169	1,501	5,864	76	420,404
2003	9,507	193,832	64,333	18,980	35,258	45,399	30,336	76,422	13,204	2,274	7,701		497,244
2004	9,929	199,012	64,959	16,536	34,105	50,455	30,285	74,794	17,655	3,091	7,185	-232	507,774
2005	9,851	215,477	69,119	18,161	33,579	54,292	28,253	70,712	31,998	10,619	6,982	-272	548,772
2006	10,216	230,956	73,028	19,149	37,529	55,157	30,082	81,367	10,487	8,144	8,720	-148	564,686
2007	9,999	216,479	67,460	17,489	35,945	54,735	23,183	67,741	14,232	6,622	9,049	-300	522,634
2008	10,421	189,147	72,134	13,394	35,989	49,925	21,514	63,480	14,232	2,106	9,212		481,554
% change FY 2000-FY 2006*	16%	86%	78%	30%	24%	68%	20%	24%	0%	555%	93%		57%
	C	C	C	C	C	S	S	S	S	S	C		

Source: DoD Greenbook, FY 2008: (National Defense Budget Estimates for FY 2008, Office of the Under Secretary of Defense (Comptroller), March 2007), Table 6-5, p. 81: Department of Defense Total Obligational Authority by Program (http://www.defenselink.mil/comptroller/defbudget/Fy2008/)

C = Combat force programs

S = Support programs

* we have chosen to use FY 2000 through FY 2006 because they are actuals, not estimates.

mixed into the actual force package where they are needed to better show the true cost of the combat force packages. For example, suppose that the force program that Table 4.2 shows that general purpose forces cost about 14 times more than strategic forces. From an analytic standpoint, it would be important information to know how much of program 8: training, medical, and general personnel is attributed to each package. Do strategic forces get about 1/14th of this category as their fair share or is something unique about them that drives support costs higher? Similarly, if program 1 were to take half of the money in program 8, leaving the rest to be divided evenly among the other programs, then program 1 would no longer be as cheap as it appeared in the beginning. Consequently, we might want to rebuild the force packages to get complete costs in them so that true cost per output could be measured. While some logic may be found in this argument, it might also be argued that displaying support costs separately allows for emphasis in the decision making process on those items that make up the support costs, whereas they might be forgotten or underfunded if embedded within the combat force activity packages.

Historically, the U.S. military has been good about logistics and transportation planning and resourcing, but the tendency is to spend first on mission personnel and weapons systems, assuming that someone else will supply the appropriate supporting personnel, communications, logistics, and transportation structures when needed. However, history teaches us that this does not always happen and a military force insufficiently provided with supporting personnel and materials does not do well on the battlefield. In World War II, the German Army planned to live off the land in its campaign in Russia, and committed to invasion on this basis. When the Russians adopted a scorched-earth policy, this did not turn out well for the German troops, who were left to freeze and starve during the winter.

A historical perspective on force structure seems to show great stability over time between combat and supporting force structure programs. Over the period, combat forces averaged about 63% of force structure. Viewing this in a little more detail indicates the changes that have occurred since 1962 and those that are predicted to occur. For example, notice the decline in cost of strategic forces since 1992. This is another manifestation of the Cold War peace dividend; more precisely, it is an indication of where that peace dividend was. Spending for general purpose forces declined in the 1990s, but the War on Terrorism indicates a substantial increase out to 2009. Also, note the increases in space, mobility, guard and reserve, and special operations funding. On the support side, research, development, training, and engineering (RDT&E) is up, supplies and maintenance is down from 1992, and training and medical costs are stable. The numbers from 2002 to 2009 are DoD forecasts and the

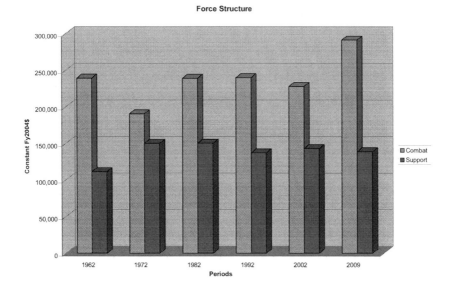

FY 2004 constant dollars
Source: Computed from DoD Greenbook, FY 2004. DoD TOA by program, FY 2004 constant dollars, Table 6-5, p. 81.

Figure 4.1.   Relative size of combat and support forces, FY 1962- FY 2009.

intelligent reader can make his own estimates of these trends, but these categories are those that defense policymakers consider and have considered for the last 40 years.

These force programs may be drawn in different ways and no structure is sacrosanct. What DoD uses internally works for its own internal processes that are driven by a sophisticated and complex planning, programming, and budgeting mechanism where these data are used in a variety of short and long range planning documents and exercises. Understanding the program and budget becomes more complex when DoD begins to translate its needs outside of DoD, as it must to obtain funding from Congress.

Defense in the President's Budget

In the President's budget, defense is summarized in the standardized unified budget functional classification code format used for all of the federal government by the President's Office of Management and

Table 4.3. Trends in Force Structure, 1962-2009

			Combat Forces			
Year	*Strategic*	*General*	*C3, I, Sp*	*Mobility*	*Guard, Rsv.*	*Sp Ops*
1962	72,996	122,910	20,779	7,446	15,261	
1972	31,355	111,243	23,678	6,877	17,291	
1982	27,170	158,490	24,791	7,520	21,210	
1992	20,189	136,725	41,231	8,973	28,503	4,409
2002	8,565	130,479	41,338	14,599	27,767	5,258
2009	8,089	167,152	57,071	18,270	34,706	6,080

			Support Forces				
	RDT&E	*Sup/Maint.*	*Train/Med.*	*Admin.*	*OT Nations*	*Undistrb.*	*Total*
1962	23,239	30,381	49,328	8,001	602		350,943
1972	22,927	36,830	71,960	7,768	10,753		340,682
1982	29,113	33,422	79,261	6,988	1,628		389,594
1992	34,387	31,727	61,286	8,424	1,422		377,276
2002	36,883	21,505	60,363	22,970	1,318	153	371,198
2009	42,733	22,763	59,080	12,857	1,325	41	430,169

Source: Computed by authors from DoD Greenbook FY 2004: National Defense Budget Estimates for FY 2004. DoD TOA by program, FY 2004 constant dollars in millions, Table 6-5, p. 81.

Budget (OMB). The National Defense Budget Function classification is 050.

Function 050 totals present a broad measure of spending for defense and may be compared to the other 16 budget functions in the unified budget display in the President's budget and in the Congressional Budget Resolution. This functional display is used to compare how much is spent on defense relative to other functions as well as to show how funding relationships change over time.

CONGRESSIONAL DEFENSE BUDGET FORMAT

Congress reviews and appropriates funds for national defense in appropriation legislation (individual bills) by title, including the following:

Military personnel: This includes among other things, pay for uniformed personnel, housing and uniform allowances, bonuses, contributions to military retirement funds, travel for permanent change of station, National Guard and Reserve pay for drill and

Table 4.4. Budget Authority by Function (On-Budget)

Federal Unified Budget ($ Billions)

		FY 2006	FY 2007	FY 2008	FY 2009	FY 2010	FY 2011	FY 2012
050	National defense*	617.2	622.4	647.2	534.8	545.2	551.6	560.7
150	International affairs	32.8	34.1	38.3	36.3	36.6	37.1	37.5
250	General science, space and technology	25.1	24.9	27.5	28.5	29.7	31.0	32.3
270	Energy	0.3	1.2	1.6	1.5	2.0	2.3	2.3
300	Natural resources and environment	38.1	29.7	30.4	29.7	30.0	30.8	31.2
350	Agriculture	25.6	19.1	19.9	19.4	20.0	20.4	120.8
370	Commerce and housing credit	14.3	11.1	10.4	9.6	13.0	8.7	7.6
400	Transportation	75.7	77.7	79.9	75.4	75.8	76.4	76.9
450	Community and regional development	31.2	16.1	10.4	10.4	10.5	10.8	11.0
500	Education, training, employment, and social services	125.9	91.2	85.5	87.7	88.8	89.5	91.3
550	Health	295.2	242.3	281.5	298.9	316.1	337.0	358.8
570	Medicare	365.4	371.9	391.6	414.8	439.0	480.5	487.3
600	Income security	351.1	361.0	376.9	388.1	402.7	421.4	419.2

Code	Category							
650	Social security	552.2	589.2	614.6	647.6	685.8	724.9	798.8
700	Veterans benefit/services	71.0	74.5	84.5	86.7	89.4	93.4	96.8
750	Administration of justice	42.7	43.8	46.1	45.6	45.7	46.8	47.5
800	General government	19.7	18.6	20.4	24.4	20.9	21.7	21.6
900	Net interest	226.6	239.2	261.3	274.2	280.8	283.7	284.9
920	Allowances	13.1	-0.3	-0.3	-0.3	-0.3	-0.3	-0.3
950	Undistributed	-68.3	-81.8	-86.3	-88.7	-82.8	-89.2	-92.4
	Total	2,841.8	2,799.3	2,941.4	2,924.6	3,048.9	3,178.5	3,293.8

Source: DoD Greenbook FY 2008, Table 1-7, p. 10. In this display, FY 2006 is the actual Budget Authority authorized DoD by Congress through September 30, 2006. FY 2007 is the current appropriation which could be modified by supplemental expenditures up to September 30, 2007 and FY 2008 is the request included in the FY 2008 President's budget. FY 2009 thru FY 2012 are executive branch predictions about the future. These are not outlays; for example actual outlays for defense in 2006 amounted to $521.8 billion (see Table 1.8), primarily due to the multiple year nature of the procurement accounts.

training. This title is intended to hold all the direct costs of maintaining uniform personnel, officers and enlisted.

Operations and maintenance: Included is funding to operate military facilities and most of the annual operating expenses of DoD.

Procurement: Funding for acquisition of military hardware assets including aircraft, ships, tanks, and weapons systems. Procurement legislation is broken down further by Congress into appropriations for specific types of hardware by military service, for example, Navy aircraft procurement (APN).

RDT&E: Included is funding for most DoD research, development, testing, and evaluation of military weapons and systems.

Family housing: Included is funding for construction of housing for military personnel in the United States and abroad.

Revolving and management funds: Included is core funding to support semi-autonomous DoD operating entities including Navy shipyards, DoD logistics operations and other revolving fund entities supported by reimbursements for services payments. For the FY 2002 budget, personnel, operations and maintenance (O&M), procurement, and RDT&E made up 96.5% of the DoD appropriation. The history of these accounts may be seen in Figure 4.2, which is in constant FY 2003 dollars and allows us to consider the long-term trends with inflation out of the picture. Notice the decline in the procurement account, the procurement holiday in the 1990s and the catch-up in the mid-2000s. It is good to remember that this is a FY 2003 budget exhibit and everything after 2003 is either an educated guess or a wish, or both.

In Table 4.5 we correlate these appropriation categories with political cycles. The period from 1976 to 1980 represents the post-Vietnam drawdown under President Carter. Notice that this drawdown is most evident in military personnel; O&M and procurement increased. The 1980 to 85 period represents the heart of the Reagan build-up, with procurement up 106%, but personnel spending only up about 11%. The Reagan build-up was clearly a procurement build. Then from 1985 to 1990, large deficits limited defense spending during the era of Gramm-Rudman-Hollings. Procurement fell more than 29% in this time period. With the end of the Cold War defense spending plummeted; personnel spending fell more than 31% from 1990 to FY 1998. Procurement was down more than 52%. FY 1998 was the low for the cycle, although spending on personnel would decline slightly in FY 1999. From 1998 to 2001, defense began to make a cyclical rebound from 13 years of decreases; personnel was still down slightly, but procurement was up more than 33%. Then with the attack on September 11, 2001 defense accounts across the board were projected to increase.

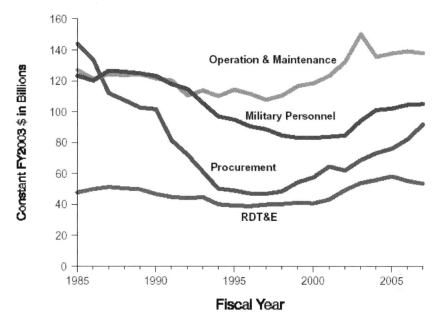

Source: Daggett and Belasco, 2002. Defense budget for FY 2003: Data summary, March 29. Congressional Research Service. Exhibit 2, p. 14.

Figure 4.2. DOD Budget Authority by Appropriation Title, FY 1985-2007.

Appropriation Account Structure

These appropriation titles are divided into appropriations accounts and further subdivided into budget activities, line-items, and program elements. Program element is used in appropriation language only to describe programs in the RDT&E accounts, for example, funds for research on a high density submarine radar antennae for the Navy. In the other accounts, appropriations are divided into budget activities and then line items. For example, the APN appropriation account is subdivided into five budget activities, including combat aircraft, trainer aircraft, modification of aircraft, aircraft spares and parts, and aircraft support activities. Each of these categories is then composed of line-items, for example, F/A-18 aircraft—a specific type of combat aircraft in the general category of APN. Congress holds hearings and debates funding down to this level, deciding upon how many F/A-18 aircraft the Navy will be allowed to purchase, and in what years, if it is a multiple year purchase contract. We should note that Congress does not always provide adequate information

Table 4.5. Real Growth/Decline by Presidential Era

FY 2003 Constant Dollar Display of Defense Appropriations Computed for Historical Periods, by Appropriation and Total

	Carter Drawdown 1976-80			Reagan Build-up 1980-85		Deficit Limits 1985-90		Cold War Dividend 1990-98		Era of Surplus 1998-01		War on Terror 2001-2007 (Projected)		1976-07
	1976	1980	Change	1985	Change	1990	Change	1998	Change	2001	Change	2007	Change	Total Change
Military personnel	117.9	111	−5.60%	123.4	10.87%	123.2	−0.16%	84.8	−31.17%	83.9	−1.06%	104	23.84%	−11.87%
Operations and maintenance	85.3	94.4	10.67%	126.9	34.43%	121.5	−4.26%	110.5	−9.05%	123.3	11.58%	140	13.38%	63.89%
Procurement	61	69.7	14.26%	144	106.60%	101.8	−29.31%	48.5	−52.36%	64.7	33.40%	91.9	42.04%	50.66%
RDT&E	26.7	26.7	0.00%	47.8	79.03%	46.7	−2.30%	40.3	−13.70%	43.2	7.20%	53.7	24.31%	101.12%
Military construction	6.2	4.3	−30.65%	8.4	95.35%	6.5	−22.62%	6	−7.69%	5.7	−5.00%	12.7	122.81%	104.84%
Family housing	3.6	3	−16.67%	4.3	43.33%	4	−6.98%	4.1	2.50%	3.8	−7.32%	4.5	18.42%	25.00%
Other	−0.5	1.1		7		−1		0.4		5		2.5		
Subtotal DoD	300.2	311	3.46%	461.7	48.65%	402.6	−12.80%	294.6	−26.83%	329.6	11.88%	409	24.09%	36.24%
DoE-defense related	5.3	6.6	24.53%	11.8	78.79%	13.4	13.56%	13.3	−0.75%	15.2	14.29%	15.5	1.97%	192.45%
Other-defense related	0.5	0.5	0.00%	0.8	60.00%	0.8	0.00%	1.2	50.00%	1.7	41.67%	1.6	−5.88%	220.00%
Total defense	306	318	3.79%	474.3	49.34%	416.8	−12.12%	309.1	−25.84%	346.5	12.10%	426	22.97%	39.25%

Source: Computed from data from Daggett and Belasco, 2002, Exhibit 6, pp. 12-13. Percent change is computed based on constant FY 2003 dollars.

FY 1976-2001 are actuals; 2002 is budget estimate; 2003 to 2007 is projected.

Constant FY 2003 dollars, in billions

Other defense computation not shown due to distortion of percentage caused by big changes to small numbers.

(Budget Authority by appropriation within DoD, the Department of Energy (DoE) and in other appropriations with defense budget authority.)

to clarify precisely what it is funding; in other cases Congress stipulates how much is to be spent and on what with great specificity.

For example, in the 1990s Congress provided an amount of money for the Air Force B-2 strategic bomber aircraft program that could have been used to either buy a few new aircraft or as support for the fleet of aircraft then in existence. In effect, Congress refused to buy the total new aircraft desired by the Pentagon, but did not say so directly. As a matter of fact, the money was only sufficient for maintenance on the then current fleet of aircraft. Thus, Congress may deliberately add complexity to the defense budget process by passing more ambiguous appropriation measures than would normally be the case, as a result of political incentives.

The level of appropriation detail varies from one title to another. In the military construction, procurement, and RDT&E accounts, appropriations bills show thousands of line items for individual weapons and military construction programs. The personnel and O&M accounts are described by fewer items. As readiness levels were increasingly of concern in the 1990s, Congress began to keep track of more separate accounts within the O&M account to provide a finer grained analysis of those accounts directly connected to military readiness. This trend began in 1993 (Tyszkiewicz & Daggett, 1998, p. 17). Table 4.6 shows the defense function (050) subdivided by subfunction (051, 053, 054) Appropriation bill and appropriation. From this, for example, we learn that the military construction bill is subdivided into the military construction appropriation and the family housing appropriation. Each of these is further subdivided as discussed above.

In Table 4.6 the defense function is also divided by subfunction which allows for a finer definition of the purposes for which defense resources are intended. For example, altogether the total national defense function (050) drew resources from eight appropriations bills in FY 2005, although most military activities (051) were funded by either the DoD appropriation bill or the MILCON bill. For FY 2005 these two bills provided about 96% of the funding for the total national defense function (051$/050$). During the long summer debates over defense, it is what is in these two bills that is of most concern to most participants. This function is divided into subfunctions as is seen in Table 4.7.

BUDGET FORMAT WITHIN DoD

Congress enacts the defense budget by appropriation titles, such as military personnel, while DoD itself organizes the defense budget into force programs, such as strategic forces. Presenting fiscal information by appropriation title focuses attention on what is being bought, that is,

Table 4.6.   National Defense Budget Authority ($ in Millions)

	FY 2005	FY 2006	FY 2007	FY 2008
Military Personnel	121,279	128,483	118,740	118,920
Operation and Maintenance	179,215	213,532	191,598	165,344
Procurement	96,614	105,371	103,211	101,679
RDT&E	68,825	72,855	75,684	75,117
Revolving and Mgmt Funds	7,880	4,754	2,241	2,454
DoD bill	**473,813**	**524,995**	**491,474**	**463,514**
Military Construction	7,260	9,530	7,480	18,233
Family Housing	4,098	4,426	3,795	2,932
Military Construction Bill	**11,358**	**13,956**	**11,275**	21,165
DoD Offsetting Receipts (Net) and Other	−1,258	−2,489	4,282	−1,431
051-Total DoD	**483,913**	**536,462**	**507,031**	**483,248**
OMB rounding/scoring	−49	57,318	−110	−275
Additional GWOT Requests			93,315	141,665
051-OMB Total DoD	**483,864**	**593,780**	**600,236**	**624,638**
Defense-Related Activities by Bill:				
053-Energy and Water Bill				
Atomic Energy Defense Activities	17,024	16,218	15,709	15,813
Occupational Illness Compensation Fund	682	1,061	1,108	1,354
Fomer sites remedial action	164	139	130	130
Nuclear Facilities Safety Board	20	20	20	20
053-Total Defense Related	**17,890**	**17,438**	**16,967**	**17,317**
054-VA-HUD-Independent Agencies				
U.S. Antarctic Log Support Act (NSF)	116	67	68	67
Selective Service System	26	25	24	22
Subtotal	**142**	**92**	**92**	**89**
054-DoD Appropriation Act				
Community Management Staff	522	538	600	689
CIA Retirement and Disab. Fund	239	245	256	263
Subtotal	**761**	**783**	**856**	**952**
054-Homeland Security				
Coast Guard (Def Related)	1,238	1,402	550	563
Emergency Preparedness and Response	39	109	49	89
Infrastructure protection and info security		619	548	538
R&D, Acquisition and Operations	363	376	296	175
Subtotal	**1,640**	**2,506**	**1,443**	**1,365**

054-Commerce-Justice-State

Radiation Exposure Compensation Trust	184	108	82	61
DOJ (Defense Related)	−53	−15	−5	−14
FBI (Defense Related)	1,240	2,289	2,473	2,538
DOC (Defense Related)	<u>7</u>	<u>14</u>	<u>15</u>	<u>14</u>
Subtotal	**1,378**	**2,396**	**2,565**	**2,599**
054-Labor-HHS-Education				
Trans-Treasury-Indendent Agencies				
Maritime Security/RRF	98	154	153	154
054-Energy and Water Bill				
Corps of Engineers-Civil Work	4	4	4	
DoE (Defense Related)	<u>10</u>			
Subtotal	<u>**14**</u>	<u>**4**</u>	<u>**4**</u>	
054-Total Defense Related	**4,033**	**5,935**	**5,113**	**5,159**
053/054-Total Defense Related	**21,923**	**23,373**	**22,080**	**22,476**
053/054-Additional GWOT Requests			130	50
050-Total National Defense	**505,836**	**559,835**	**529,111**	**505,724**
050-OMB Total Natiional Defense	**505,787**	**617,153**	**622,446**	**647,164**

Source: DoD Greenbook FY 2008, National Defense Budget Authority, Table 1-3, p. 6. In this display, FY 2005 and FY 2006 are actual amounts authorized by Congress for those years. FY 2007 could still have been amended by supplementals at the time this table was prepared. FY 2008 is an estimate drawn from the President's budget.

Table 4.7. How Function 050 is Subdivided

Function 051:	Department of Defense, Military
Function 053:	Atomic Energy Defense Activities
Function 054:	Other Agencies-defense related activities

personnel, for example, as against weapons. Presenting this data in terms of major force programs helps illustrate the purposes for which the money is being spent. Presenting it by component department indicates who spent the money.

Defense Budget by Military Department

Conversation with DoD budget participants would lead one to think that a sense of "fair share" governs the resource process, that the military departments share about equally in good times in budget increases and have to give up dollars about equally in bad times and that increases and cuts are distributed by thirds to the military departments. In some years this notion is plausible, particularly in times when the President has stated that he wishes to keep defense even with inflation, or inflation plus a percent; in those years there is a tendency to fair share the increase. However, in practice mission trumps fair share. A study of historical trends over the last 30 years illustrates this. Table 4.8 displays shares for the military departments and DoD-wide at 5-year intervals from 1980 to 2010, in current or then year dollars and as percentage shares. Table 4.8 indicates that the Navy and Air Force received larger shares of the DoD budget than did the Army over most of this period, as a result of the primacy of their positions in Cold War deterrence. In terms of shares, the Army did not catch up to Navy and Air Force until activity in Iraq and Afghanistan in 2003-2005. Note also that while the DoD budget increased substantially between 2000 and 2005, Navy and Air Force shares fell to their lows for the 3-decade period. Notice also that Pentagon planners indicate that Army will revert to its historical share of the DoD budget in 2010, assuming some favorable resolution in Iraq. For the 30-year study, Navy has

Table 4.8.   DoD Budget Authority by Military Department and DoD-Wide, FY 1980-2010 and Percent Shares

	1980	1985	1990	1995	2000	2005	2010
	Then Year Dollars, in Billions						
Army	$34	$74	$79	$63	$73	$153	$121
Navy	$47	$99	$100	$77	$89	$132	$148
Air Force	$42	$99	$93	$74	$83	$128	$143
DOD-wide	$17	$14	$22	$42	$46	$72	$82
Total	$141	$287	$293	$256	$291	$484	$494
	Percent Share by Service of Selected Year						
Army	24%	26%	27%	25%	25%	32%	25%
Navy	34%	35%	34%	30%	31%	27%	30%
Air Force	30%	35%	32%	29%	29%	26%	29%
DoD wide	12%	5%	7%	16%	16%	15%	17%
Percent change total DoD		104%	2%	−13%	14%	67%	2%

Source: Computed from DoD Greenbook FY 2008, selected years and Steven Kosiak, Historical and Projected Funding for Defense: Presentation of the FY 2007 Request In Tables and Charts. Center for Strategic and Budgetary Assessments. Washington, D.C. at www.csbaonline.org.

held the largest budget share, closely followed by Air Force. For the last 10 years, DoD-wide activities have held more than a minimal share of the DoD budget. We explain this role later. Also, it appears that for this period the military department shares have been very stable, again with the exception driven by the Army's combat role in Iraq.

If we look at the averages and maximums and minimums in Table 4.9, the military departments do have recognizable maximums and minimum that are closely grouped, but Navy and Air Force have stayed closer to their maximum longer than Army and have generally gained a larger share of the DoD budget. It is clear that substantial swings do happen over time and when the drawdown from the end of the Cold War happened all the military departments did give up some resources, yet their shares remained relatively stable until the 2005 period.

While analysts give some attention to funding by military department, this is not particularly useful information unless information about missions for each military department is also provided. For example, it makes little sense to say the Air Force or Navy needs more money based on what this display tells us, although that could be true. It could also be true that part of the Army mission has been shifted to the U.S Marine Corps or that certain logistical and support activities, which had been carried in the Navy and Air Force budgets, are now done by defense-wide agencies, and considering mission, it could be that they deserve even fewer dollars. Thus without mission information, the display of dollars by military department or service is not particularly informative. It is, however, probably easiest to understand as it describes the relative size of one component of defense as compared to another, as well as how this changes over time.

The defense-wide category describes activities that are carried on at the DoD level. Some experts have observed that left to themselves, the services would over-provide funds for their core missions, for example, artillery, tanks, and infantry for the Army; carriers, tactical air, and submarines for the Navy, and strategic and fighter aircraft for the Air Force. Before the Goldwater-Nichols Act of 1986 and its emphasis on joint

Table 4.9.   DoD Share Ranges: 1980-2010

	Avg	*Max*	*Min*
Army	26%	33%	24%
Navy	30%	35%	26%
Air Force	29%	35%	26%
DoD-wide	14%	17%	5%

Source: Computed from Table 4.8

planning and war fighting, this might have meant that logistics and communications functions might have been undersupplied as well as functions involving the support of one service by another, for example, Air Force ground support of Army or Marines or Army support in Air Force base protection. Goldwater-Nichols has helped close these potential gaps by increasing the power of the chairman of the joint chiefs, providing a structure and emphasis on joint war fighting, and including the area commanders who will actually fight the forces (the commander in chiefs or CINCs) in the resource planning process. However, that still leaves many functions that must be carried on centrally at the DoD level.

In 1998, 14 defense agencies and seven field activities comprised the defense wide component of the defense budget. These included such things as the Defense Commissary Agency, the Defense Advanced Research Projects Agency, the Defense Contract Audit Agency, the Defense Finance and Accounting Service, The Defense Information System Agency, the Defense Intelligence Agency, the Defense Logistics Agency, the Defense Security Assistance Agency, and the National Security Agency (Tyszkiewicz & Daggett, 1998, p. 21). Some of these functions are funded as revolving funds.

Revolving Funds

A revolving fund is a fund that "sells" its services within DoD to customers. The fund is expected to be managed so that after an initial endowment, it supports itself from year to year by charging its DoD customers a rate that will cover the work performed. Revolving funds can sell a product or a service. They then use the receipts from sales to pay operating expenses (shipyard wages, utilities) and purchase new stock (paper, steel, nuts, and bolts). In theory, all of a fund's income is derived from its operations and is available to finance the fund's ongoing activities. A revolving fund activity accepts an order from a customer and bills the customer when the work is done, financing the work from the working capital accumulated from payments for work done by previous customers. Funds operations are unconstrained by FY cycles, hence the term revolving fund. They do not have to worry about spending up fourth quarter funds, thus they can be managed to serve the needs of their customers. Revolving funds used in DoD include stock funds and industrial funds. The stock funds provide items such as fuel, construction supplies, clothing, medical supplies, consumable aircraft and missile parts, ordnance repair parts, and so forth. Industrial funds provide such services as equipment overhauls, shipyard services, printing services, repair depot, and public works maintenance and transportation services.

Revolving funds have a long history in the defense world, being used in the Department of the Navy since the late 1800s (Hleba, 2002, p. 108). The National Security Act Amendment of 1949 provided for establishment of revolving funds throughout DoD. While these funds generally operated in a satisfactory manner, DoD has attempted various innovations to make them more efficient. In 1992, DoD combined five industrial funds, four stock funds, and several appropriated fund support activities into a super revolving fund called the Defense Business Operations Fund (DBOF). The DoD comptroller managed the cash balances of this fund. Significant savings were expected, but never materialized. For example, rate setting had been a problem and the situation worsened. Rates were set for the future based on what had happened in the past, which seems sensible, but what if a shipyard did less work than expected (say a ship or two missed an overhaul) and the costs of maintaining the workforce was then spread over fewer units of work. Rates would go up in the following year. The next year the next customer might then look at the new rates and find that his budget did not allow for a ship overhaul because the new rates were too expensive. Suppose he cancels the work. Now the new rate setting process has even fewer units of output and the new rate will be even higher. When a cost spiral like this occurred, there was no other answer than to supplement the revolving fund with appropriated dollars.

Problems such as these and others drove DoD to abandon DBOF in 1997 in favor of four working capital funds, giving the responsibility for cash management back to each military department (Tyszkiewicz & Daggett, 1998, p. 16.) This area of complexity does not go away; its administrative locus is just shifted and instead of having one huge fund to worry about, now five smaller funds exist with the creation of the Defense Commissary Agency in 1998, each with a set of complicated rules and procedures. The bottom line for these revolving funds is that they must maintain a positive cash balance (no deficit) or their leaders may be charged with a violation of the Anti-Deficiency Act. These funds are set up to control costs by operating efficiently while paying their costs to do the work with revenues charged their customers. These are not simple functions. Management has to coordinate labor and inventories, forecast future demands for services and labor effort, and set rates that will see the work performed at a price that reflects component costs which customers are willing and able to pay out of their appropriated budget funds.

Working capital funds receive initial lump sum funding through an appropriation, called the "corpus." Thereafter they are expected to survive based on the rates they charge their customers. Supply oriented funds charge a surcharge on the items provided to recover management's costs for ordering, stocking, inventorying, and providing the goods. Non-supply funds, for example, the old industrial funds, charge a rate based on the unit cost of work plus overhead costs. Rate setting for both of these fund types has turned out to be quite complex and critical to maintaining positive cash balances. Fund customers use the rates to ask for an appropriation in the budget for work or supplies that they will purchase from the working capital funds. If the appropriation is under what is needed, customers will buy less. If enough customers do this, it could affect the rate setting process, for example, higher rates for the following year, which could lead to fewer customers and even higher rates, and so on. The working capital funds are a fact of life and whether managed centrally at DoD or at the military department level present a complex managerial challenge to do right, with accurate rate setting and good service to the customer, using modern management techniques to provide cost efficiencies in overhead tasks and cost savings to the taxpayer. In 2001, about 18% of DoD direct appropriations were managed through the working capital funds and 200,000 people were employed in them (Hleba, 2002, p. 112).

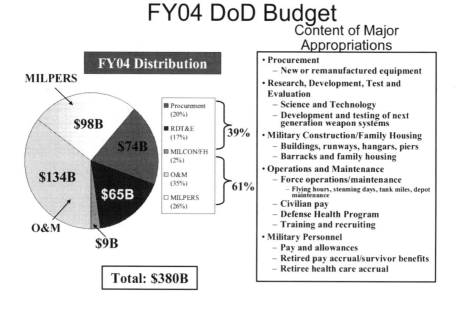

Source: Daly, 2004, p. 35.

Figure 4.3.   DoD shares by appropriation FY 2004.

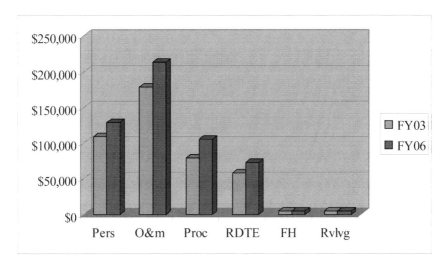

Source: DoD Greenbook FY2 008; Department of Defense BA by Appropriation Title, Table 6-8, p.118. Current dollars, in millions.

Figure 4.4. Appropriation shares FY 2003 and FY 2006.

DEFENSE BUDGET ANALYSIS

Deciding how much to spend on defense is always a difficult question that has led to some standard ways of looking at defense spending. The pie chart in Figure 4.3 shows the FY 2004 DoD TOA by appropriation title DoD uses TOA as a control on what may be spent in a FY because every DoD appropriation bill contains some funds legally to be spent in future years while the majority is spent in the named FY. The pie chart also divides TOA into operating and investment accounts as may be seen.

Single year analysis like the pie chart in Figure 4.3 is a place to start, but adding some history to a single year enables the observer to think about a richer array of concepts. In the exhibits in Figure 4.4 above we have captured DoD FY 2003 and FY 2006 actual expenditure by primary account. These years catch the central years of deployment into Iraq and ongoing activities in Afghanistan.

In Figure 4.4 the main accounts increases both on the operating and investment side, with little to no change in family housing and in the revolving funds. Table 4.10 provides a more precise numerical dimension to this display. Interestingly, the investment accounts increased more than the operating accounts between FY 2003 and FY 2006.

A logical next step is to extend the timeline of the comparison and discount the numbers for inflation in order to give a true measure of

Table 4.10.   Appropriation Percent Change, FY 2003-FY 2006

Account	FY 2003	FY 2006	% Change	Type of Account
Pers	$109,062	$128,483	17.81%	Operating
O&M	$178,316	$213,532	19.75%	Operating
Proc	$78,490	$105,371	34.25%	Investment
RDTE	$58,103	$72,885	25.44%	Investment
FamH	$4,183	$4,426	5.81%	Investment
RvlvF	$4,154	$4,754	14.44%	

Source.   Computed from DoD Greenbook FY 2008, Department of Defense BA by
Appropriation Title, Table 6-8, p. 118. Current dollars, in millions.

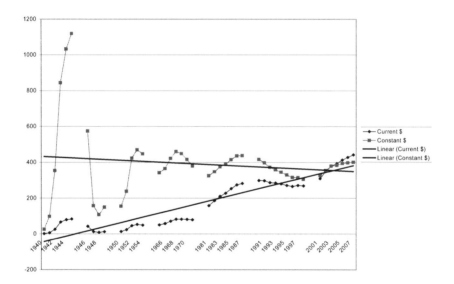

Source: McCaffery and Jones, 2004 compiled from Daggett and Belasco, 2002, Table 10.

Figure 4.5.   Defense in current and constant dollars: 1940-2007.

change in spending. This is done by converting current dollars into
constant dollars so that the impact of inflation is standardized and analysts
can see the cost of a program over time. To do this current dollars are
divided by the gross domestic product (GDP) deflator and expressed in
constant dollars for a base year. In Figure 4.5 we have interrupted the trend
lines so that periods of war and its aftermath may be more clearly seen,
starting with 1940 and building through 1945 for World War II, its

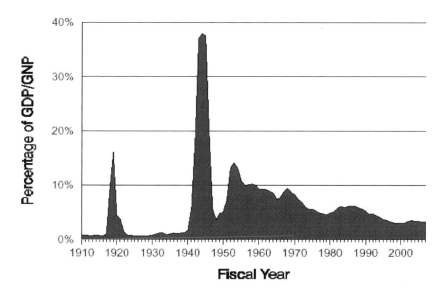

Source: Daggett and Belasco, 2002, p. 23.

Figure 4.6. Defense spending as a share of GDP, FY 1910-2007

aftermath, the build-up in Korea, then the Cold War, Vietnam and its peace dividend, then the Reagan build-up, the fall of the Berlin Wall, the peace dividend of the 1990s and finally the ramping up to cope with the War on Terrorism. While defense spending grows in current dollars seemingly at a steady pace, the constant dollar picture showing dollars spent corrected for inflation indicates the cyclical nature of defense. Defense spending reacts to crises. The graph also suggests that bad things happen when defense spending crosses below the $400 billion constant dollar line; while this is an interesting observation, it is not useful for policy guidance.

Another common measure used to size defense spending relates defense as a share of GDP (see Figure 4.6 above). This is a buying power measure.

National Defense as a percent of GDP is a good long term reference point. In Table 4.11 we see national defense spending as a percent of GDP for selected years, catching the highs and lows for those eras. Still, there is no inherent ceiling or floor to spending on defense. When threatened, a nation will pay what it must.

Defense spending as a share of federal outlays relates defense to spending on other federal functions. People who use this measure are concerned that defense and other federal functions be treated equally and that they adhere to some historical trend. Figure 4.7 shows such a display.

Table 4.11.   National Defense Spending as a Percent of GDP by Selected Fiscal Years

Fiscal Year	Percent GDP	Focal Point
1940	1.7	Low before World War II
1944	37.8	World War II top
1948	3.5	Post-World War II low
1953	14.2	Korea top
1968	9.4	Vietnam top
1986	6.2	Reagan buildup top
1999	3.0	Post-Cold War low
2006	4.0	Iraq top

Source: DoD Greenbook FY 2008, Defense Shares of Economic and Budgetary Aggregates, Table 7-7, p. 217.

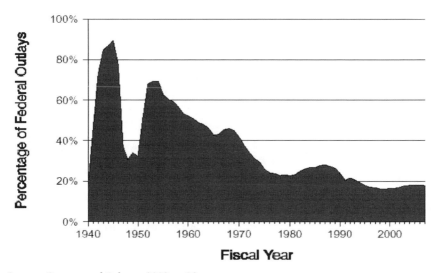

Source: Daggett and Belasco, 2002, p. 28.

Figure 4.7.   Defense as a share of federal outlays, FY 1940-2007.

However, the main thing to remember when considering defense as a share of federal outlays is that the budgetary requests for defense or any function must be tested against the need for the function. Defense may be sufficiently provided at 20% of federal spending, or, given changes in the threat, it may be woefully underprovided.

Another way to look at defense is to compare the number of personnel in uniform as does Figure 4.8.

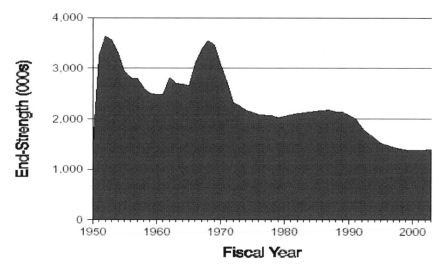

Source: Daggett and Belasco, 2002, p. 28.

Figure 4.8.   Active duty end strength, FY 1950-2003.

This graphic clearly shows increases due to Korea and Vietnam, their draw down periods and the end of the Cold War and the downsizing of the 1990s. As we have argued elsewhere, the Reagan defense build up was in procurement and not personnel. This chart shows more clearly the burden of Vietnam than do the graphics that rely on dollar measures alone.

Table 4.12 updates our end strength figure into the war on terror years. It begins at the height of the Reagan buildup, shows the impact of the fall of the Berlin Wall, and the downsizing of the 1990s and then indicates how the burden of the war on terror, Specifically operations in Iraq and Afghanistan have been assumed through FY 2006. What it shows is that the Army has assumed its burden with an end strength generally set at mid-1990 levels and Navy and Air Force end strengths have declined from 2003 and thereafter, partially as a result of cutting end strength through better management in order to afford increased recapitalization needs for new ships and planes.

Another useful analytic technique is to connect trends in support dollars and personnel levels. The next exhibit (Figure 4.9) shows operations and support (O&S) spending as a percent share of the total defense budget on the left scale (personnel and operations and maintenance). On the right scale is a dollars per capita measure of O&S spending. This reveals a steady increase in support spending for personnel over the period. The third line measures investment per capita (procurement and RDT&E) and

Table 4.12.   Military Personnel (End Strength), by Service, From Reagan Peak Until 2006, in Thousands

Fiscal Year	Army	Navy	MC	AF	Guard	Total
1987	781	587	200	607	69	2,244
1988	772	593	197	576	71	2,209
1989	770	593	197	571	72	2,203
1990	751	583	197	539	74	2,144
1991	725	571	195	511	75	2,077
1992	611	542	185	470	72	1,880
1993	572	510	178	444	71	1,775
1994	541	469	174	426	68	1,678
1995	509	435	174	400	65	1,583
1996	491	417	175	389	66	1,538
1997	492	396	174	378	64	1,504
1998	484	382	173	367	64	1,470
1999	479	373	173	361	65	1,451
2000	482	373	173	356	65	1,449
2001	481	378	173	354	65	1,451
2002	487	383	174	368	66	1,478
2003	499	382	178	375	66	1,500
2004	500	373	178	377	66	1,494
2005	492	362	180	352	69	1,455
2006	505	350	180	349	71	1,456

Source: DoD Greenbook FY 2008. Department of Defense Manpower, end strength in thousands, Table 7-5, p. 213.

reveals that on a per capita measure spending per individual increases during the FYDP period, peaks, and then decreases. The authors argue that current trends in defense show the substitution of capital for labor. This chart provides a measure of the cost of one category measured against another in percentage shares and on a per capita basis and can lead to useful insights.

However, this is not the end of the story. It appears that both capital and labor are getting more expensive. For example, in Figure 4.10 we see the decline in number of personnel by about 800,000 from 1987 to 2006 (36%) while the personnel portion of the base budget changes from 27 to 25% (OMB, 2007, p. 363).

Personnel costs have increased for a variety of reasons, including increased base pay, increased housing allowances, health care benefit increases, educational benefit increases increased moving allowances and

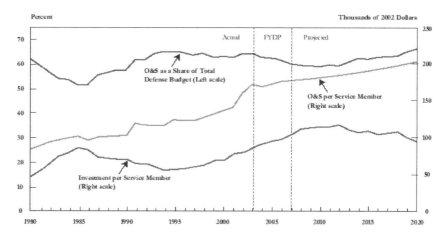

Source: Congressional Budget Office (2003, Exhibit 2.2, p. 18) using data from the Department of Defense.

Note: FYDP = Future years defense program; O&S = operation and support.

Figure 4.9. Three measures of defense spending.

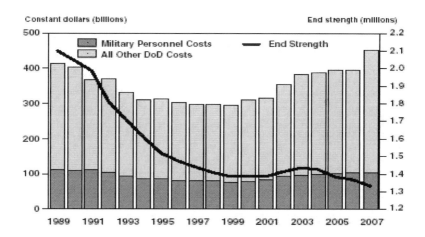

Source: OMB, 2007, Budget of the U.S. FY 2008, *Analytical Perspectives*, p. 362.

Figure 4.10. Post-Cold War end strength and dollars.

bonuses and special service pays. Figure 4.13 provides a breakdown of compensation costs for military personnel.

In Figure 4.11 it is interesting to note that the accruing costs of future benefits is almost as large as the amount dedicated to direct pay of service

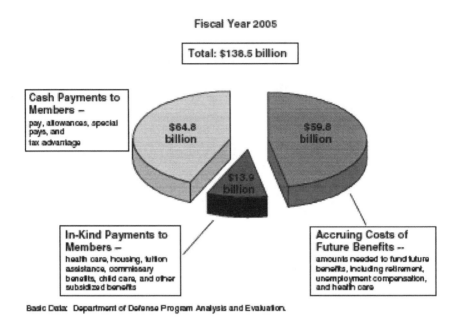

Fiscal Year 2005

Total: $138.5 billion

Cash Payments to Members –
pay, allowances, special pays, and tax advantage

$64.8 billion

$59.8 billion

$13.9 billion

In-Kind Payments to Members –
health care, housing, tuition assistance, commissary benefits, child care, and other subsidized benefits

Accruing Costs of Future Benefits --
amounts needed to fund future benefits, including retirement, unemployment compensation, and health care

Basic Data: Department of Defense Program Analysis and Evaluation.

Source: OMB, 2007, Budget of the U. S. FY 2008, *Analytic Perspectives*, p 363.

Figure 4.11.   DoD direct compensation costs, current, and future.

members. In addition to expenditures on personnel made by DoD for future benefits, it should also be noted that the "Department of Veterans Affairs spends nearly $40 billion on medical care, vocational rehabilitation, compensation, pensions, education, home loans, burial and other services for as many at 70 million veterans and their families" (OMB, 2007, p. 363).

No one best measure for defense spending exists. Nominal dollar comparisons of different functions, usually in pie charts, show the size of spending on one function related to another. This answers the question of which function received more or less funding in that particular year. This also may be expressed in percentage shares. When percent of GDP is involved the measure speaks to the question of how much we wish to afford or how much we can afford if we have to, for example, defense was 1.7% of GDP in 1939 and 37% in 1944. Use of constant year dollars answers the question of how much the defense appropriation was worth, while accounting for inflation. Converting nominal to constant is not difficult when one uses the historical exhibits in the U.S. budget and the constant dollar deflator exhibit for defense goods (defense goods are generally a little more susceptible to shifts in inflation than nondefense goods). The division of current or nominal dollars by the GDP deflator produces the constant dollar amount for the year or series of years. Many exhibits in the U.S. budget already have this computation made. This is most useful for looking at what it has cost to provide a service or function over longer historical periods while accounting for inflation.

All of the above are useful, but ultimately they all must be related to the threat. During the Cold War era, estimates were sometimes made of USSR defense spending and compared to U.S. spending on defense. U.S. estimates of USSR defense spending turned out to be very imprecise and typically were inflated. However, even budget officials in Russia often did not have accurate numbers—nobody actually knew how much was spent with any degree of accuracy. In the year 2000, as the draw-down era continued, U.S. defense spending was sometimes totaled and shown against the defense spending of the potential top five or six enemies of the United States. Table 4.13 shows estimates of defense spending for the top 25 defense spenders in terms of U.S. dollars and GDP.

Table 4.13.   Top 25 Defense Spending Nations (Current U.S. Dollars in Millions

| Country | Rank | U.S. Dept. of State: WMEAT 1998 | | IISS: Military Balance 2001-2002 | |
		Defense Expenditures (1997 Data)	*% GDP*	*Defense Expenditurs (2000 Data)*	*% GDP*
United States	1	273,3000	3.3%	291,2000	3.0%
China-Mainland	2	74,910	2.2%	42,000	5.4%
Russia	3	41,730	5.8%	60,000	5.0%
France	4	41,520	3.0%	35,000	2.6%
Japan	5	40,840	1.0%	45,600	1.0%
United Kingdom	6	35,290	2.8%	34,600	2.4%
Germany	7	32,870	1.6%	28,800	1.6%
Italy	8	22,720	2.0%	21,000	2.0%
Saudi Arabia	9	21,150	14.4%	18,700	10.1%
South Korea	10	15,020	3.4%	12,800	2.8%
Brazil	11	14,150	1.8%	17,900	2.8%
China-Taiwan	12	13,060	4.6%	17,600	5.6%
India	13	10,850	2.8%	14,700	3.1%
Israel	14	9,335	9.7%	9,500	8.9%
Australia	15	8,463	2.2%	7,100	1.9%
Canada	16	7,800	1.3%	8,100	1.2%
Turkey	17	7,792	4.0%	10,800	5.2%

Table continues on next page.

Table 4.13.   Continued

Country	Rank	U.S. Dept. of State: WMEAT 1998		IISS: Military Balance 2001-2002	
		Defense Expenditures (1997 Data)	% GDP	Defense Expenditurs (2000 Data)	% GDP
Spain	18	7,670	1.5%	7,200	1.3%
Netherlands	19	6,839	1.9%	6,500	1.9%
North Korea	20	6,000	27.5%	2,100	13.9%
Singapore	21	5,664	5.7%	4,800	4.9%
Poland	22	5,598	2.3%	3,300	2.0%
Sweden	23	5,550	2.5%	5,300	2.2%
Greece	24	5,533	4.6%	5,600	4.9%
Indonesia	25	4,812	2.3%	1,500	1.0%

Sources: Daggett and Belasco, 2002, Exhibit 15, p. 28. U.S. Department of State: Bureau of Arms Control, *World Military Expenditures and Arms Transfers: 1998,* April 2000. International Institute for Strategic Studies, The Military Balance 2001-2002, October 2001.

Note: For information on a total of 167 countries and details on this data, see CRS Report RL30931, *Military Spending by Foreign Nations: Data from Selected Public Sources.* Military spending in this tabe is defined primarily by the NATO standard definition: cash outlays of central governments to meet costs of national armed forces. This definition includes military retired pay, which is excluded in the U.S. Office of Management and Budget's definition of DoD outlays. Therefore, the U.S. outlays numbers may be highter in this table than the reported DoD outlays numbers in the rest of the report.

This analysis leads to a position where the United States is seen to have overwhelming dominance in dollars spent, but this did not seem very useful after the attack on the World Trade Center in 2001 when it became clear that terrorist groups with sufficient funding did not need submarines, intercontinental missiles, tanks, and aircraft carriers to inflict substantial damage on the United States. Still, these are the standard ways of looking at and thinking about the size of the defense budget. In most cases, the trend is the most important factor to watch.

In the 1990s when NATO expansion was under consideration, Congress required an annual report showing what allies were contributing to the collective defense effort and where potential invitees to NATO might rank. One of the analyses DoD provided was individual country defense spending as a percent of GDP compared to the aggregate GDP spent on defense of the group. The dotted line is the average; some contributed above and some below; Portugal was providing its fair share.

The report comments:

Chart III-3
Defense Spending as a Percentage of GDP
1998

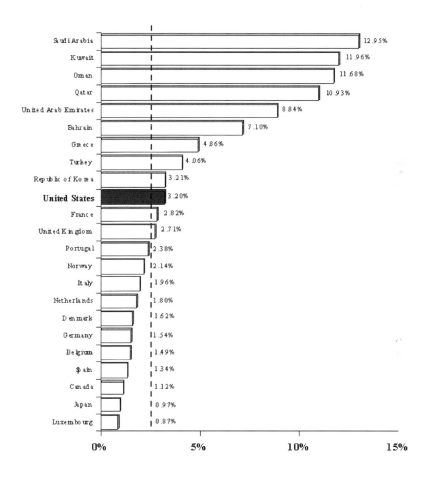

Dashed line represents the defense spending/GDP ratio at which a country's share of aggregate defense spending equals its share of aggregate GDP. Countries at this level are contributing their "fair share" of defense spending. Countries above this level are contributing beyond their "fair share," and conversely.

See Annex, Section C.

Source: U.S. DoD, 1999.

Figure 4.12.   Assessments of fair share.

The dashed vertical line shown in **Chart III-3** helps address the issue of equity among countries' defense efforts, by comparing contribution with ability to contribute. The line almost intersects the bar shown for Portugal, which signifies that Portugal's share of total defense spending (contribution) is commensurate with its share of total GDP (ability to contribute). With regard to defense spending, Portugal's is thus doing roughly its "fair share" among the countries addressed in this report. The United States and countries shown above the U.S. in this chart (the Republic of Korea, Turkey, Greece, and the GCC countries) are doing substantially more than their "fair share," with defense spending contributions in excess of their respective GDP shares by 20 percent or more. Conversely, Italy and those countries listed below it in this chart (the Netherlands, Denmark, Germany, Belgium, Spain, Canada, Japan, and Luxembourg) are doing substantially less than their "fair share." (USDoD, 1999, Responsibility, ch. 3)

It is possible to think about other ways of deciding on fair share. For example, suppose this were a neighborhood providing collective security against some outside threat. It would be possible to divide the total cost of protection in terms of each household's population as a percent of total population, or in terms of each household's property values as a percent of total property value.[1]

Figure 4.13 shows how the NATO group countries rank when spending on defense as a percent of GDP is related to standard of living. This analysis was done to see how potential new NATO members the Czech Republic, Hungary, and Poland compared to then current NATO members. It shows for all NATO nations, including the new members, how their respective defense effort (measured by defense spending as a share of GDP) related to their standard of living (measured by GDP per capita). This perspective reveals that the GDP share devoted to defense among the three new members was roughly equal to the share provided by a number of allies with higher (and in some cases, substantially higher) standards of living.

These are interesting tests to see if an ally is paying enough compared to what everyone else in the alliance is paying and compared to the country's own wealth. Of course, no amount of hectoring can make a country contribute more to the common defense. Beyond that the fundamental question is the threat. Countries may spend more or less than others as they perceive the threat differently or in respect to their local threat situation and considering their own ability to pay. The Stockholm International Peace Research Institute (SIPRI) (2002) divided countries into four "income" groups and concluded that countries in the two lower groups were increasing their spending on defense faster than the two "upper income" groups; however SIPRI noted that while the poorer counties were increasing their spending faster, they still spent less than the richer

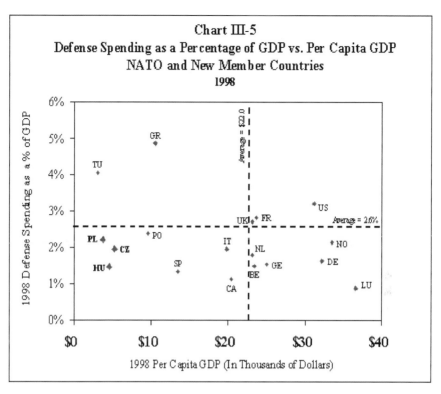

Chart III-5
Defense Spending as a Percentage of GDP vs. Per Capita GDP
NATO and New Member Countries
1998

USDoD (1999) Responsibility Sharing Report, Pursuant to FY 1999 Defense Authorization Act. Accessed at www.defenselink.mil/pubs/allied_contrib99. This series of reports can be accessed online by searching on Allied Contributions to the Common Defense. It has a plentiful supply of data on the efforts of the United States and its allies about who is providing how much for defense and whether or not this is enough and is available from 1995 through 2003 in about the same format, provided at a time when NATO was considering adding new members.

Figure 4.13.   GDP spent on defense and standard of living.

countries, both absolutely and as a share of world defense spending. In fact the world share of defense spending by low-income countries was only 8%, while spending for defense in the high-income countries constituted 70% (Doyle, 2003, p. 6). To some extent countries spend what they can on defense, based on their appreciation of the threat and what they have to lose, and their appreciation of the extent to which the action of another country will also provide defense for them as they act in their own self interest (free rider effect). Needless to say, relying on others for defense has not always worked out well.

The Deficit Context

The defense budget process is complex enough without adding another factor to the mix, but it is good to remember that typically the budget drama is played out in a context of scarce resources. When we discussed defense spending in terms of ability to pay, we did not explicitly include the role that deficit spending plays in constraining defense choices. Obviously, in a wartime situation, a country will pay whatever it has to, but defense is not that simple. A defense establishment has to be built and maintained over time, and can not be created instantaneously when weapons systems routinely take a decade to create and deploy and when it takes 25 years or more for a young officer to rise to general officer rank. It is true that "you go to war with what you have," thus those who resource defense must ensure that in the annual budget process, enough is set aside each year to make a continuous provision for an adequate defense. Having an overhang of national debt may limit necessary defense spending.

Thus, Figure 4.16 reminds us that budgeting usually takes place in an era of deficit spending and is so again in the 2000s after a short history of surplus budgets from 1998 through 2001. In this display, 2003 and after are estimates and will depend upon policy choices and the performance of the economy. The exhibit shows the impact of the Great Depression of the 1930s and World War II, as well as the Reagan era's defense build up, tax cut, and an underperforming economy in the early 1980s. In that era, the deficit as a percent of GDP rose to more than 6.0% in 1983. It took almost 15 years to get this deficit under control as a result of growth in the U.S. economy, as the first surplus appeared in 1998.

In February 2003, the President's budget predicted that the deficit as a percent of GDP would fall to under 2% by the end of the decade. The 2003 midsession review updated this prediction in July and estimated that the deficit would increase slightly from $455 billion in 2003 to $475 billion in 2004, but said: "As a share of the economy, the projected deficit remains steady in these two years, at 4.2 percent of gross domestic product (GDP). These deficit levels are well below the postwar deficit peak of 6.0 percent of GDP in 1983, and are lower than in six of the last twenty years." OMB further added: "Even more important, after 2004, the deficit is projected to decline rapidly in response to the economy's return to healthy and sustained growth. By 2006, the deficit is cut in half ... and falls from 4.2 percent of GDP in 2003 and 2004 to 1.7 percent of GDP in 2008." OMB warned that the deficits reflected, "an economy in recovery from recession, increased spending in response to the war on terror and homeland security needs, and the reversal of a massive surge in individual income tax collections." OMB admitted that the deficits were "large in nominal terms and a legitimate subject of concern," but concluded that the deficits are "manageable if we continue pro-growth economic policies and exercise serious spending discipline" (OMB, 2003c, pp. 1-2). In the FY 2008 President's budget the deficit as a percentage of GDP was estimated at 1.9%, up slightly from an actual share of 1.8% of GDP in FY 2006 (OMB, 2007, p. 223.). This is quite a good performance in estimating for OMB.

Recently, as during the Reagan years, OMB's credibility with respect to deficit projection has been very low. Generally, deficit estimates from the Congressional Budget Office (CBO) have been more accurate. Part of this is explained by the fact that CBO is a nonpartisan office that works for the leadership of both political parties in Congress. On the other hand, OMB serves only one party and one master—the President. Consequently, if a President wishes to have a lower deficit estimate he may simply order OMB to produce such numbers—and they will (Stockman, 1986). This was the case in the 1980s when OMB deficit projections were wildly unreliable.

Serious spending discipline means that appropriations will be tested against the deficit as they come up for a vote in Congress—but only when Budget Committee spending targets from the budget resolution are heeded by spending committees. However, this occurs only when Congress itself has decided to try to reduce the deficit. Congress has shown little proclivity to do so in the 2000s. During the 1990s this dynamic applied to defense spending. When defense appropriations came to a vote, legislators looked at deficit projections, budget resolution spending targets and the deficit controls Congress had enacted and concluded that the budget was too large—that we could not afford to spend more for defense. However, thus far in the 2000s Congress has not passed deficit control measures similar to those in place in the late 1980s and 1990s. There is no consensus that deficits are out of control (Posner, 2002; Meyers, 2002). Therefore, in the 2000s members of Congress have indicated that we must spend more to contain the threat, even when it is painful. An additional complication of this needs some attention. To the extent that U.S. spending is funded partially by an annual deficit and by a steadily increasing amount of debt (as it has been from 1970, except for 1998 through 2001), then who holds the debt is of some consequence. If it is all held by U.S. citizens or corporations, then well and good. In the late 1960s foreign holding of U.S. debt was about 5% but began to grow substantially in the 1970's. By 2006 foreign holdings of U.S. debt had increased to 44.2% of total debt held by the public, up from 22.2% in 1995 (OMB, 2007, p. 235). This is a dramatic change. Moreover, in 2006 12 foreign central banks owned 66% of debt held by foreigners. What this means for defense policy is that the potential exists for a foreign nation or nations holding a substantial portion of U.S. debt to successfully pursue "debt warfare" against the United States to get it to shape policies more to its or their liking. To the extent that this debt is held by foreign central banks makes it all the easier for those countries to exert undue influence on the United States, as opposed to having the debt held by individuals. Usually part of this argument is that so long as foreigners buy U.S. debt and help keep interest rates low by their willingness to buy relatively "safe" U.S. debt, they are helping the U.S. provide services more cheaply than if U.S. citizens had to "finance" all the debt. There is, however, another side to this and it is not so much that knowing foreigners hold debt would influence U.S. policymakers on any particular day, or in a sudden sharp incident, but rather that the U.S. might shape longer term policy to be more accommodating to a particular nation than it would otherwise have been, just as a homeowner might be more "understanding" of a noisy neighbor who holds his home mortgage. Thus, for example, in 2006-07 did analysts begin to discuss the consequences of China's holdings of substantial U.S. debt?

The lessons of history are not always easy to understand, but history from the early 1940s to the present seems to indicate that the U.S. government and taxpayers are willing to spend for a sufficient defense establishment, be it in an era of deficit or not, and irrespective of whether defense has to be funded by debt. The most telling arguments will be made with reference to defense spending as a percent of GDP and relative to the extensiveness of the threat. History indicates that the U.S. will spend whatever it takes to meet the threat.

CONCLUSIONS

In chapter one, we made the point that defense spending matters in terms of fiscal policy and the economy. In this chapter we have provided additional evidence to support this conclusion. At an annual spending level of $500 to $600 billion dollars (depending on the war supplementals), defense is a large piece of the federal budget. Moreover, the nature of defense budgeting is complicated by murky threat definition, the possibility of extreme violence, the need for extensive military training, the demand for weapons systems that work, and maintenance and supply systems that keep both people and weapons in the field ready to go. This is difficult business and our expectations should not be that the task of defending U.S. interests at home and abroad will be done perfectly, but rather that everyone in the defense policy and budget business, as well as in the fighting forces, will keep trying to get it right.

As we have demonstrated in this chapter the defense budget is important beyond the perspective of how data are arrayed for decision, or what

Deficit as Percent of GDP

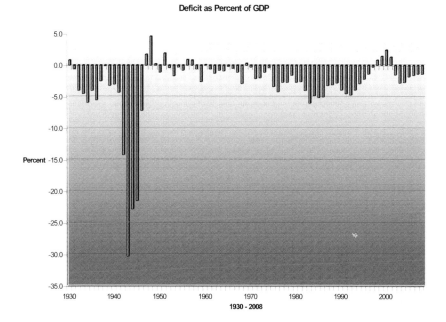

Source: Historical Tables, Budget of the United States, FY 2004, Table 1.2, p. 23.

Figure 4.14.   The deficit as a percentage of GDP.

is bought with defense dollars, or how spending is accounted for in DoD and Congress. It *is important* to understand how DoD and Congress array the budget differently for analysis and decision making. However, as much as pointing this out is the intent of this chapter, just as important is the impact of defense spending on the economy, and the extent to which the state of the economy affects the annual and total debt position of the nation.

DoD spending is an important stimulant to local, state and regional and even foreign economies. Furthermore, as noted, the trend in the U. S. is to finance war fighting and preparedness under virtually any economic condition. Perhaps the best example of this is the mobilization and debt taken on prior to and during WWII.

Both before and after World War II, U.S. Presidents and Congress have been willing to spend what is necessary to counter perceptions of threat to national interests. As we have seen in this chapter, the U.S. has been willing to pay for defense for as long as it took to do the job, from 1940

Shares of GDP

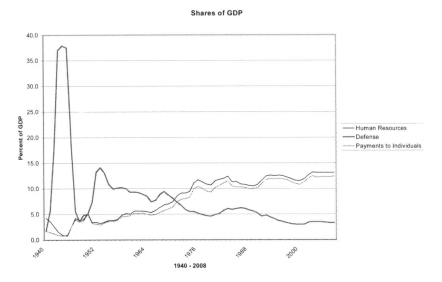

Source: Historical Tables, Budget of the United States, FY 2004.

Note: Defense and payments to individuals as a percent of GDP from Table 6.1 Composition of Outlays: 1940-2008, pp. 109-115. Human Resources from Table 3.1: Outlays by Superfunction and Function: 1940-2008, pp. 44-45. Human Resources includes Education, Training, Employment, Social Services, Health, Medicare, Income Security, and Social Security and Veteran's benefits and services. Payments to individuals is the major part of this category and includes direct payments to individuals from such accounts as Social Security, Medicare, and Veterans' benefits plus others. The two categories track each other closely after the early 1940s at a time when more payments went to state and local governments rather than individuals.

Figure 4.15.   Defense and payments to individuals as shares of GDP, 1940-2008.

through the end of the Cold War and into the War on Terrorism. Also obvious is that the United States is dedicated to supporting individuals by sending direct payments to them as a result of entitlement programs including Social Security, Medicare, Medicaid, and many other programs. The great change in federal spending by function over the past 60 years may be seen in the mapping of spending for national defense against payments to individuals. The composition of the federal budget has changed from a dominance by defense to an increasing emphasis on payments to individuals. These lines crossed in 1971.

It seems obvious that defense was a burden thrust upon the nation in World War II that continued until the 1990s and the end of the Cold War. The U.S. reacted to this burden with billions of dollars of defense spending. If the world had been a kinder, more gentle place, such sustained high levels of defense spending might not have been necessary. After all, in the United States the first instinct was to demobilize after World War II. By June 1947, U.S. military forces had been demobilized from 12 million men and women to about 1.5 million, and the Army was a volunteer body of 684,000 ground troops. The Cold War changed all this for the next 4 decades. In many respects, defense was an external bill imposed on the United States by Russia and others, or perhaps by inept diplomacy. Conversely, choices to create a Social Security system and later to add medical care entitlements, all indexed to inflation, were internal domestic policy decisions. Funding defense provided for protecting citizens as a group: this nation. Funding Social Security and Medicare "protected" citizens as individuals and in so doing, enriched the nation. While defense has varied in its needs, Social Security and health funding have created a steadily ascending obligation for the future spending—the impact may be seen clearly in the lines on the graph. The lesson here is that policy and budget choice have long-term consequences and some choices are virtually forced upon decision makers. Decisions made in any 1 year may take years to undo, if they are ever undone—and may take decades to pay for through higher taxes.

For example, the choices made in 1981 by the President and Congress to cut takes by a large amount and to increase spending for national defense had long-term consequences. Coupled with large increases in entitlement programs, these decisions led to then record peacetime deficits and deficit politics was a steady drumbeat on the national scene until 1998. The message here is that fiscal policy decision making is important and difficult, and that major decisions have large, long-term consequences. Although we may expect political systems to provide feedback and policymakers to respond to feedback as decisions are made and remade, the fact is that many fiscal policy decisions once made are virtually *impossible* to undo. Some part of the hesitancy to create a new military strategy after the fall of the Berlin Wall resulted from recognition that getting it wrong would have long term consequences. Thus, during the 1990s while military doctrine was pushed forward, building a new fighting force was not. Reflecting on this during an era that features a world war of interminable length against a host of widely dispersed terrorist threats, we are forced again to assume a new and different defense burden. This necessity is in large part what drives the debate about defense transformation. We move slowly because we want to get it right—and even when we do the costs are high. Although national defense spending is down from historical levels as a share of GDP, it consumes hundreds of billions of dollars annually. Willingness to pay for U.S. defense is not new. Rather, it has become a fiscal axiom and a signature policy of the United States of America that allies have learned to depend upon and enemies should fear.

NOTE

1. For more on why countries spend for defense and how much, see the
    annual yearbook of SIPRI. SIPRI identifies four types of determinants of
    defense spending: security, technological, economic, and political (SIPRI,
    Yearbook, 2002, p. 264). Also see the work of the International Institute
    for Strategic Studies (IISS), yearbook, The Military Balance, London. For
    comparative military capabilities and defense spending.

CHAPTER 5

THE PLANNING, PROGRAMMING, BUDGETING, EXECUTION SYSTEM

INTRODUCTION

The policy development, resource planning, and budgeting process for national defense is characterized by complexity and plurality. The Department of Defense (DoD) prepares its plan and budget using the Planning, Programming, Budgeting System, or PPBS, which was renamed by reform in 2003 as PPBES (Planning, Programming, Budgeting and Execution System). This chapter reviews and evaluates changes made to what is now termed the PPBES and also assesses other initiatives to reform the resource decision-making system employed by the DoD.

This chapter describes and analyzes the long-range policy and resource planning process employed by the DoD. The chapter also assesses the relationships in policy planning and resource allocation decision making and oversight between the defense department and Congress. However, the primary focus of this chapter is on the institutional and political dynamics of policy and resource planning for national defense within DoD.

While PPBS was discontinued for the federal government as a whole almost 40 years ago, it continues to be employed by the DoD because it

Budgeting, Financial Management, and Acquisition Reform in
The U.S. Department Of Defense, pp. 137–198
Copyright © 2008 by Information Age Publishing

meets the policy development and participatory demands of multiservice budget advocacy while also providing a long-term perspective on programs and spending. While DoD manages its internal resource management systems, this is done under the watchful eyes of Congress. Consequently, in resource planning and in budget preparation and execution, DoD continually searches for a greater delegation of authority from Congress to permit the exercise of greater managerial discretion to improve efficiency and respond to contingencies.

A number of issues related to planning and budgeting for national defense confound DoD and congressional decision makers annually. Among these are how to perform effective and competent threat assessment and the consequences of doing this job well or poorly. Another issue is how much to spend on national defense. This is determined in large part by the perceived threat. The perception of threat also must be interpreted in the dynamics of the politics of budgeting for defense. Numerous variables affect public opinion about threat and spending. Debate and consensus building for national defense budgets is part of our democratic political tradition. Budgeting for national defense is always complicated by conflicting political opinion and information, and also the need for selective degrees of secrecy with respect to identifying and evaluating the threat and budgetary responses to it. These conditions make marketing the need for national defense spending an inevitable task and part of the obligation of defense advocates working in an open political system.

Because so much of the policy framework and budget of the DoD is determined by Congress, which under the U.S. Constitution has sole power to tax and spend, analysis of resource allocation for defense cannot ignore the political context within which decisions are made and executed. Policy development and resource planning for defense is inextricably linked to constituent politics in defense budgeting. National security policy choice and implementation is made more difficult by the highly pluralistic nature of the resource allocation decision environment (Adelman & Augustine, 1990; Wildavsky 1988, pp. 191-193). Still, disagreements over policy and resource allocation should be anticipated and, indeed, welcomed in a democracy.

WHY PPBS? AN HISTORICAL PERSPECTIVE

Policy development, planning, and resource-allocation decision making for the U.S. DoD is a task of enormous complexity due to the nature and size of the defense department and the highly differentiated nature of its mission and activities. The DoD plans, prepares, negotiates, and makes decisions on policy, programs, and resource allocation using the PPBS.

PPBS was implemented in DoD originally by Defense Secretary Robert McNamara and by Charles Hitch, Robert Anthony, and others during the administrations of Presidents Kennedy and Johnson in the 1960s (Thompson & Jones, 1994). Prior to 1962, the DoD did not have a top-down coordinated approach for planning and budgeting (Joint DoD/GAO Working Group on PPBS, 1983; Korb, 1977, 1979; Puritano, 1981). Until this time, the secretary of defense (SECDEF) had played a limited role in budget review as each military service developed and defended its own budget. McNamara had used PPBS when he was President of the Ford Motors Corporation and he and Hitch, his Comptroller, had confidence that the system would be valuable for long-range resource planning and allocation in DoD. McNamara wanted PPBS to become the primary resource decision and allocation mechanism used by the DoD. McNamara implemented the system after President John F. Kennedy tasked him to establish tighter control by the SECDEF, a civilian, over the military departments and services. As a former member of Congress, Kennedy was highly distrustful of the military service planning and budgeting. He ordered McNamara to take control of DoD planning and budgeting away from the military and put it in the hands of civilian leadership. Consequently, the initial motivation for establishing PPBS had as much to do with control and politics as it did with rational resource planning and budgeting. By June 30, 1964, PPBS was operational within the DoD (Feltes, 1976; Korb, 1977, 1979; Thompson & Jones, 1994).

Hitch implemented PPBS and systems analysis throughout DoD, but most of the program analysis was done by his "whiz kids" in the Office of the Secretary of Defense (OSD) under the comptroller and the Office of Program Analysis and Evaluation. The military departments were not anxious to implement PPBS, but had to do so eventually to play in the new planning and budgeting game run and orchestrated by Hitch and his staff. After a few years, the military departments were fully engaged in learning how to compete in the new PPBS process. However, as noted, PPBS was not just budget reform—it was a new approach to analysis and competition between alternative programs, weapons systems and, ultimately, multiyear programmatic objectives. Additional reforms beyond PPBS were to be proposed by DoD under the Johnson administration.

Charles Hitch was followed as DoD Comptroller by Robert N. Anthony, a professor of management control on loan from Harvard University's School of Business, who proposed an ambitious set of changes to DoD budgeting and accounting in 1966 in what was termed Project Prime. Among other things, Project Prime would have divided all parts of DoD into mission, revenue, expense, and service centers, consistent with management control theory according to Comptroller Anthony, and required accrual accounting with reimbursable fee-for-service internal transactional payments (using negotiated or shadow prices) throughout DoD (Thompson & Jones, 1994, pp. 66-68). What Comptroller Anthony envisioned was a reimbursable accounting process similar to what was implemented in much of DoD by Comptroller Sean O'Keefe and Deputy Comptroller Donald Shycoff as part of the Defense Management Report initiatives of 1989-1992 under the Bush administration and Defense Secretary Dick Cheney (Jones & Bixler, 1992). Project Prime also included accrual accounting and budgeting for DoD. Accrual accounting is required now under the Chief Financial Officers Act of 1990, which DoD has been unable to implement successfully. Clearly, Comptroller Anthony was ahead of his time in his vision of how DoD accounting and budgeting should be organized (Thompson & Jones, 1994, pp. 67-68).

Congress did not support Comptroller Anthony's proposed changes. Key members of the appropriations committees refused to allow the change to accrual accounting and rejected Project Prime, probably because they thought it would reduce their leverage to micromanage DoD through the budget. Opposition was so strong that it was suggested Comptroller Anthony should be asked to resign. Comptoller Anthony was not asked to do so, but chose to return to Harvard and the experiment was ended (Jones, 2001b). Not until 2003 did DoD return to Congress with such a sweeping reform proposal—the Defense Transformation Act (see chapter 12, in this volume).

The post-World War II sequence of budget reforms that led to PPBS in the 1960s started with performance budgeting (PB) in the 1950s. In essence, PB (Burkhead, 1959, chapters 6-7, and pp. 133-181) attempts to connect inputs to outputs. As implemented by the President's Bureau of the Budget (BoB) under the Eisenhower administration, PB in the 1950s was characterized by indicators of cost per unit of work accomplished, focusing on workload measures rather than outputs or outcomes. The history of performance budgeting includes the Taft Commission of 1912 which recommended it be implemented and its implementation in the Department of Agriculture in 1934 and the Tennessee Valley Authority in the later 1930s, as well as having been strongly recommended by the Hoover Commission in 1949 (McCaffery & Jones, 2001, p. 69).

In 1949, Congress required that the budget estimates of the DoD be presented in performance categories. Performance budgeting was an executive branch managerial budget tool. During the 1950s under the leadership of BoB Director Maurice Stans and others, executive budgeting was transformed somewhat radically through the institution of performance measures into budgets. Many of the measures had already been in use for decades as proxies that facilitated and simplified negotiations between the Executive and Congress. However, in this first wave of performance budgeting (the second wave would hit in the 1990s) great effort was exerted to develop measures of performance and relate these to appropriations and spending. In fact, many of the measures developed in this era did not measure performance. Instead, because it was easier (and perhaps the only approach possible), workload and input cost data were used in place of real measures of performance. Still, budgeting in this era moved far from the simple line-item formats of the past. Formulae and ratios between proposed spending and actions were integrated into the Executive budget along with explanations of what the measures demonstrated and how they related to justifications for additional resources (McCaffery & Jones, 2001, p. 69).

The emphasis of budget reform shifted in the early 1960s to what was termed "program budgeting." Program budgeting (Mosher, 1954; Novick, 1969) was and is a variation of or evolution from performance budgeting in which information is collected by program categories, without much of the detail of the performance-budget construction. These categories of spending are tied to specific objectives to be achieved. Activities are grouped by department, agency, and then by mission objective and sometimes by function and projected for a 5-year period. Program budgeting was experimented with in the Department of Agriculture in the early 1960s as reported by Wildavsky and Hammond (1962) and later adopted throughout the entire federal government through executive order by President Lyndon Johnson in 1966.

The PPBS (Hinricks & Taylor, 1969; Lee & Johnson, 1983: chapter 5; McCaffery & Jones, 2001, p. 70; Merewitz & Sosnick, 1972; Schick, 1966, 1973) was intended to be a thorough analysis and planning system that incorporated multiple sets of plans and programs. Under Secretary of Defense Robert McNamara and DoD Comptroller Charles Hitch, PPBS drew upon methods from various disciplines, including economics, systems analysis, strategic planning, cybernetics, and public administration to array and analyze alternative means and goals by program and then derive benefit/cost ratios intended to indicate which means and ends to choose. Budgeting under this system was to become a simple matter of costing out the goal chosen.

In theory, the program budgets that resulted from PPBS were supposed to provide the Executive and Congress information on what the federal government was spending for particular categories, for example, health, education, public safety, and so forth, across all departments and agencies. Program budgets may best be understood as matricies with program categories on one axis and departments on the other. Thus, in the fully articulated program budget Congress could determine how much was spent on health or education in total in all departments and agencies and this would promote deliberation over whether this was enough, too much or too little.

President Lyndon Johnson thought that PPBS was so successful in DoD that in 1966 he issued an executive order to have it implemented throughout the federal government. Regrettably, although Executive branch departments prepared their program budgets and related spending to objectives, Congress largely ignored what it was presented, preferring to stick with the traditional appropriations framework for analysis and enactment of the budget (Schick, 1973). Why was this the case? Perhaps program budgets presented too much information to be used and understood by Congress. Alternatively, and as likely, perhaps Congress perceived that program budgeting would reduce the power of members of appropriations committees because the budget in this format would be determined too much by formula, thus decreasing the political spending discretion of Congress (Jones & Bixler, 1992). Although the government-wide experiment with PPBS was suspended by President Richard Nixon in 1969, this was done more for political than efficiency reasons. However, PPBS was perceived in much of the Executive branch and Congress as paper-heavy and consuming too much staff time for preparation and analysis (Schick, 1973). Still, the system continued to be used in the DoD, in part because DoD purchases substantial long-lived capital assets and since PPBS requires long-range planning as its first component, it suited the needs of the defense department.

Thus, despite criticism that PPBS was a failure in the federal government, the process remained in use by the DoD and has been modified incrementally so as to operate effectively despite some evident flaws (McCaffery & Jones, 2001; Puritano, 1981; Wildavsky, 1988, pp. 186-202). While the manner in which PPBS operates has varied under different Presidents and secretaries of defense, the basic characteristics of the system have remained in place for more than 40 years. During this period, three significant reform initiatives have influenced the PPBS: the Laird reforms, the Goldwater-Nichols Act, and the Rumsfeld transformation in 2001-2003 that renamed the process adding the word execution, that is, the system is now referred to as PPBES.

Laird Reforms

In 1969, Melvin Laird was appointed Secretary of Defense by President-elect Richard Nixon to succeed McNamara. Laird brought a different management orientation to the defense department, one more in keeping with its historical predilections, emphasizing decentralization and military service primacy. If McNamara increased scientific decision making in the Pentagon, he also installed a centralized management approach. Systems analysis, top-down planning, and benefit/cost analysis supported this centralized focus. One of the key bureaucratic players was the Office of Policy Analysis, which made use of the tools cited above to help McNamara centralize decisions in the OSD (Thompson & Jones, 1994, pp. 68-73). Laird's methods ran counter to this approach, emphasizing participatory management and decentralization of power. Beginning in 1969, Laird shifted decision-making power away from the DoD staff agencies to the military department secretaries, because there were

> many decisions that should be made by the Services Secretaries and they should have the responsibility for running their own programs. I have no business being involved in how many 20mm guns should go on a destroyer. That is the Secretary of the Navy's business. I must let the Services take a greater role. (Feltes, 1976)

Laird also pursued a process of participatory management, in which he hoped to gain the cooperation of the military leadership in reducing the defense budget and the size of the forces.

During Laird's 4-year tenure, U.S. troop strength in Vietnam fell from 549,500 persons in 1969 to 69,000 in May of 1972 (Laird, 2003). Laird was preoccupied with disengaging from Vietnam, but not to the exclusion of other issues, such as burden-sharing costs with other nations, maintaining technological superiority (e.g., B-1 bomber, Trident submarine), improved procurement, enhanced operational readiness, and strategic sufficiency and limitations on the nuclear build-up (Armed Forces Management, 1969; Feltes, 1976). He ended the selective service draft in January of 1973 and was persistent in his efforts to secure the release of American POWs.

Laird (2003) spent a lot of time preparing for and testifying in Congress and improved DoD relations with Congress. On the management side, Laird gave the military department secretaries and the Joint Chiefs of Staff (JCS) a more influential role in the development of budgets and force levels, but he also returned to the use of service program and budget ceilings (fixed shares) and required services to program within these ceilings. This concept of ceilings or "top-line" endured for most of the next 40 years and still influences DoD budget requests today, as services are expected to balance their program and budget against the TOA they are given at various stages in the planning and budget process.

Laird (2003) sought to provide a better balance between military and civilian judgment in the defense decision-making process by providing better and earlier strategic and fiscal guidance to the services and the JCS. Feltes (1976) suggests that the result of Laird's emphasis on decentralized management was that responsibility for military planning was shifted back to the military services, and the role of OSD systems analysis was de-emphasized. While no abrupt shifts were made, the Laird era was marked by a steady and persistent shift away from McNamara's emphasis on centralization of DoD decision making under the SECDEF (Armed Forces Management, 1969).

The Goldwater-Nichols Act of 1986

It may be argued that the creation of the defense department in 1947-49 never really took hold in that, by and large, the military departments continued to go their separate ways within the envelope of the DoD until the reforms of the 1960s and, to some extent, until implementation of the Goldwater-Nichols Act of 1986 (Thompson & Jones, 1994, pp. 78-79, 246). In the 1950s, Presidents Truman and Eisenhower both fought arguably losing battles to strengthen the role of chairman of the Joint Chiefs of Staff (CJCS) and the JCS (pp. 51-53).

By 1981, the sitting JCS Chairman, General David Jones (1982) was writing that the system was broken and asking Congress to fix it. The fact that General Jones as CJCS was voicing such criticisms was in itself very significant (Chiarelli, 1993, p. 71). In 1981, Jones (1982) suggested that because of the decentralized and fragmented resource allocation process driven by parochial service loyalties, there was always more program than budget to buy it; that the focus was always on service programs; that changes were always marginal when perhaps better analysis would have led to more sweeping changes; that it was impossible to focus on critical cross-service needs; and the result was that an amalgamation of service needs prevailed at the JCS level.

General Jones argued that staff to the CJCS was so small that the chairman could focus only on a few issues. The result was that the defense budget was driven by the desires of the services (usually for more programs and money), rather than by a well-integrated JCS plan. In addition, he argued that all of this undercut the authority of not only the JCS but the entire unified command structure established in the Defense Reorganization Act of 1958 (Thompson & Jones, 1994, pp. 51-53). General Jones noted this was particularly evident in acquisition, where weapons systems met performance goals 70% of the time, but schedules 15% of the time, and cost goals 10% of the time. Jones (1996) explained:

> The lack of discipline in the budget system prevents making the very tough choices of what to do and what not to do. Instead, strong constituencies in the Pentagon, Congress, and industry support individual programs, while the need for overall defense effectiveness and efficiency is not adequately addressed. (p. 27)

In 1986 Congress passed a sweeping reform plan, commonly referred to as the Goldwater-Nichols Act (for its congressional sponsors), over the ardent objections of many in the Pentagon, including Secretary of Defense Caspar Weinberger (Locher, 1996, p. 10; Locher, 2002) who thought it would break apart the DoD management system. The legislation is too complex to detail here, but among other things it strengthened the hand of the CJCS as chief military advisor and spokesman to the SECDEF and to the President, provided the CJCS with a larger staff and identified important phases in the PPBS process where the JCS would play in setting requirements and reviewing the plans of other players. It established the national command authority to run from the President to the SECDEF to the unified commanders in chief (CINCs). This increased their formal authority so that rather than using whatever forces the military services would allow them to use in their geographical area, the unified CINCs had war fighting and command responsibilities and the military service roles were to provide them with the wherewithal to do so (Thompson & Jones, 1994, pp. 51-53, 79, 223-224). This distinction clearly put the military services in the role of training people and providing personnel and equipment for the war fighting missions of the geographically based unified command CINCs. Goldwater-Nichols also created the position of vice-chairman of the JCS. Generally, the officers who have served in this spot have been strong innovators and, through various committee structures, have had a substantial impact on the resource planning process within DoD.

Goldwater-Nichols also emphasized the requirement for joint command officer duty assignment. Before Goldwater-Nichols, JCS and joint command assignments were viewed as almost career-ending assignments, thus many of the best officers tried to avoid them. CJCS Jones observed that people serving joint tours did less well in the promotion process than those who had not served such tours (Jones, 1996: 28). While implementing it has been an evolutionary process, Goldwater-Nichols has changed this perspective – such assignments now may be career enhancing. The Act also required all officers to pass certain levels of joint proficiency and upwardly mobile officers now believe a joint tour is a must.

Most importantly, Goldwater-Nichols changed the caliber of advice given to the President and SECDEF by the JCS. Former CJCS Army General Shalikashvili praised this part of the Act, "we have broken free from the 'lowest common denominator' recommendation that so often plagued us in the past" (Roberts, 1996, p. 1). Shalikashvili indicated there was still room for smoothing the role of the JCS in the planning and budgeting cycles, in the national military planning process, and in management of officers into joint billets. Nonetheless, it is clear that Goldwater-Nichols is a success, as Secretary of Defense Perry noted in 1995, "It dramatically changed the way that America's forces operate by streamlining the command process and empowering the Chairman and the unified commanders. These changes paid off in.... Desert Storm, in Haiti, and today in Bosnia" (Locher, 1996, p. 15).

On the resource allocation side, Goldwater-Nichols provided two classes of organizations, those who do the war fighting, under the unified command CINCs, and those who support them, the military departments and services and their own CINCs. The military department secretaries hold most of the DoD Budget Authority, while the service CINCs play key roles in programming, with less leverage in budgeting. Most of the combatant commands, the unified CINCs, do not have their own budgets (except for their staffs). Rather, they use the personnel and weaponry provided them by the military departments and services. However, the military CINCs must pass their budget requests through the unified command CINCs before they move upward in the budget chain of command to the Pentagon. Prior to the mid-1990s this review by the unified command staffs used to be pro forma but it has become a real review in many unified commands, for example, the review by the commander in chief of all Pacific forces (CINCPAC) of the budget of the commander in chief of naval forces in the Pacific (CINCPACFLT). The special operations forces (SOF) command, headquartered at McDill Air Force base in Florida, has its own sizable and historically increasing budget, but SOF budgets still are relatively small compared to the military department budgets.

The unified CINCs also have had an opportunity to identify requirements in the PPBS process and the CJCS has the responsibility to advise the SECDEF to certify the merit of these requirements as well as how well the budgets of the military departments satisfy the unified CINCs' needs. The JCS chairman also can submit alternative recommendations to SECDEF to meet unified the CINCs' needs in the budget. In this matter, SECDEF is the final arbiter of what the military departments get in their budgets. The unified and service CINCs both have opportunities to give input to the CJCS in PPBS planning process for development of the national military strategy, and in the final draft of the defense guidance which leads to the Program Objective Memorandum (POM) process. In the POM process, the service CINCs make inputs by providing their integrated priority lists (IPLs) that indicate their top war fighting needs (important information for the JCS and unified CINCs). Military service CINCs may indicate program deficiencies that exist and make recommendations to fix deficiencies to both the JCS and the military service chiefs. The IPLs are a part of the programming and budgeting process and are duly considered in several venues in OSD and the military departments. More detail on this is provided subsequently in this chapter.

An unresolved tension is evident here as the unified and service CINCs both have been criticized as sometimes tending to focus on short-term operational needs, war fighting issues, and the operations and maintenance (O&M) accounts that support readiness. Simultaneously, the military departments have to keep an eye not only on the short-term and immediate items and issues, but also weapons procurement and recapitalization issues, such as modernizing the aircraft or fleet inventory. Some players in the PPBS process have viewed this is a healthy tension. Others have worried that immediate issues, and some long-term needs, may be slighted. As we note subsequently, both organizationally and budgetarily, DoD developed and implemented significant changes, including those to PPBS, under former Secretary of Defense Donald Rumsfeld in the period 2001-2006. Rumsfeld and his staff pursued the goal of transforming both military and business affairs while actively employing a large part of the operating force in combat operations. This reform period may be characterized as somewhat of a return to a more centralized pattern of DoD top-down control of the type established by Robert McNamara in the 1960s.

THE REFORMED PPBES PROCESS OVERVIEW

The purpose of PPBES is to provide a systematic and structured approach for allocating resources in support of the national security strategy (NSS) of the United States. The ultimate goal of the entire PPBES process is to provide the military CINCs with the best mix of forces, equipment, and support attainable within resource constraints. Before delving into the full complexity of PPBES it is useful to review the system in summary. Once we understand how PPBES operates in general, we then review changes initiated in 2001 and 2003 to significantly modify PPBS into what is now PPBES—the result of significant reforms authorized by Defense Secretary Donald Rumsfeld under the administration of President George W. Bush. Then, when we understand the changes made during this period, we examine how the process operates in detail. An overview of the new PPBES decision cycle is provided in Table 5.1 and Figure 5.1.

PPBES has four distinct phases, with each phase overlapping the other phases (Jones & Bixler, 1992, pp. 19-31; Jones & McCaffery, 2005). Planning and assessing, for example, are continuous and all players in the process understand that decisions made in these phases will affect other phases. Because the interrelationships are so complex, players in each phase attempt to stay informed on issues in their respective phase as well as issues in other phases affecting them. However, the size of DoD and the complexity of the PPBES process render this virtually impossible. It is difficult enough for the participants in one military service to keep

Table 5.1.   Summary of the PPBES Cycle

Year 1: Review and Refinement	*Year 3: Execution of Guidance*
• Early National Security Strategy	• —
• Restricted fiscal guidance	• Restricted fiscal guidance
• Off-year DPG, as required (tasking studies indicative of new administration's priorities; incorporating fact-of-life acquisition changes, completed PDM studies, and congressional changes)	• Off-year DPG, as required (tasking studies; incorporating fact-of-life acquisition program changes, PDM studies and congressional changes)
• Limited changes to baseline program	• Limited changes to baseline program
• Program, budget, and execution review initializes the on-year DPG	• Program, budget, and execution review initializes the on-year DPG
• President's budget and congressional justification	• President's budget and congressional justification
Year 2: Full PPBE Cycle— Formalizing the Agenda	*Year 4: Full PPBE Cycle— Ensuring the Legacy*
• Quadrennial Defense Review	• —
• Fiscal guidance issued	• Fiscal guidance issued
• On-year DPG (implementing QDR)	• On-year DPG (refining alignment of strategy and programs)
• POM/BES submissions	• POM/BES submissions
• Program, budget, and execution review	• Program, budget, and execution review
• President's budget and congressional justification	• President's budget and congressional justification

Source: Secretary of Defense, Management Initiative Decision 913, 2003, p. 3.

abreast of what is happening in their own process, much less what is done in other branches of the military.

The *planning phase* begins at the executive branch level with the President's NSS developed by the National Security Council. The NSS takes its input from several federal agencies (including the Department of State, the Central Intelligence Agency, and others in the intelligence community) to ascertain the threats to the United States in order to form the nation's overall strategic plan to meet those threats, thereby outlining the national defense strategy. Subsequently, the JCS produce a fiscally unconstrained document called the national military strategy document (NMSD). The NMSD contains their advice regarding strategic planning to meet the direction given in the NSS while addressing the military capabilities required supporting that objective. As a follow on to the NMSD, the CJCS advises the SECDEF, in the chairman's program

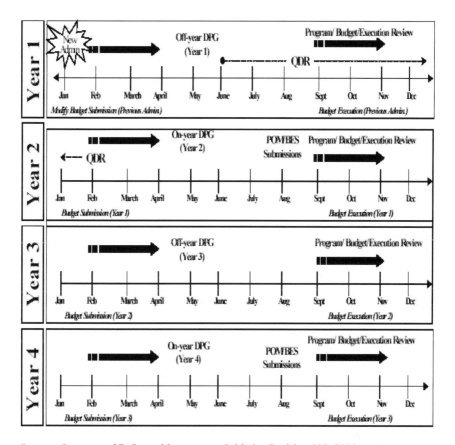

Source: Secretary of Defense, Management Initiative Decision 913, 2003, p. 4.

Figure 5.1.   Calendar of the 4-year PPBES cycle.

recommendation (CPR), regarding joint capabilities to be realized across DoD military components. The CPR provides the personal recommendations of the CJCS for promoting joint readiness, doctrine, and training, and better satisfying joint war fighting requirements in order to influence formulation of the defense planning guidance (DPG). The CPR is seen as a key joint staff input from the CJCS and his staff into the PPBES process. It is meant to help steer the DPG.

All of the above inputs are provided to the SECDEF for drafting and ultimately issuance of the DPG, and the future years defense plan (FYDP), a 6-year projection of department-wide force structure requirements. The DPG provides the military services official guidance regarding force structure and fiscal guidelines for use in preparing their POM during the programming phase of PPBES. For purposes of reporting to Congress on defense planning, the DoD also prepares and transmits a comprehensive report referred to as the Quadrennial Defense Review (QDR). In the past decade, the QDR has enhanced the FYDP and DPG for purposes of planning for the OSD and DoD. Figure 5.2 shows the articulation of the phases of PPBES for the Navy.

The purpose of the *programming phase* is for each military component to produce a POM to address how they will allocate resources over a six-year period. The development of the POM requires the services to consider numerous issues including their CINCs fiscally unconstrained IPLs stipulating programs that must be addressed during its development. The POM must also support the guidance given in the DPG and operate under fiscal constraints issued within it, for example, TOA by military department by year. POMs are developed in even numbered years and subsequently reviewed in odd-numbered years.

Woven within the POM are the sponsor program proposals (SPPs) developed by resource sponsors (e.g., the major commands, systems commands and defense agencies) to address military service objectives, and preferences of the CINCs. The SPPs must be developed within the

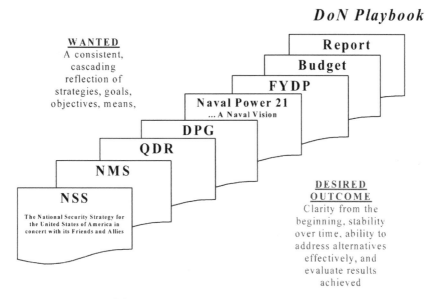

Source: Department of the Navy, 2003c.

Figure 5.2.   From National Security Strategy to Budget Execution (Navy example).

constraints of military component TOA, defined as the total amount of funds available for spending in a given year including new obligation authority and unspent funds from prior years.

Military department and service POMs are reviewed by the JCS to ensure compliance with the NMSD and DPG, assessing force levels, balance and capabilities. Following the review, the CJCS issues the chairman's program assessment (CPA) to influence the SECDEF decisions delineated in the program decision memoranda (PDM) marking the end of the programming phase. The CPA is another key steering device that the chairman uses to give his personal assessment of the adequacy and risks of service and defense agency POMs. He also proposes alternative program recommendations and budget proposals for SECDEF consideration prior to the issuance of PDM by SECDEF. The PDM issued by OSD approves or adjusts programs in each POM. The POM that has been amended by the PDM provides an approved baseline for military departments to submit their budget inputs. While the programming phase of PPBES operated as a separate cycle from the 1960s through the early 2000s, in August 2001 Secretary of Defense Donald Rumsfeld merged the POM and budget review cycles, as noted later in this chapter.

In acquisition matters, the CJCS is supported by the joint resources oversight committee (JROC) a committee led by the vice-CJCS and composed of the service vice-CJCS who review all joint acquisition programs and programs where a joint interest in interoperability is evident. The chairman then can and does make recommendations about acquisition priorities. This is another change rising out of Goldwater-Nichols and out of the Grenada operation where Army and Marine troops on the ground could not communicate with other units because the radios used were not interoperable. The JROC approves the mission need and conducts an analysis to see how well the suggested acquisition program meets these needs. The process of staffing a proposal up to the JROC decision level involves assessment and analysis by various committees ending at the Flag level, and analytic effort by JCS staff and can take four to five months. A successful program that is vetted and found to meet joint requirements then has a priority attached to it at the JROC level and is then passed into the POM and later the budget for funding.

Part of the 2003 reform was intended to accelerate and improve the acquisition process. In April 2002, DPG study #20 (SECDEF, 2002b) concluded that the resource requirements process frequently produced stovepiped systems that were not necessarily based on required capabilities and incorporated decisions from a single service perspective. The study found that the acquisition process did not necessarily develop requirements in the context of how the joint force would fight. Rather, requirements tended to be more service-focused. Moreover, duplication of efforts was apparent in the less visible and smaller acquisition programs. The study observed that the current culture aimed for the 100% (perfect) solution and this resulted in lengthy times to field weapons. In addition, the process was still found to lack prioritization of joint war fighting demands. Ongoing reform here resulted in reshaping of the JROC process so that decisions would be better set up for JROC to make its decision by two new oversight committees reporting to it, headed by flag officers and focused on functional areas. This is an ongoing part of the 2003 reform and is indicative of Secretary Rumsfeld's interest in joint operations, joint war fighting, and a quicker acquisitions process.

The *budgeting phase* begins with the approved programs in each military service POM.

Each military component costs out the items that support its POM for the budget year and submits its part of the budget as its budget estimate submission (BES). The BES in even-numbered "POM years" is a 2-year submission and is based on the first 2 years of the POM as adjusted by the PDM. The BESs are amended by the services during the POM update occurring in odd-numbered years and cover only 1 year. Every BES is reviewed by military secretariats under the authority of the military department secretaries because budgeting is a civilian function in DoD, as mandated by Congress in the 1970s. The budgets of the military department secretaries are then reviewed by the DoD Comptroller, other OSD officials, the JCS and ultimately by the deputy and SECDEF.

As noted in chapter 1, the OSD cooperates in this review with the President's Office of Management and Budget (OMB). This review attempts to ensure compliance with the DPG, the PDM and the President's NSS. SECDEF staff makes changes and provide rationale for these changes in the form of program budget decisions (PBD). Before becoming part of the President's budget, required for submission to Congress no later than the first Monday in February, PBDs are issued to allow the military department secretaries and budget staff to respond with appeals of cuts (reclamas) to SECDEF/OSD comptroller staff. Once major budget issues have been resolved, the final defense budget is sent to OMB to become part of the President's budget. This step constitutes the end of the budget proposal and review phase of PPBS. However, as noted subsequently, budget execution is a critical part of PPBS typically ignored in analysis of this system.

Budget execution consists of first gaining permission to spend appropriations approved by Congress through a separate budget submission process referred to as the allotment process. In allotment review, DoD must show how it intends to spend what has been appropriated, by quarter, month, or fiscal year (FY) for multiple year appropriations. This is always somewhat different than what was proposed in the President's budget since appropriations must now be attributed to programs and allocated into the months they will be obligated (usually by quarters). After allotment approval is received from OMB and the Treasury, DoD begins the process of separating and distributing shares of the DoD budget to the military departments and services and other DoD commands and agencies. After they have received their spending allotment authority, these resource claimants begin to incur obligations to spend, and then liquidate their obligations through outlay of money. During this process, comptrollers and budget officials at all levels of DoD monitor and control execution of programs and funding. At the midpoint of the spending year, the military departments and services typically conduct a midyear review to facilitate shifting of money to areas of highest need. At the end of the FY (September), all DoD accounts must be reconciled with appropriations and spending must be accounted for prior to closing the accounts from further obligation and outlay (for annual accounts). Financial and management audits by military department audit agencies, the DoD inspectors generals, the Government Accountability Office (GAO) and other entities follow the conclusion of execution and reporting.

PPBES REFORM UNDER RUMSFELD

In 2003, the DoD announced significant changes to the PPB system, renaming it the Planning, Programming, Budgeting and Execution System or PPBES (SECDEF, 2003a). While the basic structure of PPBS remains, it was changed in three important ways. First, the reform merged separate programming and budget review into a single review cycle. Second, it incorporated a biennial budget process. Third, it changed the cycle for OSD provision of the top level planning information to the military departments and services. The DPG that had been issued annually will now become a biennial guidance. The OSD will no longer provide the military services and defense agencies an annual classified planning document designed to help them develop their budget and program requests for the upcoming FY. The move away from developing the top-level DPG each year is part of the OSD move toward 2-year budget cycles. OSD may prepare "off-year" guidance documents reflecting minor strategy changes, according to Management Initiative Decision No. 913, issued May 22, 2003 by Deputy Secretary of Defense Paul Wolfowitz (SECDEF, 2003a).

The essence of the reform is to place the biennial issuance of the DPG document in a 2-year cycle within the 4 years that a presidential administration has to develop its national defense objectives and strategy. A series of documents has in the past guided this process, including the annual DPG, the Future Years Defense Program, the issuance of each new President's NSS, and development of the QDR for use by DoD and for reporting to Congress. The QDR consists of a comprehensive analysis of military readiness, capabilities, and force structure that helps to provide a reporting framework to permit a newly elected administration to develop its spending plan and budget. Since the early 1990s, the QDR has become the primary external and one of the major internal statements of policy by the SECDEF.

On February 3, 2003, DoD Comptroller official Dov Zakheim presented the new DoD biennial budget part of the reform with the release of the President Bush's FY 2004 defense budget request. Zakheim indicated that DoD would use the off years when budgets would not be prepared from scratch to examine how well DoD was executing its programs and dollars (SECDEF, 2003a). He noted that as of this budget (FY 2004) FY 2005 would be an "off year" in which only significant revisions to the budget would be requested from Congress. This meant that the budget process conducted during the summer and fall of 2003 to prepare the FY 2005 budget would be significantly changed. For example, DoD will not prepare the POM or budget estimates for FY 2005. Instead, OSD will use estimates for FY 2005 as they were estimated in the FY 2004 budget and FYDP, which covers FY 2004 to FY 2009. An updating mechanism has been created for the off-years, for example, FY 2005.

Military departments and CINCS may create program change proposals (PCPs) to affect the POM and budget change proposals (BCPs) to speak to new budget needs. The PCPs allow for fact of life changes to the previous year's POM; they are meant to be few and of relatively large size. Guidance for 2003 indicated the PCPs had to exceed a set dollar threshold or have serious policy and programmatic implications. For example, in 2003 the Navy submitted only three PCPs, one worth $100 million that involved 450 line items. The Navy would submit only three PCPs in 2003. For all of DoD the number of PCPs was estimated to be about 120. For the CINCs, the PCPs are a new tool provided them in the PPBE process, but like the military departments, they have to suggest offsets. For example, if a CINC wants to increase force protection in one area at a certain cost, he has to suggest weakening force protection in another area as an offset for the increase. This is meant to be a zero-sum game. Changes have to be accompanied by offsets or billpayers. As is usual with any offset procedure, claimants who submit either PCPs or BCPs take the risk that the offsets they suggest will be accepted, but the accompanying change proposals the offsets were intended to fund might not be. In such cases, the offset reveals a pot of money for a lower priority item that might be directed to another area. The BCPs were expected to be more numerous but smaller. They too would be largely fact of life changes (e.g., cost increases, schedule delays, new congressional directives) and would have to be paid for by offsets. Although the individual BCP need not be offset, the package of offsets provided by a military department has to be offset and provide a zero balance change. The FY 2006 budget request will be prepared completely anew, marking the first biennial POM and budget in the new 2-year cycle. A DPG was prepared by OSD to guide the FY 2006 to FY 2008 processes.

In April 2003 Defense Secretary Donald Rumsfeld canceled the 2005 DPG due to the budget process changes announced in February by Zakheim to concentrate Pentagon analytical resources on determining whether Saddam Hussein's ouster and the progression of the war on terror had mandated additional changes in the Bush administration national defense strategy. In addition to prioritizing how OSD believes military dollars should be spent in upcoming years, the DPG typically calls for studies on top issues and indicates new strategies to be undertaken. Rumsfeld's action violated no rules, as the SECDEF is not legally required to prepare an annual DPG.

PPBES: Year 1

Management Initiative Decision 913 sets out a 2-year budget and planning cycle within the framework of the 4 years in a presidential administration. The first year requires "review and refinement" of the previous President's strategy and plans, including only limited changes in programs and budgets, an early national security strategy, and an "off-year DPG." As stated in MID-913,

> The off-year DPG will be issued at the discretion of the Secretary of Defense.... The off-year DPG will not introduce major changes to the defense program, except as specifically directed by the Secretary or Deputy Secretary of Defense.... However, a small and discrete number of programming changes will be required to reflect real world changes and as part of the continuing need to align the defense program with the defense strategy. (SECDEF, 2003a, p. 5)

A major objective of the off-year guidance will be to provide the planning and analysis necessary to identify major program issues for the next DPG. One of the benefits of the new 4-year cycle is that it fits the PPB process into the electoral cycle. Incoming administrations usually struggle to get their people on board in the first year and significant defense policy changes usually do not come until later. The new cycle recognizes this reality. Significant events do happen in year one. The NSS is issued at about midyear and the QDR begins shortly thereafter and is issued early in the second year. These are significant guidances for defense strategy and resource allocation.

PPBES: Year 2

The second year in the new 4-year framework is more intense in that the military departments and services and OSD will conduct full program, planning, budget, and execution reviews to formalize the President's defense posture and strategy, including the resource portion of the strategy. In addition to a QDR issued early in the year, the second year will include a full, "on-year" DPG, issued in May and designed to implement the QDR results. Previously, the QDR had been issued on September 30 in the first year of a presidential administration. However, in the FY 2003 Defense Authorization Act, Congress change the QDR reporting requirement to the second year to provide new DoD leadership more time for analysis and preparation. Senior defense officials had argued to Congress that the requirement to submit a QDR in the first year was too much to ask of a new administration barely through the rigorous congressional process for confirmation of presidential appointees to head the DoD and military departments. The second year will see then a full POM and a full budget build. These result in a full FYDP build.

PPBES: Year 3

The new planning and budget process specifies that the third year be used for "execution" of the President's defense plan and budget agenda as provided in the QDR and the previous year's DPG. Year three corresponds with FY 2005 in the budget cycle and could include an "off-year" DPG if so desired by the SECDEF. This off-year guidance could task new studies, or incorporate fact-of-life changes in acquisition programs including increased costs or schedule delays as well as congressionally mandated changes. In May 2003, Zakheim indicated that no 2005 DPG was to be prepared under the Bush administration and Rumsfeld. However, the presidential elections of 2004 could change this plan. The third year is a year of refinement of objectives and metrics with only the most necessary program or BCPs considered.

Careful examination of DoD execution of dollars and plans is a critical part of the new planning and budgeting process. Traditionally, budget execution has been left primarily to the military departments. However, the revised process provides OSD with greater opportunity to examine and critique the budget execution decisions of the military departments and services. Zakheim reported in February a widespread agreement in the DoD not to return to a comprehensive annual budget and program review; rather the intent was to use the off year to measure the "burn rate" (rate of spending) in an execution review. To this end, the comptroller said the review would include asking questions such as how money is being spent, if it should be moved to other areas and accounts, and the results achieved from execution.

An important budget changed initiated by the Bush administration announced in February 2003 and subsequently by the DoD comptroller is implementation of "performance-based budgeting," to focus more on the costs of achieving desired military and programmatic outcomes, rather than concentrating budget review on the details of program administration and production. The driving military concept behind performance-based-budgeting (PBB) is the concept of "effects-based capabilities" for war fighting. The effects-based approach focuses on desired end results from a military action rather than the military action itself. Under this concept, military commanders specify the results, such as capture of territory, in addition to the amounts and types of forces needed to achieve the outcome.

PPBES: Year 4

The fourth year in the budget and planning cycle is characterized in MID-913 as the point where the achievements of a 4-year presidential administration are assessed. This year will include preparation of a full DPG to refine the alignment between presidential strategy and the DoD program and budget. As usual, the DPG will initiate and guide the cycle of military department and service POM and budget preparation, review and submission (for FY 2006). Then, the next full PPBES cycle will encompass FYs 2006 to 2011.

The PPBES and budget changes implemented by the Bush administration in 2003 are the most comprehensive since the system was established in the early 1960s. The drive to move to a refined system was impelled by many of the factors analyzed subsequently in this chapter. In short, for years the PPBS process had been criticized as duplicative, unnecessarily complicated, and wasteful of staff time and energy (Jones & Bixler, 1992; Puritano, 1981). The administration of George H. W. Bush (1988-1992) seriously considered reform of PPBS in the period 1990 and 1991 as part of its defense management report initiative, but the challenge to realign defense strategy and the overall confusion caused by the end of the Cold War understandably took precedence over these reforms. It would take a decade for a presidential administration to come back to the issue of

PPBS reform. In chapter 12, we delve more deeply into the rationale for PBB and how it is intended to operate.

PPBES IN DETAIL:
THE POLICY DEVELOPMENT AND PLANNING PHASE

In assessment of policy and planning for defense, it is important to understand that the goals and mission of the DoD are not deliberated and set exclusively within the PPBES system. Policy direction comes from the President, the State Department, the National Security Council, and other executive branch agencies, and from Congress. Under the separation of powers constitutional political system in the United States, Congress always has the authority to assert policy priorities independent of the executive branch.

Another set of factors that drives the policy and planning phase of PPBES are the treaties, international commitments, agreements, and understandings of U.S. defense obligations negotiated by policymakers over the past century and particularly since World War II. Critical to policy development and planning is the assessment of worldwide threats to United States and allied interests that are monitored constantly by a variety of intelligence agencies. In addition, a distinguishing set of factors that drive PPBES are the broad defense policy and programmatic objectives set by the President and his advisors, for example, in the areas of readiness, sustainability, force structure, and modernization.

The DoD policy development and planning may be differentiated into three semiautonomous systems: one for macrointernational security planning, a second for war-fighting planning, and a third for defense resource planning. In addition, the DoD may be viewed to operate organizationally using two or, arguably, three interdependent management control systems: one for military operations, a second for general administration, and a third for financial management and budgeting. Alignment of these semiautonomous systems has been a problem for DoD. OSD and the military departments have attempted to address realignment problems by forging improved linkage, networking and coordination, including use of better business practices as part of the "transformation in business affairs" and a corresponding transformation in military affairs, beginning in the latter part of the 1990s.

In PPBES, assessments of threats and commitments, and estimates of resources needed to meet commitments at "acceptable" levels of risk, are defined separately by the JCS in its joint strategic planning document, by each military department, and by the OSD. These independent evaluations are then combined by the OSD to produce the DPG. The

DPG indicates annually the assets, forces, and other resources needed to satisfy U. S. security obligations. The DPG covers threats and opportunities, policy, strategy, force planning, resource planning, and fiscal guidance, and includes a summary of major policy issues. The DPG provides the basis for subsequent service-branch and OSD programming and budgeting.

Policy and resource planning is accomplished within the framework of the PPBES program structure, comprised of eleven programs: support of other nations, strategic forces, general purpose forces, intelligence and communications, airlift and sealift, guard and reserve forces, research and development, central supply and maintenance, training and personnel administration, and special operations forces. The report used to amass and report all of this information is the FYDP, a 6-year projection of force structure requirements. More on the FYDP is provided subsequently in this chapter. These major programs in the PPBES program framework are cross-walked to the appropriations format employed by Congress to budget for the DoD, to warfare planning, to the programming and budgeting control structures of the service branches, and to the unified and specified command structure that combines Army, Navy, and Marine Corps, Air Force, and special forces into autonomous, geographically defined area commands. Also, as explained later, for purposes of reporting to Congress on defense planning the DoD uses a comprehensive report—the QDR.

The fact that policy development and planning is continuous suggests that perhaps planning and programming are short-term exercises. In fact, policy and programmatic planning within PPBES takes a long-range perspective of 10 to 20 years and beyond while programming has a 6-year focus as does budgeting. However, in practice, programming tends to focus on a 2-year period and much of the budget is decided annually, even though multiple-year projections are prepared for both programs and budgets.

Much of U.S. defense policy and force structure planning was developed after World War II and codified for the purposes of PPBS in the 1960s and has been changed only marginally since this time until the period 2001-2003. Reductions in force structure were made in the 1970s after Vietnam and with the adoption of the volunteer force and in the 1990s after the fall of the Berlin Wall. However, even when cuts are made or forces expanded, as in the 1980s, relative force-structure composition has been fairly stable. Of course, the mix of capital equipment in the force structure changes constantly, for example, numbers and types of ships or aircraft in the Navy and Marine Corps, or aircraft in the Army and Air Force. The most serious challenges to force structure planning have come at the end of the Vietnam War in the 1970s, the end of the Cold War in

1989, the defense drawdown of personnel and weapons from 1991 through approximately 2000, and the advent of the War on Terrorism in 2001.

Because policy planning in PPBES is long-term, some criticisms to the effect that defense policy and planning fails to take into account rapid changes in world conditions may miss the point. Policy and resource planning is often unable to anticipate short-term contingencies and risks of the type that must be accommodated by all types of organizations on a fire-fighting basis. Furthermore, many defense contingencies are anticipated with considerable accuracy by DoD analysts, but the knowledge that certain events may occur, even where probabilities may be specified with some confidence, does not resolve the problem of allocating and coordinating resources under circumstances where commitments are extensive and resource demands exceed availability.

Policy development and planning also does not insure that commitments of allies to jointly defend against certain types of threat will be upheld without considerable jaw-boning by the United States on a short-term basis, as occurred at the beginning of the Persian Gulf sea-lane protection operation in 1987 and in the late summer and fall of 1990 at the outset of enforcement of the embargo and then the first war against Iraq to liberate Kuwait. Similarly, this was a major issue in preparation and execution of the War on Terrorism for battles against terrorist regimes in Afghanistan and Iraq in the period 2001-2003. Operation Iraqi Freedom in 2003 found the U.S. at serious disagreement with two long-standing Cold War allies, France and Germany.

Given these and other constraints, long-range policy planning will inevitably fail to anticipate some threats, and will be unable to resolve, ex ante, the highly complex problems of coordinating military responsiveness to all types of contingencies. This is particularly true where the responsibility for operation is allocated among different services and nations so that mobilization efforts require significant coordination.

In attempting to anticipate the resources needed by the armed services in the future, planning within PPBES is only marginally resource constrained. For example, planning for the defense against all threats to United States within the context of Navy maritime strategy may indicate the need for 12 or more carrier vessel battle groups (CVBGs), for example. However, in all likelihood, fewer than the number of CVBGs determined to be needed in force structure planning to counter projected threats will be programmed or budgeted in the next phases of PPBES. Planning articulates the amount of resources needed to minimize threat, supposedly independent of resource constraints so that choices among alternative force structures and threat responses (and intensity levels) may be made more knowledgeably, in recognition of risks from choice, during the programming and budgeting phases of PPBES. The PPBES process may be described as the "defense funnel" in which policy and programmatic choices among alternatives narrow the resource base to fit budgetary feasibility and executability as the decision process moves toward presidential budget preparation, and congressional authorization and appropriation. Some experts also speak about it as a "wedge of risk," as fiscal constraints imposed in the process budget gradually increase the amount of risk taken, threats not met, or threats met with a lower "risk of success."

As a result of congressional action in the early 1990s, the DoD is required to report to Congress every 4 years on threat assessment, force structure planning, modernization, manpower, asset acquisition planning, readiness of forces and other matters in the QDR. Preparation of this report is an additional task, but is related to DoD development of the DPG and the 6-year FYDP.

The QDR is a mechanism to meet congressional reporting requirements, replacing the FYDP that Congress forced DoD to supply to it beginning in the late 1980s. Prior to this time DoD had refused to give Congress the FYDP out of fear of information leakage that would benefit the Soviet Union. As noted the QDR contains much of the same data as in the DPG and FYDP enhanced with military readiness assessments that are desired by Congress to assist in determining how much and on what to spend for national defense readiness. The QDR serves as the major external statement of defense strategy and policy. Publicly, DoD advertises the QDR as "the single link throughout DoD that integrates and influences all internal decision processes" (DoD, 2003a, p. 15). However, the QDR does not actually serve this purpose, except perhaps for some staff in the OSD. It is a reporting rather than a decision influencing document—but DoD cannot say this explicitly to Congress lest it stimulate unrest among members who want to know every detail of DoD planning, budgeting, and management. Consequently, DoD represents the QDR as something it is not—as a management system (which it is to some extent for OSD) to satisfy the politics of the resource planning and budgeting process. However, Under Secretary Rumsfeld OSD exerted pressure to make the QDR a management system, thus reducing the apparent duplicity referred to above with respect to representing the QDR to Congress.

The QDR was intended to become the single, fully integrated planning and management mechanism envisioned for all of DoD by Secretary Rumsfeld and OSD staff—particularly for the military departments and services. It should be noted that Section 922 of Public Law 107-314, the National Defense Authorization Act for FY 2003, amended section 118 of Title 10 of the United States Code to align the QDR submission date with that of the President's budget in the second year of an administration as we have explained above.

PROGRAMMING IN PPBES

As we have noted, an intentional process overlap has been implemented to link the separate phases of PPBES in layers, rather than sequential operations. The programming phase of PPBES is guided by the FYDP, which aggregates and translates the program elements that are the basic building blocks of projected defense asset requirements into the force program framework of PPBES. Under the 2-year budget process mandated by Congress and first implemented by the DoD for FYs 1988-1989, the FYDP period was expanded to 6 years. During the 1990s, the biennial budget process was not accepted by congressional budget appropriators, although the authorizing committees that commissioned it did use it for their review of defense programs. However, as of 2003 DoD returned to the biennial budget process as noted.

The FYDP provides a summary of requirements and alternatives for achieving force structure, readiness, sustainability, and modernization objectives. It is updated 3 times each year, in January in conformity with the President's budget, in May, and in September. The May FYDP is used to link the most recent changes determined in the POM process to the budget under preparation for the next FY. The September FYDP is roughly correlated with the budget process and the revision reflects the programmatic decisions of the SECDEF and, as such, provides the database for determining both programming requirements and budgets. The January plan update is provided to Congress (beginning in 1989) along with the President's budget proposal as required by law.

The task of programming as the second phase of PPBES is to articulate and prioritize 6-year defense resource demands into the perspective of a moving 2-year cycle. Programming is intended to integrate the capabilities of all the individual military components of each service branch into coherent packages, the results of which are summarized in POMs prepared by the individual military departments. The POMs are merged with views from the JCS on risks and military capability expressed in the joint program assessment memorandum and analyzed in detail by OSD staff before program decisions for all of defense are made by the defense resource board (DRB).

The DRB is the final arbitration and decision point of programmatic resource allocation in the OSD and is chaired officially by the SECDEF (but for the most part the chair in effect is the deputy secretary). The membership of the DRB includes the CJCS (and/or the deputy chairman and others JCS members at times), the military department secretaries, the under secretary of defense for acquisition, technology and logistics, and at times selected other senior defense officials. The members of the DRB represent their positions in the DRB face-to-face, generally without the presence of staff, and are expected to be knowledgeable and forceful advocates for their programs and areas. Once the DRB issues its PDM, programming is set and the framework for the next FY budget is established.

While programming by the military departments is a complex process that differs among the service branches (Air Force, Army, Marine Corps, and Navy), it generally comprises three phases: program planning and appraisal, program development, and program decision and appeals. Programming is considerably more cost-constrained than planning, but still places the greatest emphasis on technical capability relative to peacekeeping and war-fighting demands. The process is informed by the input of information and justifications from the CINCs of the unified and specified commands, the military service commands, and Pentagon-based armed-service-program sponsors. Committees of senior officers deliberate and make recommendations up the chain of command to the military and civilian executive staffs in the military-department-service secretariats before final program decisions are made by the service secretaries.

To illustrate the differences in programming between the services, the Navy system is more decentralized to sift program proposals through a gauntlet of separate reviews up to the chief of naval operations (CNO) and the secretary of the Navy (SECNAV). Navy programming, program appraisal, and budgeting have been and remain distinct activities performed by separate staffs. On the other hand, programming and budgeting in the Air Force has long been more centralized, with fewer separate reviews and more overlap in staff functions and responsibilities between programming and budgeting. The Army has functioned much like the Navy, with separate programming and budget staff. However, POM and budget review (as distinct from preparation) have become simultaneous since 2002 in all the military departments and services.

BUDGETING IN PPBES

Budgeting in PPBES is primarily an effort at rationing resources across and within the military departments in consonance with planning and programming decisions. Budget formulation requires the issuance of preparation guidelines, the amassing of programmatic and cost data, the provision of opportunities for program justification in hearings, the analysis of proposals for adherence to both financial and policy guidelines, and the negotiation of program priorities within the constraints of the Budget Authority projected to be available in the next 2 FYs and 4 outyears under the biennial budget process. Budgeting also attempts to respond to short-term contingencies resulting from changes in the international environment and new policy initiatives flowing from Congress, the President, and the SECDEF—presumably including the input from the JCS. The qualification about the role of the JCS is added because different secretaries of defense have had different types of relationships with the JCS and military services. Some secretaries have relied heavily on the JCS and its chairman, for example, Secretary Dick Cheney relied significantly on then Chairman of the JCS General Colin Powell in the period 1989-1992. In contrast, Secretary Donald Rumsfeld distressed the military and JCS by restricting their roles in policy and resource decision making. For example, rather than the previous process of briefings to SECDEF by the military chiefs and members of the JCS, Secretary Rumsfeld preferred to meet only with the JCS chairman. As noted, each SECDEF evolves a distinctive style of operation and management. Some are more participatory while others are less so, and some appear only to be disorganized, for example, Under Secretary Les Aspin who served at the beginning of the Bill Clinton presidency.

Budgeting is a highly constrained exercise in pricing the executability of programs within the parameters of affordability and political feasibility. Each service branch prepares its own budget request in response to directives from the President's OMB, the OSD comptroller, and comptrollers in the military department secretariats. Military department budget requests move up the chain of command from field activities to command-level comptrollers and budget offices, to the central comptrollers' offices in the Pentagon. Military and civilian budget officials analyze input from operational and system commands and the military-resource-sponsor representatives of programs within each service, conduct budget hearings, issue marks (reductions) internally, and respond to appeals from resource sponsors prior to final service-secretary decision making and submission of the formal budget proposal to the OSD.

It is important to reiterate that military department budget staff and comptrollers develop their own estimates of needs, costs, and prices to implement their own priorities and POM decisions while, at the same time, the OSD and the JCS also prepare their positions on the defense program and budget. The OSD comptroller and his budget staff compile service budget requests and analyze them in conjunction with the President's OMB defense examiners; the OMB reviews the budget internally with department staff only for the DoD as distinct from the external manner in which domestic-department budgets are analyzed by the OMB. Marks (reductions) are issued by the OSD comptroller to the military departments, another round of appeals is heard at the military-department comptroller level, and the adjusted budget requests are returned to the OSD for final decision.

Coordination and reconciliation of the multitudinous budgetary perspectives and demands of the JCS, and the separate military departments and uniformed services (and the unified and specified commands) is the responsibility of the OSD staff who prepare the materials to inform the deliberations of the DRB. Once DRB-level policy and program issues have been negotiated and resolved, DRB recommendations are issued in the form of PBD. PBDs provide the basis for preparation of the formal DoD budget once final decisions on high-profile issues have been made by the SECDEF.

The defense military component of the President's budget is issued by the OSD comptroller and the OMB each December and becomes part of the federal budget subsequently transmitted to Congress by the President. The PBDs and final secretary of defense budget decisions also guide the preparation of the national defense budget estimates that are issued annually in the spring by the OSD comptroller and, eventually, become part of the QDR.

Budget alternatives and decisions for the acquisition of major capital assets by the DoD such as missiles, weapon systems, ships, aircraft, tanks, and so forth, are reviewed and analyzed in a budget process that functions semiautonomously from the defense operating budget process. The acquisition budget is integrated with the rest of the DoD budget by the military departments under the authority of the under secretary of defense for acquisition, technology and logistics or USD AT&L. Congress directed OSD to centralize and better coordinate the acquisitions process in the 1986 Defense Authorization Act (Goldwater-Nichols Act) and then DoD Secretary Dick Cheney made this task one of the highest DoD priorities of the Bush administration in 1989. OSD and the military departments have continued to adjust to the new organizational process for acquisition decision making since this time.

Starting in July 1989, a number of changes in the acquisition budgeting and management process were implemented Under Secretary Dick Cheney, and then by Secretaries Cohen and Rumsfeld, some at the direction of Congress. A notable reform by Congress was passage in 1992 of the Defense Acquisition Workforce Improvement Act (DAWIA). The DAWIA calls for major improvement in the education and training of acquisition and contracting staff and has been implemented successfully during the past decade. However, the majority of acquisition reforms have been implemented within DoD (as part of the administrative law that guides DoD acquisition policy and procedures—the DoD 5000 series regulations).

Defense capital asset budget proposals are reviewed by the defense acquisition board (DAB), chaired by the under secretary for defense acquisition, technology, and logistics (also referred to as the defense acquisition executive or DAE) and the vice-chairman of the JCS at the OSD level, prior to the integration of capital and operating budget proposals by the DRB into the President's budget. The under secretary for acquisition technology and logistics and the DAB also are responsible, along with the OSD Comptroller, the military-department secretariats, and the service-comptrollers offices, for program budget execution and the supervision of weapons and systems procurement by the individual service system commands and program offices. In turn, the military department secretaries and assistant secretaries for acquisition (by whatever titles) and the service chiefs, assisted by their Pentagon-based staffs, the operating commanders and their staffs, the service system command officials (e.g., the Naval Air Systems Command, the Space and Naval Warfare Systems Command) and budget comptrollers are responsible for executing the acquisitions budget, orchestrating the complex and technically difficult tasks of acquisitions program and project management. Given the highly differentiated military department-based nature of program management and the extent of overlapping responsibility for budget execution in the acquisition area, a high degree of organizational complexity is the most impressive characteristic of the systems acquisition, procurement, and contracting processes.

EXECUTION IN PPBES

PPBS may be viewed as a process used only for program planning and budget preparation. However, in fact, PPBS has always included execution of the budget. This phase is so important that the Army has long referred to its system as PPBES. Renaming the system as PPBES in 2003 Under Secretary Rumsfeld cemented this recognition. Although we examine budget execution in detail in chapter 7, the basics are reviewed here.

The mechanics of budget execution require OMB apportionment to certify the release of funds from the Department of the Treasury to OSD appropriation accounts. The OSD comptroller allots the budget to the military departments and to DoD-wide functions in accordance with congressional appropriations and the allotment requests submitted independently to the OSD by the three military departments. Military-department comptrollers further allocate money internally to budget claimants or administrating offices, that is, operational and system commands, to support functions within the Pentagon and elsewhere, and down the chain of command to the three military service branch CINCs.

OSD and military department comptrollers routinely issue appropriation/continuing resolution guidance and forecasts to budget administrating offices (claimants or executors) prior to final congressional action on the budget to permit rationing of reductions, accommodation of increases, and other execution planning. OMB and Treasury apportionments are issued on a quarterly basis. Midyear budget execution reviews are conducted in various forms by each military department to monitor and adjust spending for contingencies and to spend money most efficiently. It is axiomatic in budget execution that, at times, unanticipated events will necessitate the movement of funds into and out of programs and activities in ways that may appear, on face value, to be in conflict with the manner in which the budget was proposed, authorized, and appropriated. In short, budget execution is a separate subcycle of the overall budget process and has a life of its own distinct from budget formulation.

Effective budget execution depends upon accurate and timely accounting, careful monitoring, and continuing control of spending so that underutilized dollars may be shifted to areas of highest priority and greatest need. Comptrollers at all levels in the DoD attempt to create and maintain reserves throughout the execution year. However, under budget cutback conditions, commands do not have the resources to create reserves. End-of-year spendouts to ensure that Budget Authority is obligated rather than lost to the OSD or the Treasury are a fact of life in the DoD as well as in other federal departments and agencies. Audits of accounts are accomplished on a routine basis throughout the year, and after the end of the FY to close out annual appropriation accounts (5 years after the FY), and to prepare statements of financial position for multiple-year accounts, for example, RDT&E (research, development, test, and evaluation), procurement, and construction.

Management audits are undertaken by OSD and military-department auditors and reviewers at all levels of the DoD chain of command, (for example, at the military-service-branch level and by military department and OSD inspectors general offices) and also by external agencies such as the GAO. Management audits of the DoD are performed on a continual basis. The number and scope of audits by a wide variety of internal and external agencies has led some command-level and Pentagon officials to complain that they are nearly "audited to death." Congressional oversight committees apply audit pressure extensively to the DoD to insure accountability, a difficult task given the size and complexity of DoD spending and contracting. Congressional and public perceptions that the DoD is inefficient and sometimes wasteful in budget execution drive much audit activity. In some instances, congressionally directed audits and oversight appear to be concerned as much with protecting constituent interests as with weeding out inefficiency. In practice, too many audits are conducted in an uncoordinated way so that efficiency and operating effectiveness often are impeded rather than improved by audits. No one would argue that auditing is unnecessary. However, it should be better targeted at high-payoff areas and duplication of effort should be reduced dramatically through coordination.

Budget execution involves much work that is mundane and relatively routine for DoD. However, budget execution is inevitably political as well as managerial. Incentives for strategic budget behavior present in the highly decentralized and open-ended congressional budget process produce a predictable response from the executive branch. Executive departments attempt to gain and sustain a significant degree of discretion and flexibility in execution of their budgets. One approach to gaining the ability to execute budgets independent of detailed congressional budget oversight is to reward the constituencies of members of Congress who will allow such discretion, for example, by cooperating with "pork barrel" budgeting. This trade-off of spending priority for increased flexibility is only one of the many ways in which the budget process is competitive and sometimes less concerned with efficiency than with winning and losing on specific spending proposals. Given the partisanship of defense budget negotiation, it is little wonder that DoD appropriations are seldom approved by Congress before the beginning of the FY, or that critics of defense spending are able to find instances of waste and misuse in the acquisition and management of defense resources.

DoD attempts to gain execution flexibility by requesting greater delegation of authority, advanced, and full funding for capital asset acquisition, and multiyear operational budgets. Flexibility also is sought through fund transfer, reprogramming, and supplemental budget requests. However, from the DoD perspective, the key issue is efficiency of budget execution in a highly contingent environment rather than flexibility per se. While DoD has long advocated an increase in the amount of congressional multiyear budgeting for defense, experience with biennial budgeting has not produced the results desired by the defense department. Because Congress has continued to appropriate the defense budget and to conduct oversight on an annual basis, many of the DoD efforts to gain greater budget execution flexibility appear to have been wasted. This is painful for defense department comptrollers because the DoD moved in the late 1980s to formulate a 2-year budget in response to requests from congressional authorization committees.

DoD fund transfers provide another means for increased efficiency as well as responsiveness to contingency. Transfers are governed by statute in that they involve changes to previously enacted authorization and appropriation legislation. Fund-transfer authority permits the movement of money between appropriation accounts and is provided by Congress to the OSD rather than to the individual military departments. The military departments compete for shares of DoD transfer authority requested by the OSD from Congress. Reprogramming allows the DoD and the military departments to move money between items within an appropriation account. Such changes are nonstatutory and afford more flexibility to the DoD within the limitations of item specificity. What is termed below-threshold reprogramming permits the services to make changes as needed and report them to Congress. In contrast, above-threshold or prior-approval reprogramming must be approved by the members (or their staffs) of the House and Senate authorization and appropriation committees.

Threshold criteria are established and modified annually by Congress. Generally, they specify levels of Budget Authority below which prior notification is not required. Prior notification of new starts of programs through reprogramming is always required. Other programmatic criteria may be prescribed for specific appropriation accounts, for example, language in the law or in accompanying committee reports and general provisions to prohibit reprogramming of money away from individual items of procurements. Requests to Congress for supplemental appropriations are scrutinized carefully as new appropriations and watched carefully when approved unless congressional support for the request is noncontroversial, which is often not the case.

THE NAVY PPBES PROCESS

To illustrate how the military departments and services participate in the PPBES reform process we review the practice of the Department of the Navy. As we have noted, in the past the Navy and Army have operated highly differentiated PPBS processes while the Air Force has used a more consolidated process. The trend in the Navy is toward further consolidation, as explained here and in chapter 12. After the initiatives to modify PPBES in August 2001 and May 2003 the Army also began further consolidation in review of how it manages the PPBE system.

Planning Phase

Coordination of the Navy planning process is led by the CNO assessment division (N81). The Navy operates three major planning subcycles:

- Integrated Warfare Architectures (IWAR) assessments
- CNO Program Assessment Memoranda (CPAMs)
- Programming and Fiscal Guidance

Integrated Warfare Architectures Assessments

Beginning in 1998, the Navy tool for planning became a broad-based analysis process involving 12 IWAR assessments. IWARs are comprised of five warfare and seven support areas. The five warfare areas consist of power projection, sea dominance, air dominance, information superiority/sensors, and deterrence. Seven other warfare support areas include: sustainment, infrastructure, manpower and personnel, readiness, training and education, technology, and force structure.

The 12 IWARs are assessed from the standpoint of total mission responsibility and capability to meet requirements. The assessment process attempts to answer the question "how much is enough" both in terms of quality and quantity of capability in the present and future. IWARs assessment is conducted by Integrated Product Teams (IPTs) comprised of representatives from the Department of the Navy secretariat, resource sponsors, the claimant, and the Navy fleets. IWARs are intended to:

- Provide senior naval leadership with the foundation for resource decisions by conducting end-to-end capability analysis of warfare and support areas.
- Provide linkage across Navy strategic vision, threat assessment, and resource programs.

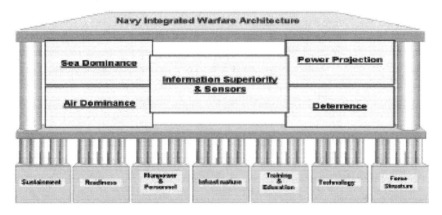

Source: Reed, 2002, p. 24.

Figure 5.3.   Integrated warfare architecture.

- Analyze current and planned programs to identify capability short-falls and surpluses.
- Identify the impact of alternate paths to reach near, mid, and long term war-fighting capabilities

Chief of Naval Operations Program Assessment Memoranda

The analysis generated by the IWAR process feeds directly into the CNO CPAM. Based on IWAR analysis, CPAMs are designed to produce a balanced program that supports Navy goals. Each of the 12 IWARs lead to an individual CPAM. CPAMs are then combined into a summary CPAM that becomes the basis for N80 programming and fiscal guidance. N80 guidance, combined with the DPG becomes the basis for development of the Navy POM. CPAMs are intended to:

- Establish balanced programs across warfare and support area capabilities and over time
- Provide senior Navy leadership with the foundation for programming and fiscal guidance
- Evaluate the impact of IWAR issues on near/mid/long term warfare and support area capability
- Recommend specific programmatic adjustments based on capability trade-offs, alternatives, and options

Navy Programming and Fiscal Guidance

The Navy programming and fiscal guidance provides Navy resource sponsors with general and specific guidance from the CNO as they develop their SPPs. The guidance is developed based on IWAR/CPAM analysis and is issued as the first POM serial report that leads to the programming phase of Navy PPBES.

Programming Phase

During the years prior to 2001, the product of the programming phase, the POM, formed the basis of the budgeting phase of the PPBS cycle. On August 2, 2001, Secretary of Defense Donald Rumsfeld issued a memo that changed the PPBS process. The memo, "Concurrent Defense Program and Budget Review" was sent to the military department secretariats, the CJCS, other military directorates, commanders, and undersecretaries. The memo states:

> This year, and in the future, we will conduct a concurrent program and budget review. The review this year (2001) will consider all program and budget

issues and be the primary venue for resolving any programmatic or budget issues arising from the QDR. It will be used to verify that programs proposed by Components can be executed within established fiscal guidance and focus on issues that arise during execution and from other fact of life changes. Issues previously resolved by the Secretary or Deputy SECDEF will not be revisited. Your submissions for concurrent review will be due October 1, 2001. We are currently in the process of developing overall guidance for the review, to include which specific exhibits will be required. All additional information will be provided to you by the Under SECDEF (Comptroller) as soon as the details are completed. (Rumsfeld, 2001)

This memo made a fundamental change to the PPBS process. Prior to August 2001, the military departments and services developed and submitted their POMs to OSD for review in May. The services would then start to build their BES based on the POM. After 2001, military departments and services are required to submit both the BES and the POM to OSD simultaneously in late-August. Navy BESs are now developed based on Tentative POM (T-POM) control numbers issued in late May. The organization responsible for coordinating and managing Navy programming is the CNO programming division (N80). N80 publishes POM serials that consist of programming instructions as well as fiscal guidance for Resource sponsors, assessment sponsors, major claimants, and others involved in the POM process. Other staff offices in the CNO executive chain also have responsibilities in this process. N81 conducts assessments of the capability plans based on the programming and fiscal guidance. N83 validates fleet requirements and programming inputs. If N80 finds that resource sponsors are not in compliance with fiscal or programming guidance, they will direct the sponsor to bring their program into compliance.

Prior to the Navy realignment that was implemented in 2002, Navy warfare resource sponsors were a part of the N8 organization. However, as a result of the realignment, resource sponsors were moved to N7, established as deputy chief of Navy operations for warfare requirements and programs to give visibility to warfare programs, training, and education. Warfare resource sponsors now have an advocate at the three-star level similar to the leadership of N8. N7 now consolidates SPPs into an integrated sponsor program proposal (ISPP) that balances the resources available to provide an equitable distribution among warfare areas based on validated fleet requirements to ensure compliance with both fiscal and programming guidance. N8 is now organized as follows:

Deputy Chief of Navy Operations for Warfare Requirements and Programs—N8
Resources, Requirements, and Assessments (N8) Programming—N80
Assessment—N81

Resources, Requirements, and Assessments (N8)

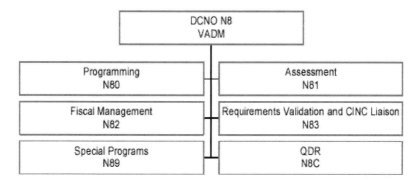

Source: Reed, 2002, p. 28.

Figure 5.4.   N8 organization.

Warfare Requirements and Programs (N7)

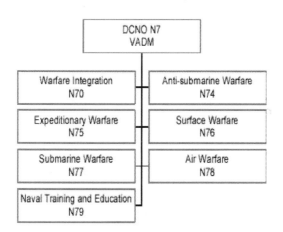

Source: Reed, 2002, p. 28.

Figure 5.5.   N7 organization.

Fiscal Management—N82
Requirements Validation and CINC Liaison—N83
Special Programs—N89
Quadrennial Defense Review—N8C

Once the Navy program is developed, it is reviewed by the Resource Requirements Review Board (R3B), the Navy focal point for deciding warfare requirements and resource programming issues. The board is chaired by N8 and consists of principals from N1 (personnel), N3/5 (plans, policy, and operations), N4 (logistics), N6 (space and information warfare), N7 (training), N09G (Navy inspector general), N093 (Navy medical), N095 (naval reserve), Navy air systems command (NAVAIR), naval sea systems command (NAVSEA), and the Marine Corps deputy chief of staff (programs and resources). Other subject matter experts advise on an as-needed basis. High interest items that involve both the Navy and Marine Corps are addressed by the IR3B that includes members of the R3B but incorporates principals from Marine Corps directorates. Major issues that remain unresolved at the R3B and IR3B are forwarded to the CNO executive board (CEB) for resolution. The CEB principals include the CNO, VCNO, N1, N3/5, N6, N7 and N8. CNO decisions on the T-POM are briefed to senior Navy military and civilian leadership at the Department Of The Navy Strategy Board (DPSB).

Prior to 2001, the Navy and Marine Corps T-POMs were combined and briefed to the SECNAV. SECNAV decisions on the T-POM were then incorporated into the T-POM and the output became the Department of the Navy POM. The POM was forwarded to SECDEF for review and the Navy began to build its budget based on the POM. As a result of Secretary Rumsfeld's August 2001 memo, the Navy now uses the T-POM to develop both its program and budget concurrently. Although the T-POM is not the final POM, is not locked, may still be changed if execution issues arise during the program/budget review phase, and is used to develop budget control numbers for claimants, it must be noted that the T-POM has received CNO approval and has been briefed to SECNAV. Thus, claimants tend to operate as if it were the final POM.

Organizations are constantly changing, thus even the Navy. By 2007, the integrated warfare architecture organization had changed and Navy now begins its PPB process with integrated sponsor capability plans (ISCP) process—planners develop capabilities within the framework of the 21 joint mission capability areas endorsed by JCS which are then balanced in the ISCP. This process defines (or constrains) the programming process. Not everyone views this in a positive light. One observer said, "In my view, going from planning to capabilities development to programming is an additional and confusing step. We're really doing PCPBE— planning, capabilities, programming, budgeting, and execution. And I still see no evidence that E has been institutionalized." In August of 2007 SECDEF Gates sent around a memo summarizing what needed to be done to leave a legacy by the end of the 2008. PPB was on that memo; it is still a work in progress.

By the same token, the N8 and N7 charts are out of date. The Navy July 2007 organization chart reveals a significant change that has taken place over the last few years. All of the N7 organization has been moved under N8, the deputy CNO for integration of capabilities and resources.[1] This is a vice admiral position (three-star). The old N7 organization was also headed by a vice admiral; now it is directed by the two-star who holds the warfare integration spot in our N7 diagram; his title is director of warfare integration and he reports to the three star who heads the old N8, the new deputy director for integration of capabilities and resources. Basically the director of warfare integration serves as a referee and integrator for the warfare barons: surface, air, submarine, and expeditionary warfare, themselves all two star positions. Our charts show the warfare requirements director (N7) of equal rank to the resources director (N8). Currently this is clearly not so. How long this pattern will remain in place is an interesting question; it clearly puts the warfare barons subordinated to the resource controllers. This may not be a stable configuration for the platform barons (air, sea, submarines) stand at important positions in the Navy for the weapons systems and personnel constituencies that they represent.

Of this one observer commented:

> N7 no longer exists and N6 has been reconstituted; the warfare barons are back under N8 with N8F serving as the referee and integrator. The N85/86/87/88 codes (subordinate positions in N8) are back to what they were 10 years ago. Since I have been paying attention, this is one full cycle.

Asked about this, another observer responded,

> In my opinion they go round and round because they each have a legitimate point of view that has to be considered, but no one can get the upper hand and keep it, so sometimes the beancounters win and sometimes the warfare requirements guys and war fighters win ... and since you need both, neither can reorganize the other out of business permanently. As often as not what comes around has been around before, and for good reason.

A third observer agreed with this insight, noting,

> The current reorganization having all under N8 seems to say that resources constrain requirements, then when the organization becomes too dissatisfied with that they split the two functions and create (as then CNO Clark called it) "a healthy tension" between requirements and resources; when people get tired of the in-fighting it goes back to the previous form. Notice the title CNO Mullen chose for the new N8—DCNO for Integration of Capabilities and Resources. A conciliatory name. It is an age-old dilemma in the Pentagon: do requirements drive resource requests or do resource

limitations constrain requirements? The answer, of course, is "yes" to both, but at any given time one side seems to have an edge. None of these organizational changes are transformational from a historical perspective, but for those participants in the Pentagon at the time it seems like something rather novel.

When money is tight for the Navy, as it is now, and as it was in the early 1990s because of the drawdown after the end of the Cold War, the beancounters tend to be ascendant. They do not always like this role; they would rather resource what the war fighters tell them they need, but there is not enough money to buy everything requested and to be seen as a "big spender" might mean being seen as being a frivolous spender by DoD, which could mean losing budget share to the other services in the SECDEF budget adjudication process. So the resource people have to be cutters, as we have demonstrated elsewhere (see the next chapter), just as there needs to be cutters in the executive branch (OMB) and in Congress (the appropriators). Still the warfare barons are very powerful; they control the hardware the Navy uses to go to war and when they and the fleet commanders ask, people listen. They tend to be ascendant in rebuilding times (the Carter-Reagan era 1979-1986) and in times of war. It is an unresolved tension, and as we indicated above, what comes around is not always new, and for good reason.

Budgeting Phase

A separate iteration of the Navy budget is developed for each of the three phases of Navy budgeting:

1.   BSO submission to the Navy Office of Fiscal Management and Budget (FMB),
2.   Navy budget submission to OSD,
3.   OSD budget review, specific program cuts (marking) and appeal (reclama).

The Navy budget process begins when FMB issues its initial budget guidance memorandum in March. These budget guidance memoranda (BGM) are written as serials throughout the FY and are issued as the need arises. For FY 2002 for example, the initial memorandum, BG 02-1 was issued on March 29th and the final serial, BG 02-1K was issued on July 17th. BG 02-1 provided BSOs with the DON Program/Budget Calendar, pricing factors to be used in preparing budget submissions, requirements for budget exhibits, and guidance supplementary to that found in the

Navy budget guidance manual. Shortly after issuing the first BGM, the Navy Budget Office (FMB) issues budget control numbers (dollar ceilings) derived from the T-POM for operating accounts that BSOs use to develop their budget submissions. Based on their control numbers, BSOs prepare and submit budget estimates and required supporting exhibits to FMB based on guidance provided in the BGM. BG 2002-1 directed BSOs to submit operating account budget exhibits for FY 2002 though FY 2005 no later than May 31st. Data required included budget numbers for the current year (FY 2002), the budget year (FY 2003), and for the two POM years (FY 2004 and FY 2005)

After budgets are submitted to FMB based on the T-POM control numbers, the Navy conducts a concurrent program/budget review. The concurrent program/budget process is a combined and reciprocal process rather than a sequential process. The reasoning behind the concurrent program/budget review is that:

- It incorporates a common perspective for program and budget formulation and execution;
- It provides a comprehensive review of pricing/executability before POM wrap up;
- Discrete program adjustments can continue to be implemented during the budget review phase;
- It allows for continued program refinement.

Prior to 2001, once the POM was locked and control numbers developed, the program was set and could not be adjusted in any significant way, except for changes mandated by OSD PDM until the following year during the POM or program review (PR). It is important to note that as a result of Navy realignment and the concurrent program/budget process, sponsor proposals incorporated into the program have been reviewed for proper allocation and compliance with guidance and consolidated as an integrated N7 input. Thus, the budget is built on the T-POM controls and the final POM or PR and budget are forwarded to OSD at the same time. The distinct separation between the programming phase and budget phase has disappeared, as Secretary Rumsfeld intended. Programming changes can be made at the same time as budget changes and in response to N80/FMB and BSO inputs to help ensure that the program is executable as a budget. The Figures 5.8 and 5.9 present the PPBS cycle before August 2001 and after.

The mechanism for making changes to BSO budget submissions also changed as a result of the concurrent program/budget review. In prior years, the mechanism FMB used to challenge BSO budget submissions

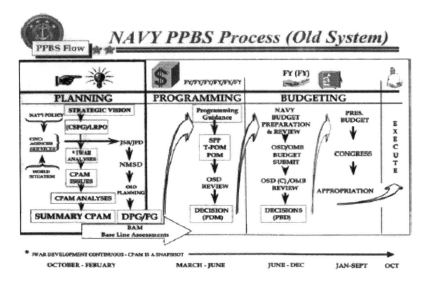

Source: Pacific Fleet, 2002d

Figure 5.6.   Navy PPBS prior to 2002.

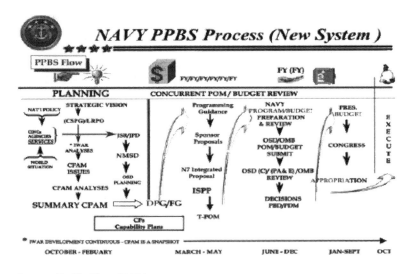

Source: Pacific Fleet, 2002d

Figure 5.7.   Navy PPBES 2002 change.

was called a mark. Marks are adjustments (usually negative) to the estimates submitted by the BSO prepared by FMB analysts. In response to marks, BSOs had the right of appeal (reclama). Reclamas provided the BSO the opportunity to respond to adjustments made in the FMB mark to restore the funding that was removed. If a reclama was submitted to a specific mark, that mark was considered a tentative decision until the mark was resolved. The method for resolving reclamas started with the respective FMB budget analyst and progressed through the department head and division director. Unresolved marks became major items to be resolved at major budget issue meetings. The absence of restoration of a mark after this meeting represented a decision to be forwarded to the secretary of the Navy.

By the summer of 2002, the old mark-reclama process had been changed to a series of issue paper alternatives that could be surfaced not only by the budget analysts or the claimants, but also by other commands like equipment or force users or contractors or industry. Once initiated these would be resolved at the lowest level. The issue paper would be posted on the FMB Web site and copies sent to all offices affected for their response. Reasons for program changes might involve such issues as reassessment of capabilities and new priorities, congressional action, contract slippage, industrial base issues, program slippage, program cost increases, or acquisition management issues. Criteria used to decide about incorporating changes would focus on favorable benefit/cost analysis, a long-term positive impact on operating accounts, and a consensus on the desirability and affordability of the capability.

The issue paper process is more dynamic than the previous mark/reclama approach. Marks focused on specific issues for which BSOs would prepare reclamas. Marks were generated at the FMB level and distributed to respective BSOs. Issue papers, on the other hand, enable all stakeholders to present an issue at the appropriate level or to view outstanding issues or comment on issues that affect them. While an issue may not specifically address a particular BSO or activity, any stakeholder is free to submit comments to reinforce the primary BSO argument on an issue.

An example of how this process operates may be drawn from information technology (IT) funding. IT is typically underfunded in the POM, and this affects all Navy commands, facilities, and activities. The subactivities are free (after coordination with the BSO) to address the specific requirements that will be affected by the issue within their region, fleet, or Type Command (TYCOM). Since, for example, IT funding issues are not limited to the Pacific Fleet claimancy (CPF) but affect the Navy as a whole, not only are CPF activities free to comment, but other claimancies also are free to comment on the issue. While claimants other than the one specifically addressed by an issue are free to comment, it must be noted that this is not a "free-for-all" and that consideration must be given prior to making comment on issues not specifically addressed to a particular claimant.

When issues are generated, they are posted to the Navy headquarters budget system (NHBS) Web site. The Web site provides stakeholders a method for generating issue papers, submitting them to FMB, making comments, and reviewing comments that have been posted. If issues are unresolved (e.g., among the FMB/N80 analyst, department head, and division directors), they are presented to the Program Budget Coordination Groups (PBCGs). PBCG membership consists of representatives of FMB, N80, N7, commander fleet forces command (CFFC) and N43 (fleet readiness). PBCGs deliberations are similar to Major Budget Issue meetings. They provide for the final resolution of issues prior to the point when the budget is forwarded to the CNO and SECNAV.

Major budget issue meetings resolve only budget issues. PBCGs may resolve both program and/or budget issues. While both program and budget issues may be addressed, the PBCG is primarily a FMB forum. PBCGs meet to review issues by budget area, that is, civilian personnel, military manpower, base operations, aircraft operations, ship operations, and so forth. The PBCG schedule is posted on the NHBS Web site. Following SECNAV review, the program, which now becomes the POM, and budget are submitted concurrently to comptroller of the OSD.

OSD adjusts the military department and service submissions, issuing PBDs. The Navy has opportunity to appeal PBD reductions, although time to do this typically is limited. This is a tactic used by OSD to limit military department and service ability to overturn PBD reductions, and it succeeds more often than not. When PBD issues are resolved with the military department and services, the budget is forwarded to the OMB where it becomes the Navy and DoD portion of the President's budget submitted to Congress.

The Navy PPBES process has produced a POM and budget for both the POM and subsequent years in even numbered years and the PR and budget in odd numbered years. Therefore, in FY 2002, the Navy prepared POM 2004 and budgets for FY 2004 and FY 2005. Then in FY 2003, the Navy conducted a PR for FY 2005 and refined the FY 2005 budget, but only changes to the POM through the program change proposal and to the budget through BCPs were considered. It must be kept in mind that while this budget is being developed, the President's budget for the upcoming FY is being considered and passed in Congress, and the current FY's budget is being executed. Thus, three budget processes must be managed in any given year, as depicted in Figure 5.8. A current year supplemental or what Congress does or does not appropriate for the upcoming FY can have consequences for the budget under construction for the following year.

Consequently, during the current year (CY), Navy staff (and its BSOs) work on three separate budgets. The Navy executes the FY 2003 budget, testifies on the FY 2004 President's budget submission and prepares the FY 2005 program/budget. Ripple effects result from the current execution year that push into both budget and program decisions for subsequent years. At the BSO level, programmers, budgeters, and execution personnel all are actively engaged in managing 3 budget years simultaneously.

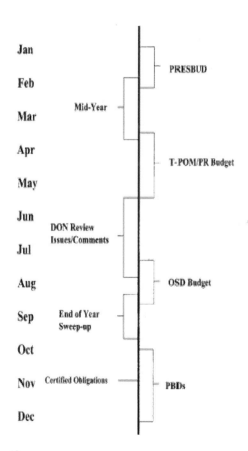

Source: Reed, 2002, p. 49.

Figure 5.8.   Budget process flow from claimant perspective.

ASSESSING RECENT DoD PPBES REFORM

During the 1990s it was clear from the beginning that the shift from a Cold War mentality to a new framework was preceding slowly. The gist of what was necessary in the post-Cold War world did not ever seem clear, despite all the discussion of asymmetric threat and successive preparation and reviews of the QDR. The terrorist attack of September 11, 2001 ended this period of doubt and confused reflection.

It is routinely acknowledged that the planning component has been the weakest part of PPBS for decades. Part of this is due to the contingent nature of threat assessment while other impediments include the sheer volume of information and absence of data coordination. In order for DoD to plan to counter threat effectively, it seems to us that a capabilities-based planning process within PPBES, rather than a theater-based approach, is one way to tear loose from the old bipolar geographic analyses that focused on the USSR, potential enemies in Asia or elsewhere. Instead, it is critical to ask what capabilities the U.S. needs to meet threats wherever they occur, especially given that the terrorist threat has a personal or group basis less geographically bound.

The deployment of U.S. forces since September 11, 2001 illustrates new concepts in joint operation, the use of special forces and the application of joint forces in unique ways, supported by traditional forces using traditional doctrines. Nonetheless, it is a new mix. Much of the transformation in military affairs that has been ongoing since the mid-1990s is driven by new threats that seem to emerge almost daily. All this points to the ties between changes in military war fighting and PPBES planning.

The programming-budgeting changes that constituted a redesigned PPBE system began in August of 2001, when SECDEF Rumsfeld announced that the POM and budget cycle would be operated contemporaneously to speed up the review and decision process (Rumsfeld, 2001). Further reforms were announced by DoD in May of 2003. By this time, military departments had already begun to transform their planning and budgeting systems. It may be observed that the comprehensiveness of the changes will take several years to implement throughout DoD. However, the cycle for adopting the innovation is likely to be shorter for military departments and services that already had begun reforms on their own. In the following section, we explore an example of this in one service, the Navy.

The Department of the Navy carried out PPBES reforms of its own in 2002-2003. Interviews with Navy FMB officials, fleet comptrollers and other DoD financial management executives revealed a number of concerns about the old consecutively phased PPBS process. Those interviewed expressed the view that the old system was characterized by the following problems:

1. Inadequate guidance: it felt the DPG issued by SECDEF to initiate the POM process which led to decisions about what to fund for the budget year was often late and unaffordable and did not provide a clear statement of SECDEF priorities.

2. Concurrent process flaws: program and budget processes appeared concurrent but were not well coordinated.

3. Continuous rework: the POM and budget were subject to disassembly, rebuilding and review every year.

4. False precision: programming for the acquisition process required excessive detail and was projected too far into the future years.

5. Revisiting decisions: decisions made during one cycle were not always recognized and respected in the next.

6. Changing rules: rules, expectations, and metrics complied with in advance by the Navy were changed later in the process to facilitate cuts.

Last, FMB executives believed that SECDEF oversight should be limited to those matters of true significance or policy compliance. Here it appears that the Navy budget-makers were trying to keep some military department autonomy, but arguing for relative spheres of decision, some appropriate to SECDEF and some to the military department.

Given these criticisms, and the August 2001 directive, but little in the way of implementing instruction from DoD, the Navy pressed forward with its version of the new PPBE system. The first step in this change was to bring together all PPBES processes under Navy comptroller oversight. This gave the comptroller jurisdiction over both the POM and budget processes. This was intended to install tighter control and coordination of the planning and budgeting segments of the process, and to establish clearer and more direct linkage to the budget process. Second, it was recognized that there was little fiscal room for growth for new initiatives. Between FY 2003 and FY 2004 the budget was scheduled to increase about $3 billion dollars, from $111 to $114 billion, but over 91% of this was fixed in military compensation growth ($1.9 billion) and inflation and civilian salary increases ($1.2 billion).

As a strategy, the Navy aggressively pursued cuts in the budget base and made some $1.9 billion in reductions, including such things as eliminating 70,000 legacy IT applications, pursuing 7 facility management initiatives that would increase efficiency, decommissioning 11 ships and 70 aircraft early and divesting 50 systems, like the MK-46 torpedo and the

Phoenix missile system. A workload validation process was undertaken by teams of senior level functional area experts to examine workload across the Navy, including maintenance, base support, R and D centers, and supply operations, to seek increased efficiencies, divesting, or contracting out, and to eliminate duplicative functionality across the department. Approximately 75 initiatives resulted in budgeted savings of almost $670 million. Cumulatively these efforts resulted in $1.9 billion that was reallocated to buy ships and aircraft, and invest in transformational weapons systems.

Meanwhile, the Navy budget process itself was undergoing a change in perspective. In the winter of 2003, FMB took a grand simplifying step. It noted that investment accounts, all tied to weapons, ship or aircraft acquisition accounted for 36% of the Navy budget. The rest could be divided into accounts that had a historical base (11% of the budget) and those that had a performance model base (53%) and were driven by formulas. These included such accounts as the flying hour program, the ship operations program, training workload, Marine Corps operations, aircraft maintenance, ship maintenance, spare parts, USMC depot maintenance, facilities, and base operations. Secondary models included military personnel (both active and reserve) and civilian personnel. These models produced cost figures, but unlike the first group did not model performance. A manager was designated for each of these formula-driven modules and a validation process was begun to verify that the models were accurate and that the correct inputs were used.

The object of the change was that in subsequent budgets these models would automatically provide their part of the budget request and that the performance models would be able to specify outputs from dollars of input. Those expenses that were driven neither by formula nor investment were termed "level of effort programs." These last items were to be based on a three year rolling average of what was actually spent and increased or decreased as the level of effort varied for the program. This is currently the scheme under which the Navy plans to operate its budget process as it goes forward with reform—first, with a more formula driven budget and, second, with intense interest in the investment accounts and, thirdly, to focus within the base to free up more money for the investment. Implicit in this would seem to be concept that the readiness accounts will be replenished by supplemental appropriations when they become unbalanced by emergency increases in military operating requirements (optempo).

It is obvious that important changes have been made in the DoD planning and budgeting process. The simultaneous execution of the POM and budget review and its consolidation into one data base is an important change. In the old system, a good POM could still be lost on the way to the final budget. In addition, sometimes the budget process ended up doing a lot or reprogramming and remaking of decisions that would have been better done in a POM exercise. For example, the 2003 POM process started by doing a pricing review of the shipbuilding budget. This is a budget drill and in the old PPB system would have been done in the budget process long after the POM had been completed. Observers comment that when such drills (repricing the shipbuilding account for inflation etc.) result in a big bill that has to be paid, it is good to have that bill considered and paid up front at the beginning of the process in the POM where large dollar changes can be made more easily. They also felt that doing the POM and budget simultaneously should result in fewer surprises and less reprogramming of changes to the POM in the budget process than there used to be. They felt that the process should be quicker, but less linear, a layered process rather than a sequential process. The routing of all products of the POM and the budget into one database was seen to be a significant move to help resolve some of this added complexity.

Second, the outcome focus of the process is an important change. Secretary Rumsfeld has emphasized outcomes and the Navy approach illustrates this concern in two ways. The procurement accounts are focused around the outcomes each weapon system bought will provide and the performance models for steaming hours and flying hours are also outcome focused. As has been stated above, this covers almost 90% of the Navy budget. Nevertheless, Congress still appropriates by line item and DoD has to be able to translate capabilities into budget items and make winning arguments for those translations. The fact is that budget lines (line-items) make it easy for Congress to buy things and what has not changed is where the power of the purse is located. In the words of one DoD budget player,

> there are a lot of changes, but what has not been changed is the Constitution. Changes will end when they bump into things that are Constitutional. The appropriation process is still a congressional process and changes in the Pentagon process have to be responsive to the needs of Congress. The menu of changes the Pentagon can pursue is not unlimited.

Third, the new process put SECDEF into the process at the early stages, "in the driver's seat," in the words of one budgeteer. Decisions in the new PPBES are intended to reach the secretary before the decision has become a foregone conclusion, while options are still open, and while important and large-scale changes can still be made. When SECDEF inputs come at the end of the stream of decisions, some decisions that could be taken get preempted simply because they might cause too much breakage in other programs or because everyone has already become committed to the likely outcomes of the decision. Former Secretary

Rumsfeld had a clear interest in transformation, but not all communities within the defense establishment were equally committed or committed at all to Rumsfeld's vision. As we have noted, inserting SECDEF in the decision process early stands up so long as history proves the decisions SECDEF make are right. While this is true whether SECDEF input is early or late, inserting SECDEF early in the PPBES process puts a larger burden of proof on SECDEF. Veteran observers see these changes as an evolving process, cautioning officers bound for the Pentagon in a couple of years not to bother memorizing the new process until they get there since it has changed significantly since 2001 and will continue to change.

Last, the new emphasis on execution seems an important change, but it is too early to speculate on how this will turn out. It seems clear that no one wants to be viewed as decreasing military effectiveness in the name of saving dollars. Through 2003 a continuing theme of administration critics was that the United States was trying to do Iraq "on the cheap" with not enough troops and not enough of the right kind of troops. If the new emphasis on execution becomes a code word for efficiency and this is parsed into "doing things on the cheap," then the emphasis on execution will not have important or long lasting effects. As we have indicated in analysis of the durability of defense financial management problems in chapter 10, this is a difficult problem to resolve, but one that must be addressed.

The 2003 budget process within DoD was dramatically changed. The 2003 process exemplifies incrementalism triumphant. Only changes to the POM and the budget were brought forward in 2003. This is a dramatic change from past. Aaron Wildavsky in developing the concept of incrementalism may have ignored defense, but DoD appears to have gone to school on Wildavsky. The result of the 2003 budget process is that unless a budget change proposal is explicitly approved, then a unit's budget is the same as it was the previous year; in Wildavsky's terms, the base is reappropriated. Thus, if a unit does well in the on-year cycle (second and fourth year), it may carry some "fat" through the off-years. This would seem to intensify the struggles during the on-year processes, making the stakes higher. Success is rewarded for 2 years and failure is doubly penalized, that is, remember to change in the off year, off-sets have to be offered up, so the only way to get better in the off-year is by giving up something else. In the off-year cycles, only changes to the base are explicitly considered, both in the PCPs for the POM (big dollar numbers, but fewer of them) and BCPs for the budget (more, but smaller dollars). However, there is an interesting twist to this. Changes may come from anywhere someone has an issue, for example, the military services, combatant commanders, and assistant and undersecretaries of defense.

CONCLUSIONS ON PPBES AND DEFENSE BUDGET REFORM

Budget processes normally focus around ownership. For example, the concept of a claimant or a budget submitting office identifies who will submit a budget. They alone control what goes into this submission. This new process seems to empower friends and neighbors to examine how a neighbor is managing his property and submit a program or budget change proposal if the neighbor is not doing the right things, by for example putting in a budget change for more frequent mowing of the lawn or a program change for construction of a two car garage. Suppose an Under SECDEF believes the Army should provide more force protection to an Air Force base and submits a program change proposal to do so; since these must have offsets, who will be designated as billpayer is a good question. The under secretary is unlikely to have any money; thus Army might have to pay that bill or Air Force. No matter who the billpayer is, they will have to make an adjustment to their budget, just as you would were a neighbor able to dictate that you have to paint your house more frequently and pay for it by decreasing your entertainment budget. Thus, this new system is incrementalism with enhanced pluralism. Disinterested neighbors with good ideas have the opportunity to insert them in other people's budgets. Some observers will say that some of this has always gone on, but now the process is formal and invitations have been extended to players at various levels within DoD. Then, during the second and fourth years of this new PPBES cycle, zero-based budgeting is invited, based on the QDR and the NSS. This seems like a scenario doomed to fail because complex organizations have difficulty in adapting to radically different routines, for example, incremental and zero-based procedures. For DoD, it must be remembered that the full year cycles are largely incremental also, given no dramatic change in the threat. This new process is likely to work most satisfactorily only at the top and only if top-level players are somewhat restrained about their intrusions into the domains of other players.

It must also be recognized that while the arguments we have advanced above are interesting in theory, the fact is that DoD has been grappling with how to express and budget for significant war fighting needs and how one command is a little fat here or a little lean there pales in significance when set beside $100 billion supplemental requests, some of which is for this year's expenses, some making up for last year's expenditures, and some providing for future years. Meanwhile, in the main budget struggle the Navy and Air Force find their modernization budgets for ships and planes woefully underfunded, the Air Force by as much as $20 billion a year for the next 5 years. Thus while our observations are accurate, they are not as significant as the real fiscal problems faced by DoD in carrying out current operations without compromising its ability to fight future battles.

While this places our argument in context, we submit that it is important to have a budget system that works and therefore it is also important to know where there are problems in reformed systems. For example, as might be expected, changes in the PPB process this significant created turbulence in the process for lower echelon commands. In 2003, the new PPBES process as it was actually implemented was not functioning satisfactorily from the view of PACFLT and the TYCOMs. One example of the type of problem encountered was that by August 2003 the control numbers for the FY 2004 budget (beginning October 1, 2003) had not been issued to PACFLT and the TYCOMs by the FMB. Prior to the reconfiguration of PPBES by DoD to merge program and budget review, control numbers typically would be available from the final POM review to guide fleet and TYCOM budget execution planning—by April—April 2003 in this case. However, with the merger of program and budget decision making, because major resource allocation decisions are made to complete the final POM (e.g., trade-off reductions in OMN to fund procurement), the time to completion of programming was pushed out and, consequently, final budget control planning numbers could not be developed and issued by FMB—to their consternation. And, with no budget control numbers available to fleet and TYCOM comptrollers until August, these officials experienced real problems briefing and responding to questions from their commanders on what could and could not be afforded in terms of operational support capability in the upcoming FY. This was a source of stress and in some cases embarrassment for comptroller staff facing questions from frustrated commanders (Admirals) who wanted to know what actions they could order subordinate commanders, including those of operating forces, to perform and support in the next year. These kinds of problems continue to surface.

Such problems may be anticipated with any comprehensive change in PPBES or other major system directed from the level of the OSD. The lag time for full and satisfactory implementation of DoD-level macro changes in planning, programming, and budgeting is probably 2 to 4 years, although many wrinkles will be worked out by the military departments after the first new cycle has been completed. However, it is understood by seasoned observers of such change that the solutions and new processes developed by the military departments will differ by service and therefore some degree of incompatibility between different service solutions is inevitable despite the intention of DoD decision makers to prevent this from happening. DoD prefers uniformity but this is not possible, and probably not desirable, given the highly differentiated resource management systems and processes used by the respective military departments and services.

In defense of diversity it may be observed that any system developed by DoD should serve the needs of its constituents, that is, the military departments and services. From the view of the OSD, diversity in implementation of reform is an annoyance at best, and a direct violation of authority at worst, to be illuminated and eliminated. However, the power of OSD is not such that it can mandate what the military departments and services, as semiautonomous operating entities, will do in implementation of any DoD directed reform, or congressionally mandated reform for that matter, for example, the CFO Act or the GPRA.

This conclusion is not based on examination of reforms (e.g., Management Improvement Decision 913) in formal lines of authority and management control in the DoD, nor on how differences in old and new systems are depicted in highly detailed, multicolored graphics and tabular "wiring diagrams" of the type typically used in Pentagon level briefing slides. Rather, it is based upon examination of how process changes actually are incorporated into existing systems in the real world of comptroller-based financial management in a highly decentralized and diverse organization—the U.S. DoD. It is axiomatic to observe that no matter what reforms successive defense secretaries wish to implement, what will be done in reality is more a matter of evolution of existing military department and service based systems rather than transformation of a single centralized system. In practice, DoD cannot be centrally managed and controlled, and any assumption that such control is possible is a combination or wishful thinking and fantasy.

REFORM OF PPBES: ISSUES AND CONSEQUENCES

We have explained that reform of PPBES has been underway in DoD since 2001, and especially beginning in 2003. Defense Secretary Donald Rumsfeld changed the process in August 2001 and appointed committees to recommend further changes. Changing the PPB system is not a new idea. One change recommended by PPBS observers for more than a decade has been to simplify the process (Joint DoD/GAO Working Group on PPBS, 1983; Jones & Bixler, 1992; Puritano, 1981). Other options have included increasing the assets and time devoted to threat assessment, improving the quality of integration of threat information, and reducing the time devoted to programming so that this cycle consists of essentially the "end game" of the current programming process, and multiyear budgeting. While the 2001 transformation to merge POM and budget review represented a laudable incremental reform, changes made in 2001 and 2003 are far more sweeping.

The rationale for combining POM and budget review was essentially that too much time was being spent on POM preparation and approval than was worth the effort. Placing more emphasis on the end game of programming seemed a good idea. The end game is the phase in programming at which major decisions are made, for example, about weapons systems acquisition and force structure changes. Shortening the programming phase was recommended to the Rumsfeld study committees to eliminate the time consuming process of preparing the POM de novo for each POM cycle from the field level up. Not only had so much time been devoted to POM preparation in the past, but critics also felt that this time was, essentially, wasted because most of the POM was composed of virtually the same information used for previous POM submissions. The criticism was that, much like zero-based budgeting, building the POM from the bottom up for each cycle created great workload without producing much that was new for decision makers.

The key to effective programming is to make the right decisions on major asset acquisition and force structure changes. The other important task accomplished by programming is to align the various assets, including manpower, so they are budgeted in a coordinated way, making assets available together, coinciding with the time they are needed for war fighting. As of 2002, this important part of programming was essentially merged with the budget review phase of PPBES so that budgeting has become a longer-range process, focusing more on a multiple year period rather than the one year cycle that is used by Congress with considerable inefficiency.

Moving to multiyear budgeting does not represent a radical change to DoD for two reasons. First, as we have explained in this chapter, DoD and the military departments and services already manage at least 3 budget years simultaneously. Second, because many budget accounts for national defense already are funded on a multiple year basis (RDT&E, and weapons acquisition accounts for example), moving all of them to a multiple year structure does not require serious change in the way that DoD budgets for national defense. The degree of change required of Congress is another matter. While it would be desirable to have Congress move to a multiyear budget process for defense, this was not a necessary condition for change to PPBES by DoD. DoD can move to multiyear budgeting and Congress can continue to appropriate annually portions of the multiyear budget it receives from DoD. The point is that DoD can and has decided to reform PPBES itself, and then attempt to gain congressional support and approval for the change by demonstrating that the reform produces better decisions and reduces the costs of decision making.

WHY HAS REFORM OF PPBS TAKEN SO LONG?

Strategy in budget gaming is an inevitable result of the highly participative manner in which Congress reviews and acts on the defense plan. To reduce the extent of budgetary gaming behavior and to build greater trust into budget advocacy, it appears desirable for Congress to consider reforming its own budget making rules and oversight procedures, as explained above.

A similar conclusion was drawn by Secretary Rumsfeld's ad hoc study committees and high level DoD staff regarding reform of PPBS in 2001 and 2002. Reform of the system had to go on hold during 2002 and early 2003 due to the necessary preoccupation with winning the war in Iraq. However, those who studied PPBS for Rumsfeld characterized the system as exhibiting considerable redundancy, excessive procedural complexity, and involving too many players. However, until 2002-2003 attempts to reform the system appeared to have worked against the incentives that have nurtured and protected PPBS from radical transformation. The PPBS process expanded continually since the 1960s to include more participants in program and budget decision making because resource competition is the crux of the political process. As such, more and more players were successful in finding roles that permitted them to compete for a piece of the budget action, just as has been the case in the congressional budget process (Art, 1985; Lindsay, 1987). Ironically, the nature of the reform disincentive appears to be the same for the DoD as it is for Congress.

The reason why PPBS has served the policy development and planning needs of DoD resource decision making for so long while it floundered in other federal agencies is that the process was developed specifically for defense, where many program and budget decisions are made on the basis of assessing alternatives and where quantitative data are available and amenable to the application of the tools of economic cost/benefit analysis, systems analysis, operations research, and other sophisticated analytical methodologies. Many defense inputs and outcomes may be constrained so as to be measured quantitatively with greater confidence than is the case for domestic programs. On the other hand, PPBS may have survived with only minor and incremental revision for so long in DoD because it is expensive to change to another system or for reasons related more to political competition between the OSD and the military service branches than to analytical utility. However, as explained earlier in this chapter, comprehensive reform of PPBS was authorized by Under Secretary Rumsfeld in 2003 and implementation of these changes continues to this date.

It is too early to assess in any meaningful way the results of the 2003 and ongoing reform to PPBES. As with any reform, it will take time for DoD to internalize the changes in process throughout the military departments and services. Change at the OSD level is far more easily accomplished. We anticipate, however, that many if not all of the changes, will be implemented because they make sense in terms of reducing unnecessary and duplicative workload for the military departments and throughout DoD. Still, even though PPBES has been reformed significantly we believe that it should be replaced by a better system. We address this further in chapter 14.

CONCLUSIONS

The PPBS process has been employed by presidential administrations served by Defense Secretaries from McNamara in the 1960s and Laird in the 1970s to Caspar Weinberger and Frank Carlucci in the 1980s, to Dick Cheney, Les Aspin, William Perry, William Cohen, and Donald Rumsfeld in the 1990s and into the new millennium. DoD has managed the constant evolution of PPBS while keeping its basic structure relatively stable. However, the pace of evolutionary change quickened Under Secretary Rumsfeld.

In February and May 2003, the SECDEF announced changes to PPBS. The new system was termed the planning, programming, budgeting, and execution (PPBES) system and was announced to "revolutionize" internal DoD planning and budgeting, increase effectiveness and add additional emphasis to budget and program execution (SECDEF, 2003a). It has been noted that for decades the Army had used the term PPBES to give proper emphasis to execution.

The change to PPBES was announced as having resulted from the directions given by then Deputy Secretary of Defense Paul Wolfowitz to a senior executive council to study and recommend improvements to the overall DoD decision-making process Wolfowitz approved a management initiative decision to implement the changes recommended to PPBS. No legislative changes were required and Congress would continue to see the same budget and QDR as it has in the past.

As explained earlier in this chapter, the 2003 reform determined that DoD would move from an annual POM and BES cycle, to an essentially multiple year cycle, beginning with an abbreviated review and amendment cycle for FY 2005. However, the military department budgets itself continue to be prepared and submitted annually to OSD and Congress. Under the revised process, DoD is supposed to formulate resource plans (along with the POM) on a 2-year cycle with use the off-year to focus on fiscal execution and program performance. The 2-year programming and resource execution cycle is designed to guide DoD strategy development, identification of needs for military capabilities, program planning, resource estimation and allocation, acquisition, and other decision processes. This change was intended to more closely align the DoD internal PPBES cycle with external requirements embedded in statute and administration policy, including the QDR as amended.

Under the new system, the QDR is intended to continue to serve as the major DoD statement of defense strategy and policy. This distinction is noteworthy as it reflects the revolution in business affairs initiative undertaken in DoD in the late 1990s and continued into the new millennium. And, from the OSD perspective, the QDR also constitutes the single link throughout DoD to integrate and influence all internal decision processes, for example, preparation of the FYDP and DPG. Section 922 of Public Law 107-314, the Bob Stump National Defense Authorization Act for Fiscal Year 2003, amended section 118 of Title 10 of the United States Code to align the QDR submission date with that of the President's budget in the second year of an administration as noted earlier.

As a result of the 2003 process modification, the off-year DPG will be issued at the discretion of the SECDEF. The off-year DPG will no longer introduce major changes to the defense program, except as specifically directed by the DoD Secretary or Deputy SECDEF. As noted previously in this chapter, DoD announced that no DPG would be issued in 2003 for FY 2005.

In addition, rather than preparing the POM during the off-year, according to the reform, DoD is supposed to use PCPs to accommodate real world changes, and as part of the continuing need to align the defense program with the defense strategy. As envisioned by the 2003 reform, DoD is supposed to use BCPs instead of a BES during the off-years. BCPs are intended to accommodate fact-of-life changes including cost increases, schedule delays, management reform savings, and workload changes as well as changes resulting from congressional action. After FY 2005, budget execution reviews have supposed to provide opportunity to make assessments of current and previous resource allocations and evaluation of the extent to which DoD achieved its planned performance goals. Performance metrics, including the OMB program assessment rating tool (PART) used in the George W. Bush administration from 2001 to 2008, might have provided the analytical underpinning to ascertain whether an appropriate allocation of resources exists in current budgets. However, PART was never as fully applied to the defense budget as it was for domestic agency budgets. Still, to the extent performance goals of an existing program established in the PPBES process are not being met, recommendations are to be made to change or even replace the program with alternative solutions including making appropriate funding adjustments to correct resource misallocations and imbalances. Under the reform the QDR has been used to demonstrate how DoD has complied with the requirements of the Government Performance and Results Act (GPRA) of 1993 as enforced by OMB. Submission of the QDR to Congress has been intended to satisfy GPRA requirements for DoD.

In chapter 1 we remarked that no SECDEF could alone manage an enterprise as complex as the DoD. And in fact, it is important to point out that in the past and presently, input to program and budget decisions in DoD is provided by the Deputy SECDEF and staff, the position in DoD that bears a large part of the responsibility for actually attempting to manage the DoD. In addition, the under secretary comptroller, the under secretary for acquisition, technology and logistics, and under and assistant secretaries for other OSD functional areas including program analysis and evaluation, policy, force management and personnel, legislative affairs, health, reserve affairs, and others, all provide views and analyses to guide program and budget decision making. Figure 5.9 gives some indication of how complex the DoD resource allocation is and why it is impossible for any one SECDEF to "run" the department. Concentrating just on the lower half of the figure which represents the internal Pentagon processes, we can see a constellation of events, processes, and committees which must be mastered for each annual budget process. At certain points the decisions of a few people are critical. For example, CPR and CPA refer to assessments made by the CJCS in the POM process (CPR) and in the budget process (CPA) where he has the opportunity to contribute opinions on both program decisions (PDM) and budget decisions (PBDs). In the end game, committees remain critical to decisions. In the middle 2000's, Secretary Rumsfeld used the SLRG, the senior leaders resource group (pronounced "slurg"), consisting of himself, and the civilian secretaries, assistant secretaries and undersecretaries as well as the JCS and the service chiefs, a group that consisted of 18 participants sitting at the table with 6 more arrayed in chairs against the walls of the room; these included the secretary, the deputy secretary, the three service secretaries, the five undersecretaries, the 6 JCS, the general counsel, and the assistant secretary for public affairs. The additional 6 included the director of administration, the assistant secretary for networks, the director of program analysis, SECDEF's senior military assistant, and his senior special assistant. Others may be called in as the issues warrant.

Of forming the SLRG, Rumsfeld said:

Concurrent POM and Budget

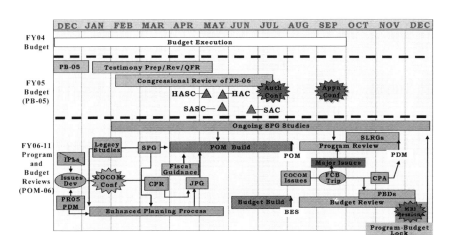

Source: Peter Daley, 2004. *Joint Perspectives on Defense Budget Issues: Briefing to the Naval Postgraduate School.*

Figure 5.9. The SLRG.

it has had as much effect as anything else we have done. Everyone got to know each other, we know our strengths and weaknesses, we know what's important, and we learn from each other. There's no one smart enough to know what ought to be done in this department on big things like that, you have to be informed by others. And you have to have a process where people are confident, they can talk, they can take risks, they can speculate on things and raise questions, and it has been just an enormously important part of what's happened. (Barnett, 2005)

Rumsfeld's SLRG had a set of well known rules for behavior attached to it including:

1. Secrecy is crucial: both civilian and military leaders need to be able to speak freely without worrying about setting off bureaucratic skirmishes in their own organization or others

2. If you do not show up, you do not get to fight: a vice chief may substitute for a chief, but that is about it.

3. No SECDEF, no SLRG: early on a subordinate chaired the meeting and the outcome was not to Rumsfeld's liking, hence the rule.

4. One fight at a time, and fights go on as long as they have to.
5. When the war fighter is involved, the war fighter is invited. When this happens and a CINC is invited then the SLRG is called the SPC, or senior planning council.

The SLRG not only addressed major budget issues (MBI in Figure 5.9), but it also addressed the joint acquisition process and was the place where senior military and civilian officials began to compare and critique each other's acquisition strategies (Barnett, 2005). The SLRG was in effect Rumsfeld's personal vehicle for directing the Pentagon; for him, the SLRG replaced the DRB we have talked about earlier and its name changed slightly as it was expanded to participate in different phases of the PPBE process, to SLC when a war fighter was involved for example. We have said that no one man can run the Pentagon, but that does not stop determined executives from trying.

The discussion above notwithstanding, while former Secretary of Defense Rumsfeld and his staff, and the military department secretaries made significant PPBES changes, much of the process still operates now as it did under previous administrations and PPBS.

In addition, we have explained that defense budgeting is part managerial and part political. While some observers might conclude otherwise, from our perspective no amount of budget process or PPBES reform will reconcile the different value systems and funding priorities for national defense and security represented by opposing political parties, nor will it eliminate the budgetary influence of special interest politics. Value conflict was evident in the early 1980s when public support combined with strong presidential will and successful budget strategy produced unprecedented peacetime growth in the defense budget. And despite the implementation of deficit control reforms since 1985, constituent and special interest pressures made it difficult for Congress and the DoD to realign the defense budget. While we applaud the changes made in 2001-2003, reform of defense budgeting process does not mean that producing a budget for national defense will be much easier politically in the future than it has been in the past. As we have explained, threat perception and politics drive the defense budget, and to some extent the budget process itself. We concentrate on the political dimension of budgeting for defense in chapter 6.

We may observe that a sequence of annual budget increases for national defense in the early and mid-2000s have not brought relief to many accounts within the DoD budget. At the same time, requirements of fighting the War on Terrorism have intensified the use of DoD assets and the costs of military operations. Because the need for major asset renewal has been postponed for too long, new appropriations have gone and will go in the future largely to pay for new weapons system acquisition, and for war fighting in battles against terrorism. What this means is that accounts such as those for O&M for all branches of the armed services will continue to be under pressure and budget instability and restraint will remain a way of life for much of DoD. This places a heavy burden on the PPBES process, on DoD decision makers, analysts, and other resource process participants to achieve balance in all phases of defense budgeting and resource management.

With respect to the future of PPBES, despite some apparent weaknesses in application, and given the promise of the reforms implemented in 2001-2003, it is likely that the system will continue to produce budget requests for defense on a regular and reliable schedule. It will continue to provide a structured context for policy negotiation and decision making for the service branches and the OSD. It will supply the framework for integrating the views of the JCS and the military departments and services and a template against which the views and recommendations of other defense and security agencies may be compared, for example, those of the National Security Council, the State Department, Central Intelligence Agency, Defense Intelligence Agency, and others. In addition, it is noteworthy that other agencies of the federal government have recently adopted PPBES as their resource management and decision system of preference, for example, the Department of Homeland Security, the National Aeronautics and Space Administration, and the National Oceanic and Atmospheric Administration (NOAA). However, as indicated in this chapter we believe it would be in the best interests of DoD to completely replace PPBES with a longer range budgeting process. More on this change with respect to the need for capital budgeting in DoD and the rest of the federal government is provided in chapter 14.

Analysis of the manner in which PPBES operates and should operate provides a basis for assessing criticisms of defense policy setting and budgeting/resource management as well as policy and process reform proposals such as those implemented by the Bush administration in the early 2000s and the new administration elected in 2008. We continue to wonder whether the most recent reforms will be permanent and will be taken up by succeeding presidential administrations, whether and to what extent to they have actually improved DoD planning, programming and budgeting, and what the likelihood is for further PPBES reform. One answer may be provided in the aftermath of the presidential and Congressional elections of 2008. Any administration in office in 2008 and beyond will need to have confidence that the defense resource decision system used by DoD can match the capability requirements of an environment in which the U.S. continues to wage a long-term world War on Terrorism while responding to the inevitable contingencies that arise to confront national defense resource planning and budgeting.

NOTE

1. A short footnote on Navy organization: the Navy is headed by the CNO (four-stars) and a vice-CNO (also four stars). Then there are five line vice admirals (three stars) that report to CNO, for communications, strategy, logistics, manpower, capabilities, and resources; the sixth line position is the director of naval intelligence, a two-star position. Then there are a series of staff positions reporting to CNO, of various flag ranks, including such functions as naval nuclear power (4 star), naval reserve, test, and evaluation, and so on. It is a complex and constantly changing environment. (Also our division of line and staff is not the same as the Navy would use, where line officers can hold command of a ship at sea while staff officers can not because they have chosen specialty careers such as engineering, or logistics.)

CHAPTER 6

CONGRESS AND THE DEFENSE BUDGET

From the Cold War To the War on Terrorism

INTRODUCTION

In earlier chapters, we provided templates and detailed analyses of the federal and defense budget processes. The purpose of this chapter is to follow-up this material with analysis of how Congress influences national defense policy and resource management through deliberations and decisions made in the budget process. The chapter focuses on policy setting and budgeting in the period 1989 through the late 2000s, an era in which defense policy and the budget were reshaped beyond all expectation.

National defense policy and budget strategy suffered two great shocks as a result of historic events that occurred on May 1, 1989 and September 11, 2001. The first was the dissolution of the USSR and the end of the bipolar Cold War world with its heavy emphasis on strategic nuclear deterrence, strategic air and sea power, forward deployment to surround the enemy, and the acquisition, manpower, and training programs geared to meet a single, powerful adversary. As this threat dissolved so did the need for some amount of defense spending in the view of Congress and

Budgeting, Financial Management, and Acquisition Reform in
The U.S. Department Of Defense, pp. 199–267
Copyright © 2008 by Information Age Publishing
All rights of reproduction in any form reserved.

the executive, particularly after the elections of 1992. By the middle of the 1990s, defense budgets had come down by about one-third from peak Cold War levels and military strategists were struggling to think beyond the doctrine and force structure that had guided and serviced American defense posture for more than 40 years—from the rise of the Iron Curtain at the end of World War II to the fall of the Berlin Wall.

At the beginning of the twenty-first century, defense planners continued to grapple with what it meant to be the primary military power in the world. A moderate political consensus emerged in the late 1990s to recognize that some parts of the force structure had been underfunded and cut back too far at the end of the Cold War. With the election of President George W. Bush, plans for a sizable increase in defense spending began to emerge. However, in the aftermath of the shock of the terrorist attack on the United States on September 11, 2001, this buildup was accelerated as a new enemy shifted the threat scenario to a fundamentally different perspective.

In this chapter, we chronicle the congressional defense policy and budget dialogue during the period of decline and retrenchment in the 1990s followed by the war on terrorism expansion to meet an even more menacing worldwide threat in the first years of the new millennium. Our purpose is not merely to describe the ebb and flow of defense spending. Rather, we examine this period in the content of a theoretical assumption grounded in the work of noted budget authority Allen Schick. His thesis is that budgeting by the U.S. Congress fits a model of improvisation under conditions of policy uncertainty. Relative to our previous definition of the budget process as incremental, following the tenets of Wildavsky's theory of the budget process, Schick's hypothesis provides another perspective on the ways in which congressional policy definition and budgeting, and the budget process itself, may be examined and explained. We ask whether the theory of improvisation contravene that of incrementalism. Does one hypothesis provide a better model to guide inquiry than the other? Alternatively, is it possible that a combination of the two hypotheses provides a more complete contingency theory of budgeting? First, we must examine the evidence in detail. Second, we draw conclusions based upon this assessment. We begin by reviewing the legislative process for budgeting national defense.

THE CONGRESSIONAL BUDGET PROCESS AND DoD

The defense budget begins as the product of executive branch deliberations and is presented in the budget of the United States that the President annually transmits to Congress. Congress then reviews this proposal and decides what to fund. The Constitution provides sole authority to Congress to tax and spend. No such authority is available to the executive.

Congress approves DoD programs and the funding to carry them out annually. The President's budget is merely the starting point, that is, only a proposal. A number of decision points occur during the congressional review and approval process; these are sometimes confused by the news media and the public. Essentially, the annual congressional budget process may be divided into three phases: passage of the Budget Resolution (not a law), passage of the Defense Authorization Act (law), and passage of the Defense Appropriation Act (law).

The Budget Committees and the Budget Resolution

The Budget Resolution (BR) represents the budget plan of Congress. Typically, the media calls the BR "the budget" but, only appropriation law actually provides money that may be spent by the executive including DoD. The BR is a commitment by Congress to itself as to how much it will tax, spend and borrow in the appropriation process; it is a resolution passed by both Houses of Congress but it is not sent to the President for signature. The BR is not law and, as such, it is meant to be used as a planning tool to guide congressional budget negotiation and enactment. The BR provides prospective limits for what will be spent on defense in the current year and what might be spent on defense in the next several years. Deficit avoidance strategies have been built into the BR and related acts since 1985 and have been specified in the congressional BRs that projects out 5 to 7 years as guidance. While the amounts detailed in the BR in the out-years after the upcoming fiscal year (FY) under consideration for approval are seldom realized as such, Congress watchers, including the President and Office of Management and Budget (OMB), can look at the trend lines and draw their own conclusions about what Congress plans to do.

For example, the 1997 BR included a decrease in defense spending for FY 1999. While this was the judgment of Congress in 1997, the enlightened observer would have viewed this decrease with some skepticism, given DoD problems in meeting its readiness goals after the drawdown budget years of the early 1990s and the flat budget years of the mid-1990s. Thus, while it provides useful information, the BR is not the defense budget. It does indicate what is likely to be the top limit for defense spending in the current year, and the BR out-year profile indicates how Congress plans the prospects for future year defense spending.

It is important to note that the Budget Committees do their work in each house of Congress as committees of the whole, that is, without subcommittees. When each house of Congress has passed its own version of the BR, the differences between the two versions have to be negotiated in a joint conference committee. The leadership of both houses sit on the conference committee and it is here that final deals are struck that determine the outcome of the BR.

According to the guidelines provided in the Congressional Impoundment and Control Act of 1974 as amended, the BR is supposed to be passed by April 15th, but Congress rarely meets this deadline. Typically, it is June or July before Congress passes the BR that provides taxing and spending targets for all appropriation committees and subcommittees. It should be noted that in some years Congress has failed to pass a BR at all. When the BR is passed, the media tend to refer to the amounts approved as the "defense budget." In reality, after passage or during the period of consideration of the BR, attention turns to defense authorization and appropriation legislation.

The Authorization Committees and Authorizing Legislation

The authorizing committees for defense are the House and Senate Armed Services Committees (HASC and SASC). These committees do their business in both the full committees and in subcommittees. They are critically important in terms of overall federal spending because they have authority over entitlement programs (e.g., Social Security, Medicare, Medicaid) and changes to these programs, some of which may affect defense (e.g., related to veterans benefits, although veterans benefits changes typically are budgeted under appropriation legislation annually). The authorization legislation is important because it approves defense programs, focusing especially on both existing and proposed new programs.

While the intent of the BR is to provide a guide to overall federal and defense spending, the attention of the authorizers is on specific programs. However, authorization legislation may include a variety of budgetary and nonbudgetary policy directions for DoD. For example, the authorization bill may include the annual military pay raise, authority to begin development of a new weapons system, or authority for new military construction projects, such as a barracks or family housing. Such measures also require appropriations from the appropriations committees of Congress. Authorizers also may suggest policy positions on treaties, arms limitations, nuclear weapons research and the use of American troops to the President. Critically, without the annual approval of authorization legislation, DoD is not permitted to begin spending money on new or existing programs, even if it has provided money for them in defense appropriations legislation. To begin new programs and continue existing ones, DoD needs passage of both authorization and appropriation laws. However, these two pieces of legislation differ in some important respects. For example, the graphic below shows how the FY 1999 authorization and appropriations bills differed.

Table 6.1. Comparison of Appropriation and Authorization Bill Structure

Title	Authorization	Appropriation
Military Personnel	70.6	70.6
Operation and Maintenance	93.5[a]	84.0
Procurement	49.5[b]	48.6
RDT&E	36.0	36.8
Military Construction[g]	4.9	NA
Family Housing[g]	3.5	NA
Revolving/Mgt Funds	1.5	.8
Other Defense Programs	NA	11.8[a, b, c]
Related Agencies	NA	.4[d]
Other Defense-Related	1.0[e]	NA
General Provisions	NA	−2.4
Scorekeeping Adj.	NA	0.0
Trust Funds	.3	NA
Receipts/Other	−2.3	NA
Atomic Energy Defense Activities[f]	12.0	NA
Total DoD	270.5	250.5

Source: Tyszkiewicz and Daggett, 1998, p. 42; Budget Authority in billions of dollars.

a. O&M differs mainly because defense health and drug interdiction are included in O&M in authorization bill, but in "other defense programs" in appropriation bill.

b. Chemical agent and munitions destruction is included in procurement in the authorization bill, but in "other defense programs" in the appropriations bill.

c. Also includes Office of Inspector General

d. Includes CIA Retirement and Disability System Fund, Intelligence Community Management Account, National Security Education Trust Fund, payments to Kaho' olawe Island Fund.

e. Includes Selective Service System and defense-related civil defense activities of FEMA, funded in VA-HUD-Independent Agencies Bill.

f. Funded in Department of Energy Appropriation Bill.

g. Funded in Military Construction Appropriation Bill.

h. Note that appropriators do not have to fund all that is authorized. In 1999, it appears they approved more RDT&E than was authorized.

As shown in Table 6.1, authorization legislation contains specific dollars but, these amounts are only advisory guidelines to the appropriations committees, numbers that appropriators typically ignore. Authorization legislation does not provide Budget Authority (BA) to actually spend money but approval of programs is required of authorizers. Thus we begin to understand the basis for competition between these two powerful committees in both houses of Congress.

To maintain their oversight and policy and program approval leadership role, the authorizing committees have gradually expanded their scope of interest and authority to cover virtually all aspects of defense policy and budgeting, and their role in this respect has continued to expand since the late 1950s. This is evident in the Table 6.3 indicating the spread of annual authorization requirements in the period 1959 to 1983 (more recent data demonstrating this trend may be found in Candreva, 2005; and Jones & Bixler, 1992). Each year the authorization committees and subcommittees (e.g., for national defense) hold sessions in which the "posture" of defense policy is presented in testimony by the secretary of defense, the chairman of the Joint Chiefs of Staff (JCS) and the heads of the military services. The authorizers also hear testimony on various policy and program issues from the secretaries of the military

Table 6.2.   Addition of Annual Authorization Requirements

Year	Public Law	Programs Added
1959	86-149	Procurement of aircraft, missiles, and naval vessels
1962	87-436	RDT&E for aircraft, missiles and naval vessels
1963	88-174	Procurement of tracked combat vehicles
1967	90-168	Personnel strengths of each of the selected reserves
1969	91-121	Procurement of other weapons
1970	91-441	Procurement of torpedoes and related support equipment; Active duty personnel strengths of each component of the armed forces
1973	92-436	Average military training student loads of each component of the armed forces
1975	94-106	Military construction of ammunition facilities
1977	95-91	National defense programs of the Department of Energy
1980	96-342	Operation and maintenance of DoD and all its components
1982	97-86	Procurement of ammunition and "other" procurement
1983	98-94	Working capital funds

Source: Tyszkiewicz and Daggett, 1998, p. 36, from Senate committee on armed services, defense organization. S. Prt. 99-86, 1985, p. 575.

departments and numerous other defense officials. Such hearing begin in February each year and typically continue through the summer months until the committees deem themselves ready to complete their work at the subcommittee level and then in full committee on the authorization bill. Precise language about how programs may and may not be executed and managed is provided directly in authorization bills and also in subcommittee reports accompanying the authorization act, and in joint conference committee reports. As is the case with the BR, each house of Congress approves separate authorization bills annually, and the differences between the two then have to be negotiated in a joint conference committee. After approval by the conference committee the authorization bill with identical content must be approved separately in floor votes by both houses before it is sent to the President for signature into law.

Appropriations Committees and Appropriation Legislation

The third major step in the congressional budget process is appropriation. This role is performed by the House and Senate Appropriations Committees (HAC and SAC) and their subcommittees for national defense. Appropriation committees assess spending proposals contained in the President's budget submission plus alternative spending proposals that may be solicited by them from DoD and the military departments or may come from measures initiated by the members of the committee themselves on behalf of their constituents. After subcommittee hearing each year that last from February to September (or later in some years) they approve spending for programs as BA by separate appropriations and by title (type of appropriation). Generally, defense appropriations acts provide BA to DoD in lump sums by separate and specific appropriations accounts, for example, $6,535,444,000 for Navy aircraft procurement (APN). Precise language about how money may and may not be spent is provided directly in appropriation bills and also in the committee reports accompanying the appropriation act; typically by line item amount. If appropriation bills and committee reports do not specifically change the DoD budget request, then whatever the justification was presented by DoD for various programs, accounts, and items is taken to be binding and Congress assumes DoD will carry out program as it presented in the President's budget. Most of the actual funding for the Department of Defense is provided in the two separate pieces of legislation annually, in the defense appropriation bill and the military construction bill. While Congress attempts to pass both authorization and appropriation legislation before the beginning of the FY, sometimes this schedule is not met, requiring Continuing Resolution Appropriations (CRAs) to allow DoD to continuing operating. More on CRAs is explained subsequently. Table 6.3 shows the details of congressional appropriation approval for a year in which the defense appropriation was passed on time before October 1, the beginning of the federal government FY.

In 2000, the defense appropriation bill was passed on time. As can be seen, it was approved by voice vote in the subcommittees in each chamber of Congress. The full appropriations committee passed it by voice vote; in the House and the Senate committee the votes were unanimous. Substantial majorities passed the bills on the floor of each chamber in early June. The conference committee report was issued about a month later, on July 17 and passed each chamber within 10 days. The President signed the bill into law on August 9, 2000. The bill was early because there was substantial agreement on what needed to be spent for defense, the overall federal government budget was in a surplus position that year, and a presidential election would take place that fall. Because of its size, the defense appropriation bill is usually passed late. From 1970 through 2007 the defense appropriation bill was signed on or before the start of the FY only 10 times, in 1976, 1977, 1988, 1994, 1996, 2000, 2001, 2004, 2005, and 2006. In 2007 the defense appropriation for FY 2008 was held up and not passed until late in the fall due to pre-2008 election politics and congressional policy concerns related primarily, but not exclusively, to the war in Iraq. Part of the reason the defense appropriation for FY 2008 was late was due to Democratic Party leadership initiatives in both houses of Congress to increase nondefense discretionary spending above what President Bush had requested. Bush threatened to veto such appropriations if approved at higher spending levels and Democrats, who controlled both houses of Congress after the 2006 elections, held the defense appropriation "hostage" and would not pass it until the President negotiated compromises on domestic spending increases.

Without an appropriation, little or nothing new can be done by DoD. As explained subsequently, without an appropriation the CRA passed by Congress essentially puts spending on "automatic pilot" at the level approved for the previous FY. Thus, the annual appropriation act is the true defense budget and is vital to stable operation of DoD and the military departments and services. It contains the dollars to fund people, operations and maintenance, weapons procurement and other important areas of expense. As noted, both the authorization and appropriation legislation, once passed out of conference committees and sent to the President for signature into law contain dollar figures for defense programs, but the appropriation bill provides the only authority (BA) to spend as specified in the Constitution, that is, "real money," The approved authorization bill figures typically viewed by DoD as ceilings

Table 6.3. Defense Appropriation Bill in 2000

Bill No.	Subcommittee Approval		Committee Approval		House Passage	Senate Passage	Conf. Report	Conf. Report Approval		Public Law
	House	Senate	House	Senate				House	Senate	
Defense	5/11/00	5/17/00	5/25/00	5/18/00	6/7/00	6/13/00	7/17/00	7/19/00	7/27/00	P.L. 106-259
H.R. 4576	(vv)	(vv)	(vv)	(28-0)	(367-58)	(95-3)	H. Rept. 106-754	(367-58)	(91-9)	8/9/00
S. 2593			H. Rept. 106-644	S. Rept. 106-298						

Source: Library of Congress, 2000.

for the appropriations and programs, unless they are exceeded by the amount enacted in the appropriation acts. However, the appropriations act BA is not enough to initiate spending. Appropriations must be matched what has been approved for DoD in the authorization act. Additionally, critical information on congressional intent must be gleaned by DoD from careful reading of the subcommittee and conference committee reports that accompany the defense appropriation act. As noted, usually language is inserted in an act (and therefore has statutory authority) about how money is to be spent, and it also appears in the committee reports.

Congressionally approved reports provide direction and cues about how money should be spent and usually such directives are not subtle. For example, in 2000 the conference committee took funds from the C-17 program and put them in a revolving fund so that the Air Force could not spend them on fighter aircraft. The conference committee also cut $1 billion from Navy LPD-17 ship construction program but in its report allowed the secretary of the Navy to move up to $300 million from other budget accounts to cover unexpected increases in this program (Congressional Quarterly Almanac, 2000, pp. 2-51). More subtle cues also may be provided directly by individual members or the full subcommittee in hearings where legislators indicate what they think DoD ought to do. Again, this is a normal part of the politics of the budgetary process.

Statutory language is law and must be obeyed. Report language does not have the force of law, but when DoD fails to heed nonstatutory report directives, inevitably some official will be called before a defense appropriations committee to explain why. Questions and suggestions made in committee hearings by members of Congress and their staffs are not binding and DoD need not fully implement them when they run counter to the needs of the services. Still, disregarding congressional "suggestions" may result in lengthy interrogations at hearings, requests for reports detailing how money was or is to be spent, and, sometimes, a tightening of thresholds for reprogramming dollars in budget execution if Congress loses faith in DoD stewardship. When DoD deviates from what both authorization and appropriation committees have explicitly recommended, these actions must be justified to Congress. And even if members of Congress forget, their staffs do not. This is a key part of committee and member staffs' jobs, that is, to "assist" in controlling the management of DoD to satisfy constituent interests.

Consequently, multiple sources of information must be considered by DoD to understand what is required to meet congressional directives and intent with respect to how the defense budget is spent, with greatest emphasis on the conference committee reports that accompany the defense authorization and appropriation acts. Defense insiders routinely deal with "fences, floors, and ceilings" with respect to congressional actions and directives, meaning that Congress has required specific amounts of money within an overall appropriation to go exclusively to certain programs (a fence), that not more than a specific amount be spent (a ceiling), and not less than a specific amount be spent (a floor). DoD program and financial managers treat what are called "congressional special interest items" with great care because they know that if they do not, they will be called to account for their errors.

ADDITIONAL FACTORS THAT INFLUENCE DEFENSE SPENDING

A number of additional factors influence how DoD can spend its BA. Some of the most important of these are explained below.

All Defense Spending Does Not Come From the Defense Appropriation

In any specific year, the Defense Appropriations Act provides most of the funding to be executed by DoD, for example, approximately 90% (the exact percentage varies by year). Other funding for defense is appropriated in the Military Construction Appropriations Act, the Energy and Water Development and Appropriations Act (mainly for nuclear weapons activities and base closure), the Housing and Urban Development and Independent Agencies Appropriation Acts (mainly for the Federal Emergency Management Agency, support of National Science Foundation logistics in the Antarctic, and the Selective Service), the Commerce, Justice and Department of State Appropriations Acts (mainly for defense related activities of the FBI and for maritime security programs). Further, these appropriations may be enhanced by supplemental budget appropriations, as we analyze in chapter 7. In addition, creation of the Department of Homeland Security (DHS) shifted funding for some national defense related program into the DHS appropriation.

Competition Between Authorizers and Appropriators

Once the defense appropriation bill is approved, it would appear to most observers that Congress has done its job on the budget unless it has to act again in response to a presidential veto or a request for a supplemental appropriation. So, at this point we might conclude that we have the defense budget. However, as explained below, this is not quite the case. Authorization and appropriation acts differences must be reconciled before DoD can begin to spend. Adding confusion to the congressional process, sometimes the authorization bill is passed after the appropriation bill; this is not the intended process but it happens. Programs that are appropriated, but not authorized, may not be executed and no money may be spent until the differences in authorization and appropriation language and controls are negotiated between DoD officials (typically by comptrollers) and Congress (typically by committee staffs).

It was not always this way. Until the mid-1960s, if a program gained an appropriation but had not been authorized, it was assumed that the appropriation included an implicit authorization for surely, the logic ran, those who voted to appropriate the program meant to authorize it too, otherwise they would not have appropriated it. However, different members serve on the two committees; only a few members serve on both. Thus, prior to the late 1960s were treated as though the appropriation included implicit authorization. During the Vietnam war members of Congress who opposed this engagement attempted to stop funding for it thorough both the authorization and appropriation committees, as well as in floor votes. Similar differences of opinion found there way into the deliberations for both committees over the years so that by the 1980s the authority of the two committees was treated as distinct from one another. In a strict legal sense the concept of implicit authorization was found lacking in terms of the ability of Congress to apply control over executive branch spending, and perhaps deservedly so given the size of the both the defense authorization and appropriation bills and the number of programs included in them.

It is an understatement to observe that not all members of Congress know all the provisions of the defense authorization and appropriation bills when they are passed. Rather, members concentrate on areas that affect their constituents. It is up to committee staff and Congressional Budget Office (CBO) to keep track of the extent to which appropriators are staying within the guidelines of the BR and any other applicable restrictions established by Congress itself, for example, those to reduce the budget deficit. Moreover, given the lack of membership overlap between the authorizing and appropriations committees and subcommittees for defense, it is hard to argue that the appropriations bill carries the implicit will of the authorization committees. The law is clear on this matter. Title 10 of the U.S. Code, the body of law that governs the executive branch including the military departments and the DoD, stipulates that even when appropriated all funding must be authorized before it may be spent.

Moreover, the rules of the House and Senate each prohibit appropriating funds for programs that have not been authorized. However, some of this clarity is dissipated in actual practice when appropriations bills often provide funds over and above amounts provided in authorization bills and approve money for activities that were not mentioned either in the authorization bill or the report that accompanied it. Legal opinion and judicial action has consistently held that appropriations bills may provide more or less money than has been authorized for a particular program. Government Accountability Office (GAO) has recommended to Congress that the piece of legislation passed most recently prevails over earlier legislation. This is customarily, but not always, the appropriation bill, and GAO advised that more specific provisions should prevail over less specific provisions (GAO, 1991).

What makes this whole congressional budget process workable is that the authorization and appropriation laws themselves are rarely in conflict over the same item or issue because each year the authorization bill aggregate totals are provided in categories that are more general rather than by discrete items. The language of the reports accompanying the bills carries the details clarifying what and how many programs or units are funded within a general category, and rather than being tested in court, this language is subject to the brokering and clarification that goes on between members of Congress and their staffs and DoD representatives during the legislative budget process and later in budget execution after the appropriation legislation has been signed into law.

In 1998, a particularly acrimonious debate over defense authorization led to a letter from Senator Robert Smith of New Hampshire explaining the consequences of not passing the Defense Authorization Bill (Congressional Record, 1997, S11817-8). Smith commented on a variety of items that would impede the good management of the DoD and defense programs, ranging from delaying the construction of family housing and other military construction projects to the absence of authority to expand counternarcotics programs in South and Central America, to the absence of authority to accelerate advance procurement programs, the lack of authority to increase military personnel retention bonuses. What is threatened by the mismatch between authorization and appropriation bills is found primarily in the acquisition area in absence of provision of funding for new program starts and expansions or decrements to existing programs. While the dollar amounts this involved are small relative to the size of the defense appropriation in most instances, they are the marginal dollars that may prevent new program starts from implementation, keep good programs from expanding, and protect programs that have outlived their usefulness from being reduced or eliminated.

Establishing the Will or Congress on Defense Issues

The authorizing committees provide DoD a forum for advocacy. These committees look both at macro and micro defense needs, examining defense force structure, overseeing DoD operations, and looking for ways to fund projects that will benefit their constituents.

To illustrate how this happens, the following excerpt is based on a conversation with a staff member of the Senate Armed Services Committee. It traces the Defense Authorization Bill cycle through to passage. The comments may not hold for all committees or all years or all staffers, but they do provide insight into a process that is not normally revealed publicly.

The Authorization Cycle from the Staff Perspective

The cycle starts with State of Union address. The first round of hearings is pretty cut and dried; the same witnesses appear before the committee each year: secretary of defense, the chairman of the joint chiefs, the secretaries of Army, Navy, Air Force, etc.

The subcommittee hearings are different every year, dependent on issues. Staff work as facilitators, call 5 or 6 people, ask them what they would say if asked certain questions, then decide who to use, form panels of witnesses and script the hearings so that the issues are brought out and both sides aired. Staff suggest to members the questions to ask and what the answers should be so that expected issues are highlighted.

The hearings are valuable. They educate the Senator on a complicated issue; he learns when he gets truth, when he is being hoodwinked. The hearing builds a record to take new action. It is also used for oversight: has the agency done what it was asked to do last year? The hearing involves briefing books, statements by members and witnesses, questions, answers. Most of the time it is hard for staff to get ten minutes with the Senator so the briefing around the hearing is a two hour block of time to go over in detail what the issue is about, educate the Senator so that in mark-up he will remember this issue and be prepared to make a decision on it, or so that he can be put back into the picture by referring to the discussions at the hearing when the issue was first considered.

Markup follows hearings. This is usually in June, by subcommittee and by full committee. The subcommittees are given a number by the committee chairman that they have to attain. Staff sit where witnesses sat, answer questions, give advice. Members decide. It can take from 20 minutes to 4-5 hours. Staff have to agree on markups and what has been decided. Go late on Thursday night, put a report together, put a bill together, proof with members and staff, send to the government printing office and file with clerk of the Senate. Staff position is that bill is perfect when it comes out of committee and no amendments are needed, but amendments are always offered because Senators have to have a record for reelection.

Staff tries to fend off amendments, instead give them language in the report, or a sense of the Senate resolution, or promise to take it to conference. First couple of days serious amendments are offered, after that it is cats and dogs and the committee/staff tries to kill them. Staff talk around, try to find out who is going to offer an amendment, without giving them any ideas.

Floor debate on our bill is usually scheduled right before August recess; otherwise they would talk about defense issues all fall. Armed Services bill is a bill that will pass, so people try to add amendments. A Senator who gets an amendment accepted has a three-for: he gets to talk about it when it is accepted, again when the bill passes, and when it goes to conference with the House. He can talk about it three different times.

After floor passage, the bill goes to conference with the House. Members tell staff to negotiate a potential solution on 90% of issues, take them back to the member to be blessed. House are specialists, really tough task to prepare Senate member to go head to head with guy who is expert. When you get agreement, the staff member has to speak up and say "we understand the agreement is ..." in order to make sure all agree on what has been agreed. Have to build in enough time in the conference process so people can talk, ventilate issues, but not too much time. At this point also the appropriations bill is right on the heels of the authorization bill, like passing notes through a hole in the wall between committees. If authorizers propose too much for a program according to appropriators, authorizers may change the shape of program a little to spend less.

Big ticket issues drive the conference. Most of the other issues (80-90%) can be settled by staff and blessed by members.

Precision in conference committee work is very important. When bill is in the hearing stage in committee, if errors are made they can be fixed on the floor (or if "bad" provisions inserted), floor mistakes can fixed in conference committee work with the House, but the product of the conference becomes law; has to be perfect, language has to be clear, precise, correct. To fix it requires passage of another law.

Next the bill and the conference report go back to Senate. The Conference Report is not amendable and must be voted up or down.

My job is to make the Senator look good: I might say "here are the pros and cons; I'll give you a recommendation if you ask me." As the Soviet threat vanished, the tendency to pork and earmarking seems to be increasing.

The product of the Senate Armed Services Committee is the Defense Authorization Bill. The sense here is that a getting a bill passed is a very significant achievement, like winning the Superbowl. (Author Interviews, 2007).

In the period from the mid-1990s and continuing through 2008, when the Senate Armed Services Committee was chaired by Senator John Warner (R-VA) during the time Republicans controlled the Senate (or when Warner served as the "ranking" senior minority party member),

Senator Warner was particularly effective in initiating pay raises and pension benefit changes for the military services, helping to mitigate the consequences of defense budget shortfalls imposed by the Clinton administration. During the administration of George W. Bush, Senator Warner led the Senate on defense issues including war related supplemental authorization and appropriation deliberations and decision making. Warner also remained a firm advocate for Navy shipbuilding and other programmatic initiatives. For servicemen and women, the authorization committees provide the first indication of how quality of life issues such as pay, benefits, and family housing will be treated. Contractors and manufacturers also are very interested in authorization committee work, for they too get early indication of when and how much a program or weapons contract might be changed.

While most programs in defense are authorized annually, the number of uniformed military personnel to be employed is handled differently. A permanent personnel end-strength level is authorized, then the appropriators fund the dollars to support the level of employment authorized. From time to time the end strength numbers are changed up or down by the authorizers as needs warrant, based on recommendations from DoD and the executive branch. The appropriators must fund this authorized strength in the annual appropriation bill. The authorizers may also approve annual pay increases, but sometimes the appropriation process results in DoD being given new money for half the pay raise and being asked to fund the other half itself, by finding a billpayer among current accounts.

Additional Opportunities to Influence Defense Spending

Members of Congress have many opportunities to cast votes for or against defense spending, and defense legislative liaison personnel have many locales in Congress upon which to focus their energies. As many as 22 stages of congressional action occur where votes are cast on parts of the defense budget, including subcommittee and committee votes to report out each bill (5 in each chamber), the final floor vote on the bill, and the vote on the final conference committee report for the bill (6 in each chamber). Other appropriations bills that have defense dollars in them would increase the number of voting opportunities, for example, the military construction or the energy bills.

Table 6.3 presents this information somewhat differently. The legislative liaison groups representing each military department, and some Office of the Secretary of Defense (OSD) level staff, have to monitor votes in as many as 38 steps throughout the legislative year, beginning with Budget Committee action, proceeding through the authorizing committees (armed services) and the appropriations committees. DoD often has a supplemental bill to worry over and since Congress usually does not finish appropriations on time, each session usually sees one or more CRAs. Any decisions reached at any phase of the total process could be critical for selected defense programs and thus must be monitored and, in many cases, responded to with appeals and lobbying efforts. Only committee members are allowed to vote in committees, and subcommittee members on subcommittees, but all members vote on final floor action and on accepting both committee bills and conference committee bills (reports). Thus, every Congressman has at least 10 opportunities to vote on aspects of defense bills; this expands to 13 or more separate voting opportunities when supplemental appropriations and one or more CRAs are employed by Congress.

For example, in 2001 Senator Diane Feinstein (D-CA.) made 15 roll call votes on defense issues in the Senate from April through December, which she listed on her Web page. The votes included two BR votes, one reconciliation vote that was tied to a defense reduction, four supplemental appropriation votes, two votes on the military construction appropriation, five authorization votes and one appropriation bill roll call vote (Feinstein, 2003). The Senator does not list all the actual roll call votes; only the final actions are public. But the Senator's list was easily accessible on her Web site so that her constituents could see how she voted on important defense issues.

If one were to step inside each stage and make a count based on individual votes taken, the voting opportunity of legislators is enlarged significantly when amendments to bills are considered or as items are voted up or down in the subcommittee and committee mark-up process. Moreover, the critical vote on a bill could come on one of dozens of amendments that are offered, if the coalition that can successfully pass an amendment to a bill can hold together for the vote on the main bill. Although we only

Table 6.4.   Milestone Votes on the Defense Budget

	Budget Resolution	Armed Serv.S	DoD Approp.	Milcon Approp.	DoE Approp.	DoD Suplmtl.	CRA
Opportunities for Votes on Defense							
Subcommittee		H, S	H, S	H, S	H, S		
Full committee	H, S	H, S	H, S	H, S	H, S		
Floor	H, S	H, S	H, S	H, S	H, S	H, S	H, S
Conf. Rpt. approval	H, S	H, S	H, S	H, S	H, S	H, S	

Source: McCaffery and Jones, 2004, p.150.

show three in the exhibit, it is also useful to note that tracking the total defense budget in Congress involves tracking five appropriations bills, as well as the BR. Also, a supplemental appropriation often provides a possible venue for change, as the current year supplemental can affect in some way current-year programs in the following year.

CONGRESSIONAL APPROPRIATION PATTERNS

Aaron Wildavsky's (1964) incremental model of budgeting includes the assumption that agencies will be advocates for their programs and, as a result, the executive branch will tend to ask for more that it had last year and, taken together, Congress will tend to reduce these requests (pp. 13-35, 63-84). The Wildavsky argument also assumes that Congress will be a marginal modifier, reducing the budget request some, but not a lot, in the aggregate. LeLoup and Moreland's work on the Department of Agriculture budget found this to be true. From 1946 to 1971, they found that Congress reduced the annual request by 2% on average. From 1980-1989, Congress changed the Agriculture request an average of 5.9%, more, but still what might be called marginal (McCaffery & Jones, 2001, p. 132). Table 6.5 shows the defense appropriation bill from 1980 to 2001. Congress reduced the request in 14 of the 20 years. We note that the House and Senate never arrived at the same number for defense, thus a conference committee was always a necessity and even where numbers were close, the two chambers could have funded dramatically different defense profiles.

On average, during this period the President asked for an average of 6.03% more for defense each year over these years and Congress cut this request by an average of 2.02%. The outcome is a modest 4% year over year change. However, these averages mask the Reagan build up of the early 1980s (81-85) where the President asked for increases up to 25.75% over the previous year's appropriation. These increases were supported in 1981 and 1982 by Congress and then cut substantially, but the result was still a substantial increase. The defense drawdown really began in 1987 when the outcome was 2% below the previous year's appropriation. The 1990s clearly show cuts to defense appropriations, with the President asking for decreases in 1993, 1994, 1996, and 1997. Congress supported the President and cut more in 1993 and 1994 (made the decreases larger), but added money back to the defense budget in 1996 and 1997 (diminished the President's cuts). It may be noted that from 2001 on the President requested substantial increases in defense spending through FY 2009, and Congress tended to cut these in some places and add on other spending in others. This was particularly true for large supplemental appropriations.

**Table 6.5.   Defense Appropriations 1980-2001:
President's Request, House, Senate and
Final Enactments and Change From Request**

Fiscal Year	Request	House	Senate	Enacted	Change From Request
1980	132,321	129,524	131,661	130,981	−1,339
1981	154,496	157,211	160,848	159,739	+5,242
1982	200,878	197,443	208,676	199,691	−1,187
1983	249,550	230,216	233,389	231,496	−18,054
1984	260,840	246,505	252,101	248,852	−11,988
1985	292,101	268,172	277,989	274,278	−17,823
1986	303,830	268,727	282,584	281,038	−22,792
1987	298,883	264,957	276,883	273,801	−25,082
1988	291,216	268,131	277,886	278,825	−12,391
1989	283,159	282,603	282,572	282,412	−747
1990	288,237	286,476	288,217	286,025	−2,211
1991	287,283	267,824	268,378	268,188	−19,095
1992	270,936	270,566	270,258	269,911	−1,025
1993	261,134	251,867	250,686	253,789	−7,345
1994	241,082	239,602	239,178	240,570	−512
1995	244,450	243,573	243,628	243,628	−822
1996	236,344	243,998	242,684	243,251	+6,907
1997	234,678	245,217	244,897	243,947	+9,268
1998	243,924	248,335	247,185	247,709	+3,785
1999	250,999	250,727	250,518	250,511	−488
2000	263,266	267,900	263,932	267,795	+4,529
2001	284,501	288,513	287,631	287,806	+3,305

Table by Stephen Daggett, Foreign Affairs, Defense, and Trade Division, CRS.

Source: For FY 1950-74, Department of Defense FAD Table 809, issued October 21, 1974; FY 1975-82 and FY 1989-99, annual Appropriations Committee conference reports; FY 1983-88, Department of Defense Comptroller, annual reports on congressional action on appropriations requests (FAD-28 tables); FY 2000-01, House Appropriations Committee.

To simply indicate that the President asked for increases on average of 6% over the period while Congress cut on average of 2% does not do justice to the complexity of the story, nor does it adequately describe the turbulence of the period. In Figure 6.1 this turbulence can be seen clearly. The defense drawdown is apparent in the 1990s. In 1993 and 1994, what Congress appropriated was below the previous year's base, and so was the President's budget request. Appropriations stayed at or slightly above the

Table 6.6.   Percent Increase From Prior Year Appropriation: 1981-2000

Fiscal Year	Pres. Request	Final Approp	Pres Req/ App CY-1	Cong. Change	Total Change
1980	132,321	130,981			
1981	154,496	159,739	17.95%	3.28%	21.24%
1982	200,878	199,691	25.75%	−0.59%	25.16%
1983	249,550	231,496	24.97%	−7.80%	17.17%
1984	260,840	248,852	12.68%	−4.82%	7.86%
1985	292,101	274,278	17.38%	−6.50%	10.88%
1986	303,830	281,038	10.77%	−8.11%	2.66%
1987	298,883	273,801	6.35%	−9.16%	−2.81%
1988	291,216	278,825	6.36%	−4.44%	1.92%
1989	283,159	282,412	1.55%	−0.26%	1.29%
1990	288,237	286,025	2.06%	−0.77%	1.29%
1991	287,283	268,188	0.44%	−7.12%	−6.68%
1992	270,936	269,911	1.02%	−0.38%	0.64%
1993	261,134	253,789	−3.25%	−2.89%	−6.15%
1994	241,082	240,570	−5.01%	−0.21%	−5.22%
1995	244,450	243,628	1.61%	−0.34%	1.28%
1996	236,344	243,251	−2.99%	2.84%	−0.15%
1997	234,678	243,927	−3.52%	3.79%	0.27%
1998	243,924	247,709	0.00%	1.53%	1.53%
1999	250,999	250,511	1.33%	−0.19%	1.13%
2000	263,266	267,795	5.09%	1.69%	6.78%
Avg. change			6.03%	−2.02%	4.00%

base from 1995 through 1998, although in these years we can see that Congress added some money to defense while the President sought both decreases and small increases (a reduction in spending overall when calculated in constant dollars). Since 2001 Congress has appropriated substantial increases for defense in response to presidential requests. This trend continued through FY 2009 but there was no guarantee that such increases would continue under another President and Congress and, as of 2008, most defense analysts were projecting reductions in defense spending both from the proposals of the President and from Congress. Still, our main point is that Wildavsky's assumptions on incrementalism

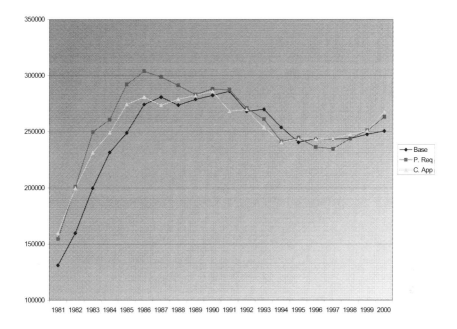

Figure 6.1.   Turbulence in defense funding.

and the roles of the President and Congress tend to hold in most but not all circumstances.

HISTORICAL BACKGROUND ON
CONGRESSIONAL BUDGETING DYNAMICS

Allen Schick has asserted that the federal budget process in the 1980s could best be characterized as a case of improvisational budgeting, where the nature of the annual budget process could not be predicted (Schick, 1990, pp. 159-196). According to Schick, leaders in Congress and the executive branches adopted whatever strategies and process steps they thought necessary during the 1980s to approve the annual appropriation bills. In most years, appropriations bills were passed late; CRAs were a common occurrence and some years were funded by omnibus continuing resolutions. In 5 of 10 years from 1980-1989, more than half of the appropriation bills were funded for the whole year by a CRA; this included all 13 in 1987 and 1988. Over the course of the decade, 63 of a potential 130 bills were funded for a whole year in a CRA (p. 181).

Summit meetings were held between key presidential advisors and congressional leaders with varying degrees of success; some summits succeeded and some did not. On the whole, argues Schick, outcomes from these summits were minimal and the problems budget makers faced after the summits and in succeeding years were much the same as those faced before budget summits were held.

In this environment, budgetary gimmickry was pervasive. Such deceptions ranged from a simple shifting of paydays in the DoD to meet spending caps to overly optimistic GNP (gross national product) and revenue estimates. The budget process usually began slowly with Congressional BRs passed late, an average of 53 days late from 1980-1989 (calculated by authors, 2007, based on Schick, 1990, p. 174). Much of the appropriation "end game" was deliberately moved into the new FY where the sense of fiscal urgency helped decision makers reach compromises. In this turbulent environment, Schick observed that CRAs that were usually funded at the cost to continue current levels of operations actually gave agencies more money than they had had the previous year and more money than was asked for by the President's budget (p. 181). They became real appropriation bills rather than placeholders.

Another change of consequence concerned the President's budget which, in Schick's view, had lost its status as an authoritative statement of what the President felt was needed to fund the executive branch. Instead, according to Schick the President's budget had become an opening bid in some sort of extended budget auction where true needs and motives were difficult to perceive, all the better to facilitate bargaining over key programs during the remainder of the session. Consequently, it is not surprising that timeliness was a victim of improvisational budgeting. The process began late, with late BRs, and failure to adopt both authorization and appropriation bills on time. In some years, the appropriation bill preceded the authorization bill, making the guidance-giving role of the authorization bill problematic.

The view that the President's budget is merely a starting point for Congress, or that it is unnecessary (e.g., dead on arrival), is disputed by those who view the budget as the primary document and point in the budget process where the President has the opportunity to state his policy and program priorities—that will inevitably differ from those of many members of Congress (McCaffery & Jones, 2001, pp. 97-108).

INCREMENTAL AND IMPROVISATIONAL MODEL DIFFERENCES

A simple incremental model of the budget process would include aggressive agencies asking for more for their programs, OMB cutting them back somewhat to include all priorities within available resources and Congress acting as a marginal modifier. In the classic period of American budgeting, this might have meant the House as masters of detail and guardian of the purse cutting deeply while the Senate acted as a court of appeal and restored some, but not all, of the cuts made by the House. Outcomes would show the Presidents budget request being the best predictor of the outcome of the struggle. This picture changes somewhat for the 1990s, but it held true for the period 2001 to 2008.

The budget process of the 1990s, as was the case in the 1980s, was characterized by improvisational budgeting routines that included an abundant supply of CRAs, appropriations passed before authorizations, budget summits between Congress and the President, and one government shutdown. Some of this improvisation may be seen from the funding outcomes. In 6 of the years in this decade the DoD appropriation bill signed into law was less than the President's budget request, with cuts ranging from $19B in 1990 to $488 million in 1998. In four of the ten years, DoD appropriations enacted exceeded the President's budget request, with increases ranging from $6.9 billion in 1995, to $9.2 billion in 1996, $3.7 billion in 1997 and $4.5 billion in 1999. From 2001 to 2008 Congress tended to appropriate roughly what the President requested for defense, in both regular and supplemental appropriations.

The beginning of the decade of the 1990s was marked by the end of the Cold War and the dissolution of the USSR. From 1990 to 1994, the President's budget request was less than the previous year's request. Conversely, a pattern of increased spending, in order to restore a military over-tasked and underfunded, emerges from 1995 to 1999 with each year the President asking for more than the previous year. In addition to a larger request, the Congress added to the Presidents budget request every year from 1995 to 1999, with the exception of 1998 when the request was cut by $0.3 billion.

In the later years of the decade, the policy changes early in the decade and the Internet economy produced a surplus for the federal budget, but the budget struggles in these years make it abundantly clear that harmony was not restored to the budgetary process by the accumulation of a surplus. After the election of George W. Bush and the terrorist attacks of September 11, 2001 total defense appropriations rose dramatically and this trend continued into 2008.

Congressional Budgeting From 1990 into the New Millennium: Recurrence of Themes and Trends

While Schick's analysis pertains to the 1980s, his conclusions still seem accurate. BRs continue to be late and, in some years, no BR has been

Table 6.7. Participant Changes in President's Budget for Defense in the 1990s

Changes From Presidents Budget Defense Authorization	Number of Years	Average % Change by House Committee	Average % Change by Senate Committee	Average % Change by House	Average % Change by Senate	Average % Change in Appropriation
Decrease	4	(2.90%)	(2.46%)	(3.24%)	(2.29%)	(2.55%)
(Decrease: > 10 billion)	1	(7.79%)	(5.83%)	(7.79%)	(5.83%)	(6.06%)
(Decrease: $2-8 billion)	2	(1.39%)	(1.56%)	(2.13%)	(1.82%)	(1.78%)
(Decrease: < $1 billion)	1	(0.37%)	(0.44%)	(0.29%)	0.80%	(0.11%)
Increase	4	3.05%	2.87%	3.05%	2.78%	2.78%
(Increase: > $10 billion)	1	4.72%	4.83%	4.72%	4.22%	4.22%
(Increase: $2-8 billion)	3	2.51%	2.24%	2.51%	2.31%	2.31%
(Increase: < $2 billion)	0	—	—	—	—	—
Same or no change	2	—	—	—	—	—

Source: Buell, R. C., 2002.

passed, rather each chamber adopted a resolution to guide itself irrespective of what the other chamber did. Appropriations have been late and the Defense Authorization Bill sometimes preceded the defense appropriation bill. Partisanship remains high, sometimes bringing the budget process to a halt. In 1994 and 1996, the budget process was on time, but this was the result of a greater policy failure. In 1994, the failure of the Clinton health plan to gain serious consideration allowed for a timely budget process; in 1996, the budget process was timely because congressional leaders were exhausted from the 1995 struggle and were facing an angry public willing to hold both parties accountable for the most extended budget struggle in modern history.

The turbulence of the 1990s culminated in the bitterly contested presidential election of 2000 with the nation divided into red and blue states and many Democrats believing the election had been stolen. When President George W. Bush was elected in 2000 and the House and Senate elections gave his Republican Party majorities in both houses of Congress, conditions were ripe for increased consensus on defense policy and budgets. A budget surplus and recognition that defense had been cut too deeply in the 1990s provided rationale for both large tax cuts and increases for defense. However, the 9/11 attacks put the nation into a new war on terrorism that was to result in significant increases in defense appropriations and spending.

The narrowness of the majority in the Senate following the elections of 2000 made the House, where the Republican majority was far greater, the driver of defense spending increases. However, the Republican defense hawks were defeated by the Republican tax cut hawks and President Bush gave policy priority to the tax cut in 2001. If effect the debate over the size of the tax cut and its timing, fought out in the first half of the year, took both policy decision space and dollars away from a potential defense recapitalization. Despite the fact that the President had suggested adding as much as $48 billion to defense during the 2000 campaign (Congressional Quarterly Weekly, 2001, p. 337), he submitted a placeholder defense budget in April contingent upon a review of defense needs by former defense secretary Donald Rumsfeld. Bush asked for more for defense in June, but mostly for operating expenses, not recapitalization. As debate over the tax cut proceeded, Congress put off any wholesale defense changes (Congressional Quarterly Almanac [CQA], 1975-2001, pp. 7, 3), and Rumsfeld said that the administration's plan was to stick with the Clinton budget of $310 billion for FY 2002. Secretary Rumsfeld's attempt to lead the Pentagon into a new era of force and business transformation at times angered the service chiefs and some in Congress as he proceeded with transformation plans that focused on technology and less on conventional weapons. The outcome for FY 2002 was a presidential request of $319.5 billion and a defense appropriation of $317.6 billion, an increase of 10.7% over FY 2001 at $298.5 billion, with most of the increases going to the operating accounts rather than recapitalization.

Before final action on the FY 2002 defense act could be taken, the terrorists struck New York and Washington, D.C. with devastating effect. The response from the President and Congress was swift and purposeful, including an extraordinary emergency supplemental of $40 billion introduced on September 14th and passed on September 18th. President Bush vowed to respond to the terrorist threat with whatever was necessary for as long as was needed to find and punish those responsible for the attacks. Congress responded on both sides of the aisles with unanimous support for further increases in defense spending to counter the terrorist threat. The divisiveness of the 1990s was replaced quickly, and for a limited time, with unanimity on policy and budget direction. Consensus prevailed as the President mobilized forces and initiated a war on the Taliban regime in Afghanistan. The President requested and Congress appropriated both defense budgets and budget supplementals to fund what turned out to be a victorious war in Afghanistan. The Taliban leadership was routed and replaced by a regime friendly to the West and committed to the end of terrorism in their country and abroad.

The terrorist attack that struck the World Trade Center also struck the Pentagon, causing more than 100 casualties and extensive damage to one side of the building. That day produced many heroes, in New York City, in Pennsylvania where passengers took over a terrorist-hijacked flight at the cost of their own lives, and at the Pentagon. Secretary Rumsfeld himself, hearing the explosion, ran to the impact area and helped casualties from the Pentagon. In late August, rumors had begun to circulate that Rumsfeld was bogged down in his transformation efforts, had lost the confidence of Congress and the service chiefs and would be the first cabinet member to resign. After 9/11, then Secretary Rumsfeld went from being an average Secretary of Defense to what might be termed a "Secretary of War" during the recovery period after 9/11, and during the campaigns in Afghanistan and Iraq. The initial success in Afghanistan was followed by a prolonged build-up, both diplomatically and militarily for the war on the regime of Saddam Hussein in Iraq. At this point, the consensus of support for the presidential policy began to weaken, but it did not break.

This trend continued until the 2008 elections. Despite the political unpopularity of the war on Iraq and growing dissent that was reflected in some serious head-scratching and debate in Congress, in the end congressional support for this war was unwavering through 2008. A defense supplemental appropriation of roughly $76 billion was passed in April, 2003 to fund DoD for costs it had incurred in the successful campaign to oust Saddam Hussein from power. From the fall of 2001 on as the nation went to a war footing, funding for defense came from a stream of sources including the regular appropriation bills and several supplemental appropriations, including those of July 24, 2001 ($6.74b), September 18, 2001 ($3.4b), May 21, 2002 ($14 b), and April 2003 ($62.6b). The defense shares of these bills are in parenthesis. In addition, on September 8, 2003 President Bush requested an additional $87 billion for the war on terrorism and rebuilding of Iraq.

As the war in Iraq dragged on, Congress and the public became more and more convinced that it could not be brought to a successful conclusion. The Democrats basically ran on an antiwar platform in 2006 and gained a majority in both houses of Congress, somewhat to the surprise of the President. Secretary Rumsfeld was a casualty of this upheaval in American opinion and was asked to resign, which he did eventually, and was replaced by Secretary Robert Gates.

Meanwhile the budget process had slipped back into improvisational routines. In our view, these improvisational routines occurred because of three huge policy decisions. The first was the Gingrich House Republicans and their Contract with America from 1994 to 1996 that eventually provided a roadmap to a balanced budget and forced President Clinton to follow. The second event was the victory of the tax cut Republicans in the summer of 2001 which limited defense recapitalization and other domestic discretionary spending. The effect of the tax cut was exacerbated by the puncturing of the dot.com stock market bubble and the recession that followed the events of September 11, 2001. The third event was the proactive war policy which was immediately successful in Afghanistan, but then relapsed into continuing conflict in 2006 and thereafter, and the part of the war on terrorism which was increasingly controversial in Iraq. By 2007, according to opinion polls a majority of the public did not believe U.S. efforts in Iraq would succeed.

In conclusion, since the mid-1990s the appropriation process had to work around a series of security threat altering events. It did this by using improvisational means. The fourth event was the attack on September 11, itself, but in this instance the policy apparatus reverted to a fairly orderly if late process; it was almost normal. Thus the period 2001 to 2008 was not like the decade that preceded it in terms of the pattern of both presidential and congressional actions.

BUDGET RESOLUTION TURBULENCE

An important indicator of budgetary turbulence and improvisation may be found in the treatment of the BR. This BR is developed by the House and Senate Budget Committees and begins the budget process in Congress as we have explained. The BR was established to precede the authorization and appropriation bills and set limits on spending, both as to totals and to functional areas. The limits or targets are derived annually by the committees using the President's Budget as a starting point together adjusted by estimates from the CBO as to GDP, inflation, unemployment, and revenue and spending projections and from testimony presented to the Budget Committees by various administrative officials such as the director of OMB, the secretary of treasury, and various agency officials such as the secretary of defense. The BR is an advisory guidance for Congress and does not have the force of law; however a point of order can be raised on the floor by any member of Congress when a bill exceeds its BR limit (rules differ between the House and Senate with the Senate tending to enforce this order far more that the House of Representatives). In such a case, a 60% supermajority is required to defeat the point of order. In effect, once set, BR targets become substantial guides for the budget process. This provides discipline to the budgetary process. The appropriators generally have to meet the BR targets with each spending bill, although this as not typical practice between 2001 and 2007 and in some years during this period no BR was passed by Congress.

During discussion and amendment on the floor, a point of order also may be raised against any amendment that would send a particular bill over the target. While the point of order can be voted down, the 60% supermajority of votes necessary to do so is not easy to gain. All things considered, the BR functions as a useful starting pointing and continuing reference for score keeping.

Notwithstanding its importance, the BR has been subjected to the same turbulence as the rest of the budgetary process. It has usually been late, and by a significant margin. It was on time once in 1999 and once it was early as the Democrats rushed to get out ahead of their President in 1993. In 1998 no BR was ever passed. In 1990 a "deeming" resolution was passed in the spring when the two chambers could not agree and an official BR was passed in the fall. The BR is not a law but rather a compact by which Congress agrees to limit its spending tendencies by providing freedom within the amounts provided, but sanctions when they are exceeded. As noted, turbulence over the BR continued during the 2000s, often evidencing the differences in party preferences on major policy issues including defense. When the BR was not passed this still indicated the extent to which Congress had taken the BR process seriously. In fact, during the years it was not passed differences between the preferences of the two political parties were most extreme and irresolvable in the BR.

With all this turbulence, the temptation is to denigrate the importance of the BR. This would perhaps be premature. The BRs of 1995 and 1997 had great symbolic value as they set out pathways to a balanced budget by 2002. In 1997 particularly, congressional Republicans resistance to developing its own pathway toward a balanced budget forced President Clinton to make several changes in his budget estimates to reach this goal. In 1993, congressional Democrats thought that they were making an important statement about their readiness to do budget business in a new era. In 1998, the BR was never completed because some participants thought the guidance provided by the 1997 Resolution was good enough and because the House and Senate Republicans could not agree on policy directions. Still, this was a bitter and prolonged battle and indicates that the BR meant something to the participants, even if they could not agree on the numbers. Thus, it can be argued that the BR emerged from the period from the early 1990s through 2008 as an important and useful budget mechanism to sometimes mediate conflicting political policy preferences.

Two comments seem relevant. First, progress is not linear. While incrementalism may indicate that budgets change by a relatively small percent in respect to the previous year's base, the budget process itself is not predictable; what will lead to a good or bad budget year is not easy to discern. The fireworks around the BR do provide a hint, but only a hint of what is to come. Consensus and cleavage count. When consensus exists, the budget year is smooth; cleavage results in a disrupted budget process. Sometimes the cleavage is between the President and Congress, sometimes between the parties in Congress, sometimes between the two chambers and sometimes within one chamber. Second, when cleavages exist, surpluses matter, but not much. The year 1998 was a year of surplus; it was also a disaster of a year for the budget process.

Table 6.8 makes it clear that passing the BR on or before the April 15 deadline is difficult; Congress did so only twice during the decade of the 1990s. Not depicted is that Congress failed to pass the BR on time during the period 2001 to 2007. Excluding calendar years (CYs) 1990 and 1998 when Congress failed to agree on a BR, and subsequently when it failed to pass a resolution during some years in the 2000s, the BR was passed late, often in July.

Delays in establishing spending levels in the BR would seem to inevitably lead to late appropriation bills. However, late defense appropriation bills did not always result from late or no BRs and late defense appropriation bills have followed in 2 budget years when the BR was approved on time, in CYs 1993 and 1999.

It is important to remember that the BR is constructed by budget function, and the number it carries is a guide or limit to total spending for defense, the 050 function in the President's budget. Thus, the BR does not directly correspond to the defense appropriation bill. For FY 2002,

Table 6.8.   Budget Resolution:
Adoption Dates Compared to April 15 Deadline

Calendar Year	Adoption Date	Number of Days After Deadline
1990	None	DNA
1991	22 May	36
1992	21 May	35
1993	01 April	−15
1994	12 May	26
1995	29 June	74
1996	13 June	58*
1997	5 June	50
1998	None	DNA
1999	15 April	0

* Submitted May 14, 1996 following late approval of FY 1996 appropriations.

Source: Buell, R. C., 2002, p. 55.

the defense function was actually funded by 8 different appropriation bills, notwithstanding that almost all of it (99.9%) coming from the defense appropriation bill (92.8%), the military construction bill (2.9%) and the Department of Energy bill (4.2% for defense related atomic energy purposes). Nonetheless, the BR allows us to see how much spending is planned for the total defense function. This may be compared to the display by function in the federal budget to provide a rough judgment about how good a guide for the budget process the BR is. We have done this for the 1990s and find that agreement existed between the President's budget and the congressional BR in 1993 and 1996.

This was basically a matter of party in both cases. In 1993, the Democrats controlled Congress and the Presidency and wanted to usher in the Clinton era by illustrating how well they could govern, hence the agreement between the President's budget and the BR. In 1996, both parties were exhausted by the 1995 debacle, mainly a struggle between the Gingrich-led Republican majority in the House and the President, and just wanted to get to the next election in the fall. This represented agreement by default. Without going into the details of the battles that have occurred annually since then, it may be observed that generally a similar pattern has characterized the years from this time through 2007.

The BR needs to be further interpreted before it is a good guide for the appropriating process. This must be done by the appropriations committees in the House and the Senate. These committees are given a sum of discretionary BA by the BR and it is their job to divide it among the 13 appropriation bills. The individual bill targets are called 302b targets. The 302a amount is the total amount provided in the BR to the appropriations committees. Table 6.10 is a press release issued by the Senate Budget Committee, which shows the appropriation bill targets by chamber, the FY 2003 appropriation enacted and the FY 2004 President's budget request. At the end of the process, the appropriation bill should be on or under the target. While the trend seems to be toward more frequent and more public disclosure of data like this, it is not always publicly accessible on a timely basis.

By 1999, it became obvious that the information technology boom was going to produce a budget surplus much faster than anyone had predicted and Congress again chafed under the "unrealistic spending caps set in 1997" (CQA, 1990-1999, pp. 2-3), and again disagreed with President Clinton on several policy areas. Eventually eight appropriations bills would be passed separately, four of these on the last day of the FY. President Clinton vetoed four of the other bills and threatened to veto a fifth. After eight CRAs, the final five appropriations were wrapped in an omnibus legislation package and passed on November 29th. As might be expected, this bill included a series of gimmicks including a shifted payday for federal employees, and exceeded the budget caps by $31 billion (pp. 2-3).

Budget process problems continued in 2000. The year began with a BR that was timely (passed April 13, 2000), but the year soon became

Table 6.9. Budget Resolution Funding Levels in the 1990s

Calendar Year	Pres. Budget	Budget Res.	Authorization
1990	306.0	None	288.3
1991	291.0	290.8	291.0
1992	280.5	277.4	274.3
1993	263.4	263.4	261.0
1994	263.7	263.8	263.8
1995	257.8	264.7	265.3
1996	254.4	254.4	266
1997	265.3	269.0	268.3
1998	271.8	None	271.5
1999	280.8	288.8	288.8

Source: Congressional Quarterly Almanac, 1990-1999.

Table 6.10.   FY 2004 302b Appropriations Targets

FY 2004 302(b) Total Budget Discretionary[1] 302(b) Allocations
(Budget Authority in Millions of Dollars

	FY 2003 Enacted[2]	FY 2004 Pres. Request	House 302(b) Allocations[3]	Senate 302(b) Allocations[3]
Agriculture	18,096	16,981	17,005	17,005
Commerce, Justice, State	39,201	37,673	37,914	36,989
Defense	364,243	371,819	368,662	368,637
District of Columbia	509	421	466	545
Energy and Water Department	25,856	26,801	27,080	27,313
Foreign Operations	16,227	18,889	17,120	18,093
Homeland Security	21,267	27,482	29,411	28,521
Interior	19,463	19,555	19,627	19,627
Labor, HHS, Education	132,069	137,558	138,046	137.601
Legislative	3,343	3,804	3,512	3,612
Military Construction	10,546	9,115	9,196	9,196
Transportation, Treasury	27,259	27,462	27,502	27,502
VA, HUD, Independent Agencies	86,717	89,481	90,034	90,034
Total	765,796	787,071	785,565	784,675

1. Total discretionary is the sum of general purpose, mass transit, and highway categories.
2. FY 2003 enacted excludes the FY 2003 emergency wartime supplemental.
3. House and Senate 302(b) allocations columns reflect Budget Authority (BA).

Source: U.S. Senate Committee on Appropriations (2003), FY 2004 302b allocations.

engulfed in a hotly contested presidential election that ended up bitterly disputed. The budget process reflected the growing separation of the country into red (voted Republican for President) and blue (voted Democrat) states. For the budget process, enough of a consensus on defense held that Congress and the President could pass the regular DoD and Military Construction bills in a timely manner. The President signed them on July 13th and August 9th, 2000, respectively. This was the high-water mark for the year. No other appropriations bills passed on time.

Normally, Congress and the President hurry to pass appropriations bills in an election year. The election of 2000 gave control of the Presidency, the House and the Senate to the Republicans. The budget was in a surplus position and until September 11, 2001, the biggest policy issue was the size of the tax cut. The terrorist attack of 2001 did not wreck the budget process or destroy budgetary consensus; these were already in a state of some disrepair, as we have seen. In fact it might be suggested that the budget process after the terrorist attack showed some reversion to order as all 13 appropriations bills for FY 2002 were passed separately, if late. The passage of the supplemental in response to the September 11, 2001 attack in 7 days in September of 2001 was the first indication that U.S. politicians would pull together in time of crisis. Then, beginning in November, all 13 appropriations bills would be passed in a little over 2 months, in a period which included two major holidays and the aftermath of the attacks on the World Trade Center and the Pentagon. Eight bills were passed in November with the earliest being interior and military construction on November 5; two were passed in December, and three on January 10, 2002. The fall of 2001 was definitely not a time of business as usual for the legislative process and passage of the regular appropriations bills by the end of January cannot be criticized.[1] The consensus which enabled this basically would hold for defense in the coming years, but not for other policy areas.

However in 2002, the budget process ground to a halt and the routines of improvisation returned. The defense and MILCON bills were passed on October 23, 2002, but these would be the only separate appropriations bills completed that calendar year. After eight CRAs,[2] the other 11 appropriations bills were bundled into a consolidated[3] appropriations bill and passed on February 20th, 2003. The first wartime supplemental was passed on April 16, 2003.

By 2002 the new consensus was clear, and generally speaking this consensus was sustained through 2007 when it dissipated over the content of the FY 2008 budget. It meant funding defense first and disagreeing on the rest. Defense was the first appropriations bill to be passed in 2002, 2003, and 2004. It passed on time or nearly on time from 2005 to 2007. In 2003 and 2004, the defense bill was passed before the FY had begun, while the majority of the rest of the appropriations bills were not passed at all. In contrast, in the 1990s, the DoD appropriation was passed on time only twice in 10 years, and then only barely, on the 30th of September in 1994 and 1996 (McCaffery & Jones, 2004, p. 104). The use of consolidated bills for nondefense matters indicated an inability of policy makers to develop policy consensus in those areas. Thus, proposals from Congress for omnibus appropriations for FY 2008 were not greeted warmly by President Bush who vowed to veto them.

Taken altogether, changes in the annual congressional budget process have been striking. First, the BR began to operate on a hit or miss basis. The congressional Budget process, as reformed in 1974 and afterwards,[4] provides agents and a process to set overall goals and targets. In addition, the BR sets sectoral allotments (defense, education etc.) and often carries sense of Congress provisions (e.g., that Congress ought to pursue a balanced budget) and stipulations about process (paygo rules, firewalls).[5] However it has become a highly uncertain event. In 2002, 2004, and 2006 the House and Senate could not agree on a BR. In 2003 and 2005 they did. In 2002, the Senate failed to complete a BR for the first time since 1974.[6] In 2004, the Senate again could not bring itself to agree with the more conservative House budget policy outlines, although it did produce and pass its own BR (Heniff, 2005, p. 5). In 2006, the Senate attached a "deeming" resolution to the spring defense supplemental for use as its BR.

Some critics might argue that since BRs do not get passed every year, they must not be very important. This does not seem to be case. Had it not failed in recent years, the case could be made that the BR has become increasingly important.

Beyond the BR there is also some other process improvisation. First, some of the usual budget process controls including caps on spending by appropriation enacted in the 1990s and pay-as-you-go financing for entitlements elapsed in the 2000s (PAYGO reemerged in 2007 by congressional rule rather than law) and with the more tentative nature of the BR spending discipline could not be reinstituted in Congress through 2008. The Paygo provisions and spending caps were self-disciplining mechanisms that came into the budget process with the Budget Enforcement Act of 1990 and gave discipline to the congressional budget process during the 1990s. Thus, overall, since 2000 not only did the process become more uncertain, it became less disciplined. Although efforts have been made to reinstitute some of these self-control congressional provisions, the jury is still out as of early 2008.

The House made efforts to reform the committee structure so that by 2005 the two chambers were operating with a slightly different committee structures. The Senate went from 13 appropriations subcommittees and 12 bills while the House went to 10 subcommittees and 11 bills. Of the 12 Senate bills, six have corresponding bills in the House with identical jurisdictions; the rest do not. How this process will work out is unknown as this is written, but patterns of the recent past indicate that Congress is adept at improvising, with consolidated appropriation bills or by attaching what used to be a separate bill to the conference committee report for another bill at the conference committee vote stage. What this asymmetrical structure does is make it hard to compare bills in the House and Senate, thus further decreasing the transparency of the budget process.

Another change which has had to be absorbed has been the creation of a homeland security department and appropriations bill affected the structure of Congress. At first, everyone wanted a part of homeland security and the White House identified 88 committees and subcommittees of Congress that exercised some authority over homeland security policy in 2002; in the House this included 10 of 13 appropriations subcommittees. This was rationalized somewhat in the appropriations process with a homeland security appropriations bill and homeland security subcommittees for the HAC and SAC. Opposing forces have also been in play; the House for example utilized 10 appropriations subcommittees in 2005 while the Senate has changed to a 13 bill-12 subcommittee structure. All of these changes have impacted the appropriations process and have had to be digested.

TURBULENCE IN AUTHORIZATION AND APPROPRIATION COMMITTEES

Another way to analyze how Congress provides coordination and control for its disparate elements is to examine the timing between passage of the authorization and appropriation bills. Traditionally, the authorization bill, at least in the eyes of the authorizers, provides critical guidance for the appropriators, particularly on the macroitems in the budget, for example, how big should the military be, how many carriers should the Navy have, what is the right balance between strategic and tactical air, does the pay and benefits system sufficiently reward uniformed members of the services, and is the retirement system adequate?

When the authorization bill is passed later than the appropriation bill, the timing of these guidances is impaired. This happened 6 out of 10 years in the 1990s. The defense authorization was properly passed before the appropriation only twice during this decade, during CY 1996 and 1999. In 1990 and 1998, the authorization and appropriation bills were passed on the same day. This simultaneous timing allows the authorization bill to provide timely guidance for the appropriators on defense policy questions. Nonetheless, it appears that the authorizers cannot be depended on to provide timely guidance for the appropriators. To some extent this failure occurs because the authorizers also see themselves as giving the President guidance over treaties, strategic deterrence, and peacekeeping activities, notification of Congress when American troops are about to be committed and so on. In some years the authorization bill stalls due to issues with the President, thus sometimes the authorizers are caught between an internal guidance role to the appropriators and an external guidance role to the President. Their average outcome is hard to interpret, but it should be remembered that guidance on a thorny issue successfully given 1 year might last for many years.

However, even when the authorization bill is late for a good reason, this is still a source of turbulence for the appropriators who might look to the authorizers for guidance as well as for the authorizers who want the appropriators to pay attention to their guidance. All in all, the turbulence that Schick found in the 1980s continued and worsened in the 1990s and 2000s, which is not shown in Table 6.11

TURBULENCE IN THE INABILITY TO
BUDGET RESULTING IN THE CRA

One of the most telling indicators of legislative turbulence from the 1990s to 2008 was need for frequent usage of CRAs. When Congress fails to pass an appropriation bill before the FY begins, it must enact a CRA to fund agencies until the full appropriation is passed. The funding level may be set at last year's rate or any rate that Congress desires. These are usually passed quickly and are not controversial. They usually preclude new program starts and new initiatives. In defense this sometimes penalizes procurement programs that get off to a late start and thus may run into cost overruns due to delays in production or contract execution.

In every year except 1994 and 1996, CRA were necessary. From 1990 through 1999, 39 CRAs were passed, covering 524 days or 14% of the days of the decade. In 1995, CRAs were necessary for 219 days, into April of 1996. No CRAs were needed for the 1994 and 1996 budget processes. However, CRAs were passed from January to April in 1996 to conclude the 1995 budget process. The budget process did not get incrementally worse or better, rather very good years followed very bad years and the

Table 6.11.   Defense Authorization and Appropriation Timeline Comparison

Calendar Year	DoD Author	DoD Approp.	Auth. On Time?	Approp. On Time?	Auth. Before Approp?
1990	5 Nov	5 Nov	N	N	Same
1991	5 Dec	26 Nov	N	N	N
1992	23 Oct	6 Oct	N	N	N
1993	30 Nov	11 Nov	N	N	N
1994	5 Oct	30 Sep	N	Y	N
1995	10 Feb 1996	1 Dec	N	N	N
1996	23 Sep	30 Sep	Y	Y	Y
1997	18 Nov	8 Oct	N	N	N
1998	17 Oct	17 Oct	N	N	Same
1999	5 Oct	25 Oct	N	N	Y

Source: Buell, R. C., 2002, p. 52.

Table 6.12.   Continuing Resolution Appropriations in the 1990s

Calendar Year	# CRA(s)	Range of Days Covered by CRA(s)	Total # Days Covered
1990	5	3 to 10	35
1991	2	28	65
1992	1	22	22
1993	3	8 to 21	60
1994	0	0	0
1995	10	1 to 67	219
1996	0	0	0
1997	6	1 to 23	48
1998	5	2 to 9	16
1999	7	1 to 21	59
Total	39		524 days or 14.35% of the decade

Source: Buell, 2002, p. 53.

period of 1993 through 1997 was marked by oscillation from bad to good and back. This was the reflection of a bitter policy fight between the President and Congress on a number of fronts, not the least of which was defense. The need for CRAs for defense was largely absent in the 2000s until the FY 2008 budget was considered in 2007, and a series of CRAs were passed due to pre-election year politics.

Process Inversions

The events described above characterize an appropriations process under stress, one that may be said to be in some sense "inverted." Unusual events are occurring and usual events are happening out of the usual order. This has been manifested in several ways. First the DoD appropriations bill in the 1990s was one of the last bills to be taken up on the floor; it was large and Congress wanted to give it due deliberation; the process usually began with smaller bills like that for District of Columbia or legislative affairs. After 2001, the DoD appropriations bill went first, and sometimes it was almost the only bill passed. Moreover, this did not mean the authorization bill was accorded the same level of precedence.

The task of the authorizing committees is to build programs and authorize changes, particularly in personnel levels and benefits and in procurement. Authorizers like to think of themselves as giving guidance to the appropriations committees, therefore a normal or "good" process should exhibit a pattern where the authorization bill precedes or parallels the appropriations bill. This happened in calendar 2001, after the 9/11 terrorist attack. In subsequent years the authorization bill has been dramatically behind the appropriations bill. For comparison, in the 1990s the authorization bill preceded the appropriations bill twice, was signed on the same day twice, and was within 10 days twice and 20 days twice in 10 years; thus 80% of the time in the 1990s, the authorization bill could be said to be close enough to the passage of the appropriations bill to provide guidance (McCaffery & Jones, 2004, p. 164). Since 2000, only in calendar 2001 was this same profile evident; this was a year when Congress wanted to make a statement of business-as-usual.

A second sign of inversion of the budget process is that the DoD appropriations bill has been passed on time in 2003 and 2004; it was passed on time twice from 1990 through 1999. This is an inversion in Congress' normal way of doing business.[7] A third sign of inversion is the failure to pass the rest of the appropriations bills; these are considered must pass legislation because the government must be funded, but in this era after 2001 the majority of appropriations bills have been passed as consolidated appropriations bills, used because Congress could not resolve policy conflicts, despite the consensus on defense. Later we argue that the use of consolidated bills and turmoil around the BR have resulted in a diminished budget process and less efficient spending outcomes as the prevalence of earmarks has increased and aggregated to a level where important sums of money are allocated without a useful priority setting

Table 6.13.   Defense Appropriations Precede Authorization

Fiscal Year	Authorization Conference Rpt	Appropriations Conference Rpt	Days Behind
2001	10/12/2000	7/27/2000	77
2002	12/13/2001	12/20/2001	−7
2003	11/13/2002	10/16/2002	28
2004	11/12/2003	9/25/2003	48
2005	10/9/2004	7/22/2004	79
2006	12/21/2005	12/21/2005	0
2007	9/30/2006	9/26/2006	4

Source: Linwood B. Carter and Thomas Coipuram Jr. Defense Authorization and Appropriations Bills: A Chronology, FY 1970-FY 2006. CRS Report for Congress, 98-756C May 23, 2005. Washington, DC. FY 2006 and 2007 data from Library of Congress Web site, Thomas.gov

process. We argue that this new budget era may have led to significant opportunity costs. A fourth sign of inversion concerns the dramatic changes in defense appropriation supplementals. These used to be small; now they are large. Moreover they are now provided inside and outside of the appropriations bill. When inside the DoD appropriations bill, they are designated emergency funds and not necessarily called supplementals in scoring, but DoD spends them the same way. Also supplementals have been supplied for defense and homeland security issues (i.e., these portions are spent outside of DoD) and some defense appropriations bills have also carried appropriations for other agencies. In 2004, Congress included in the regular appropriations bill $25 billion in emergency funds for early FY 2005 costs of operations in Iraq and Afghanistan and another $1.3 billion for such activities as wildfire firefighting in the Western states (mostly California), humanitarian assistance for the Sudan, Iraq embassy security and conventional security funds for Boston and New York. In the 1990s these would probably have been done in a separate supplemental bill. Also the $40 billion 9/11 supplemental was passed swiftly, but spent out slowly and only a small portion of this went to DoD ($14 billion in 2001 and $3.86 billion in 2002 (Carter & Coipuram, 2005, pp. 24-25), although it was all intended for national defense and homeland security. Another complication involves the contingent emergency designation. In 2002, Congress passed the summer supplemental for $28.9 billion, but designated $5.1 billion as contingent emergency spending, leaving it to the President to accept some, all, or none of the $5.1 billion. Within two weeks of signing the original bill, the President turned down all the $5.1 billion in contingent emergency funds (CQA, 2002, pp. 2-40).

A fifth inversion occurred in 2003, when a supplemental was tapped to provide money for the main defense appropriations bill. The April 2003 supplemental created an Iraqi Freedom Fund of about $16 billion. Meanwhile Republican appropriators shifted about $3 billion from the Bush defense budget request to domestic discretionary spending, but this left the defense bill short and later in the summer appropriators shifted $3.5 billion out of the IFF to fill out the defense appropriations bill. The result was that the funding meant for reconstruction of Iraq in the wartime supplemental ended up increasing domestic discretionary accounts and ongoing operations within DoD normally funded by the main appropriations bill.

Also in 2003, there were two supplementals, one in the spring and the other in the fall. The first paid for ongoing operations; the second appears to have contained funds for future operations. The result of all of this turbulence makes it hard to track supplemental and emergency funds and the boundary between normal DoD operations and emergency and wartime operations (normal appropriations and emergency and supplemental appropriations) has become blurred as has the boundary between defense appropriations and total spending for defense, homeland security, and foreign aid and assistance efforts in support of the war on terrorism. And finally, some things happened out of sequence. In passing the consolidated appropriation bill in February 2003 to clean up the 2002 appropriation process, Congress provided DoD with $10 billion in emergency funds mostly for intelligence matters; this normally would have been included in a supplemental bill, for example the one which would be passed in April of that year. In addition, the consolidated bill in 2003 also carried mandatory funds (mainly for the health and human services accounts) of $397.3 billion. Since this included a small increase in the mandatory programs, the consolidated bill also had to pay for it with a Paygo provision cut of .0065 across the board for most discretionary accounts, including those funded in the consolidated bill and in the previously passed appropriation bills. All of this illustrates a period of great stress for the budget process. In Table 6.14 below we have mapped these years of turbulence.

Over the 9 years, BRs were passed only 5 times. Congress had the opportunity to pass 91 appropriations bills on time; only 10 were produced on time. Forty-three bills were wrapped up in 6 consolidated appropriations bills. While the fall 2001 process was understandably late, the process in 2000 and 2002-2003 were not significantly different. The failure to pass annual BRs and the increasing failure to coordinate the defense authorization and appropriations bills indicates a lessening of Congressional policy control mechanisms.

While the terrorist attack of 2001 had a great impact, the appropriations process was already in some disarray. This does not mean the policy system was "frozen." For example, in 2000, the consolidated appropriation bill allowed President Clinton some victories despite strong Republican opposition; these included a tax break for distressed urban and rural communities and a loosening of immigration rules. It also allowed a multitude of earmarks and resulted in spending that exceeded the spending caps set 3 years earlier and the BR targets set that year. The consolidated bill also allowed special interest policy provisions; Senator Stephens was able to gain a delay and review of a Department of Commerce order to limit the pollock and cod fishery off Alaska to protect the endangered Steller sea lion.

If anything, consolidated or omnibus bills allow too much policy to be made in too short a time and with insufficient public scrutiny. On a more positive note, dysfunctions in the budget process did not mean that the policy making apparatus ground to a halt. Judging by the budget process, 2003 looked like a bad year, but the Republicans, in control of the Presidency and both houses of Congress for the first time in 50 years, were able

Table 6.14. Budget Process Chaos

A	B	C	D	E	F	G	H	I	J	K
Calendar Year	CBR Passed	Approp. Bills on Time	Late Approp. Bills	Late and in Consol. App. Bill	Total # of Bills Passed	Date of Last Bill	Days Late	Vetoes	CRAs	DoD Au > App.
1998	no	1	4	8	6	10/17/1998	17	0	5	yes
1999	yes	4	4	5	9	11/29/1999	48	4	8	yes
2000	yes	2	8	3	11	12/21/2000	82	3	21	no
2001	yes	0	13	0	13	1/10/2002	102	0	8	yes
2002	no	0	2	11	3	2/20/2003	143	0	12	no
2003	yes	2	4	7	7	1/23/2004	115	0	7	no
2004	no	1	3	9	5	12/8/2004	69	0	3	no
2005	yes	2	9	0	11	12/30/2005	91	0	3	no
2006	no	1	1	10	3	2/14/2007	171	0	4*	yes

A = calendar year
B = did budget resolution pass?
C = number of appropriation bills passed on time.
D = number of appropriation bills passed late.
E = number of appropriation bills passed in consolidated bill
F = total number of appropriating bills passed (on time + late + consolidated bill)
G = date of last bill
H = how many days after the start of current fiscal year
I = number of appropriation bills vetoed
J = number of Continuing Resolution Appropriations needed to fund missing appropriations
K = for DoD, did authorization precede appropriation.
* = the CRA passed on 2/14 included the funding for the remainder of the fiscal year for ten appropriation bills. Only Defense and Homeland Security passed as normal appropriation bills. Only defense was timely (9/29/2006).

Source: CQA and Library of Congress Web site for calendar years 1999-2006.

to pass significant pieces of legislation, including the second major tax cut in 2 years, and the biggest overhaul of Medicare since its creation, the passage of a "wartime" supplemental for Iraq in April, in 3 weeks, as well as initiatives to fight AIDS, and "do not call" and "do not spam" laws. What the budget struggle does indicate is that even when solidly in control, the Republicans could not get all they wanted. After the elections of 2006, the Democrats seized control of Congress with an ambitious agenda to end the war in Iraq against the inclination of a stubborn President. Irrespective of how this comes out, what is predictable is an even more turbulent budget process.

PORK IN THE BUDGET WARS

Each year the appropriation bills fund projects of dubious merit, inserted at the request of lobbyists, state and local governments, and various contractors who profit from federal spending. These projects tend to be location-specific and highly visible. In the nondefense area, they take the form of courthouses, highways, airports and the like and are traditionally referred to as "pork." Mechanically, a project is inserted in an appropriation bill as an adjustment to a specific line-item; hence, it may also be called an "earmark." Pork, according to the Webster's dictionary, is a government appropriation that provides funds for local improvements "designed to ingratiate legislators with their constituents" (Utt, 1999, p. 1). Some realists argue that pork is a characteristic of democracy in action, that pork projects are part of the price of passing legislation; votes for a bill are bought by inserting items of benefit to particular members or groups of members so that they will support the bill. In this way, the interests of the many are protected by making a side-payment to the few in order to provide the necessary votes to pass the more general measure.

Utt and Summers (2002) observe that pork, "is really all about electoral insecurity and efforts to buy friendship and public affection with the hard-earned dollars of American taxpayers" (p. 1). They worry that pork has become the province of expensive lobbyists who persuade state and local governments to hire them to get projects inserted in federal budgets, but the projects that get funded may not match state and local priorities, thus the lobbyists set themselves up as knowing more about local preferences than either the state-local or federal legislators and distort the priorities of all by pursuing whatever money is available. Utt (1999) argues that proof of these distortions may be found in transportation funding in FY 1999 budget when half of the highway earmarks went unfunded because local governments refused to put up their matching share of the money.

Pork is easily recognized. In FY 2003 federal funds were spent on Hawaiian Monk seals ($82,500); waste management in North Carolina ($489,000); Alaskan groundfish surveys ($661,000); hoop barns in Iowa ($225,000); and the Walla Walla basin habitat ($750,000). In FY 2002, money was spent to help Blue Springs, Mo combat teenage Goth culture ($273,000); provide a statue of the God Vulcan in Birmingham, Alabama, ($1.5 million); fund a tattoo removal parlor in San Luis Obispo County, California ($50,000); and to study how thoroughly Americans rinse their dishes ($26,000) (Riedl, 2003).

Defense also has it share of pork barrel projects inserted into its appropriations, almost always against the will of DoD. Sometimes these items are related directly to defense, sometimes they are not. Further, it is important to recognize that much pork spending comes at the request of DoD and contractors. And significant amounts of pork funding takes place outside of defense. Schlecht (2002) found these bits of nondefense pork spend in the FY 2002 defense appropriation:

Examples of Pork in FY 2002

$1.3 billion for environmental restoration.
$7 million for HIV prevention in Africa.
$150 million for the Army breast-cancer research program.
$85 million for the Army prostate-cancer research program.
$19 million for international sporting competitions.
$3 million for aid for children with disabilities.
$1 million for math-teacher leadership.
$4.5 million for a cancer research center.
$2.6 million for the Pacific Rim Corrosion Project.
$2 million for the Center for Geo-Sciences.

Schlecht (2002) concedes that some of these programs may be worthy of funding, but that none of them can be said to be directly related to the defense of, "our country or the safety of our soldiers on the modern battlefield," and finds that, "Such pork-barrel spending wastes billions of taxpayer dollars and undermines the combat effectiveness of our military" (p. 2). As a percent of a $2 trillion budget, the amounts cited above may seem trifling, but these dollars accumulate into the billions of dollars, as we will explore later, dollars misused or applied to priorities that do not serve well all or even large numbers of the people of country.

One of the watchdogs of wasteful spending is the Washington, D.C. based institute, Citizens Against Government Waste (CAGW). This group was founded in 1984 by Peter J. Grace, Chairman of President Carter's Commission on Cost Control (Grace Commission), and by newspaper columnist Jack Anderson. CAGW has grown to a membership of more than one million. CAGW refers to itself a public interest lobby against government waste. Every year it issues a summary of pork projects called the "Pigbook." This document may be accessed online from 1996 forward. For FY 2003, CAGW reported as follows:

- Appropriators stuck 9,362 projects in the 13 appropriations bills, an increase of 12% over the previous year's total of 8,341.
- In FY 2002 and 2003, the total number of projects increased 48%.
- The cost of these projects in fiscal 2003 increased 12% to $22.5 billion.
- The total cost of pork increased by 22% since fiscal 2001.
- Total pork identified by CAGW since 1991 added up to $162 billion (Pigbook, 2003).

CAGW found that Alaska led the nation with $611 per capita ($393 million), or 18 times the national pork average of $34, with the runners-up being Hawaii at $283 per capita ($353 million) and the District of Columbia with $262 per capita ($149 million). CAGW observed that the common thread in the top two states is that they are represented by powerful senators and appropriators, then Senate Appropriations Committee ranking member Ted Stevens (R-AK), and former Senate Appropriations Committee Chairman Daniel Inouye (D-HI). Senator Stevens has generally been regarded as the "king of pork" by Congress-watchers, a title he inherited from Senator Robert Bird (D-VA) who also reveled in provision of pork for his state and others for decades.

In addition to analyzing by state and by member, CAGW provides a summary for each policy area. For defense, CAGW found that total pork increased 25% in FY 2003 over FY 2002, from $8.8 billion to $11 billion. The number of projects also increased 22% from 1,404 to 1,711 (Pigbook, 2003). Undoubtedly, these projects are of value to someone (sometimes only to the contractor and the labor that builds them), and perhaps of some value to all. The question that must be answered lies in opportunity cost, that is, in lost value the projects displaced and not funded or the value of leaving the tax dollars in the hands of taxpayers. It is also clear that the totals are of some consequence.

Definitions of Pork

CAGW takes a process approach to defining pork: "all of the items in the *Congressional Pig Book Summary* meet at least one of CAGW's seven criteria, but most satisfy at least two:

- Requested by only one chamber of Congress;
- Not specifically authorized;
- Not competitively awarded;
- Not requested by the President;
- Greatly exceeds the President's budget request or the previous year's funding;
- Not the subject of congressional hearings; or
- Serves only a local or special interest (Pigbook, 2003).

For CAGW, process is important. A project that is requested by the President, subjected to congressional hearings, is specifically authorized and appropriated, serves a general interest and is competitively awarded might avoid the label of pork.

Our observation is that the definition of pork depends on the principle of "where you stand depends on where you sit." In other words, pork is a two-sided coin. One side reads "unnecessary and wasteful spending" while the other side reads "local economic development."

Pork Funding

Where does the money come from for these pork projects? Ultimately, of course, it comes from the taxpayer, but in the immediate case Congress finds ways to provide the money, sometimes out of thin air. A study by Wheeler (2002) of the FY 2002 military construction bill found that the military construction subcommittees of the HAC and SAC used two gimmicks to provide dollars for such projects as museums, gyms, warehouses, water towers, day care centers, and so on. First, they assumed that the value of the dollar would rise against foreign currencies, thus making projects in foreign nations cheaper, by about $60 million. Second, they wrote into the MILCON bill a 1.127% across the board cost reduction in all projects in the bill. This appeared to save $140 million and provide space for additional projects while keeping the bill under the $10.5 billion cap allocated the House and Senate Appropriations Committee chairmen for the bill. Wheeler found that California was the number one beneficiary of these tactics, gaining $25 million for the 8 projects added that were not requested by the President, followed by Texas and West Virginia. Wheeler notes:

> the top eleven recipients of added military construction projects consisted entirely of senior Democrats and Republicans in the House or Senate who just happened to be the sitting Chairmen or Ranking Minority Members of the Appropriations and Armed Services committees and subcommittees

who handled the Department of Defense and its military construction budgets. (Wheeler, 2002, p. 104)

Writing in the fall and winter of 2001 and in the aftermath of the 9-11 terrorist attack, Wheeler was concerned that spending on pork would divert funds from reaching necessary defense items. He also studied the defense appropriation bill for FY 2002 and found Congress using gimmicks to reduce O&M funding in order to make space for projects they favored. For example, Wheeler cites the following sections inserted into the 2002 appropriation bill as being largely phony, that Congress assumes will occur or tries to force to occur, but that the appropriation committee members are "acutely aware" are not going to happen, since they have been told they would not happen by DoD and OMB. The provisions were:

- Section 8095 in the same general provisions title of the bill reduces O&M by $240 million to "reflect savings from favorable foreign currency fluctuations."
- Section 8102 takes out $262 million to restrict travel of DoD personnel.
- Section 8135 extracts $105 million "to reflect fact of life changes in utilities costs;"
- Section 8146 takes $100 million to improve scrutiny and supervision in the use of government credit cards.
- Section 8123 is the granddaddy; it reduces O&M by $1.650 billion for "business practice reforms, management efficiencies, and procurement of administrative and management support."

Wheeler adds,

these mandated reductions involve unrealistic assumptions about DoD's ability (and willingness) to adopt management reforms within the current fiscal year; they make unjustified economic assumptions regarding foreign currency exchange rates; they assume that in a time of war DoD is going to reduce travel costs, and they pretend that defense contractors can and will adopt efficient business practices despite decades of obstruction to full and open competition. (Wheeler, 2002, p. 19)

On the accuracy of reduction projections OMB warnings were eloquent: First, on November 28, OMB officials wrote:

these reductions were based on unrealistic assumptions about achievable FY 2002 savings—primarily from reductions in consultant services, headquarters staff, and A-76 studies. The real effect of the House's deep O&M reductions would be to undercut the President's plan to address readiness shortfalls and competitive sourcing, and reduce funds available for military operations and support. (OMB, 2001a, p. 4)

Then on December 6, OMB noted:

> The Committee has made reductions to Operations and Maintenance programs, based on unrealistic assumptions of how much savings could be achieved through reductions in consultant services, foreign currency fluctuation account balances, and travel. These reductions would undermine DoD's ability to adequately fund training, operations, maintenance, supplies, and other essentials. They would seriously damage the readiness of our armed forces and undermine their ability to execute current operations, including the war on terrorism. (OMB, 2001b, p. 3)

Congress exercised its Constitutional prerogatives and kept the provisions, and the projects, in the bill.

Pork barrel spending also manifests itself as overinvestment in certain weapons systems produced in the home state of a powerful committee chair, floor leader, or long-term defense authorization or appropriation committee member. Here, the pork involves overproduction of missiles, or aircraft or whatever. They have a clear defense benefit, but too many are bought; that is what makes them pork. Examples of pork during the 1990s are more abundant than could be economically documented here. Buell (2000) found that some players did stand out in the struggle for defense pork. For example, in 1994 Senator Byrd (D-WV), successfully added $21.5 million to beef up the National Guard C-130 squadrons in his state. This increase was not requested in the President's budget. Senator Inouye (R-HI) successfully added millions to the defense authorization and appropriation bills almost every year during the 1990s. Some of it was slated for assisting in the clean up and clearing of unexploded mines and ordnance from Kahoolawe, a small island that the Navy used as a bombing range for over 50 years. He also successfully added $250,000 for Hawaiian Cruise lines to ferry troops around the island chain in 1997 and $19 million for two reservist-manned amphibious transport ships to shuttle troops from Pearl Harbor to training grounds on the big island of Hawaii.

Former Senator Trent Lott (R-MS) was very successful at adding additional ships for construction at the shipyards in Pascagoula, including an extra Aegis Destroyer in 1995 and 1997, a helicopter carrier and amphibious assault craft in 1999 and an LHD-7 in 1995 and 1998. He was not the only member of Congress benefiting from the addition of ships. Rep. Livingston (R-LA) added $974 million for the LPD-17 to be built in his state in 1995 and Senator Cohen (R-ME) added $2.16 billion for two new Arleigh Burke destroyers to be built in Bath, Maine. In 1997, Rep. Newt

Gingrich (R-GA) added nine C-130J to the one requested by the Air Force, adding $529 million to the budget. Of the 36 C-130s built during the decade, only five had been requested by DoD in their budget (Buell, 2000).

Calendar year 1993 had some classic examples of pork, with the addition of ten new Apache helicopters for $273 million, not requested, and the initial order for a Wasp class helicopter carrier for $50 million, also not requested. The addition of high-speed cargo vessels for a new Army prepositioning force was increased from $291 million to $1.5 billion in a strategic shift for the Army to find new missions for themselves. Similarly in 1990 a billion dollars was added to the defense bill for the production of more M1 tanks after the Army requested zero funding for that program in their budget request.

Sometimes pork battles are not about the weapon system per se, but about in whose district they will be built, that is, who will be the beneficiary. For example, in 1995, an argument developed over who would get to build a new submarine. The contenders were the Seawolf submarine being built in Groton, Connecticut and the new Virginia class being built in Newport News, Virginia. Arguments also occurred over which programs to fund, the LPD-17 in Louisiana versus LHD-7 in Mississippi, or an extra Aegis Destroyer (DDG) in Massachusetts versus Mississippi.

Other types of pork emerge in the budgetary process as typified by the 1999 omnibus CRA, which was filled with additional funds for things such as satellite TV for rural areas, Canyon Ferry Reservoir funding and dairy policies for Midwestern states. In CY 1995, Congress added $10.4 million for a new physical training center at Bremerton, $99 million for a Navy berthing wharf at North Island (although ships could use the wharf at Long Beach) and $370 million to research the Russian nuclear demise. This year also had one of the more notable pork additions to defense bills with the inclusion of $16 million in support of the 1996 Olympic Games in Atlanta, Georgia.

How may pork be identified? In addition to the CAGW process tool, we have provided a method to think about pork outcomes, in terms of its cost, where it is spent, to whom the benefit accrues and to what extent the good provided is a public good and can not be provided by the private sector. Table 6.15. above suggests some analytic criteria, that is, if it is a lot of money for a local district where only the district benefits and the good is private in nature—it is high on the pork scale. The essential problem with pork is that it is funded by tax dollars drawn nationally and dedicated to projects of somewhat dubious merit or low priority where the benefit could be quite local. In some cases, the project is a substitute for something the private sector could do. Many defense functions are close to pure public goods and can only be supplied by government. Others are

Table 6.15.   The Pork Badness Scale

What does it cost? (A lot or not much?)

Where are dollars spent? (In one locality or nationally?

What is the benefit? (Local or national?)

What kind of good is purchased? Private goods (a private automobile) or public goods (Navy ship)?

not. Annual defense funding exhibits a mix of true public goods of high merit and some not so meritorious projects that support local interests.

Earmarks and Pork

Study of changes to the DoD appropriations bill indicates that the pork question is more complex than is normally thought for defense. Some projects clearly benefit local areas and are at the request of the local Congressman. Some clearly are of national interest, even if at the request of local representative. What is commonly called pork gets into DoD bills because Congress keeps an eye out for things that will benefit them and their re-election records; however DoD also campaigns for pork. The battle of the budget does not stop inside the Pentagon when the President's budget is printed. When the President's budget is presented to Congress, the battle for congressionally inserted pork has just begun, but the struggle within the Pentagon has been going on for some time, at least since the spring of the previous year with the beginning of the military department budget process and goes on through the congressional session as the services campaign to get things included in the appropriations bill right up to the vote on the conference committee report on the appropriations bill. DoD and the military services may appeal on provisions in the appropriation bill through conference report deliberations, if the treatment of an item in one chamber differs from that in the other. Then too, DoD testifies, responds, informs, and educates members of Congress all during the congressional budget process. Inside DoD, various unfunded priority lists are created and maintained and when it appears Congress is interested in funding something close to a unmet priority list (UPL) item, DoD representatives will attempt to get the item "torqued" toward the UPL list item and fully costed with language inserted that gives DoD some flexibility in how it administers the program. These ideas may be surfaced at various committee venues, conversations, and in responses to questions asked by Congress or, as politics in Washington go, at a variety of social settings. Thus, while Congress appears to take the blame for all these earmarks, it is clear that that they are not necessarily totally congressional; DoD may find a sponsor to carry a needed project.

Pork projects are not necessarily taken from secretary of defense's UPL, either. Pork may also result when a Congressman visits an obscure base and is given the routine command brief, a brief which just happens to include a need which may have been denied in the DoD budget process. This need will help the base and the local area, so it is no wonder it strikes the Congressman as a good idea, one which he may take up and pursue at the committee stage in hearings and markups. Thus, for a base commander who is knowledgeable, there are wrestle back provisions in the budget process. He may lose within DoD, but what is to prevent him from salting his commander's brief for a visiting Congressman with a "nice to do" project that did not make it past the next step upward in the administrative budget process? If the Senator or Congressman wants to take on an issue as a personal item and add it to the appropriations bill later, how is the base commander to protest? In the Pentagon, this item may not be anywhere close to an important priority and did not make it in the budget request, but if it appears that the Congressman has a good chance of getting it included in the appropriations bill, DoD will work to see that it is fully costed (if a new building, that maintenance and personnel costs are included) and that to the extent possible the item is changed to help include some item that is a DoD priority. From the DoD perspective, these earmark plus ups are of a lower priority than the items Congress is pushing out of the appropriations bill to make room, thus from DoD's perspective Congress is substituting inferior for superior goods. It is probable DoD will have to ask for some or most of these excluded items in the following appropriations cycle, or perhaps in the supplemental bill.

We have said nothing yet about defense corporations, yet it is clear from our earlier discussion of "iron triangles" that the defense corporate world is an interested and eager participant in this part of the policy process, especially in procurement and RDT&E account and in the development, fielding, production, maintenance and continued sales of new weapons systems. DoD represents, after all, where the big money is to be earned in doing business with the government. Here pork manifests itself when Congress holds production lines open when DoD has suggested it already has enough of a certain weapon system or when Congress inserts additional copies of an aircraft in the appropriation bill even though DoD has not asked for any in its request. Naturally the affected corporations and their subcontractors and suppliers are only too happy to let individual Senators and Congressmen know how important these systems are to the national defense effort. These efforts start when the budget is still in the military departments in the late summer, before the Program Objective Memorandum (POM) and budget have been finalized. For weapons systems, it is particularly important to win in the POM, since the POM controls the future year defense plan and may well be the difference between 5 fat years and a series of years so lean as to put the company out of business (or into someone else's hands). With stakes this high, it is a small wonder that any decisions are taken on merit. In our judgment, the majority of the DoD appropriation is presented and passed on merit, but there is a large and seemingly growing part of it that seems to respond to the rules of pork, inserted in appropriations bills through earmarks.

HOW CONGRESS ACCOMMODATES THE PRESIDENT'S BUDGET

Another place to study for policy inversions in the budget process is how Congress treats the President's defense budget request. Conventional wisdom suggests that Congress is a marginal modifier of presidential budget requests and will support wartime budget efforts of the President. The closest analogous historical period to the current era is the Vietnam era. Table 6.16 shows a record for that time of what the President requested and how he was treated in Congress.

The bills for FYs 1965 and 1974 were constructed and passed in a peacetime environment. Calendar 1965 marked the first war year; note the high levels of support in Congress for the war effort in 1966 and 1967. Thereafter, Congress began to make substantial cuts to the DoD bills. Conventional wisdom says that consensus around defense disappeared in the late 1960s; what this profile illustrates is that Congress modified the President's budget before Vietnam (FY 1965) and afterward (FY 1974); supported the war in the beginning and became disenchanted later (FY 1969 et seq.). The first Nixon bill was in 1969 but Johnson signed the 1968 appropriation bill as a lame-duck President who chose not to run again because of the war. On average, the President received about 97% of what he requested during this period.

More recently, the President seems to have been even more successful. On average, the President received over 99% of what he requested from 2001 to 2005. Congress added funds to the last Clinton request in 2000.

Three of the bills shown were passed before the start of the FY. In the Vietnam era, all the bills were late. It appears that whatever the lack of consensus that has dragged down the BR process, in the main appropriations struggle there has been consensus around defense spending levels, which lasted until consideration of the FY 2009 budget.

Conventional wisdom suggests that Congress is a marginal modifier of executive budgets, that three months debate over a BR and countless

Table 6.16.   Wartime Budget Request: Vietnam

Signed	FY	P. Req.	C. Gave	Pres. Rcd	Cong. Impact
8/19/1964	1965	49.014	47.22	96.34%	−3.66%
9/29/1965	1966	47.471	46.752	98.49%	−1.51%
10/15/1966	1967	45.248	46.887	103.62%	3.62%
9/29/1967	1968	57.664	58.067	100.70%	0.70%
10/17/1968	1969	71.584	69.936	97.70%	−2.30%
12/29/1969	1970	77.074	71.869	93.25%	−6.75%
12/28/1970	1971	75.278	69.64	92.51%	−7.49%
12/15/1971	1972	73.543	70.518	95.89%	−4.11%
10/13/1972	1973	79.594	74.372	93.44%	−6.56%
12/20/1973	1974	77.2	73.7	95.47%	−4.53%
Averages				96.74%	−3.26%

Source: Congressional Quarterly Almanac, 1964-1973.

Table 6.17.   Wartime Budget Request: War on Terror

Signed	FY	P. Req	C. Gave	Pres. Rcd	Cong. Impact
8/9/2000	2001	284.5	287.8	101.16%	1.16%
1/10/2002	2002	319.5	317.6	99.41%	−0.59%
10/23/2002	2003	366.7	355.1	96.84%	−3.16%
9/30/2003	2004	372.3	368.7	99.03%	−0.97%
8/5/2004	2005	392.8	391.1	99.57%	−0.43%
Averages				99.20%	−0.80%

Source: Congressional Quarterly Almanac, 2001-2005. (We chose to end this analytic table with FY 2005 because of the complexity of the intermixture of the base budget and supplemental bills, including bridge supplementals, in later years, as we discuss later.)

hours of hearings, testimony, and debate over the appropriations bills beginning in February and ending in November or December, or more recently January all result in very small increments of change to what the President has requested (Jones & Bixler, 1992; McCaffery & Jones, 2004). However, a closer examination of the budget process reveals that conventional wisdom may be incomplete. In 2004, Congress added 2672 earmarks worth $12.2 billion to the defense appropriations bill. Congress appeared to give the President 99.57% of what he requested. However, Congress cut $1.9 billion from the President's request and pushed out

$12.26 billion in funds requested by the President and replaced it with earmarks. In fact, in 2004, the President received from Congress only about half of what he wanted for defense. Speaking about budget reforms, Allen Schick commented:

> The present role of the President is informal and political, and arises out of the fact that he can veto appropriations and reconciliations bills, as well as other budget-related measures passed by Congress. The President already exerts considerable influence on congressional budgeting, and in some years he is the dominant player. The exuberant hopes of 1974 that the BR would be a declaration of congressional independence from the White House have been dashed by the realities of American politics. Yet, even as a political partner, the President does not get all that he wants. (Schick, 2005, p. 12)

Long ago, Aaron Wildavsky (1964) defined the concept of budget base, "[the] base is the general expectation among the participants that programs will be carried on at close to the going level of expenditures" (p. 17) Wildavsky also noted,

> Budgeting is incremental, not comprehensive. The beginning of wisdom about an agency budget is that it is never actively reviewed as a whole every year.... Instead, it is based on last year's budget with special attention given to a narrow range of increases or decreases. Thus the men who make the budget are concerned with relatively small increments to an existing base. (Wildavsky, 1964, as cited in McCaffery & Jones, 2004, p. 15)

Some research completed since Wildavsky's work has focused on incremental change and generally found the executive getting most of what he wanted in the budget with the legislative branch playing the role of marginal modifier (LeLoup & Moreland, 1978; Meyers, 1994, pp. 1-60).[8]

This is not an accurate picture of the recent budget process. In 2004, after the congressional process had finished, of $24.3 billion in year to year change in defense appropriations, the President got half and Congress got half. If the cut is scored with the money Congress controlled, then Congress directed $14.1 billion of the $26.2 billion requested, or 53.8%. Also, when cost of living adjustments and fuel and medical care cost increases are counted, they may well soak up the rest of the year to year change in the President's share. Thus it is possible that Congress is "dictating" almost all of the programmatic year to year changes, but doing it in a piecemeal fashion, earmark by earmark.

For students of congressional budgeting, the conclusion rendered above is both good and bad news. It is good in the sense that all the effort exerted by Congress actually leads to important outcomes, but bad because of the nature of the earmarking process. Earmarks set funds for a specific purpose, use, or recipient. In some cases, these represent useful additions or specifications for budget spending, but the process lacks transparency and this means that it is hard to know what the earmark does and who benefits.

Earmarks as Policy and Process Inversions

Earmarks are inversions in the budget process in theory because they replace the preferences of the many with the preferences of the few. Clemens (2005, p. 10) analyzed the FY 2005 defense appropriations bill and found that most earmarks were added while the bill was in subcommittee, committee, or conference committee. In the Senate version of the appropriations bill, Clemens attributed 65% of the earmarked funds to members of the appropriations subcommittee on defense, 11% to other appropriations committee members and 24% to Senators not on the appropriations committee. Committee leaders did better than members. States with representation on the appropriations committees did better than states without; this was particularly true in the House, where states with representation on the subcommittee averaged $316 million and states with no representation on the committee (either full or subcommittee) averaged $22 million (Ashdown, 2005, pp. 11-12). Clemens worried about this process because of the lack of review to determine the merit of these earmarks and because "very few of them are discussed in detail in the bill, contributing to poor transparency in the appropriations process" (Clemens, 2005, p. 2). A further irony is that the process shows Congress as very active in changing presidential budget requests on a piecemeal basis while backing off from a harmonized plan to guide its changes, the BR.

Some may believe that the fact that earmarks escape control at the BR level is all right because earmarks do not amount to much money. This was perhaps true in the past; it is not so now. The FY 2005 $12.2 billion earmarked in the defense appropriations bill adds up to a significant amount of program space. If the earmark total belonged to a country, it would have ranked 14th in the world in defense spending in 2000. In fact, the U.S. Congress earmarks more money in defense than Israel, Canada, Australia, and Turkey each spend on defense (McCaffery & Jones, 2004, p. 72).

On a total budget basis, CAGW[9] tracked earmarks and finds a considerable growth in their number since 2001 when there were 8,341 earmarks worth $20.1 billion to 2004 where there were 13,997 earmarks worth $27.3 billion; this a growth of 67.8% in number and 35.8% in dollar value in 4 years. This amount is more than the budgets for such agencies as commerce, energy, interior, state, EPA, NASA, and NSF (measured against their 2004 outlays) (Budget of the U.S, FY 2006, p. 76). Speaking of the power of the BR to allow Congress to set the increase at the macro level for domestic discretionary spending, Allen Schick (2005) said, "Through the budget process, Congress has effectively decided the annual amount of increase in discretionary appropriations" (p. 9).

Our perspective is from the microanalytic side. Our analysis shows Congress not only dictating the amount of change permissible under the budget caps, but also going inside the different appropriations bills and, using the earmarking process, pushing out items requested by the President and replacing them with items desired by some or a few members of Congress.[10] When Congress cannot find enough items to throw out of an appropriations bill, it uses across the board reductions in one or another accounts, based on supposed savings or improved methods, or reduced inflation expectations or some such pretext.

Table 6.18 shows total earmarks for FY 2000 through FY 2005 compared to total discretionary spending for the same period.

Concentrating only on the discretionary side of the budget for FY 2001 through FY 2005 it appears that Congress "owns" about 35% of the year to year change and owns it in a process that is by and large neither public nor transparent and may or may not uphold the general interests of American society. This sounds harsh, but despite all the information that is readily available either published or on the Internet, it is very difficult for outsiders to look at congressional documents reporting congressional changes to the President's budget request to determine whether they are good changes or not. Any statement will provide unlimited opportunities for puzzlement.[11]

Table 6.18.   Earmarks as Share of Discretionary Dollars

	Disc. $	*Y/Y Change*	*Earmark $*	*Cong. Share*
FY 2005	964.8	69.4	27.3	39.34%
FY 2004	895.4	70	22.9	32.71%
FY 2003	825.4	91.1	22.5	24.70%
FY 2002	734.3	85	20.1	23.65%
FY 2001	649.3	34.5	18.5	53.62%
Average				34.80%

Source: Budget of the United States for Fiscal Year 2006, Historical Tables, Table 8.1, Outlays by Budget Enforcement Act Category: 1962-2010, p. 125; Earmarks from Pigbook for years cited.

Earmarks are not new. Clemens (2005) traces earmarks in the 1969 and 1979 defense appropriations bills where there are many fewer earmarks, but substantial sums of money ($5.6 and $8.9 billion, respectively (p. 5). Senator John McCain (R-AZ) has been on an antipork crusade for some years,[12] but neither he nor CAGW or Taxpayers for Common Sense seem to be making much of an impact. Congress attempted to pass earmark reform in 2007 but in view of most critics, by the time the legislation was passed it was so watered down and weakened that it was essentially inconsequential. Further, soon after the law was passed Congress failed to abide by the rules it had just enacted.

Wartime events have underscored the complexity of the earmark issue. From 2003 on the Pentagon has moved more slowly than Congress has wanted in delivery of such force protection items as armored Humvees and other armored vehicles and Congress has intervened, for example, by increasing funding for Humvees by $5.2 billion more than the Pentagon requested from 2003 through 2007 (Morrison, Vanden Brook, & Eisler, 2007, pp. 1-2). Compared to total DoD procurement spending for this period of about $470 billion, the $5 billion was not much, but it was helpful in that it did speed armored Humvees on their way to the battlefield, and it seemingly helped change the mindset of the acquisition bureaucracy early in the process. A House staffer met with steel suppliers with result a 7 month acceleration of armored Humvees to the field in 2004 as reported by Eisler (2007, p. 2). Of this episode, the staffer's boss, Representative Duncan Hunter (R-CA) said, "The acquisition bureaucracy was not in the war...We have an acquisition bureaucracy (at the Defense Department) that just doesn't respond quickly enough to the needs of the war fighter" (Eisler, 2007, p. 2).

Similarly, Senator Mary Landrieu (D-La.) originally sought an earmark to help a company in her state that produced armored police vehicles (Textron's ASV) which proved highly resistant to explosive devices. Landrieu pleaded with the Army to use the vehicles, but the FY 2004 budget included no money for the vehicles: "They send me an executive budget with this zeroed out," Landrieu recalls, "I hit the roof." She kept the program alive by inserting earmarks in appropriations bills. Says Landrieu, "They were getting ready to shut the line of these vehicles down completely.... It was on its last breath and I literally was so determined (that) I said I was not going to allow it." For the next 2 years Landrieu helped insert $700 million for the vehicles in the DoD appropriations and kept the program alive even though the Army budgeted nothing for it until FY 2007. Looking back, Landrieu says, "I can say in my life there's one really great earmark" (Morrison, 2007, pp. 1-2).

Defense experts warn that congressional intervention imposes a solution on the process, rather than letting the acquisition process work to reveal the best solution and that when lawmakers add money for current needs to the DoD budget, it has to come from somewhere, and it may be that it is shortchanging the development of future weapons systems to pay for immediate needs (Kelly, 2007).[13] Thus Rep. Roscoe Bartlett (R-MD) said, "In this case, defending our troops (in Iraq) is the urgent thing; the important thing may be developing weapons systems to protect them in our next engagement" (Kelly, 2007, p. 10a). We may note that the same tension exists within the DoD programming and budget process; future weapons may be sacrificed when the emphasis shifts to what the war fighter needs now.

Observers outside of DoD are often frustrated with the pace of change as well as the decisions that the Pentagon makes. Thus, in some respects the earmarking process may function as a pressure relief valve while also doing some good in the short run. However history has taught us that defense is too complex an arena to be managed by committee, let alone by numerous congressional committees, and the decision process in Congress needs to honor the long run as well as current pork and constituent demands.

It is no coincidence that an era of big appropriations bills has led to an increasing use of earmarks. Passing large bills often is a case of taking the bad with the good; in this case the bad is extensive earmarking. Defense is often considered veto-proof because of its size and mission, but its size and status are also an invitation for earmarking. The large consolidated appropriations bills were also an invitation to earmarking due to their size and to the fact that they were the best compromise going between a bad bill and no bill at all. The result seems to be a degradation of budget process outcomes, because earmarking adds up to sizeable amounts of money through aggregation of smaller sums over multiple years, thus escaping both the discipline of executive budget review at the department and OMB level, and congressional budget review in the appropriations process. Anyone sensitive to the concept of opportunity cost cannot help but be critical of this outcome.

With the earmarking process, there is no choice of spending $27 billion more on education or environmental protection or defense, rather bits and pieces add up to an unanticipated, but large, total, as if some manic grocery shopper were filling his cart with all sorts of things that might be tasty in the short-run, but would not add up to a well-balanced diet.

Part of the absence of discipline in congressional budgeting is due to policy conflicts and part is due to process problems. Almost all of it has been seen before, in the Vietnam era, or the budget debacle of 1995-1996. What has not been seen before is the consistent use of large consolidated appropriations bills accompanied by a growing tendency to earmark funds. Earmarks have been around before, inside and outside of the defense bill, but the size and critical nature of the defense bill makes them difficult to stop. The same things happen in the large consolidated omnibus appropriations bills. These outcomes are a setback for rational budgeting and efficient allocation of resources for public purposes. It is clear that the earmarking process needs to be brought within greater budget process discipline but prospects for a good outcome in this respect are not good because it contravenes the incentive of lawmakers to serve their constituents, even if wastefully. The earmarking process does not appear to be a triumph of the rank and file over the leaders or a repeal of committee decisions by floor events (as were, for example, the passage of the GRH Deficit Reduction Act In 1985 and the Reagan tax cut legislation in 1981 via the reconciliation process). Rather, it appears to be a result of a slackening of the discipline of the "cutters," those committees and individuals interposed in the appropriations process to control the appetites of the spenders. This is a serious weakening of the appropriating process. If it is allowed to continue in obscurity, then it may not necessarily be self-correcting. If the "guardians of the public purse" are the ones who benefit most from earmarking (appropriations subcommittee and committee members and other powerful congressional leaders) then there is little reason to suppose that they will fight to restore the budget process to a more rational and open pattern.

What also may be appearing in the improvisational patterns of the budget process is a reflection of an increasingly divided country where what used to be just a matter of dollars has become a matter of principle and that these matters of principle are transmitted to the national scene from increasingly safe districts whose representatives need not compromise to stay in office. For example, Abramowitz and Saunders (2005) analyzed election and attitudinal data to come to some interesting conclusions. First, they observed the rise of safe districts and noted that in the House election of 2004 "over 95 percent of safe Democratic districts were won by Democratic candidates and over 95% of safe Republican districts were won by Republican candidates (p. 17).

This also shows up in presidential elections:

states have become much more sharply divided along party lines over the past several decades. While the 2000 and 2004 presidential elections were highly competitive at the national level, the large majority of states were not competitive. Compared with the presidential elections of 1960 and 1976, which were also closely contested at the national level, there were far fewer battleground states in 2000 and 2004 and the percentage of electoral votes in these battleground states was much smaller. The average margin of victory at the state level has increased dramatically over time and far more states with far more electoral votes are now relatively safe for one party or the other: red states have been getting redder while blue states have been getting bluer. (Abramowitz & Sanders, 2005, p.11)

Abramowitz and Sanders (2005) attribute these changes not to electoral redistricting efforts, but to fundamental changes in American society:

> Over time, red states, counties, and congressional districts have been getting redder while blue states, counties, and congressional districts have been getting bluer. This increasing geographical polarization is not a result of redistricting. It is a result of fundamental changes in American society and politics. Internal migration, immigration, and ideological realignment within the electorate are producing a nation that is increasingly divided along partisan, ideological, and religious lines. (p. 19)

It may be argued that the majority party has little need to compromise over budget issues so long as they reflect the values of their districts, districts which are becoming increasingly divided into two camps. Moreover, when party control of the levers of government is split, this also means that compromise will be harder to attain, since what is compromised may be seen as a matter of principal and not dollars. This could be a partial explanation for the absence of spending discipline in the federal and congressional budget process over the last decade.

When people in the "red" and "blue" states have significantly different beliefs and these beliefs are rooted in values related to political partisanship and ideology, then the art of political compromise becomes more difficult to pursue. This clearly affects the budget process which rested on a slowly moving consensus about the ends of society from the end of World War II until the mid-1960s and then again from the Ford Presidency (1974-1976) until perhaps the mid-1990s, ending with the Republican's Contract With America in 1995. During this time, the belief systems imbedded in American society were generally agreed upon by both parties, and in the budget process members of Congress could decrease conflict by claiming issues were just a matter of dollars.

How this relates to pork is interesting. It may be argued that with strong party control, pork is no longer necessary to secure elections. Given the size and frequency of pork dollars recently and their persistence despite well-intentioned efforts to rectify the situation by some leaders and some followers, pork persists, as if to say, "we won and now we will enjoy the spoils." In any case, pork is a fact of life and so is the earmarking process despite valiant efforts to the contrary. In conclusion, as we have observed elsewhere, pork barrel spending is a two-sided coin. On one side, it is wasteful spending, on the other local economic development. On the definition of pork, clearly where you sit determines where you stand (Jones & Bixler, 1992, pp. 11-12; see also Wildavsky, 1988, p. 42).

TURF BATTLES IN DoD AND THEIR IMPACT IN CONGRESS

Military departments compete with each other for missions and this can result in a public squabble both in DoD and Congress. The source of competition is mission responsibility. The military service that has the mission will get the money, and this may mean control of that mission and all the weapons it takes to perform that mission for the next decade.

For example, the Air Force core mission involves tactical fighters and strategic bombers; the Navy core mission involves carrier groups, Marine deployment groups, and ballistic missile submarines. The Army core missions revolve around infantry, artillery, and helicopters and armored vehicles. In the 1990s, turf battles were seen in appropriation struggles over the Apache and Comanche helicopters, M-1 Abrams tanks and THAAD missiles. The Air Force concerns were focused on the B-1 and B-2 bombers, the F-15 Eagle, F-16 Falcon and F/A-22 Raptor fighter jets as well as the C-17 cargo plane. The U.S. Marine Corps is defined in part by its amphibious warfare capabilities using the LHD-7 and LPD-17 ships in addition to the AAAV (Advanced Amphibious Assault Vehicle) and other assault vehicles.

Turf battles arise where one service challenges another either for a peripheral mission or for a core mission. For example, thematic debates during the 1990s concerned such big ticket items as the V-22 Osprey, SDI, B-2 bomber, Seawolf submarine, aircraft carriers, tank production, F-18 E/F, and other costly weapons programs. Such weapon systems essentially define a branch of service and the cancellation of any one of these money-generating programs would put into question the purpose or mission of the respective branch of service using or defending that program.

The debates between military services over funding for major weapons programs are a source of heated discussion with service reputation, readiness and constituency employment implications hinging on the outcomes of the contested items. The emerging trend toward joint weapon systems is the natural result of spreading the support for a weapon system across services, resulting in a more stable platform made less vulnerable to cuts due to its widely distributed support across services and districts alike.

In 1999 a turf struggle occurred over funding for the Navy F/A-18E/F Super Hornet and the Air Force F/A-22 Raptor. Each represented billions in potential new procurement dollars, in addition to representing its respective service's future strategic requirements. Since both were required for their respective services, they were both funded with the

Super Hornet getting an increase from $1.35 billion to $2.1 billion and the Raptor received an increase of $331 million.

The B-2 bomber-funding story spanned the entire decade with continuous struggles between congressional program supporters and opposition for further production in the White House. President Clinton threatened to veto the defense bill over wording regarding whether B-2 production would continue. Although it seemed a relic of the Cold War, it defined some of the Air Force's mission capabilities and represented billions of dollars in government funding to the various firms in states supporting its continuation. For these reasons, funding continued to be provided for the B-2 and it emerged as one of the classic turf issues of the decade.

Turf battles are not new. They go back to the beginning of the nation, as shown in chapter 2. As such, examples abound. In 1993 the U.S. Army was attempting to redefine its mission with the introduction of funding for 12 prepositioning ships, competing for turf against the U.S. Marine Corps amphibious role. Although this position was opposed by the Marine Corps, the gambit was ultimately funded, but the Marines were compensated with funding for a new Wasp class helicopter carrier to the amount of $50 million. The U.S. Navy had a turf battle of its own in 1993 when it faced the cancellation of their A/F-X stealth attack plane in favor of the U.S. Air Force F/A-22 Raptor which was reportedly to be operable from an aircraft carrier; it is under modification to do so in fact. While it remained unconvinced of the Raptor's seaworthiness, the Navy was placated by the addition of $1.47 billion for Super Hornet enhancements. In 2007 a turf battle erupted over control of the highly successful Predator and Global Hawk aerial drone programs with the Air Force pushing to become the executive agent for drones that fly above 3,500 feet. This move was opposed by the Army, Navy, and Marine corps who feared that it would make the Air Force responsible for the acquisition and development of unmanned aerial vehicles for these latter services, including the Army's Sky Warrior program (Financial Times, 2007).

A large part of what the military services and DoD fear from turf battles that spill over into Congress is loss of control and a product which will not quite fit their missions as well as if they had sole control over its development.

Another turf battle broke out with the Marines and Army on one side and the Air Force on the other. This was over competing aerial surveillance systems. The Air Force and Marines wanted a system called Angel Fire which was a high definition real time system that could be replayed quickly to show a soldier on the ground what was just happening or what had just happened in the street in front of him. The tactical advantages seem obvious. The Army favored a Global Hawk that surveyed a broader area but was more difficult to use and produced a video that had to be studied by analysts before dissemination. This turf battle began in late 2005 and ran through 2006; while events seemed calm on the surface, a furious e-mail campaign was conducted with members of Congress and their staffs being contacted and urged to save the Air Force and Marine program (Angel Fire). Eventually, funding for both programs was continued with a spokesman noting that the two systems served different missions (Vanden Brook, 2007). Turf battles also happen within military departments, even within a weapons community. For example, the development and deployment of a new Army rifle was halted because the Army weapons community could not agree basically on the process. New weapons begin with the statement of a requirement and then approval within the department and then at the joint level. This can take a long time and Army tried to shorten this by backing the new rifle into an old requirement stated in 1995; this tactic was meant to speed up the process, but eventually it raised too many problems and the new rifle program was scrapped and the Army continues to use the M-16 rifle family, a 4 decade-old weapon system. Part of the problem here was the acquisition process itself. To start up the process of acquiring a new rifle, the Army Infantry Center would have had to write a requirement for a new rifle, a project that was nowhere on the horizon. Ultimately a superior weapon systems failed to get off the ground and resulted in a DoD inspector general audit, an argument over who should have control between the Army and DoD (depending on dollar cost of the program) and accusations flying back and forth in the Army between the requirements bureaucracy and the Army requirements setters. An Army colonel familiar with the case proclaimed,

> There was not a common vision between the two. We have a broken process. When you don't have a requirement and acquisition process with a shared vision, you are not going to get anything, and you are going to waste a lot of money.

General Jack Keane, Army Vice-Chief of Staff, still bitter about the new rifle's demise said, "To have the best means you have to regenerate to get the best every so often.... In this case, our bureaucracy failed our troops" (Cox, 2007, p. 38). The concluding chapters of this book analyze in greater detail a number of the issues related to defense acquisition and the design attributes of the DoD acquisition processes. In any case, turf battles erupt almost every year in DoD and typically they spill over to Congress.

It would seem that to a great extent in defense budgeting, the phrase, "What goes around, comes around" applies all too often. We might ask, however, whether any of this is new? The answer is emphatically, "No!" In the 1790s, members of Congress complained about making lump sum appropriations with no accountability mechanisms in place and, "that is to say" clauses in appropriation bills that directed money to be spent on specific items or locations (McCaffery & Jones, 2001, pp. 54-55). Members of Congress are elected by their constituents in part to "bring home the bacon." In this context, the appropriations for national defense provide great opportunity. Pork barrel spending and protection of turf are standard elements of resource competition in democratic political systems. As long as specific constituencies want their piece of the federal budget and defense portion of the pie, such dynamics will be a prominent part of the defense budgeting game. As we have noted elsewhere, some degree of strategic misrepresentation is taken for granted in budgeting in all contexts (Jones & Euske, 1991). To think it otherwise is naive relative to the norms, values and history of our budget culture.

CONCLUSIONS

From the 1980s to the present, the normal complexity of the budgetary process has been embellished with improvisational behavior by the participants. This has made it hard to find stability in process from one year to another. It is true the same products were produced ultimately, including the BR, and the defense authorization and the appropriation bills, but the process that produced these outputs was and is marked by great diversity from one year to the next. The President (of whichever party is in office) is always in the lead due to his responsibility for formulating the annual budget proposal, but ofttimes Congress hears a different melody and, like a headstrong dance partner, begins to veer off in a different way; the result is improvisational budgeting.

This is evident in the government shutdowns in 1990 and 1995, budget summit meetings in 1990 and 1997, omnibus appropriation bills in 1996, 1998 and 1999, CRAs in all but 2 years including 21 in the year 2000 alone, and widespread use of the veto power averaging about 7 a year—including 21 in 1992 and 11 in 1995, the veto of a CRA in 1990, and the first use of the item veto in 1997. It was highly apparent in the budget stalemate of 2007. Sometimes it appears as if the only worthwhile decision-forcing mechanism is the end of the *calendar* year. Negotiators seem to take important policy and budget issues more seriously the further into the year they get. Moreover, spending and taxing choices are high-visibility decisions that can be the basis for election campaigns. This means that legislators must make sure they get their position in front of their relevant public. It also means that while some debates are just a question of dollars and thus easy to compromise, other issues raise large philosophic questions and are not easily resolved. These issues include such things as nuclear disarmament, space-based weaponry, theater-based antimissile systems, and strategic bombers. These raise serious questions about the U.S. defense profile and often result in serious and prolonged debate. With these complex and expensive systems, the traditional budgetary compromise is not easy to attain.

A summary of congressional budgeting in this period would consider both the good and bad news about the effectiveness of the congressional budget process:

The Good News

- 1994 marked the first time since 1948 that Congress had cleared all 13 appropriation bills before October 1.
- 1997 marked the first surplus in the annual budget for federal government since Lyndon Johnson was President. The budget remained in surplus for four years in succession (FY 1998, 1999, 2000, 2001).
- In 1996, all bills were passed on time for the second time in three years, but only because the 1995 process ended in April 1996 and neither party could afford another budget debacle, especially in an election year.
- Deficits returned by 2001, as a result of tax cuts and a slow economy and defense appropriation bills were again late, in 2001 and 2002. However, by FY 2007 the annual deficit was in decline after years of a prosperous economy.
- In late 2001 through 2007 Congress supported the President's war on terrorism initiatives with significant defense spending increases passed in the annual defense appropriations acts and sizable supplemental appropriations.

The Bad News

- 1998 marked the first time since the modern budget process was established in 1974 that Congress failed to produce a BR with the House and Senate unable to agree on a spending plan.
- Despite endeavoring to put an end to CRAs by shifting the FY from July 1 to October 1 in 1973, the period from the 1990s through 2007 saw plentiful employment of CRAs. They were employed in every year except 1994.

- The Defense Authorization Bill was signed before the defense appropriation twice during the 1990s and once during the 2000s and thus it is hard to make the case that the authorization bill is needed to guide the appropriation bill.

- The most serious breakdown in budgeting occurred in 1995 when the failure of the President and the Congress to agree on budget priorities culminated in two separate government shutdowns, totaling 28 days. A government shutdown was again threatened in 2007 as negotiations over the FY 2008 budget dragged on towards the end of the calendar year.

- Excluding 1998, and several years in the 2000s Congress failed to agree on a BR, the average BR was passed 37.5 days after the 15 April deadline with delays ranging from 0 to 74 days. CRAs spanned 524 days over the decade of the 1990s, representing 14% of the entire decade spent deliberating over late appropriations, and reappeared in 2007.

- Because authorization and appropriation legislation was passed satisfactorily in years when no BR was approved by Congress, it is difficult to argue that the BR is a necessary part of the congressional budget process.

Based on this summary it appears that when the path to definition of defense policy is not clear, improvisational budgeting is an inexorable result. However, improvisation does not result only from dissensus; it also typifies consensual decision making. When consensus was present, as it was after the attack of September 11, 2001, improvisation also occurred and Congress showed remarkable ability to move quickly in support of the President in waging the war on terrorism through approval of funding for enhanced defense spending and the wars in Afghanistan and Iraq.

The good news in the years following 2001 was consensus on the need to fight terrorism on individual battlefronts such as Afghanistan and Iraq, and more generally in a war where U.S military action to counter terrorism increased around the globe. The bad news in this period is that the consensus developed through 2002 began to deteriorate as the presidential and congressional elections approached in 2004. Defense issues that had been treated in bi-partisan manner prior to this point became increasingly politicized, dividing Republican and Democrats. The good news is that even with this partisan split, overall support for defense and the war on terrorism persisted. The bad news is that the gap in funding for replacement of aging military weapons platforms remained after the spending increases of the 2000s. While the tactical aircraft needs of the 21st twenty-first century were addressed by funding for the new F/A-22

Air Force fighter and the Joint Strike Fighter (JSF) to be flown by all military services in various configurations, and while the Navy was funded to complete construction of a new aircraft carrier and several tactical submarines, the gap in overall weaponry to replace obsolete and used up assets including ships, tanks and other infantry weapons, bomber aircraft or alternative delivery systems including nonpiloted and space base weaponry remained significant.

The ultimate bad news about the period from the end of the twentieth century to 2008 was that neither Congress nor the executive had exhibited to any significant degree any discipline to control the growth in federal spending. Also, the consensus about funding the war on terrorism, and the war in Iraq in particular, begin to deteriorate in 2007 in negotiations over the preelection year FY 2008 budget. Further, by 2008 neither branch of government had moved to begin to tackle the impending financial crisis resulting from demographics as no plans for insuring the fiscal solvency of the nation's major entitlement programs were developed and planned for in any way in the budget process. In this case, the President blamed Congress, and Congress would not get started on this mammoth but crucial task until the 2008 elections were over.

The budget process employed by Congress to determine defense policy and budgets is highly disaggregated, disjointed, at most times conflictual, competitive, and inefficient. National interests compete with local interests and in many instances the compromises struck result in less than optimal allocation of scarce resources. However, as has been demonstrated in this chapter and in actions taken subsequent to the terrorist attacks of 9-11, Congress also is capable of responding to contingency in a relatively rapid fashion despite a considerable degree of visible and not so apparent decision making and budgetary dysfunctionality. We would observe that such dysfunctionality is the price to be paid for reaching agreement in policy and budget negotiation in democratic political systems.

Finally, what may we conclude with respect to alternative theories of budgeting? Is Schick's improvisational a better model than Wildavsky's incrementalism to explain congressional budgeting in the period 1990 through the 2000s? Schick's argument has considerable merit. In fairness to Schick, he did not explicitly propose his theory as a better model than Wildavsky's. The competition between these alternative models is one we have posed.

Support for the Wildavsky model is evident. The overall budget of the federal government continues to increase by relatively small increments in the period from 1989 to 2008. In addition, while defense spending did not decrease or increase consistently over this period, the cuts and increases to the defense budget during this period generally were seldom huge relative to the size of the base in any particular year. We admit that some critics would argue this point about the rate of increase in defense spending under President George W. Bush, particularly if supplemental appropriations are considered as part of the total as they should be. They would also point out that the cumulative decreases from 1990 to 1998 and increases from 1998 to 2008 were substantial, Still, they might concede that defense and federal budgeting for the entire period appears to fit Wildavsky's model. Both policy decisions and spending decisions in this period were made on the margin in adjustments to the base, even when sizable supplemental war appropriations were approved. Therefore, it would appear that quite a bit of evidence exists in support of Wildavsky's thesis. However, the congressional budget process and budgetary behavior during the period examined also was very highly improvisational. It would appear that Schick too is right. The budget process innovations in this period were not incremental—they deviated from the base operations defined in contrast to how the budget process operated in periods of greater defense policy and budgetary consensus, for example, the 1950s, or even in the more contentious and inflation plagued period of the 1970s.

How may we reconcile the perspective of these two scholars? It is our view that both scholars are right. The key conception supporting our conclusion relates to the difference between a focus on substance versus process. Wildavsky based his theory on the evidence that budget numbers and the programs in budgets are modified incrementally from the base by Congress in annual budget deliberations, that is, on the substance of budgets. On the other hand, Schick observed the entire congressional budget process over more than a decade and found significant deviation from standard practice and process. Our examination of congressional budgeting supports Schick's assertion that to accommodate dissensus (ironically a term Wildavsky coined to describe budget disagreement), a pattern of significant improvisation and innovation emerged, that is, in process. However, our analysis adds to Schick's theory. We conclude that congressional budgeting since the 1980s has been improvisational in responding to contingency, whether the budget share for defense was increasing or decreasing, and despite whether the annual budget was in deficit or surplus. Consequently, we conclude that the combination of the theories of incrementalism and improvisationalism lead to a more sophisticated understanding of federal and defense policy setting and budgeting. Finally, we believe that despite what is written in law, budgetary life as it is practiced indicates there is no model budget process. Rather, there is a responsibility to produce appropriations to continue the stable operation of the federal government and the Department of Defense, and Congress modifies its budget processes in some way each year to accomplish this end.

NOTES

1.   In addition to the attack on 9/11, there were also anthrax attacks on October 15 and November 16 and in one of the Senate office buildings, half of the offices were closed for 96 days. Even so, Congress' days in session were comparable to previous years (Senate 173, House 142). CQA 2001: 1-7 and 1-12.

2.   Streeter describes two kinds of CRAs. The first is the interim or partial CRA which funds an agency or group of agencies until their appropriation bill is passed. Interim bills provide funding set by a rate or formula based on the previous FY (the one which just ended September 30); for example, if the CRA were for DoD for one month, it could be set at 1/12th of the appropriation for the previous FY. Full-year CRAs last the entire (or rest of) the year and are "generally similar to the regular bills" (Sandy Streeter, "The Congressional Appropriations Process: An Introduction," CRS Report for Congress, 97-684, Washington, DC, December 6, 2004, p. 15. Retrieved June 29, 2005, from accessed through www.opencrs.com on.

3.   The Library of Congress uses the term "consolidated" rather than omnibus to indicate that the consolidated bill did not include all the appropriation bills. In 1986 and 1987 all 13 bills were included in an omnibus CRAs that lasted for the whole FY (FY 1987 and FY 1988) (see Allen Schick, *The Capacity to Budget*, Washington, DC: Urban Institute Press, 1990. p. 181). In recent years it would be correct to say that since no appropriation bills were passed on a timely basis in 2001 and 2002, the first CRAs for those years were omnibus CRAs. Streeter's analysis of CRAs does not include the consolidated appropriations bills as CRAs; for FY 1985 thru FY 1988, the second Reagan term, 41 of a possible 52 appropriations bills were enacted as full year CRAs (over the 4 year period only 11 appropriation bills were passed) (see Streeter, "Congressional Appropriations Process," 2004, p. 16.). Streeter classifies the consolidated appropriations bills of FY 2001, 2003 and 2004 as "Omnibus (or Minibus) Measures. FY 2003 saw 11 appropriations packaged into one Omnibus; in FY 2004 seven in one, and FY 2001 two minibus measures, one of two bills and one of three (Streeter, 2004, p. 12).

4.   Major reforms included the Balanced Budget and Emergency Deficit Control Act of 1985 (commonly called "Gramm-Rudman-Hollings" and the Budget Enforcement Act of 1990 (BEA). There reforms were modified or extended through less comprehensive measures in 1987, 1993, and 1997. For FY 1991 though FY 2002, there were statutory limits on discretionary spending (e.g., defense, education et al.) and paygo provisions for direct (mandatory) spending (e.g., social security) and revenue legislation. Bill Heniff Jr. and Robert Keith, "Federal Budget Process Reform: A Brief Overview," CRS Report for Congress," CRS RS21752, Washington, DC, updated 8 July 2004. Accessed through www.opencrs.com on June 29, 2005.

5. See Bill Heniff, "Congressional Budget Resolutions: Selected Statistics and Information Guide," CRS Report for Congress, updated January 25 2005, Washington, DC, p. 14 and pp. 24-25.

6. The Senate passed built and passed BRs in 1998 and 2004, but it did not go to conference with the House and pass a joint resolution. This means the Senate itself had its own resolution to guide its appropriations work, thus 2002 remains unique.

7. McCaffery and Jones, 2004, p. 164.

8. For example, this is the picture LeLoup and Moreland defined. Others, including Roy Meyers penetrated within the base to show active decision making taking place there, but incrementalism with a dominant executive has proved a durable organizing concept.

9. See "2005 Pig Book Exposes Record $27.3 Billion in Pork" CAGW, Washington, DC, Press Release, April 6, 2005, June 30 2005.

10. Ashdown argues that at least part of the earmarks included in the 2005 DoD appropriation bill were paid for in the spring 2005 emergency supplemental bill; he cites the O&M cuts made in the fall of 2004 and then replenished in the emergency bill in 2005.

11. For example, plus ups in 2003 in the FY 2004 defense appropriation bill for $1 million for forward osmosis water filtration devices, $13.5 million for Building 9480 renovation, $1.4 million for contaminant Air Processing System, $.6 million for Sooner Drop Zone extension and $.6 million to repair the jump tower at Kirtland Air Force Base and so on for about 3,000 more items. The DoD comptroller can find all these and some Congressmen and their staff know about each of them. Allen Schick (Capacity to Budget, 1990) warned that baselines were the new dividing line between those who could know about the budget and those who could not and could not test the adequacy of baseline construction. "The baseline has become a dense barrier separating budget insiders who know what the numbers mean, from outsiders including seasoned budget observers who cannot figure out what is being done to the budget unless they uncover the assumptions behind the numbers" (Schick, 1990, p. 99). Earmarks have always been an arcane area. Now their totals are big enough to matter and they are still arcane.

12. Senator McCain has taken on this issue since July, 1997; see the list of 95 press releases and talking points about pork. Retrieved July 4, 2005, from //mccain.senate.gov/index.cfm?fuseaction=Issues.ViewIssue&Issue_id=27.

13. In 2007 Senator Ted Stevens (R-AK) directed the Navy to build an experimental ferry it once rejected to serve a little used port in a remote area of his home state at an estimated cost of $84 million. The ferry route follows the same route of one of the two "bridges to nowhere" spotlighted as wasteful spending in 2005 and officially ended by the state of Alaska in 2007 despite the $452 million inserted into the 2005 transportation bill. The Navy did not request the ferry. The version forced on the Navy by Stevens was more than double the original cost the Navy was considering because of its unique design specifications (Kelly, 2007).

CHAPTER 7

SUPPLEMENTAL APPROPRIATIONS FOR NATIONAL DEFENSE

INTRODUCTION

Supplemental appropriations provide emergency adjustments to the current year, usually for national defense contingencies and natural disaster emergencies. Supplemental appropriations have been critical for the Department of Defense (DoD) in the recent past, for example, to pay for the costs of wars in Afghanistan and Iraq. Recently, supplementals have adopted some of the complexities of the regular appropriation process. For example, both the President and Congress may suggest when a supplemental is a dire emergency and thus beyond spending discipline and when it must be offset. Some supplementals have paid for nonemergency activities; others have resulted in funding decreases and still others have resulted in spending in future years. Compared to the normal appropriations process, supplementals are usually passed expeditiously. Defense supplementals are generally precisely priced, while disaster supplementals tend to be lump sum estimates.

In the 1990s, the major focus of supplementals has been to meet emergent needs resulting from disasters and defense concerns. As the budget year unrolls, unpredictable events occur. Some of these are met by

Budgeting, Financial Management, and Acquisition Reform in
The U.S. Department Of Defense, pp. 269–319

transfers or reprogramming of dollars already appropriated, but others are of such a magnitude that only an additional appropriation will suffice. Supplemental appropriations fill these needs. This essay is about the new complexities of supplemental appropriations. Supplementals are an old tool, but deficit-limiting procedures mainly stemming from the Budget Enforcement Act of 1990 (BEA) have changed the use of supplementals (Doyle, 1991, 1992). We examine some of these changes by focusing on defense supplementals to explore the relationship between departmental administration and supplemental appropriations. Our data is drawn from interviews with budget analysts and comptrollers in the DoD and the Department of Navy and from congressional records.

PURPOSES OF SUPPLEMENTALS

Supplemental appropriations may provide for natural disaster aid to states, regions, and directly to individuals for blizzards, floods, drought, fires, and hurricanes. Supplementals provided assistance in the Northeast blizzard of 1978, the Mount St. Helens eruption in 1980, Hurricane Hugo, and the Loma Prieta earthquake in 1989, Hurricanes Andrew, Iniki, the Chicago flood, and the Los Angeles riots in 1992, the Northridge (Los Angeles) earthquake in 1994 and flooding in the Dakotas in 1998.

Supplementals met emergent military missions in the 1990s beginning with the Gulf War in 1991 and Bosnia in 1996 and thereafter. In addition to military missions, the DoD has also been tasked with providing aid in time of disaster to foreign nations. For example, from 1990 to 1997, U.S. Naval forces participated in 50 major disaster missions. The duration of 12 of these operations lasted more than 2 years. Refugee support missions averaged over 3.6 annually, peaking at eight concurrent operations in 1994. From 1990 to 1993, 17 natural disaster missions were carried out. Between 1993 and 1997, 15 peacekeeping missions were undertaken. During the 1990s, the Navy evacuated embassy personnel from Liberia, Somalia, and Zaire; assisted refugees from Iraq, Cuba, China, Panama, and Rwanda; helped fight disease in Venezuela, helped people recover from storms in Antigua, the Philippines, Guam, Bangladesh, and the Bahamas; provided earthquake relief in the Philippines and Guam; aided the drought-stricken in Micronesia and Somalia; coped with volcanic eruptions in the Philippines and Italy and was involved with peacekeeping operations in Bosnia, Somalia, Liberia, Ecuador, and Haiti (McGrady, 1999, pp. 1555-1556). Supplemental appropriations provided the wherewithal to help fund some of these missions; those of a smaller scale were absorbed within the DoD budget. Although not much is made of it, in the

1990s DoD became a sort of emergency response force, fulfilling a variety of police, fire, and emergency rescue functions when tasked to do so by the President with the concurrence of Congress.

Agencies other than defense also executed supplementals in international programs. As a result of Hurricane Mitch in November of 1999, USAID (United States Agency for International Development) administered $587 million dollars worth of supplemental-funded relief efforts for water and sanitation projects in Honduras, road and bridge repair in Guatemala, housing rehabilitation in the Dominican Republic, and water and sanitation projects in Nicaragua. The inspector general's office for USAID monitored these projects for completion on time and on budget, to guard against fraud and abuse, and to ascertain if the host government followed USAID competition requirements in bidding and contract management. According to USAID Inspector General Everett Mosley, his auditors helped contractors and grantees set up and recalibrate work plans and create fund accounting systems (Mosley, 2001). This is an example of where a supplemental aimed at disaster relief also carried with it capacity-building technology, and a value system and technical assistance to get the work done on schedule and without losing funds to corruption, fraud, and misuse. Thus, funds from supplementals have been spent both in and outside the United States as policymakers have reacted to events suddenly thrust upon them, to provide direct aid and sometimes to help the other country build technical capacity that would be of benefit to it in the future.

THE SUPPLEMENTAL PROCESS

Before the passage of the Congressional Budget and Impoundment Reform Act of 1974, the federal government regularly used supplementals for provision of funding for day-to-day agency operations, like pay raises for federal employees, but in recent years supplementals have been mainly used "to provide funding for unanticipated expenses-although there is sometimes an argument about the whether the requirements should have been anticipated or not" (Tyszkiewicz & Daggett, 1998, pp. 42-43). The Congressional Budget and Impoundment Control Act of 1974 attempted to control the use of supplementals, by providing that expected supplementals be included in the Presidents budget and that an allowance be set aside in the budget resolution for anticipated supplementals. The BEA of 1990 imposed spending caps on federal spending which meant that a supplemental that exceeded the cap could only be passed if it could be matched with an offsetting spending reduction or revenue increase, or if it were deemed to be a dire emergency supplemental, in which case it would simply be funded out of deficit spending.

During the 1990s, Congress and the President tried to offset supplementals with rescissions or cancellations of Budget Authority provided in earlier appropriations bills that had not been obligated or had been deemed unnecessary. An examination of the 1990s indicates that rescissions made an impact on supplementals, but only in 1995 were rescissions large enough to offset supplementals. In fact, in that year, rescissions exceeded the amount appropriated for supplementals. The supplemental in 1995 was enacted for $6.42 billion; rescissions enacted amounted to $18.94 billion (Congressional Budget Office (CBO), 2001, p. 10). Rescissions were enacted in every year from 1990 through 1999, amounting to $51.97 billion. Supplementals enacted amounted to $137.99 billion, thus the net cost of the supplementals for the decade was $86 billion (CBO, 2001, p. 10). Since rescissions did not pay for all supplemental costs, the dire emergency provision also had to be used. During the 1990s, most supplementals were passed under the dire emergency clause and thus avoided spending cap discipline. No such escape clause existed prior to 1990, but in every year from 1991 through 1999, part or all of the supplementals passed were designated emergencies.

The executive branch controls the timing of when the supplemental will be submitted to Congress and its initial size. The President's Office of Management and Budget (OMB) may also suggest "bill-payers" (the bill is an emergency, but not a dire emergency) or ask for dire emergency designation. Congress must concur. Emergency designation is a shared power. Congress can add projects and funds to both supplementals and regular appropriations and suggest that they be considered for emergency designation. Congress has routinely done this over the decade of the 1990s, originating about 33% of the dollars designated as emergencies (CBO, 1999, p. 4). The President has to officially accept Congress' designation of the funds as emergency before the dollars may be released for obligation.

Between 1993 and 1995, Congress was unwilling to grant the full amount of supplemental appropriations requested by the President. After 1996, Congress generally provided more than the President requested. While Congress's overall supplemental spending in the 1990s ($137.99 billion) was slightly larger than the Presidents requests ($131.6 billion), Congress was also willing to offer up more rescissions than the President: the amount of total rescissions enacted by Congress was almost 3 times that requested by the President, about $52 billion compared to $18 billion (CBO, 2001, p. 10).

In the 26 years from 1974 through 1999, 61 supplemental appropriations were enacted, totaling more than $430 billion dollars. This is an average of $7.05 billion per supplemental and $16.5 billion per year. The largest emergency supplemental was for the Persian Gulf War in 1991 and totaled $42.6 billion. Two other supplementals were passed that year totaling just over $1 billion (Godek, 2000, p. 36). Conversely, in 1974 a supplemental was passed for $4.75 million. Two more small supplementals were passed in 1974, one for $8.77 million and another for $8.66 million (Godek, 2000, p. 36). Twenty supplementals were passed from 1974 through 1979, including five in 1978 alone. More recent practice due to attempts to hold down spending to curb the deficit and, perhaps because of fewer natural disasters, has seen fewer bills passed each year; only one supplemental was passed in 1995, 1996, 1997, and 1999. Starting with the Gulf War supplementals, emergent military missions often made the defense portion of the supplemental a key policy issue where the debate focused not only on funding but also on the Presidents responsibility to inform Congress before engaging in certain peacekeeping missions. In 1991, 1992, 1999, (and 2000) defense needs constituted more than half of the supplemental bills. Over the 26-year period, the Gulf War supplemental was the only bill of 61 that was dedicated 100% to defense. Twenty of the 61 bills had some money for defense in them; the defense portion amounted to 19.1% of total supplemental spending over the 26-year period and 6.5% if the Gulf War supplemental is set aside. Compared to cumulative defense spending from 1974 through 1999, cumulative defense supplemental amounts are negligible, amounting to about 1.38% of total defense spending over the 26-year period (Godek, 2000, p. 37).

SUPPLEMENTALS PASS QUICKLY

Supplementals allow the federal government to show that it is willing and able to respond to urgent needs, both at home and abroad, and that it is capable of doing so quickly. This is a useful corrective to the regular appropriation process, which is often in a state of gridlock and where appropriation bills are usually late. Supplementals have a clear seasonal distribution. In most years, Congress introduces and passes supplementals within a 4-month period, usually in the late winter and spring (Godek, 2000, p. 47). From 1974 through 1999, 38 of the 61 supplemental bills passed during this period were introduced in February 12, March 13, or May 13. Eight bills were introduced from August through January. The last of these occurred in 1983; since then, no supplemental bills had been introduced and passed in the August through January time frame until the September 2001 supplemental stimulated by the terrorist attack on the United States.

**Table 7.1.   Supplemental Size Analysis
1974-1999 ($ Millions)**

Fiscal Year	Supplemental Amount	DoD Portion	DoD Budget	% of Supplemental	% of DoD Budget
1974	179		82600	0.000%	0.000%
1974	4.75			0.000%	0.000%
1974	8.77	2.14		24.401%	0.003%
1974	8.66			0.000%	0.000%
1975	638		90500	0.000%	0.000%
1975	143.2			0.000%	0.000%
1975	15070	256.3		1.701%	0.283%
1975	10300			0.000%	0.000%
1976	18000		97,567	0.000%	0.000%
1976	9400			0.000%	0.000%
1976	872			0.000%	0.000%
1976	2140			0.000%	0.000%
1977	200		110,362	0.000%	0.000%
1977	28900	101.1		0.350%	0.092%
1978	6800	514.8	117,349	7.571%	0.439%
1978	80.5			0.000%	0.000%
1978	250.2			0.000%	0.000%
1978	300			0.000%	0.000%
1978	7800	423.8		5.433%	0.361%
1979	13800	2900	126,880	21.015%	2.286%
1980	40		144,502	0.000%	0.000%
1980	7.6			0.000%	0.000%
1980	3800			0.000%	0.000%
1980	16900			0.000%	0.000%
1981	20900	6900	180,443	33.014%	3.824%
1981	11800			0.000%	0.000%
1982	2300			0.000%	
1982	5000		217,179	0.000%	0.000%
1982	14200	435		13.063%	0.200%
1982	5400			0.000%	0.000%
1983	24300		244,972	0.000%	0.000%
1983	7000	469.8		6.711%	0.192%
1983	4,600			0.000%	0.000%
1984	290		265,584	0.000%	0.000%
1984	6180	332		5.372%	0.125%
1984	1150			0.000%	0.000%
1985	13020		294,853	0.000%	0.000%
1985	784			0.000%	
1986	1500		289,625	0.000%	0.000%

1986	5300			0.000%	
1986	1700			0.000%	
1987	9400	720	287,960	7.660%	0.250%
1988	672		292,497	0.000%	0.000%
1988	709			0.000%	
1989	3300	2400	300,067	72.727%	0.800%
1990	4300		303,946	0.000%	0.000%
1991	42600	42600	332,228	100.000%	12.823%
1991	3700			0.000%	
1991	6900			0.000%	
1992	1100		299,115	0.000%	0.000%
1992	11100	4100		36.937%	1.371%
1993	4000		276,109	0.000%	0.000%
1993	3500			0.000%	
1993	5700			0.000%	
1994	13855	1497	262,246	10.805%	0.571%
1995	3100		262,862	0.000%	0.000%
1996	5051	982	265,014	19.442%	0.371%
1997	8900	2100	266,217	23.596%	0.789%
1998	6100	2800	272,370	45.902%	1.028%
1998	20800	1859		8.938%	0.683%
1999	14500	10900	288,117	75.172%	3.783%
	$430,354	$82,293	$5,971,164	19.122%	1.378%

Source: Godek, 2000. Table 4.1, p. 36.

In general, supplementals are passed quickly: 50.8% were introduced, passed, and signed in 2 months and 86.8% (53 of 61 bills) were passed within 4 months over the 26-year period. The average duration was a little over 3 months. Only eight took longer than 4 months. Over the 26-year period, only three supplementals took 8 months or longer. One of these, the supplemental passed in May of 1980 had been in process for 12 months. This was the longest duration supplemental. At the other extreme, seven bills took 1 month or less, from introduction to signing.

This may not seem particularly quick, but the passage of normal appropriation bills can take up to 10 months from introduction to passage, or longer. The great budget debacle of 1995-1996, saw the last Omnibus Appropriation Bill for 1996 passed in April 1996, 14 months after the Congressional budget process for that year had begun and 7 months into the then current fiscal year.

Table 7.2. Supplemental Timing

Supplemental Tracking Profile 1974-1999

CY	FY	Public Law No.	J	J	A	S	O	N	D	J	F	M	A	M	J	J	A	S	O	N	D	J
1974	74	93-305																				
1974	74	93-321																				
1974	75	93-554																				
1974	75	93-624																				
1975	75	94-06																				
1975	75	94-17																				
1975	75	94-32																				
1976	76	94-157																				
1976	76	94-252																				
1976	76	94-266																				
1976	76	94-303																				
1976	76	94-438																				
1977	77	95-13																				
1977	77	95-16																				
1978	78	95-240																				
1978	78	95-255																				
1978	78	95-330																				
1978	78	95-332																				
1978	78	95-355																				
1979	79	96-38																				

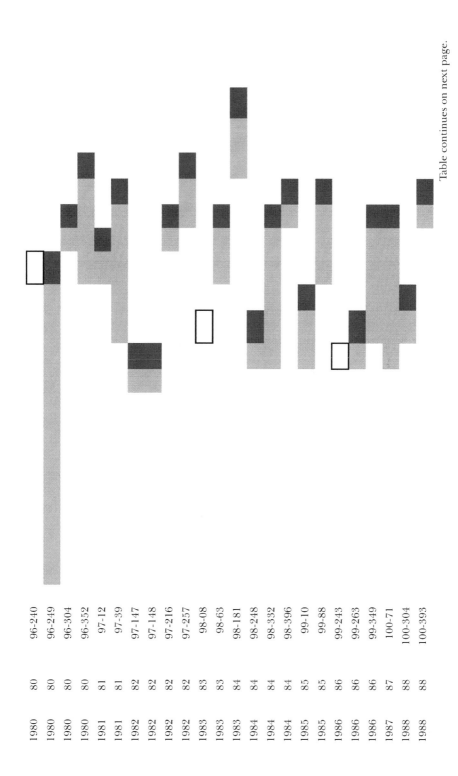

1980	80	96-240
1980	80	96-249
1980	80	96-304
1980	80	96-352
1981	81	97-12
1981	81	97-39
1982	82	97-147
1982	82	97-148
1982	82	97-216
1982	82	97-257
1983	83	98-08
1983	83	98-63
1983	84	98-181
1984	84	98-248
1984	84	98-332
1984	84	98-396
1985	85	99-10
1985	85	99-88
1986	86	99-243
1986	86	99-263
1986	86	99-349
1987	87	100-71
1988	88	100-304
1988	88	100-393

Table continues on next page.

Table 7.2. Continued

Supplemental Tracking Profile 1974-1999

CY	FY	Public Law No.
1989	89	101-45
1990	89	101-302
1991	91	102-28
1991	91	102-27
1991	91	102-229
1992	91	102-302
1992	92	102-368
1993	92	103-24
1993	92	103-50
1993	93	103-75
1994	94	103-211
1995	95	104-6
1996	97	104-208
1997	96	105-18
1998	97	105-174
1998	98	105-277
1999	99	106-51

Month columns: J J A S O N D J F M A M J J A S O N D J

NEW FY

NEW FY

Introduced/Congressional Action

Conference Report Approval/President signed

Introduced, Approved, and signed by President

Source: Godek, 2000. Table 4.1, p. 36.

Table 7.3. Supplementals Pass Quickly (*N* = 61)

Within 1 Month	*Within 2 Months*	*Within 3 Months*	*Within 4 Months*
7	24	14	8
7/61(11.4%)	31/61 (50.8%)	45/61 (73.7%)	53/61 (86.8%)

Note: Percentages are cumulative.

Source: McCaffery and Jones, 2004, Table 7.2.

SUPPLEMENTALS AND THE DEFENSE FUNCTION

When the President decides to evacuate embassy staff, rescue American citizens, deploy troops in an emergent peacekeeping action, or deploy defense forces in a humanitarian enterprise, he creates an unbudgeted expense for the defense budget. The President authorizes action by the DoD, which executes the mission using whatever tools and personnel are at hand. If the mission is small and very limited, the expense may be absorbed within the DoD budget. For larger, more extensive, missions, DoD executes the mission and pays for it by "borrowing" against its fourth quarter supporting expense budget, the O&M account. DoD then depends on Congress to pass a supplemental appropriation early enough in the summer so that whatever was borrowed can be replaced, and the original fourth quarter spending program can be executed as planned.

Conventionally, if the mission happens in the fall or winter of the year, the DoD comptroller, with OMB's permission, waits to submit the supplemental until after the budget process has begun in the following spring so that Congress does not confuse the supplemental needs with those asked for in the regular appropriation bill. This usually means that since the regular Congressional budget process begins in February, the defense supplemental may be submitted in late March or April for events that happened the previous fall or winter. Usually DoD and OMB agree to wait with the supplemental until some weeks after Congress has started hearings on the President's budget request for the next fiscal year. This prevents confusion between paying for what has already been spent in the supplemental request or what will be spent in the rest of the current fiscal year and the needs presented in the President's budget for the next fiscal year. Normally the Budget Committees and the authorizing committees do not have jurisdiction over supplementals, but as will be seen later in this chapter the authorizers can bring up questions about the supplemental at hearings about the main DoD budget bill. Also recently the Budget Committees have sometimes left space in the budget resolution for upcoming wartime supplemental bills (for example, in

2007 the Senate budget resolution set aside $99.6 billion for expected supplemental war costs, while noting that true emergency costs need not be provided for under the budget resolution spending cap). Planning for the supplemental is particularly prudent when the President/DoD have announced in the fall that a sizeable supplemental will be presented in the spring. Also, generally OMB and DoD have agreed to what DoD should include in the supplemental, thus the wartime supplementals that included items that many in Congress thought should be in the base budget were agreed to by OMB, and not just by DoD. Thus the development and passage of supplementals is neither so simple nor quick as it would seem. In general, dire emergency supplementals go directly to the floor of Congress; for example, the ($10.5 billion) supplemental for Hurricane Katrina was introduced in the House, was referred to the appropriations and then the Budget Committees, discussed and approved on the House floor (in 28 minutes), sent to the Senate where it was passed and sent back to the House for agreement, cleared for and sent to the President, where it was signed and became law, all on

Table 7.4.   Legislative Route of the Defense Supplemental in 2005

H.R.1268: Title: An Act Making Emergency Supplemental Appropriations for Defense, the Global War on Terror, and Tsunami Relief, for the Fiscal Year Ending September 30, 2005, and for Other Purposes	
3/11/2005	Introduced in House
3/11/2005	The House committee on appropriations reported an original measure, H. Rept. 109-16, by Mr. Lewis (CA).
3/16/2005	Passed/agreed to in House: On passage Passed by the Yeas and Nays: 388-43 (Roll no. 77).
4/6/2005	Committee on appropriations. Reported by Senator Cochran with an amendment in the nature of a substitute and an amendment to the title. With written report No. 109-52.
4/21/2005	Passed/agreed to in Senate: Passed Senate with an amendment and an amendment to the Title by Yea-Nay Vote. 99-0. Record Vote Number: 109.
5/3/2005	Conference report H. Rept. 109-72 filed.
5/5/2005	Conference report agreed to in House: On agreeing to the conference report Agreed to by the Yeas and Nays: 368-58, 1 Present (Roll no. 161).
5/10/2005	Conference report agreed to in Senate: Senate agreed to conference report by Yea-Nay Vote. 100-0. Record Vote Number: 117.
5/11/2005	Signed by President.
5/11/2005	Became Public Law No: 109-013.

Source: Library of Congress Web site at Thomas, http://thomas.loc.gov/cgi-bin/bdquery/ z?d109:HR01268

September 2, 2005. The 2005 defense supplemental went through the appropriation committees and took a somewhat longer path.

DoD and Incremental Costs

With the BEA of 1990, the DoD was directed to report only the incremental costs of carrying out contingency or emergency operations. This was clarified in 1995 and extended in February, 2001 to incremental costs "above and beyond baseline training operations, and personnel costs" (DoD Financial Management Regulations, pp. 23-26.). Thus, the military departments may identify incremental costs related to military and civilian pay, clothing and equipment, reserve activation, operational training, supplies and equipment, facilities and base support and airlift, sealift and inland transportation. Before this detailed guidance, each service had great latitude in developing incremental costs.

Historically, DoD has tended to absorb the costs of very small missions. The incremental cost doctrine has increased this tendency, thus in some missions an incremental cost bill could be specified, but DoD decides to absorb it, because the cost of the mission is small enough that absorbing it does no serious harm to executing the defense program planned for in the original defense budget. In some instances, diverting personnel and ships from training exercises for a small short-term rescue mission does not cost much and the actual activity might be close enough to the training mission to provide a training benefit, albeit reduced. Moreover, it is a fact of life that the President is not going to ask for, nor will Congress provide, funding for such small missions. For the larger and more elaborate peacekeeping and humanitarian missions, incremental costs are priced and the OMB comptroller, in essence, submits the bill to Congress as part of the annual supplemental appropriation request. It is clear that the incremental pricing requirement has resulted in more precise pricing of defense supplementals. It may also have resulted in a slight decrease in total costs, as DoD absorbs small incremental costs rather than asking for their reimbursement, especially small operations which can be considered similar to baseline training exercises.

DoD Supplementals: Pay-As-You-Go and Dire Emergency Status

DoD supplementals were impacted by the pay-as-you-go provisions of the BEA of 1990 and subsequently. BEA required that spending above the allowable spending caps be deficit neutral, meaning that either a revenue source had to be found or funding taken from another program through a rescission. This program, then, would become the bill-payer. Alternatively, the President could ask that the supplemental be classified as a dire emergency; if Congress concurred, the supplemental amount would be relieved of the bill-payer necessity and the allowable spending cap would, in effect, be increased. In addition, beginning in 1991, the emergency designation was used for some funds in regular appropriations bills.

Consequently, by the end of the decade, the result of this process was rather large supplemental appropriations bills, with a mix of emergency and nonemergency funding, and regular appropriations with emergency funding in them. Designation of part of a supplemental as emergency is a further complication, as is the mixing of the emergency designation into regular appropriations bills. Moreover, sometimes the emergency funding lasts for more than 1 year, thus a supplemental with emergency funding could affect not only the current year, but the next year as well. In fact, this is what happened in 1999 when the emergency supplemental enacted on May 21, 1999 carried funds not only for 1999, but also for 2000. Most of the nearly $2 billion for 2000 was for military pay and retirement (CBO, 1999, p. 2). These are items that would seem to fit well into the normal appropriation bill; moreover, they also are an ongoing expense which become part of the budget base, as opposed to a one-time emergency. Using emergency supplementals in this manner tends to confuse an already complex budget process.

Critics have argued that the obvious solution is to put the second year of emergency funding in the regular appropriations bill. Others observe that when spending caps are set unrealistically low, spending is deemed emergency spending simply to get around the caps, and that caps that are more realistic would lead to fewer or smaller supplementals and less emergency spending. An example of cap avoidance using emergency spending designation occurred in 1999 when $21.4 billion of emergency spending was included in regular appropriations, a further confusion of the process. The budgetary politics of the 1990s was a history of spending caps first imposed from 1990 through 1995 with the BEA of 1990 and then extended in 1993 and 1997, for a 5-year period through 2002. The net effect of these caps was to put the discretionary sector of the budget under very tight discipline, so that budget growth was capped at less than inflation for most agencies.

From 1995 to 1996, discretionary Budget Authority fell by $12 billion and the caps set in the 1997 budget agreement for the years out to 2002 were quite tight (CBO, 2001, p. 12). Without a concomitant decrease in the discretionary tasks of the federal government, these tight caps were an open invitation to escape onerous and tight discipline through the supplemental appropriations escape hatch. The caps, a surplus in 1998, and electoral positioning for the year 2000 election resulted in large supplemental appropriations, of $13.3 billion in 1999 and $22 billion in 2000.

With an ordinary supplemental, one not deemed a "dire" emergency, the President and Congress have to find funding sources to pay for the bill. This could mean sources inside or outside the defense budget. A bill-payer outside the defense budget, say surplus money in the food stamp account, means that the supplemental would be funded outside defense. Sometimes DoD is required to come up with a portion of the supplemental from DoD funds. For example, in some years Congress has authorized a pay raise and funded only half the increase, asking DoD to find money for the rest of it. It is an interesting irony that a supplemental can result in a decrement for someone when a supplemental is not designated as a dire emergency. In this case, some departments and programs will be used as bill-payers and will find their programs decreased, not supplemented. If the money comes from an under executing procurement or mandatory account and goes into permanent salaries as a pay raise, not only will the salary account have to be increased in the following years, but the original funds will probably have to be restored to the procurement or mandatory program account. The net effect in future years may well be an increase all around.

In the Navy budget, offsets might come from delaying property maintenance or ship repair and overhauls, categories where spending could be delayed into the next fiscal year without too much damage to program execution. Notice that in these cases DoD and Congress are, in effect, cost-sharing the supplemental. Congress uses this technique to drive down the cost of the supplemental. Thus, even though the supplemental has been accurately priced, Congress may refuse to pay the whole bill and require a contribution from DoD, which, on its part, may borrow against the future by delaying construction or maintenance programs already funded in the budget into the future. These items won approval through the executive budget process and in Congress, but now they may be forced to recompete for approval. In effect, the supplemental has forced a cut in currently approved programs and put them in jeopardy again. For DoD, the cost-sharing solves one problem in the present, but creates future budgetary needs.

Congress, for its part, has a series of choices to make, that is, fully-fund or cost-share, find bill-payers to provide rescissions or declare all or part of the bill a "dire emergency." A mixed approach is possible. For example, the 1999 supplemental was a hybrid; it was offset by $1.7 billion mainly from the food stamp account and the rest of the $15 billion was called a dire emergency.

Conventional wisdom suggests that dire emergency status is only used for supplemental bills, but this is not true. In order to escape spending cap restrictions, all of the supplementals passed in 1991 were designated as dire emergency spending, but so was an additional $1 billion of regular appropriations. Using the dire emergency designation to avoid spending cap discipline in the regular appropriations bills set a precedent that was embraced throughout the decade. The use of the dire emergency designation for regular appropriations occurred every year, ranging from $314 million in 1992 to $2.1 billion in 1997 (CBO, 1999, p. 2). This may be seen as both an evasion of budget discipline, and as a safety valve, particularly since the amounts were generally a very small percentage of the total budget. Whatever the justification, since the mechanism was available, Congress decided to use it.

In 1999, $21.4 billion was designated emergency spending in the regular appropriation bills, indicating a rather large breakdown in budget discipline, resulting from a budget surplus and rather tight spending caps. An additional $12.9 billion was designated as emergency spending in the supplemental bill for that year, but the argument can be made that this is proper because the primary purpose of supplementals is to fund emergencies.

Logically, it would seem improper to include emergency designated funds in regular appropriation bills, which are, after all, for the following fiscal year. Since most appropriation bills are passed after the start of the fiscal year, Congress does have an opportunity to address emergency needs within current appropriations bills. This may help explain the rise of this practice as well as why supplementals are generally a spring and summer phenomenon.

While the emergency designated sums are very small compared to the total budget, the emergency designation device is rather widely used among appropriation subcommittees and in agency appropriation bills. For example, in 1999, while defense was the largest beneficiary of the emergency designation, 12 appropriation subcommittees had some emergency money within their purview (only the District of Columbia failed to benefit) and 18 cabinet level agencies gained dollars through the emergency designation process, with NASA (National Aeronautics and Space Administration), Veterans Affairs, and the Department of Education being overlooked in this year. NASA was the only agency never to receive emergency designated money in the 1990s (CBO, 1999, p. 3). How this trend develops needs to be monitored.

Both the President and Congress originated emergency designated money from 1991 to 2000. Over the period, about $146 billion was designated as emergency money in both supplementals and regular appropriations. The President took the lead in asking for the emergency designation from 1991 through 1995, asking for 81% of the emergency funds during this period, with Congress leading from 1996 through 2000, asking for 56% of the funds designated as emergency money (CBO, 1999, p. 10). This is a reflection of the great struggle between the Republican Congress and President Clinton over the budget in the latter years of the decade.

The temptation is to say that Congress and the President learned over the decade how to get around the spending cap discipline by designating some funding in regular appropriation bills as emergency funding, but the fact is that $1 billion of regular appropriation money was designated as an emergency in 1991, thus illustrating that the President and Congress understood the mechanism in 1991 and used it.

Additionally it may be noted that when supplementals pass early in the session, as most of them do, and with the tendency in the 1990s to pass only one supplemental, and considering that Congress does not get serious about its budget work until August, September or later, the ability to designate dollars in regular appropriations bills for emergency purposes presents a convenient safety valve. However, it complicates our understanding of the supplemental appropriation process when emergencies are found in regular appropriation bills and nonemergencies in supplemental bills (e.g., maintenance, spare parts, and family housing funding in the defense supplementals in 1998 and 1999). It also impairs the transparency of the budgetary process.

SUPPLEMENTAL HISTORY

Supplemental spending bills are not new. The use of supplemental appropriations began with the second session of the first Congress in 1790 and continued in the 1800s (CBO, 2001, p. 1). They frequently included additional appropriations for agencies that had overspent their appropriations and as such became known as deficiency appropriations. This practice became so routine that the House Appropriations Committee divided them into general deficiency bills and urgent deficiency bills. In the 1870s when antideficiency legislation was passed, critics of this process accused Congress of underfunding the regular appropriation bills to appear frugal stewards of public funds, and then, "after elections were over, make up the necessary amounts by deficiency bills" (CBO, 2001, p. 1).

The Antideficiency Act of 1905 attempted to control deficiency spending by giving the Treasury Department the authority to apportion funds to agencies to reduce the need for supplementals. This power was further refined by the Budget and Accounting Act of 1921 and the Antideficiency Act of 1950 that encouraged agencies to set aside reserve funds for emergencies, and limited supplemental appropriations to legislation enacted after the Presidents budget had been submitted and for emergencies relating to the preservation of human life and property (CBO, 2001, p. 2). However, the issues surrounding supplementals have remained contentious. In 1966, the joint committee on the organization of Congress again raised the issue of lawmakers projecting an image of economy by underfunding the regular appropriation bills with the tacit understanding that they would later pass supplemental appropriation bills (CBO, 2001, p. 2).

In the 1970s, supplementals primarily affected mandatory accounts like unemployment benefits and increased food stamp funding. In 1977, discretionary programs were supplemented with a $9.5 billion program for job training to counteract the recession of 1973-1975. Most of the other supplementals for the 1970s were for federal pay raises, programs that were authorized after their appropriation bill had been passed, and expenses related to natural disasters such as blizzards, floods, droughts, and forest fires. Supplementals declined in the 1980s, to a low of .1% of Budget Authority in 1988, mostly as a result of the struggle with the deficit and the requirement for offsets to pay for supplementals. Mandatories were again high: two-thirds of the supplementals in this decade were for mandatory payments, primarily for support of farm commodity programs due to worse-than-expected conditions in farm commodity markets. Supplementals also were applied to the food stamp program, unemployment insurance, and various higher education programs.

In the 1990s, only 6.3% of all supplemental appropriations went to mandatory accounts (CBO, 2001, p. 10). Most of the discretionary supplemental appropriations in 1991 were for military operations Desert Storm and Desert Shield. Domestic spending dominated discretionary supplemental appropriations from 1993 to 1998, but defense spending re-emerged as the largest category in 1999 and 2000 because of peacekeeping missions in Bosnia and Kosovo.

Academic research on supplementals has been sparse. In his *classic Politics of the Budgetary Process* and in subsequent revisions, Aaron Wildavsky mentions supplementals, suggesting that, "Congressmen get headlines for suggesting large cuts, but they often do not follow through for they know that the amounts will have to be restored by supplemental appropriations" (Wildavsky, 1984, p. 23; 1997, p. 50). Wildavsky observes that appropriations committee members may even talk to agency officials about areas they can cut and then restore later in supplementals, but he does not pursue this issue further. However, Christopher Wlezien does, examining the relationship between appropriations and the supplemental process based on data from 1950 through 1985. Wlezien found that regular appropriations and supplemental appropriations were linked in a two-stage process with certain accounts underfunded in the regular appropriations bills and replenished through the supplemental process. Wlezien associated these results with party identification and found that Republican Presidents were more likely to engage in strategic underfunding and make it up later with supplementals during the period under study.

Wlezien (1996) explained that the two-stage process he discovered in the appropriations process from 1950 to 1985 occurs in a single stage in the early 1990s, with the politics of bargaining between the President and Congress, "largely confined to regular appropriations" (p. 62), primarily as a result of changes in the budget process. Observation of the 2000 and 2001 supplementals seems to indicate supplementals and main appropriation bills each have an arena for institutional bargaining; with supplementals the bargaining seems to be about size, dire emergency designation, finding bill-payers, and pork. Supplementals have also been the locus of bargaining about the limits of executive power, with Congress suggesting that the President must get advance approval from Congress before engaging in peacekeeping operations and then presenting Congress with a bill for a fait accompli.

While these are topics for future research, we agree with Wlezien that the old two-step process has largely disappeared. Largely, but not completely; the defense spending supplementals of the latter years of the 1990s deserve a closer look; it appears that some of the items funded were continuing expenses and that the pattern was for Congress and the President to hold to a tight cap for defense spending and then add money through the emergency supplemental process - the old two-stage process.

NONDEFENSE SUPPLEMENTALS

In the 1990s, after the DoD, the Federal Emergency Management Agency (FEMA) was the second largest recipient of discretionary supplementals, receiving substantial amounts of supplemental Budget Authority in every year of the decade except 1991 to total $22.1 billion. That amount primarily represents appropriations to the FEMA disaster relief account to pay for relief efforts in the wake of hurricanes, floods, earthquakes, and droughts. From 1992 on, the Small Business Administration also received funding for disaster loans, as did the Commodity Credit Corporation for aid to farmers for crop losses. In 1999, funds were also provided to compensate farmers through price supports for certain commodities (CBO, 2001, p. 17).

For disaster assistance that involves longer-term aid, like loans, loan guarantees, or other forms of financial assistance other than direct federal emergency aid, passage of the supplemental may take longer and reflect different party viewpoints on what is the best way to handle longer-term disaster relief. Thus, generalizations about supplementals are difficult to make other than that they are small compared to total budget spending and result from unforeseen events. Their symbolic value is high. It was clear after the terrorist attacks in September 2001, that the American people expected their government to protect them and help with disaster relief. This was obvious in the quick passage of the $40 billion supplemental in aid of the World Trade Center disaster in 4 days that September and the relatively quick passage of the $65 billion Operation Iraqi Freedom supplemental in 2003.

NONEMERGENCY SUPPLEMENTALS AND OFFSETS

When DoD had to offer up offsets in a nonemergency supplemental, the usual practice was for the military departments to share equally, each coming up with one-third of the offset. If the costs of the mission are equal, this is fair for all. However, the military departments have different structures and in joint missions, one service might be able to accomplish its mission more easily, and cheaply, than another service. If they all are asked to offset the bill on an equal shares basis, a tradition within defense because all share the pain equally, one military department may help pay for expenses in another.

For example, since the Navy is forward deployed, its costs may be less than Army and Air Force, but, if it has to offer up a full third of the offset, the result could be that Navy money will flow to pay Air Force or Army bills. Suppose the total cost of the supplemental is $6 billion, with the Navy share $400 million in incremental costs and the Army and Air Force costs each being $2.8 billion. If Congress decides that DoD has to offset this cost and the DoD comptroller asks each military department to offer up $2 billion in offsets, the Navy would gain $400 from the supplemental but contribute $2 billion to the offset for the supplemental. The Navy might have to cut its fourth quarter operational tempo—flying and steaming hours—so that the Army and Air Force can more fully fund their fourth quarters. Much of this depends upon how DoD leaders, including the comptroller, decide to allocate shares of the cost burden. For example, the decision might be made to share the costs equally or a differential cost pattern might be imposed, taxing one military department more heavily than the others because budget execution reports indicate that some slack is evident in its budget due to program underexecution.

Recovering true incremental costs is sometimes difficult when the incremental cost is small. For example, if Navy flying hours increase 3% over the normal yearly allotment because of incremental costing related to emergencies, the DoD comptroller tends to urge the department to absorb it, because it is so small. This tendency is even more apparent when it comes to incremental maintenance costs due to emergency operations. As one source said,

> If flying hours go up 30,000 in a year due to a contingency operation, that is a small number on a 1,000,000 hour flying hour program. Because we are out there flying and steaming anyway, it is hard to argue for these small percentage increases.

He added that absorbing costs was all right for the ships already deployed, but the nondeployed units would not be getting enough training dollars to get to the right readiness level unless the incremental costs of the emergency operation were put back in the budget.

While an emergency tasking cost might seem quite small as a percentage of the overall defense budget, it is a much larger percentage of the bill-paying budget account. This is usually the fourth quarter O&M account where the opportunity exists to change the pace of spending. For example, in 1995 while Congress was debating a $2.5 billion supplemental, the Navy Comptroller Office calculated that this would be about 1% of the DoD budget, 3% of the DoD O&M budget, the usual bill-payer for newly emergent military missions, but 10% of flexible O&M and 40% of fourth quarter flexible O&M. What this means is that if the money was not restored, a sizeable amount of planned and budgeted program would go unexecuted, most of it in force operations and direct support.

Each quarter in the O&M account mandatory expenses of a contractual or semicontractual nature must be paid and these cannot be arbitrarily foregone. Amounts have to be paid that support troop exercises or training missions that have to go forward and be executed. Money cannot

Table 7.5. Flexibility in O&M Budget: Department of Defense FY1995 Operations and Maintenance Budget: $91 Billion

No Flexibility	*$44 Billion*	*Limited Flexibility*	*$20 Billion*	*Flexible*	*$27 Billion*
Civilian pay/benefits	23	Natl guard/Reserves	6	OPTEMPO*	10
Health program	10	Contract services	5	Depot maint.	5
Environmental costs	4	Drug interdiction	1	Support/transport	12
Utilities/rents	3	Recruiting/training	3		
Mobilization/other	4	Support activities	5		

* *Note:* Optempo represents the series of accounts that provide funding for direct operations such as flying and steaming hours.

Source: McCaffery and Jones, 2004, p. 198.

simply be "robbed" from these accounts (it may be borrowed but it has to be repaid within the same fiscal year).

In 1995, the flexible accounts comprised $27 billion of the $91 billion O&M budget. If these were spread evenly over the four quarters, DoD would spend about $6.75 billion in the fourth quarter. The $2.5 billion supplemental would comprise about 37% of this amount. In these accounts where flexibility exists, the military departments that advance the money out of earlier quarters count on getting it back in the fourth quarter in time to execute it as planned.

This is not always a certainty. In 1998 the supplemental bill was passed early in the spring, but the Office of Secretary of Defense allowed only about two-thirds of it to flow through to the Department of Navy while it conducted a rebudgeting drill to see if the Navy really needed the other one-third. In 1999, to help with budget execution, DoD created a revolving fund to pay for emergency actions. This appeared to be a promising approach, but it too had problems in estimating proper reimbursement and getting the funds returned on a timely basis to the correct source.

Observations

The BEA of 1990 changed the supplemental process by requiring incremental costs and imposing deficit control measures that involve spending caps, bill-payers and/or dire emergency designation. These changes appear to have changed the nature of supplementals as well as the supplemental process. Insofar as the old two-step supplemental process is concerned, the budget ceilings set aside all a functional area could receive for a year, so it would be a waste of time to under fund an area and expect to come back later and fund it in a supplemental; another program might have used up the space under the ceiling in the meantime and a sequestration (across-the-board cut) could be ordered for an appropriation that breaks the ceiling subsequently. In 2005, the defense appropriation bill broke the ceiling on discretionary spending and resulted in an across the board cut, although the bridge supplemental was held harmless from this be declaring it a dire emergency, as we discuss later.

One such across-the-board cut was ordered as a result of the supplemental that was passed on April 10, 1991 (PL 102-27), going into effect on April 25, 1991 and cutting all nonexempt accounts (by .0013%). Supplementals that are passed before July 1 of the year and break a spending cap result in an immediate sequestration. Supplementals passed after July 1 result in a sequestration against the spending cap for the following year. The April 1991 supplemental was the only supplemental in the 1990s found to have breached a spending cap (CBO, 2001, p. 5). Thus, the BEA enforced a pay-as-you-go discipline after 1990 for supplementals not designated as dire emergencies. This complexity remains with us. In 2005 the defense appropriation bill was the last one passed and it broke the discretionary spending caps and resulted in a 1% across the board cut on it and on the other appropriations bills already passed. However the supplemental part of the bill was designated a dire emergency and sheltered from the cut. Then, in 2006, as is discussed later, Congress cut from the requested defense appropriation and used the defense cuts to fund their domestic priorities, while making up the defense cuts in the dire emergency supplemental part of the defense appropriation bill. This kind of behavior is sometimes called smoke and mirrors because Congress is using procedural magic to evade the discipline of agreed upon budget process mechanisms.

Irrespective of the funding technique used, the supplemental becomes a focus of attention and coercive deficiency strategizing becomes less likely, if not simply impracticable.

Thus we suggest that supplementals are not like other appropriation bills. An appropriation bill is a forecast of what is to come, a promise that X units of work will be accomplished with Y dollars for personnel and supporting expenses. However, no matter how carefully budget reviewers labor over the numbers, the budget remains a forecast and the corresponding appropriation is similarly imprecise. Supplementals, be they defense or disaster related, are different. Disaster-focused supplementals are not like regular appropriation bills because they are basically lump-sum appropriations focused on a goal, remediation of the disaster. The funding mix may resemble what was done for similar disaster situations in the past or it may be set up as what is recognized as an initial endowment with more funding to follow, once the magnitude of the disaster is measured. Moreover, the symbolism of governmental response to the crisis is important. The funding provided may or may not be enough for the task; what it is designed to do is get aid to a problem quickly, in the current fiscal year, in order to start the healing process. Rough estimates of what will be necessary are made, but a disaster supplemental cannot have the careful budget quality numbers that go into an appropriation request.

Defense focused supplementals are different from both appropriation bills and disaster-related supplementals. A defense-focused supplemental is unlike an appropriation request because the defense supplemental is a statement of costs for services actually performed. It is accurate and precise, much more precise than an appropriation bill can be. Even the best appropriation estimate cannot have the accuracy of the defense supplemental; after all, one is a forecast, the other is a bill for the incremental costs of services rendered. Some defense supplementals have funded ongoing military activities that last the duration of a fiscal year, or more. These are more like an appropriation bill in their duration, but better than an appropriation bill in their precision. Moreover, the defense supplemental with its itemized pricing and focus on incremental cost is at the opposite end of the scale from the lump-sum disaster supplemental; the former is retrospective and pays for work already done while the latter is prospective in providing aid in a disaster situation.

In sum, supplementals result from unbudgeted and largely unpredictable events. They are small compared to total budget spending, but they pay for what is needed in defense areas and have great symbolic value both in defense and disaster relief. It was clear after the terrorist attacks in September 2001, that the American people expected their government to protect them and help with disaster relief. One way government can show it is listening is by passing a supplemental appropriation. This was obvious in the quick passage of the $40 billion supplemental for relief of the World Trade Center disaster in 4 days in September of 2001. Events in recent years have only added complexity to the description of supplementals. Since 2001, DoD has gained its funding in a stream of regular appropriation bills and supplementals, including those of July 24, 2001 ($6.74 of $9 billion), September 18, 2001 ($3.4 billion of $20 billion), May 21, 2002 ($14 billion of $21 billion), and April 2003 ($62.6 billion of $75 billion). None of these was completely defense, but all had some defense money in them. July 2001 saw passage of a "normal" supplemental; the rest have been reactions to wartime emergencies, including the cost-or-war supplemental in April 2003 and thereafter. These wartime supplementals need separate study because they depart from the model presented thus far in this chapter.

Wartime Supplementals

So far we have argued that supplementals fund the current year for emergency operations, that disaster supplementals are an estimate and defense supplementals precise because they are paying back for events that have already occurred; thus the theory is that an emergency occurs and the President sends DoD to help. DoD borrows against its fourth quarter funds and the supplemental pays back the fourth quarter budget in time for DoD to execute its fourth quarter budget as originally planned. While this was true of the humanitarian, disaster-relief, and peacekeeping activities of the 1990s, it was not true of either the Vietnam era or the Iraq/Afghanistan supplementals. Thus, our simple scenario for supplementals in peacetime (and peacekeeping), does not work in time of war.

In essence, the DoD appropriation (as well as that in Department of Energy for nuclear purposes) builds a defense establishment to deter war or to go to war. Recent history suggests that when that machinery is put to use, the costs of war are at least partially funded through a supplemental, at least initially and then folded into the base budget. This was true of World War II, Korea, and Vietnam. For example, two supplementals were passed in 1941, the first in March and the second in October before the United States entered the war. Wartime supplementals were also passed in 1942 and 1943.[1] The October 23, 1941 supplemental was the Lend-Lease Act which allowed the United States to provide ships and other supplies to Great Britain while still remaining technically neutral (Evans, 2006). This was of great symbolic significance to our friends and enemies alike and resulted in some real material help for Great Britain. World War II was pursued subsequent to the attack on the Pearl Harbor, under the aegis of a Declaration of War in which the U.S. pledged its full treasure to seeking a successful outcome (Miller, 2006). The magnitude of the effort for World Ware II dwarfs anything since that time, as can be seen in Figure 7.1. No conflicts since then have been fought under a declaration of war, but they all have involved supplemental funding. In general, supplementals helped finance the early stages of these conflicts and then ongoing costs were included in regular appropriations bills as soon as cost projections could be made (Daggett, 2006, p. 2). Korea is a perfect example of this; the wartime supplementals passed for FY 1951 beginning in September of 1950 tripled the size of the DoD budget in 1952.

In Vietnam, the Johnson administration used a mixed model of a supplemental for FY 1965, a budget amendment for FY 1966 and regular appropriations and supplementals for FY 1967 and FY 1968. In the early 1990s funding operations in Somalia, Southwest Asia, Haiti, and Bosnia were provided in supplementals; from 1996 on funding for further activities (Southwest Asia, Bosnia, and Kosovo) was provided through the regular defense budget and appropriations bill after an initial supplemental. Perhaps the least typical occurrence was the funding for the Persian Gulf War of 1990-91. There was a $42 billion supple-

Table 7.6. Korean War Funding: Regular and Supplemental DoD Appropriations During the Korean Conflict, FY 1952-FY 1953 (Billions of Then-Year Dollars)

Fiscal Year	Regular Appropriations	Supplemental Appropriations	Total Appropriations
1951	13.0	32.8	45.8
1952	55.2	1.4	56.6
1953	44.3	0	43.3

Source: Daggett, 2006. p. 3.

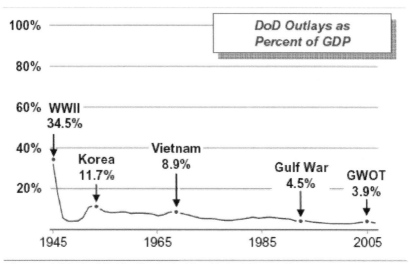

Source: Historical Table, Budget of the United States Government

Source: Department of Defense, 2007. FY 2007 Supplemental Request for the Global War on Terror. Washington, D.C., February, 2007, p. 2.

Figure 7.1.   War funding as a percent of GDP.

mental for FY 1991. Most of this was paid back by allied nations. Also as Daggett (2006) observes, "Costs declined rapidly after combat operations were over, so additional funds were not needed, either in supplemental or in regular appropriations bills" (p. 6) It should be noted that this conflict did change the focus of U.S. defense posture in the 1990s and included creating and maintaining bases around the Persian Gulf as well as increased deployments for carrier groups to the eastern Mediterranean and Persian Gulf, thus there were continuing costs that were built into the base budget.

From the perspective of growth in spending and persistence over time and for impact on DoD, the closest comparison to the current era appears to be Vietnam. For example, the fiscal history of the Korean War is short but dramatic. Three supplementals were passed in FY 1951, the first in the fall of 1950. They totaled almost $33 billion. In FY 1952, there was no supplemental per se for the war, but there was a $1.4 billion deficiency appropriation supplemental to cover overspent war accounts passed on June 28, 1952. In FY 1953 there were no supplementals and costs were included in the regular budget (Daggett, 2006, pp. 3-4). Vietnam was an example of a much longer and complex fiscal history. Robert Buzzanco (1996) observes:

In order to pay for the growing commitment to Indochina while keeping political attention at home focused on the Great Society, McNamara's budgets always underestimated the amount of money needed for Vietnam. The defense secretary always assumed, for budgeting purposes, that the war would end on 30 June of the following year (the end of the fiscal year at that time), which meant that the defense budget included no provisions for additional military funding for Vietnam. (p. 238)

Then says Buzzanco, when funds proved too limited, the administration would return to Congress for supplementals:

By underfunding the military in Vietnam in his budgets and using supplemental requests to make up for the shortfalls, the defense secretary was providing political cover for the administration, which wanted to avoid debate on the costs of the war, and placing the burden for the economic consequences of escalation on the armed forces, which were always asking for more money, and the Congress, which approved the additional expenditures. (p. 238)

Thus President Johnson could build both his Great Society (including Medicare, voting rights, the War on Poverty, and education legislation) and pursue the war in Vietnam. Buzzanco adds that this strategy masked the true costs of war, led to exorbitant defense spending, and caused military leaders to focus on domestic political battles rather than the best alternatives for Vietnam spending and by early 1968 "would create one of the gravest crises in civil-military relationships in contemporary times" (Buzzanco, 1996, p. 240).

General Taylor observes:

as the nature of the war developed into one of attrition, there was an intense feeling frustration created in the minds of Congressmen. They wanted to assure that the American fighting man was provided everything he needed, but his needs were often in direct competition with domestic programs. This change in Congress was a subtle one, with no real turning point, but the budget hearings slowly required more and more information concerning the appropriate use of resources. (Taylor, 1974, p. 11)

General Taylor's comments are drawn from an official Army study using official records conducted by key people. General Taylor was assistant director and then director of the Army budget office from 1966 to May of 1971. He says that initially participants thought Vietnam would be like Korea, with no requirement for stringent accounting controls or budgeting and that a sufficient amount of funds would be available and by the time the rules were clarified, it was almost too late on the accounting side, for transactions had already occurred which were impossible to document in the normal fashion (Taylor, 1974, p. v).

This could easily have been written of 2006 or 2007. For his part McNamara would argue that "no Congress in the midst of military operations has ever had as much detailed information as this Congress has been receiving." He further said that the information Congress was getting was vastly more precise than what had been submitted for Korea and World War II, that Congress still had not lost control of the war, that DoD was "so uncertain of the requirement as to be unable to document it" and that it was "absolutely impossible for us to predict the actions of our enemy 21 months in advance (the length of the budget cycle)" (Taylor, 1974, p. 27). In 1965, U.S. troop deployments to Vietnam went from 3,500 in March to 200,000 (Taylor, 1974, p. 19) by the end of the year; no ordinary budget process would support this kind of change. However, it is important to note that the FY 1968 defense budget did change the process. It assumed that the war might extend beyond the end of the fiscal year (June 30, 1968) and that the current level of activity would be experienced (Taylor, 1974, p. 29). Thus by 1967-1968, the costs of the war were included in the budget base and it was even anticipated that the war might last longer than 1968. As a result, it was anticipated that no supplemental would be needed, although, in fact, one was passed, albeit smaller than those of 1966 and 1967. Table 7.7 displays these Vietnam era supplementals.

By 1967 Congressional observers thought that the cost of carrying on the war in Vietnam had become $20-25 billion. (The exact amount was classified information.) In fact, the administration's budget for 1968 said

Table 7.7.   Vietnam Era Supplementals

Calendar Year	DOD $ Approp.	Y:Y (%) Change	Supplemental Bill $	Ratio (%) $S/$DoD
1964	46.7		1	2.14
1965	46.88	0.39%	‘ 0.7	1.49
1965	46.88	0.00%	1.7	3.62
1966	58.06	23.85%	13.13	22.61
1967	69.93	20.44%	12.169	17.4
1968	71.9	2.82%	6	8.33
1969	69.64	−3.14%	1.272	1.82

Source: Congressional Quarterly Almanac, 1964-1969. In 1964 while there was no defense supplemental; there was a general supplemental that affected 13 agencies, totaling $1.33 billion; about $1 billion of this was for DoD pay increases.

that the cost of war was now in the base budget and that a supplemental would not be needed. Notwithstanding, a supplemental was necessary, one that is large in historical terms and a smaller one was sought the next year. Reviewing Table 7.7 in this light, it is clear that the supplemental process as it was used understated the cost of activity in Vietnam and that if supplementals were meant to indicate the total cost of war, then they should have been much higher from 1966 through 1968. For example, in 1967 if $20 billion were in the base and if an additional $12 billion needed to be passed as a supplemental, then the wartime effort was closer to 37%[2] of the total appropriated for defense in that year than 17%. We raise these issues only to point out that these were estimates and not precise and that the same phenomenon would occur from 2003 on, large supplementals, but also an increase in the base, and some difficulty in keeping the two activities, base budget and supplemental for war, separate, although the Bush administration appears more forthcoming than the Johnson administration. Not everyone agrees on this point. Tom Friedman (2007) says, "From the start, the Bush team has tried to keep the Iraq war "off the books" both financially and emotionally." Our point is that compared to Vietnam the Bush administration has not tried to keep the costs of war secret, even if it is sometimes hard to tell what is a direct cost of war and where the indirect costs change into costs which should more properly be attributable to reshaping the Pentagon rather than costs of war. Either way, this is another complication of our initial model for supplementals. If the original model for supplementals held (payback for something already done), the net dollar effect would have been to increase outlays in the current year (original budget plus supplemental), but not to increase the base for the following year (because no ongoing activity). This was not true in the Vietnam era and not true in the current era either. Base budgets increased to reflect the costs of war even as more supplementals were passed. We see this as a function partly of size and partly of duration. Pricing the cost of 1,000 Marines to help rescue flood victims in the Philippines for 7 days is a lot different from pricing 150,000 troops in continuous, if sporadic, combat for 5 years. In the latter case equipment replacement, modernization, innovation in response to the enemy's efforts (improvised explosive devices or IEDs) and other direct and indirect supporting costs can not be ignored.

A further complication occurred after 1969 when the costs of war were in the base budget, and there were no more war supplementals requested. But there were supplementals submitted to Congress and opponents of the war attempted to attach policy guidance about Vietnam to them notwithstanding that they carried no funding for Vietnam in them. Supplementals in 1969 and after funded federal pay and retirement, including DoD, and there were supplementals for other agencies. In 1971 there were three such supplementals, but none for Vietnam. In the early 1970s Vietnam spending was a subject for debate during the adoption of these other supplementals as those who were opposed to the war, attempted to get at in other ways. For example, the July 1969 supplemental debate focused on objections to Vietnam spending and the adoption of a total spending limit for 1970. The July 6. 1970 supplemental set ceilings for federal expenditure for 1970 and 1971. The June 1973 supplemental banned bombing in Cambodia and provided that no funds could be used to support combat. Thus during this time, supplementals not intended for DoD became focal points for effort of policy control of DoD activities for the current and future years. This is another complication of our initial simple and linear model for defense supplementals. Finally, let it be emphasized that the supplementals did not carry the full cost of the wartime activities, another complication to our model and one that is repeated in the current era. From 1964 to 1974, the DoD base budget increased about $28 billion (about 58%). When hostilities for the U.S. ended in Vietnam the defense budget did not drop to its 1964 levels. In fact, the appropriations for 1974 were for $73.7 billion. The major driver here was personnel costs, which doubled in the period, even though the force levels declined by 12%. Personnel spending increases made up $27 of the $28 billion, primarily as a result of a change to an all-volunteer force.

Current Era Supplementals: Large and Frequent

Current era supplementals are as large as or larger than the Vietnam era and the base budget has expanded significantly. Additionally, from 2004 on, it became the usual practice to pass two supplementals, a bridge supplemental in the fall with the DoD appropriation bill and another, and larger, supplemental in the spring. These were both for the current fiscal year. Thus in any calendar year or fiscal year there were two wartime supplementals. The first bridge supplemental was passed as a separate section in the main appropriation bill (e.g., Title IX in the FY 2006 and FY 2007 appropriation bills) and was intended to ensure that DoD did not run out of money before the end of the fiscal year in 2004. It later years it was intended to bridge DoD over until the spring wartime supplemental became available.

In the summer of 2004, the first bridge supplemental was clearly aimed at ensuring DoD did not run out of money in the current fiscal year. The part of the appropriation bill that would contain the supplemental was labeled a dire emergency and thus put that part of the DoD appropriation bill outside of any discretionary spending caps. This would become common practice with the later bridge supplementals. The supplemental was to be "available immediately to pay for shortfalls in the final months of fiscal 2004." It passed with the FY 2005 regular appropriation bill, which the President signed on August 5, 2004. The Pentagon and Congress disagreed on the amount of money needed. The conference report was filed just hours after the Government Accountability Office (GAO) said the Pentagon would face a $12.3 billion shortfall before the end of the fiscal year. DoD officials had estimated they would be short by $2.8 billion and proposed to make that up by reprogramming. Congressional appropriators had insisted all along that the shortfall would be greater. While Congress was more generous with funds than was the DoD leadership, Congress added controls over where the money could be spent. Congress restricted all but $2 billion of the $25 billion to specific accounts, thus DoD was not given carte blanche about how to spend the money. Moreover, the administration did not get all that it wanted; transformational programs in the Army and Air Force did not fare well and lawmakers worried about a procurement shortfall and warned that there were more big-ticket items in the pipeline than DoD would be able to afford. Congress also supplied specific remedies to battlefield weaknesses that the Pentagon had not requested, adding $1.5 billion not requested by DoD to deploy more armored vehicles and replace other equipment (Congressional Quarterly Almanac (CQA), 2004, pp. 2-14). In 2005, a $50 billion bridge supplemental was added to the FY 2006 appropriation bill to tide DoD over until the new supplemental was submitted in February 2006. During the summer, appropriators warned that Army would be out of money by august. Nonetheless the appropriation bill and its bridge supplemental was the "last and arguably most contentious" appropriation bill passed that year, with the conference report cleared on December 22 (CQA, 2005, pp. 2-14).

The supplemental carried additional equipment for Army, air National Guard, and Army reserves. Also $8 billion was earmarked for equipment lost in combat including humvees, trucks, and radios and $1.46 billion for testing and fielding new equipment to thwart roadside bombs (IEDs) (CQA, 2005, pp. 2-18).

The FY 2006 appropriation bill carried an across the board cut of 1% applied to all discretionary spending (other than Veteran's Affairs and emergency spending, like the DoD bridge supplemental and disaster related funds). This cut was necessary because the discretionary appropriations total exceeded the agreed upon budgetary caps. Although it resulted in a $4 billion cut from the DoD appropriation bill, the cut did not affect the bridge supplemental because it was dire emergency spending. Part of the discussion in 2005, was the Army's attempt to fund the first 2 years of its transformation efforts (about $10 billion) through emergency supplementals for the war in Iraq. Proponents called transformation "too urgent a task to risk the delays and pitfalls of the regular budget process." Defense expert Anthony Cordesman observed: "You have immediate real time requirements that are being validated in the field in Iraq and Afghanistan. You are not talking about theory." Cordesman said that essentially what Army was doing was playing catch-up with the fact that the Army's force structure did not change quickly after the Cold War (Schatz, 2005, p. 510). At first glance it may seem odd that the Army did not change more quickly at the end of the Cold War, but remember three things: Army and DoD did draw down about 33% in personnel levels in the early 1990s, DoD planners had to be prepared for the eventuality that the Russian experiment might not work out and central Europe could again become a battleground, and Desert Storm/Desert Shield in 1991 reinforced cold war planning and tactics since it was basically won by air power and the M1A1 tank and its gun, which allowed U.S. tanks to stand out of range and destroy Iraqi tanks. Army and DoD did move to a new base force profile, but stopped well short of the transformation urged later by Secretary Rumsfeld.

The 2006 supplementals were interesting for three reasons. First, in the main defense appropriation bill process, Congress cut the defense request by about $4 billion in order to provide more than the President had requested for some domestic programs. This move did not affect defense because "appropriators made up the difference in the emergency section of the bill" (CQ Weekly, 2006, p. 3331), in the bridge supplemental. The appropriation bill (including the bridge supplemental) was signed on September 29, 2006. Note here that this bill was passed "on time" before the start of its fiscal year; also note that it is common for appropriation bills to fund the next fiscal year, but the fact that the supplemental was passed in the current year and funded activities at least through the next 6 months of the following fiscal year is a departure from our model because it is funding future costs, not past costs. Second, in February, the main supplemental for DoD was sent to Congress as a separate bill, but merged by Congress with hurricane relief spending. The majority of the bill was for defense ($70.4 out of 94.6 billion) and all of it was designated emergency spending, meaning it would escape any spending cap discipline and no bill-payers would need to be found, other than future taxpayers who have to pay down the national debt. Third, a deeming resolution was included setting discretionary spending limits for the Senate because Congress could not complete the normal budget resolution process (CQ Weekly, 2006, p. 3336). The Senate cleared the supplemental on June 13 2006 and thus could use the deeming resolution in the supplemental for its deliberations on all of the upcoming appropriation bills (including the defense bill). Last, the deliberations over the supple-

mental took longer than expected. There were some efforts to give the
president policy guidance over the war and some fiscal conservatives in
the house objected to the inclusion of items not related to war and hurricane relief, but "enactment of the measure was never in doubt because it
included money for troops in wartime." The House voted for the bill 351-
67 and the Senate 98-1. The President signed it on June 15. By now the
wartime supplementals had become miniappropriation bills, with the
President submitting a bill and Congress changing it substantially
through a hearing, markup and floor debate process, offering debate over
policy guidance, using the emergency provision to escape spending caps,
inserting add-ons and earmarks into the bill and in the bridge supplementals using some smoke and mirrors to shift money from defense to
other domestic programs while making defense whole in the bridge supplemental. The main difference is that the independent supplementals
were passed rather quickly in the spring and the bridge supplementals
lent an air of immediacy to the main appropriations bill as it supported
troops in the field in time of war. They were all "must pass" legislation. In
form these supplementals were somewhat like the main appropriation
bill, with justification, familiar line items, and comparison to the past and
the future. Also it could be argued that by 2005, most of the supplemental
costs were for future costs, as opposed to past costs, and thus depart from
our model of peacetime supplementals, although we acknowledge that
some of those future costs would certainly be payback for past actions, for
example, in vehicle and aircraft replacement. Table 7.8 and Figure 7.2
show the current history of these wartime supplementals.

In the sense that these wartime supplementals were for future costs of
war, they became miniappropriation bills for a specific purpose, not for the

Table 7.8. War on Terror Supplementals

Year	DoD $ Approp.	Y:Y (%) Change	Main $ Supplmntl.	Bridge Supplmntl.	Ratio (%) (S/Dod)	Ratio (%) W/Bridge
FY 2001[3]	297	3.13%	14		4.71%	
FY 2002	328	10.44%	17		5.18%	
FY 2003	375	14.33%	69		18.40%	
FY 2004	377	0.53%	66		17.51%	
FY 2005	400	6.10%	101	25	25.25%	31.50%
FY 2006	411	2.75%	115	50	27.98%	40.15%
FY 2007	435	5.84%	93.4	70	21.47%	37.56%

Source: Computed from DoD FY 2007 Emergency Supplemental Request p. 1, taken from
defense appropriation acts, FY 2001-FY 2007.

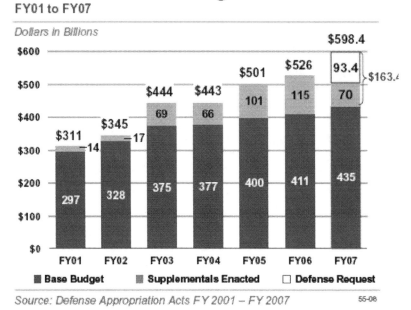

Source: Department of Defense, 2007. FY 2007 Supplemental Request for the Global War on Terror. Washington, D C., February, 2007, p. 1.

Figure 7.2.   GWOT war spending.

whole function of defense, but rather targeted at a specific goal: winning the war on terrorism which soon came to mean winning the war in Iraq. These supplementals also had other characteristics of appropriation bills.

Wartime Supplementals Complex

Like the normal defense appropriation bill, the wartime supplementals are complex. For example, in the 1990s a supplemental might simply have asked for reimbursement of O&M costs, but the supplemental presented to Congress in the spring of 2007 was very complex, with some money for standard operations of DoD, some money for allies, some procurement money to rebuild force structure (reconstitution), some money to defeat IEDs, and even a little pot of money that commanders in the field could tap to buy supplies and or use to provide for humanitarian relief (Commander's Emergency Response Program or CERP). For further explanation, please see the FY 2007 supplemental justification materials. This complexity also manifested itself in the budget execution process. One U.S. Marine Corps (USMC) officer observed that while in a financial management billet in 2006 he had about 300 accounts to monitor in the execution of these supplementals for Iraq and any account that showed a 10% variation from the previous month was flagged and had to be explained. His estimate was that meant about 10% of the accounts had to be explained each month, irrespective if the variance were up or down. The lesson here is that nothing about the wartime supplementals is simple. Figure 7.3 is testimony to the complexity of these wartime supplementals.

Wartime Supplementals: Normal Budgetary Logic and When it Applies

Normal appropriation bills often use historical comparison as a basis for analysis. This practice has carried over into the wartime supplementals.

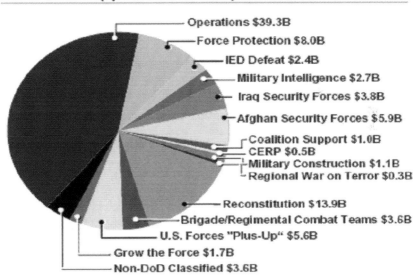

FY 2007 Supplemental Request: $93.4B

- Operations $39.3B
- Force Protection $8.0B
- IED Defeat $2.4B
- Military Intelligence $2.7B
- Iraq Security Forces $3.8B
- Afghan Security Forces $5.9B
- Coalition Support $1.0B
- CERP $0.5B
- Military Construction $1.1B
- Regional War on Terror $0.3B
- Reconstitution $13.9B
- Brigade/Regimental Combat Teams $3.6B
- U.S. Forces "Plus-Up" $5.6B
- Grow the Force $1.7B
- Non-DoD Classified $3.6B

Source: Department of Defense, 2007. FY 2007 Supplemental Request for the Global War on Terror. Washington, D.C., February, 2007, p. 11.

Figure 7.3.   Wartime Supplementals Complex.

Table 7.9.   Supplemental Baseline Comparison

Force Protection ($ in Billions)	Title IX[1]	FY 2006 Supplemental	Total	Title IX[2]	FY 2007 Supplement	Total	Percent Change
Body armor	0.1	0.3	0.4	1.9	1.6	3.5	775%
Protection equipment	1.5	3.2	4.6	1.5	5.7	7.2	57%
Armored vehicles	0.1	0.3	0.4	—	0.7	0.7	75%
Total force protection	1.7	3.8	5.4	3.4	8.0	11.4	111%

1. Title IX, FY 2006 Defense Appropriations Act (PL 109-148)

Numbers may not add due to rounding

2. Title IX, FY 2007 Defense Appropriations Act (PL 109-2890

Source: Department of Defense, 2007. FY 2007 Supplemental Request for the Global War on Terror. Washington, D.C., February, 2007,p. 23.

For example, the 2007 supplemental justification was presented with comparisons to the bridge supplementals for both 2006 and 2007 and the main 2007 supplemental. Table 7.9 shows the force protection section of these supplementals.

This part of the supplemental request reflects the changing conditions of the battlefield, with the total for body armor increased 775% over the 2006 levels. Nonetheless there was some money for force protection in the bridge supplemental in 2006, in the regular supplemental for 2006, in the bridge for 2007 and in the regular supplemental for 2007. Some would argue that this is a good case for including this kind of funding in the base budget because it was a known and continuing need, had a track record, and would lend itself to analysis which might lead to an increase or decrease in the category. (In fact, in 2005 Congress took the initiative in providing money to uparmor Humvees and for body armor which the Pentagon had not requested.)

Others would point to the percentage growth in the category and argue that this is precisely what a supplemental needs to fund, a need whose dimensions were either unknown or changing so fast that it could not wait for the regular appropriation bill. Budgeteers would argue that such a huge percentage increase would be deserving of scrutiny and perhaps reduction, a normal part of the budget process. War fighters might argue that troops are currently at risk and the task at hand is to meet the need and not worry about 'bean counter' concepts like "percentage rate of increase." At the end of the day, the 2006 and 2007 supplementals did

Table 7.10. War on Terror Supplementals: 2006 and 2007

($ in Millions)	Title IX[1]	FY 2006[1] Supplemental	Total	Title IX	FY 2007 Supplemental	Total	Percent Change
Military Personnel	6,144.5	10,278.8	16,423.3	5,386.5	12,144.7	17,531.2	7%
Operations and Maintenance	29,238.0	29,992.0	59,230.0	37,582.2	37,162.9	74,745.1	26%
Procurement	6,462.7	13,911.1	20,373.8	16,603.9	23,077.8	39,681.8	95%
Research and Development	13.1	112.1	125.2	123.5	734.9	858.4	586%
Military Construction	—	214.8	214.8	—	1,854.0	1,854.0	763%
Iraqi Freedom Fund/JIEDO	1,360.0	1,958.1	3,318.1	1,970.7	2,638.4	4,609.1	39%
Defense Health Program	—	1,153.6	1,153.6	—	1,073.1	1,073.1	−7%
Iraq and Afghan Security Forces	—	4,915.1	4,915.1	3,200.0	9,748.7	12,948.7	163%
Working Capital Fund	2,516.4	516.7	3,033.1	—	1,320.5	1,320.5	−56%
Subtotal	**45,734.7**	**63,052.3**	**108,787.0**	**64,866.8**	**89,755.0**	**154,621.8**	**42%**
Non-DoD Classified and Non-GWOT	2,765.3	2,975.0	5,740.3	5,133.2	3,627.2	8,760.4	53%
Total	**48,500.0[2]**	**66,027.3**	**114,527.3**	**70,000.0**	**93,382.2**	**163,382.2**	**43%**

Numbers may not add due to rounding

1. Reflects FY 2006 enacted amounts.
2. Does not include $1.58 billion for Non-GWOT activities.

Note: The amounts included under the Title IX columns were emergency supplementals included in the regular appropriation bill for that fiscal year.

Source: Department of Defense, 2007. FY 2007 Supplemental Request for the Global War on Terror. Washington, D.C., February, 2007, p. 97.

look a lot like the regular defense appropriation bills and not like the stereotype of the 1990s defense rescue and peacekeeping supplemental. Table 7.10 from the 2007 supplemental request presents these 2 years by the usual DoD appropriation category.

As opposed to our model for supplementals, both years had funds for military construction, research and development, and procurement; these are investment accounts as opposed to readiness accounts aimed at current operations. Spending on investment accounts is another indication of the complexity of wartime supplementals. Our model suggested that supplementals paid back for actions that were already over. As can be seen from Table 7.10, current wartime supplementals have amounts in them devoted to long-term tasks. Also these are of significant size and of high growth. For example, in both years the procurement account is third largest in dollar size. For FY 2007, the procurement accounts primarily went to reconstitute the force, for such things as armored vehicles, trucks, and helicopters that were either destroyed or worn out, but some of those funds went for such things as the next generation strike fighter (the Air Force had asked for two in the request (FY 2007 Supplemental, 77)), with the argument made that as the current generation planes were "used up" in Iraq there was no production line open to buy current models of these, and that reconstituting therefore meant stepping up to the next generation of equipment.

Costs May Be Differentially Borne

Currently the majority of the costs of war are borne by the Army. Stanford's Larry Diamond has observed, "America is not at war. The U.S. Army is at war" (Friedman, 2007). Thus while the Navy and Air Force are arguing for next generation ships and planes, Army recapitalization and transformational initiatives may come to grief under the sheer weight of wartime activities in Iraq. However, in fact, spending in Iraq also squeezed Navy and Air Force accounts because there simply was not enough money left over to pursue modernization activities, be they transformational or simply recapitalizing by buying more modern airplanes and ships. This illustrates another departure from our simple model of supplementals: the burden of the activity may be unequally borne within DoD, but with the passage of time it eventually impacts all DoD programs and financing. Remember also that one of the unintended outcomes of Vietnam was the elimination of the draft and the change to an all-volunteer force. Our model simply suggests that the national

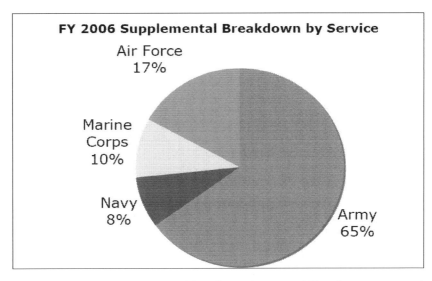

Source: FY2 006 Supplemental Breakdown by Service (From: Office of Management and Budget, FY 2006 Estimate #3 Emergency Supplemental Appropriations) in Evans, 2006. 37.

Figure 7.4.   Supplemental shares by service: FY 2006.

command authority tasks DoD to do something and the military department that carries out the task is reimbursed for that task on an incremental cost basis, that is, is reimbursed for services rendered. Wartime supplementals are not so simple.

Wartime Supplementals May Have Long Term Consequences

Our simple model of supplementals assumed an event or events that were quickly concluded and reimbursement was made for incremental costs of activities already concluded, over a known time period where cost factors could be derived, accumulated, and submitted for reimbursement. This is not the case with the current wartime supplementals (nor was it the case in Vietnam). In some respects, these supplementals are like a rolling declaration of war, with a promise to pay incrementally over time. As the graphic below indicates, the U.S. involvement with Iraq may last years with all that that means for treasure and blood. In fact, it already has. The first Gulf War did not end when conflict ceased in Kuwait in February,

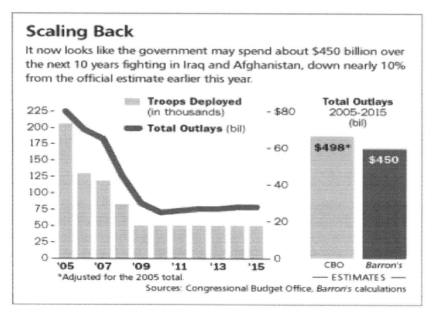

Source: From Barron's Online, "Leaving Iraq, Gaining a Rally," January, 2 2006, from Evans, 2006, p. 39.

Figure 7.5.   Supplementals have long term impacts.

1991; the United States kept troops in the region at sea and on land bases, and patrolled a no-fly zone over Iraq and no one knows when the current effort will end.

By 2005 it was clear that the effort in Iraq was going to be sizeable and ongoing and that large supplementals would continue to be necessary and that despite the large supplementals, the DoD base budget also was growing, albeit not as fast as it did during the Vietnam build-up from 1964-69. Congress became restive with this situation and took secretary of the defense (SECDEF) to task for not including known costs of war in the base budget. In a sense, Congress was encouraging DoD to do what SECDEF ultimately came to do during the Vietnam era: include the known and predictable costs in the base budget and ask for the unknown and emergency items in supplementals. The dimensions of this disagreement surfaced in February 2005 in the Senate Armed Services Committee when Senators Levin (Democrat-Michigan) and McCain (Republican-Arizona) interrogated Secretary of Defense Donald Rumsfeld, Chairman of the Joint Chiefs of Staff General Richard Myers, and Under Secretary of Defense Comptroller Tina Jonas concerning

Army troop strength, shown at 512,000 in the 2005 base budget, but at 482,000 in the FY 2006 budget request. Senator Levin argued that since Army troop strength was a known number that it should be in the 2006 budget and not funded through 2005 supplemental. Also in question was the extent to which SECDEF was including costs of transforming the Army (modularity) in the supplemental when some thought that those costs should have been in the base budget because they were not strictly costs of war. Finally, Senator McCain would argue that the way DoD was doing things impaired Congress' constitutional authority over the power of the purse because it diminished transparency in the budget and supplemental requests. What follows is a key passage from the hearing:

Senator Levin:	The Army has given us core end strength of 512,000. Why doesn't the budget request fund 512,000? Why is that only 482,000 in the '06 budget request?
Secretary Rumsfeld:	… we need to look at the budget and the supplemental together …
Senator Levin:	No, the real question is, since it's a known cost, why don't you put it in the budget? Why are you hiding the cost of 30,000 troops in the Army?
Secretary Rumsfeld:	There's nothing hidden. It's all right there. It's either in the supplemental or in the regular budget.
Senator Levin:	We don't have the '06 supplemental yet. We only have the '06 budget request that says 482,000, although the '05 active duty end strength, by your own chart, 512,000.
Secretary Rumsfeld:	Right
Senator Levin:	If you know '06 is going to be 512,000, why aren't you putting the 512,000 in your '06 budget request? That is the question … that's a short, direct question.
Secretary Rumsfeld:	The, well, I can't answer it briefly. I'm sorry, I'd be happy to submit something in writing …
Senator Levin:	It would be better for the record then.
Senator Levin:	What did you ask for modularity in the '05 supplemental?
Ms. Jonas:	There's $5 billion that they're requesting.
Senator Levin:	Thank you. Is there any reason to doubt that that modularity need is going to continue in '06, general?

General Myers:	I think that's correct, and I think in the '06 budget, you'll see it in the regular budget.
Ms. Jonas:	'07
Senator Levin:	But why not '06? Why isn't it in the '06 regular budget? It's a known amount.
Ms. Jonas:	… we could have waited and put them in the baseline budget, but General Schoomaker suggested it was urgent, and we agreed.
Senator Levin:	I'm not suggesting you wait. They're in the '05 supplemental. I'm suggesting you put them in the '06 regular budget. You know what they really are, don't you?
Ms. Jonas:	I'm not sure that they know fully. We have a good idea, but we didn't know for sure.
General Myers:	Let me give you a short answer on that … it is still being refined. To insert a large number in the '06 budget with the uncertainty surrounding them would have perturbated a lot of the '06 budget. As the secretary said, we start the '06 budget prep a long time ago…. So that's why the decision was made, I think, to put them in the '07 budget, give us time to work them, and work big numbers, billions of dollars, into the '07 budget, and handle it in the supplemental in the 2 years where it's still being developed.
Senator Levin:	Thanks.
Senator McCain:	Mr. Secretary, what Senator Levin is trying to get to is part of your presentation. The normal budget cycle is 30-33 months, and the supplemental is 9 months. And there are many of us who feel that the supplemental which is earmarked for combat operations in Iraq and Afghanistan has been expanded to a significant degree to other programs such as the modules that Senator Levin just talked about, which are—in the view of many of us, should be in the normal authorization process so that we can exercise our responsibilities of oversight … I can certainly see things from your point of view, where it would be a lot easier. But we're going to have to make a decision at some point in the congress as to exactly what should be included in supplementals and what shouldn't (Senate Armed Services Committee, 2005a).

SECDEF Rumsfeld argued that the budget and the supplemental had to be considered as a package: "The only way you can look at this budget is to look at the supplementals with it" (Shalal-Esa, 2005). He argued that there were moving parts that made it hard to estimate costs and that the Army was transforming as units returned from Iraq and some of those costs were in the supplemental. He said that the costs were either in the base budget or in the supplemental and seemed to reject the idea that he was hiding anything. History again provides a lesson. In the 1990s the United States operated no-fly zones in northern and southern Iraq to attempt to limit the damage Saddam Hussein could do to his countrymen from the air. The Air Force chief of staff called these steady state operations and said that they were planned for and counted as part of the Air Force's routine operating requirements, yet they were funded by over $12 billion in supplemental dollars during the 1990's (Tirpak, 2003, pp. 46-52). Also during the Kosovo conflict the Clinton administration discovered that there were additional incremental costs beyond what the supplementals for Kosovo and Bosnia were providing. As a result the Air Force asked for supplemental funding to repair and restore infrastructure (GAO, 2000d, p. 2) In 2000, a mixed funding model was used, where "partial war costs were submitted in the annual budget and others were funded through emergency supplemental appropriations (Evans, 2006, p. 15)," with the President asking for $2 billion to protect readiness in the current fiscal year in the FY 2000 supplemental and adding another $2.2 billion to the defense budget request for FY 2001 (DoD News Release, 2000). Thus there was plenty of precedent for the way Secretary Rumsfeld approached his supplementals; what was different was their size.

On April 12, 2005, the Republican policy committee of the Senate issued a policy paper on the issue that illustrated the complexity of this issue. The paper was issued with two summary statements:

> one being that Congress should fund operations in Iraq and Afghanistan through emergency supplemental appropriations (because funding it through the regular appropriations process would unnecessarily inflate the defense base), and the other being that Congress was correct to scrutinize the Administration's request, since it arguably included items that do not meet the definition of emergency spending. (Evans, 2006, p. 24)

Still the fundamental rationale for supplementals was clear. Former Pentagon Budget Chief, Dov Zakheim, defended the use of supplementals, saying that current troop levels were far higher than expected the previous year, proving the wisdom of using supplementals to fund such operations (Shalal-Esa, 2005), and within the Pentagon key military leaders argued that "the circumstances of ground war can change quickly and a supplemental allows for more accurate cost estimates and quicker access to needed funds (Evans, 2006, p. 31)." On October 19, 2005, the Pentagon restated its position: "The objective is to ensure that all war-related costs are being captured in supplementals to include resetting the force for damaged and destroyed equipment" (England, Memorandum for Service Secretaries, CJCS, VCJS, and Service Chiefs, 2005). Of Vietnam, Taylor (1974) comments on attributing cost that it was common knowledge that the prevailing trend was, "When a doubt exists, charge it to Vietnam" (p. 25.

By the summer of 2006, it was clear that action in Iraq was impacting the structure of the U.S. military in terms of troop strength and equipment replacement. What Congress became more and more uneasy with was that some of the supplemental dollars were either buying equipment or funding transformational objectives which should have been put in the base budget. And if SECDEF was happy with his approach, others were not. Democrats complained that supplemental spending had hidden the true scale of war costs (Robinson, 2006). Senator McCain argued that the continual use of supplementals "Distorts understanding of the defense budget and removes from our oversight responsibilities the scrutiny these programs deserve" (Weisman, 2005, p. A42). House armed services readiness subcommittee Chairman, Republican Joel Hefley, also questioned the current reliance on supplemental spending to cover predictable costs. "My theory has always been that you put in the supplemental things that surprise you" (Klamper, 2005).

Even with the large amounts of money going into Army budgets, evidence was surfacing that it was not enough. Some bases were operating almost around the clock, while others were paring back expenses as best they could. Retired Army General Paul Eaton said that because of spending in Iraq, the Army had a $530 million budget shortfall for posts (Army bases) in 2006 (Friedman, 2007).This is not a new phenomenon. Speaking of the Vietnam years in the late 1960s Major General Leonard Taylor reports:

> The result is that U.S. Army funds were supporting the major portion of the war while in the continental U. S. badly needed repair and maintenance projects were being deferred from year to year at nearly every installation. (Taylor, 1974, p. 86)

In May of 2006, the Army ordered a series of belt-tightening moves while waiting for the supplemental. These included:

1. Stopping orders for noncritical parts and supplies.
2. Canceling nonessential travel, training, and conferences.
3. Postponing civilian hiring.

4. Plans were made to furlough civilian employees and halt depot maintenance.

When the supplemental was passed on June 15 the Army was spared from laying off contract employees, halting recruiting, and freezing promotions. Speaking about a delay in the 2007 supplemental, Acting Army Secretary Pete Geren told the Senate Armed Services Committee:

> We don't have to imagine what would happen if the supplemental is delayed. We experienced it last summer.... We had to lay off temporary workers, we had to lay off contract workers, we had to cut back on recreation opportunities for children in the summer, we had to reduce a wide range of services, and we also had to cut back on many essential contracts, on programs that were supporting procurement and refit opportunities around the country. The impact on the soldiers, on training and on soldiers' families was significant. We had to slow everything down. (Mathews, 2007, p. 4)

Critics tend to believe the Pentagon is awash in money. The typical attitude was expressed by a House staffer discussing the 2007 supplemental debate:

> This is not the first time the military has said it will run out of money. They've got about a bazillion dollars in the defense bill and the ability to reprogram money. I imagine they can keep things rolling for awhile. It isn't like they won't get paid. That's what the administration would like us to believe, but there's not any danger of the military running out of ammo. We've got a giant game of chicken going between the House, the Senate, and the White House. (Mathews, 2007, p. 4)

To some extent the budget process is always a game of chicken, and experts do disagree about the level and even the need for programs, but to Pentagon planners anything that has made it into the budget has survived fierce competition with items that were not funded, and reprogramming means not doing things that were of high enough priority on someone's list to make it through the appropriating process. Reprogramming means that someone's valued program is cut. Everything in the Pentagon budget is tied back to an intended expense. To displace money from one item to another means a shortage in the first account. Some things do not get done and the cool discussion of reprogramming masks the real impact on programs and people of reducing or even canceling programs on short notice and within the budget year. Also it is clear that a war run on supplementals is a partially funded war. Robert Buzzanco (1996) comments that as the economic situation worsened in the mid-1960s SECDEF McNamara tried to conceal further the true costs of the war:

> began to postpone equipment overhauls, stretch out the purchases of new weapons systems, cannibalize materiel stocks in Europe, and extend repair and maintenance cycles by 50 percent. Such measures outraged the military hierarchy and helped make it impossible for service and civilian leaders to coordinate policy on Vietnam. (p. 240)

In 2007 some of the same partial-funding cleavages had sprung up around Iraq. Elsewhere we have discussed strains on the Army and the other services, but the partial funding syndrome applied to the National Guard as well. For example, a Congressional commission chartered to study the effect of the war on the National Guard and the Reserves, reported, "Right now in the United States, 88 percent of the Guard is not combat-ready when it comes to equipment ... (and) ... that's worse than the worst days of the so-called hollow force in the late 70's and early 80's." The story went on to report that around 200,000 National Guardsmen had served in Iraq and Afghanistan, despite these shortages and despite the fact that they received less training than professional soldiers (Dreher, 2007).

It was also clear that by 2007 a majority in the country had lost faith in the mission in Iraq and the supplemental process illustrated that fact as a Democratic Congress provided more money than the President requested, but also provided guidance in the supplemental about when and how to end the war in Iraq. In some ways the supplemental process began to resemble the main appropriation process. The proposed supplemental request amounted to $103 billion, with $99.6 billion for costs of war. Prior to Congress's return at the end of February, dates for hearings (Senate Feb. 27) and committee markups for appropriators were announced in advance (House March 5, Senate March 19). Also announced were intentions by Rep. Murtha to put limits around the supplemental as a way circumscribing how the war in Iraq would be pursued. In March the supplemental took center stage in both the House and Senate. In simpler times, supplemental proposals were short, with brief justification and quickly worked their way through Congress. The supplemental for 2007 was accompanied by 101 pages of text, charts, and graphics, all indicating the complexity of the ongoing effort. Congress tried to find ways to limit the involvement in Iraq while still supporting the troops. Said Rep. David Obey (Dem-Wisconsin and Chairman of the House Appropriations Committee): "Congress has the power of the purse and we're trying to exercise it in a way that does not damage the troops" (Rogers, 2007, p. A2). As this was written, it was unclear how this would come about, other than that it would be a complex and politically explosive issue, as a majority of the American public (55%) (p. A2) strongly opposed the President's decision to boost troop levels in Iraq. By late March Congress had inserted a series of benchmarks and deadlines for withdrawal from Iraq in the supplemental that some observers thought were bought with $20 billion in earmarked funding. This included $74 million for peanut storage, $25 million for spinach growers and $283 million for dairy farmers, according to the *Wall Street Journal* (2007, p. A-10). On May 1, 2007, President Bush vetoed the supplemental bill saying, "This legislation is objectionable because it would set an arbitrary date for beginning the withdrawal of American troops without regard to conditions on the ground; it would micromanage the commanders in the field by restricting their ability to direct the fight in Iraqi; and it contains billions of dollars of spending and other provisions completely unrelated to the war (Bush, 2007). In order to override a veto, both chambers must reject the President's veto, 66% voting to override. On May 2, 222 members so voted, while 203 voted for the veto (to sustain), thus did the attempt to override the veto fail and the long debate over the war and the supplemental dragged on into the summer. As was said earlier, the supplemental process usually "must fund" legislation in a wartime situation, proved vulnerable, and it also proved that the supplemental process, at least in this year, was looking like the main budget process, predictable only in its disorder.

What lessons may be drawn from a comparison of supplementals in the Vietnam and Iraq war periods? First, supplemental spending efforts are roughly comparable in size. Second, to date, the Bush administration has been more forthcoming about the costs of war; basically the supplementals have funded it and supplemental amounts are a matter of public record. Third, as the effort has continued some blurring is occurring between what should be in the supplemental and what should be in the base budget appropriation. Ultimately, despite base budget increases and large supplementals, the result appears to be either underfunding for the war effort or underfunding the rest of the defense requirement. This last point is more complex than it seems at first glance. Consider the troops in Iraq, for example; what they use and expend in the way of gasoline and ammunition should clearly be in the supplemental. Clearly their transportation to and from Iraq should be in the supplemental. Now for the complexity: how much of their training should be attributable to Iraq (their time and equipment costs: some, ½, all); how much of the maintenance of the base they train at or how much of the time of the instructors? This is not a new problem. Taylor remarks about Vietnam:

"If we moved one of the two divisions from Fort Hood, we would have reduced the troop operations by one half, but would not be able to reduce the base operations by one half. The reason is that some costs continue whether there are 10,000 or 20,000 soldiers at an installation. For example there are the same miles of roads to maintain and essentially the same number of buildings to repair … the cost of fire protection remains the same…The base services such as commissary, laundry and dry cleaning can reduce some costs … but costs cannot be cut in half solely because the military population has been cut in half. (Taylor, 1974, p. 38)

The Army introduced a line of accounting to identify costs both indirect and direct associated with Vietnam. This appears to have led to an over-attribution of costs. Taylor remarks (1974):

At the training centers, it was difficult, if not impossible, to identify each individual who was destined to go to Vietnam upon completion of training. Additionally it was equally difficult to specifically identify accurately all of the costs on an installation and pinpoint whether or not it should be included in the Vietnam account. (p. 25)

When helicopters are damaged and repaired, clearly the immediate repair should be charged to Iraq, but what if the repair is accompanied by renovation that extends the service life by a few years—should that be charged to Iraq? The same question may be asked of other types of equipment, including aircraft that wear out and can not be replaced by the same type because the production line has long been closed (hence the Air Force asking for two Joint Strike Fighters in the 2007 supplemental), is the supplemental a good place for that kind of expense? It seems that defense is not a separable good when the mission extends over a longer period of time; this is part of the tooth to tail argument where for every person in combat there are troops in support as well as a training, equipping, and supporting infrastructure, to include medical care, family housing, and such. It is no wonder then that during both periods the base DoD budget has increased without any commensurate change in the threat. In sum, what goes into the wartime supplemental is a judgment call and there is some logic to Rumsfeld's approach to it, that the supplemental and the main appropriation bill are a package.

SUMMARY

In this chapter we have explored how supplemental appropriations have been used in the defense area. These are our findings. First, compared to the normal appropriation process, supplementals are usually passed expeditiously. Second, contrary to popular perceptions, supplementals do not always result in supplements. In fact, a supplemental package may result in a net decrease due to offsets. Third, some supplemental bills may also supplement future year budgets, an act commonly thought to be in the province of future year appropriation bills. Fourth, supplementals are commonly thought to be a tool the executive branch uses to supplement Budget Authority to meet a current year need, but in practice, Congress may substantially increase the size and scope of the supplemental bill and may take the lead in proposing the supplemental. Fifth, although supplementals are commonly thought to be for emergency supplements for the current year, some supplementals are aimed at nonemergency uses. Sixth, when supplementals are designated as dire emergency bills, they escape the control of budget caps; Congress and the President have found this mechanism such a convenient idea that the emergency designation has slipped into use in regular appropriation bills. Seventh, though small compared to normal appropriation bills, supplementals especially in defense, fund 100% of the need. This makes them a very efficient vehicle. Eighth, the wartime supplementals, both in the Vietnam era and the War in Iraq differ from defense humanitarian and peacekeeping activities: they are large, complex, increase the base, decrease budget transparency, affect all services no matter who the primary responder is and have long term consequences. Finally, supplementals have great symbolic importance for they show an immediate governmental response to a current year crisis.

On a cautionary note, we observe that there is a downside risk to such immediacy. The huge supplementals passed quickly after 9/11/2001 and after Hurricane Katrina in September 2005 appear to have led to wasteful, inefficient, and even fraudulent fiscal behavior. The continuous deployment of huge sums of money in Iraq for contractors in support of the military effort and for reconstruction seems also to have led to misuse of money. When it comes to money, good intentions are not enough; proper controls must be put in place for appropriate execution. If not, the cost of symbolic actions will be higher than anticipated, although the return on the investment may still be positive. In 2007 a GAO audit discovered more than 22,000 cases of fraud stemming from Katrina relief efforts (Cohen, 2007).

We have gone on to make the distinction between peacekeeping supplementals and wartime supplementals. The normal "peacekeeping" supplementals are usually small, passed quickly, and not controversial. However, wartime, supplementals have been large, frequent and complex and the usual distinctions between appropriations and supplemental have been blurred. Clearly supplemental activities have resulted in increases in the budget base. These supplementals have also affected current and future years and are not simply payback for tasks already accomplished. Like the disaster supplementals, they are more an estimate than a rigorous costing out, and it is clear, just as with disaster supplementals, more money will follow. For these reasons, they are not as transparent as the humanitarian and peacekeeping supplementals of the 1990s and this lack of transparency may hamper Congressional oversight of the main appropriation bill because Congress may not be able to see how the supplementals and main appropriations fit together. This could result in either underfunding or overfunding of defense. Moreover, even though it may look as if one service or military department is garnering the lion's share of the supplemental, it still may not be enough to properly fund that service for that mission. Also, these wartime supplementals may lead to longer term commitments, whereas the 1990s supplemental missions[4] were usually quickly over, and they may or may not lead to reduced funding after the conflict was over. After Vietnam, defense spending remained high because our main opponent was still in the arena (the USSR), and because the all-volunteer force was much more costly than had been predicted. Whenever, Iraq is settled, the war on terrorism will still have its requirements and all the services will face modernization costs to prepare for peer competitors in a modern state-to-state conflict.

Politics also has been at play in the wartime supplementals. The administration has used supplementals to keep the apparent cost of the deficit down. DoD did not put some items in the base budget that should have been there (Army troop strength) and Congress has made cuts in the main DoD appropriation bill that it later funded in the supplemental in order to free up space under spending caps for nondefense domestic spending. For example, in 2005 funds were shifted out of the defense appropriations area into nondefense domestic areas and the resulting shortages in defense were then made up by taking from the Iraq Freedom Fund, passed as a supplemental; thus the result was funding of continuing DoD operations out of a supplemental. This mixing of the base and the supplemental is something Congress would later chastise Secretary Rumsfeld for. Given the obvious usefulness of supplementals and considering their complexity, it would seem that supplementals deserve a little more attention than they currently earn as a useful tool in the budgetary process.

NOTES

1.   See for example, Defense Aid Supplemental Appropriation Act, 1941, Mar. 27, 1941, ch. 30, 55 Stat. 53; Defense Aid Supplemental Appropriation

Act, 1942,Oct. 28, 1941, ch. 460, title I, 55 Stat. 745; Defense Aid Supplemental Appropriation Act, 1943, June 14, 1943, ch. 122, 57 Stat. 151.

2. We divided the supplemental by the total of the appropriation bill and the supplemental bill to get the percentage the war effort comprised of total defense spending in this year. As we have described elsewhere, the DoD appropriation does not included total defense spending, but it is close enough for the point we want to make, which was that the war effort was not totally funded by the supplemental.

3. The FY 2001 supplemental was the DoD share of the supplemental passed in response to the attacks on the United States on 9/11/2001. The U.S. then began a bombing offensive on the Taliban in Afghanistan on October 7, 2001.

4. Efforts in Bosnia, Kosovo, and the No-Fly zone enforcement in Iraq are examples of missions that lasted longer than a year, but their fiscal consequences were relatively low and the mission requirements could be predicted and easily absorbed within the current DoD capability envelope, thus we chose not to treat them as wartime supplementals. After FY 2001, these efforts continued at a steady pace and in FY 2002 were directly appropriated into the annual budget for each military service in the O&M and Personnel accounts, according to Evans (2006, p. 15). See also Office of the Secretary of Defense, Justification for FY 2004 Component Contingency Operations and the OCOTF, February 2003.

CHAPTER 8

DEFENSE BUDGET EXECUTION

INTRODUCTION

Defense budget managers face conflicting cultural and legal norms in their quest to execute budgets. Their culture tells them to spend up all the money they have, but not to overspend it or they might ruin their career, be the recipient of a substantial fine, and win a go-to-jail card. Moreover, they are limited in their effort to come out even (spend it up, but do not overspend) by law which limits where different appropriations may be spent. Defense managers must make sense out of a welter of laws, rules, norms, and cultural imperatives.

To understand the Department of Defense (DoD) and its problems, we begin with some simple concepts in budget execution and then turn to an examination of the issues and problems that still confront defense financial managers. In some cases we use older data in this chapter to illustrate our points. In these cases the points illustrated are the same regardless of the time frame.

Legal Framework

Executing public budgets occurs in a rule-based environment. Managers commit public money and must follow rules that guarantee funds are spent on intended purposes and fraud, waste, and misuse is avoided. Table 8.1 describes the legal basis for budget execution. The scope of the list is indicative of the complexity of budget execution.

Budgeting, Financial Management, and Acquisition Reform in The U.S. Department Of Defense, pp. 321–379
Copyright © 2008 by Information Age Publishing

For almost all budget managers this body of law comprises a set of rules that guides their daily actions. For example, they know that they are not to commit funds before they have been appropriated, that they may not spend in excess of an appropriation, and they may only spend the appropriation on items for which the appropriation is made. Breaking these rules leads to fines, possible imprisonment, and the certainty of career termination.

FLOW OF FUNDS:
BUDGET AUTHORITY, OBLIGATIONS, AND OUTLAYS

Defense dollars are provided by Congress in terms of Budget Authority (BA). BA allows agencies to accrue obligations for the provision of goods and services, through hiring personnel, placing orders, signing contracts, making loans, providing grants, and so forth (31 U.S.C 1501). Outlays occur as payments are made for the contracts, purchases, and personal services and as grants and loans are executed. An outlay means a payment has been made, by cash, check, or electronic funds transfer, to someone who has delivered goods or services to the government or who has received funds from government as a grant, loan, or other payment.

BA represents government intention to spend. Outlays represent actual spending. Budget accounts for a FY include both BA and an estimate of outlay for that year. When this is the appropriation bill, the BA is a promise that x dollars will be spent on function y, no more, and probably not much less. The budget outlay number is an estimate that includes historical wisdom about how fast money in such a function is normally spent out. For example, if this is a new function, then people must be hired, equipped, and trained, and perhaps not all the money will be paid out (result in outlays) in the first year, even though it is all promised or obligated. If this is a construction project where final plans have to be drawn and contracts let, perhaps very little of the money will be actually be obligated or outlayed (spent) in the first year. This pattern of BA is true for all federal agencies. The current appropriation bill creates BA for future years and what is spent this year is comprised of BA gathered from past years as well as this year's budget appropriation.

Calculating the outlay rate from a given amount of BA is done using historical experience and trend lines, but the calculations are imprecise and experts differ. For example, in 1997 on a proposed FY 1998 defense appropriation of $265.3 billion in new BA, Office of Management and Budget (OMB) estimated $259.4 billion in outlays compared to Congressional Budget Office's (CBO) $265 billion, a difference of $5.6 billion in outlays (Tyszkiewicz & Daggett, 1998, p. 9). The budget resolution had to be adjusted to the CBO outlay estimate. In the Gramm-Rudman era

Table 8.1.   Legal Basis for Budget Execution

1. Article 1, Section 9, of the Constitution of the United States, which states that, "No money shall be drawn from the Treasury, but in Consequence of Appropriations made by Law" and upon which the apportionment and Treasury warrant process is based.

2. Title 31, United States Code (U.S.C.), Section 1301, "Application of Appropriations, which restricts the expenditure of funds to the purposes for which they are appropriated."

3. Title 31, U.S.C., Sections 1341, 1342, "The Anti-Deficiency Act," which states that no Federal officer or employee may authorize Government obligations or expenditures in advance of or in excess of an appropriation, unless otherwise authorized by law, and that no Federal officer or employee may accept voluntary services except as authorized by law.

4. Title 31, U.S.C., Section 1512, "Apportionment and Reserves," which, provides the legislative basis for the apportionment process by requiring, except as otherwise provided, that all appropriations and funds available for obligation be apportioned.

5. Title 31, U.S.C., Section 1514, "Administrative Division of Apportionments," which requires establishment of administrative control of funds designed to restrict obligations against an appropriation or fund to the amount of the apportionment or reapportionment, and that the agency head be able to fix responsibility for the creation of any obligation in excess of an apportionment or reapportionment.

6. Title 31, U.S.C., Section 1517, "Prohibited Obligations and Expenditures," which prohibits making or authorizing expenditures or obligations in excess of available apportioned funds, or amount permitted by regulations under Section 1514, and requires the reporting of violations of this section to the President and the Congress.

7. "The Budget and Accounting Procedures Act of 1950," which defines the legal basis for the issuance of appropriation warrants by the Secretary of the Treasury, who is responsible for the system of central accounting and financial reporting for the Government as a whole.

8. "Congressional Budget and Impoundment Control Act of 1974," that establishes the fiscal year to commence on 10-1 (Title 31, U.S.C., Section 1102), and prescribes the rescission and deferral process (Title 2, U.S.C., Sections 681-688).

9. Title 31, U.S.C., Section 1535, "Agency Agreements," (commonly referred to as the "Economy Act") which authorizes a Federal agency to place reimbursable agreements for work or services with other Federal agencies.

10. Section 111 of the Energy Reorganization Act of 1974, as amended, Public Law 93-438, which cites provisions and limitations applicable to the use of operating expenses, expenditures for facilities and capital equipment, new project starts, and the merger of funds.

11. Section 659 of the Department of Energy Organization Act of 1977, Public Law 95-91, which allows the Secretary, when authorized in an appropriation act for any fiscal year, to transfer funds from one appropriation to another, providing that no appropriation is either increased or decreased by more than 5 percent for that fiscal year.

12. Annual authorization and appropriation acts, which may contain specific guidance on Department funding as well as limitations on reprogramming, restructuring, and appropriation transfer actions.

13. OMB Circular No. A-34, "Instructions on Budget Execution," of 10-18-94, which provides instructions on budget execution including apportionments, reapportionments, deferrals, proposed and enacted rescissions, systems for administrative control of funds, allotments, and reports on budget execution.

14. Treasury Fiscal Requirements Manual, Volume I, Section 2040, which prescribes the procedures to be followed in the issuance of Treasury appropriation warrants

Source: Treasury Fiscal Requirements Manual, Volume I.

(1986-90), when outlay was a controlling limit, a difference of $5 billion in outlay estimates by the experts would (and did) cause severe turbulence in the budget process because they were required by law to hit precise deficit target.

Spend Out Rates

The rhythms of defense spending have allowed for the development of historical spendout rates for defense appropriations. Appropriations have different legal obligation periods and after an obligation has been made, all accounts have a funds availability period of 5 years in which to outlay or spend out the funds to pay for the obligation (Hleba, 2002, p. 22). Since some programs may take years to complete, the total consumption of an obligation may take years to finish. Historical records indicate that military personnel funds are almost all spent out at the end of the FY (95%), while the Navy shipbuilding account spends out about 10% of its funds in the first year. In Table 8.2, we present some selected DoD budget accounts, indicate their legal period of availability, and compare the estimated obligation rate and outlay rates as computed by DoD from historical experience.

The military personnel account has a 1-year period of obligation and a 5-year period for outlays. Most of personnel is obligated in 1 year and spent out by the end of the second year. The shipbuilding and military construction accounts have 5 years for obligation and 5 more years for outlay; in theory money from these accounts could still be spent out 10 years later. Thus, a financial manager could inherit a problem account where the problems occurred many years before he arrived on scene, particularly if he is member of the uniformed services and is serving on a three-year rotation in a collateral duty assignment.

Figure 8.1 indicates selected appropriations and their obligational and expenditure availability periods. The military construction account, in this example, could legally be open for expenditure for 10 years.

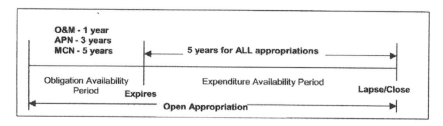

Source: Hleba, 2002, p. 22.

Figure 8.1.   Expenditure availability period.

Table 8.2. Defense Account Spend-Out Rates

		Selected Obligation and Outlay Rates (Percents)						
		BY	BY+1	BY+2	BY+3	BY+4	BY+5	BY+6	Total
Military personnel (all)	Obligation	100							100.00
(1 year)*	Outlay	94.95	4.24	0.021	0.08				99.29
O&M (Army)	Obligation	100							100.00
(1 year)	Outlay	68.80	24.40	4.10	1.50	0.50	0.20		99.50
RDT&E (DoD composite)	Obligation	90.22	9.78						100.00
(2 year)	Outlay	54.64	37.1	7.21	1.38	0.38	0.13	0.09	100.93
Procurement (Marine Corps)	Obligation	81	11	8					100.00
(3 year)	Outlay	30.00	42.00	16.00	8.60	1.50	1.00	0.40	99.50
Aircraft procurement (Navy)	Obligation	84.00	12.00	4.00					100.00
(3 year)	Outlay	18.00	42.00	14.00	4.20	1.50	1.00	0.60	81.30
Shipbuilding (Navy)	Obligation	67.00	16.00	7.00	5.50	4.50			100.00
(5 year)	Outlay	9.63	17.27	20.84	19.73	10.21	7.79	2.56	88.03
Military construction, Army	Obligation	84.00	8.00	4.00	2.00	2.00			100.00
(5 year)	Outlay	4.10	37.10	35.60	13.30	3.80	3.00	2.60	99.50
Base closure (BRAC)	Obligation	77.56	12.62	5.03	2.56	2.23			100.00
(No year)	Outlay	12.45	42.18	28.51	9.32	3.42	2.10	1.38	99.36

By = 2003. Totals may not add to 100 due to rounding.

Source: Compiled from DoD Greenbook, 2003: FY 2004/2005 Budget, FY 2004 DoD obligation and outlay rates. February 2003.

* = legal period of obligation

These differences in spending rates have consequences for those who oversee budget development and appropriation management. For example, if a new budget analyst encountered a MILCON (Military Account, Navy or MCN) account and found it had only obligated 16% of its funds in the first year and outlayed even less, his first thought might be that the program was greatly overfunded and that the account offered an opportunity for making some cost-savings to fund other projects or programs. However, this would be the wrong conclusion to draw. Instead the analyst would have to look at the total shape of the program and measure planned execution against actual execution to see if the rate of completion were proceeding at the correct pace. What analysts and program managers actually do is measure obligation rates to make sure that all the money that is appropriated is obligated in a timely manner. The actual outlay rate is not perfectly predictable; events in far off places can interrupt project accomplishment and thus although the money is obligated, it may be slow to be actually paid out.

DoD alone in the federal government uses the concept of Total Obligational Authority (TOA), which is the total amount available to be spent on defense programs in the FY. TOA is the sum of all BA granted by Congress plus amounts from other sources authorized to be credited to certain accounts, plus unobligated balances of funds from prior years that remain available for obligation.

For example, in FY 1998, BA equaled $254.9 billion, while TOA was $256.8 billion, an increase of $1.8 billion, with the difference coming from BA from previous years, transfers and offsetting receipts (Tyszkiewicz & Daggett, 1998, p. 6). On a total basis, the small difference between BA and TOA would seem to be almost insignificant; however, it masks a greater turbulence. For FY 1999, DoD estimated about 67% of outlays would result from BA provided in the FY 1999 defense appropriation bill, while 33% of outlays occurring in the FY would result from BA provided in prior year appropriation bills (p. 9). This percentage tends to vary with fluctuations in the procurement accounts, for example, for ships, aircraft, missiles, tanks, and other hardware.

By direction of Congress, appropriated funds with 1 year obligation periods (personnel, and operations and maintenance (O&M) normally must be obligated in the first year they are provided, otherwise they expire. Obligation occurs when a formal commitment is made. This could happen the first day of the FY or the last, but when it is made, then the funds are said to be obligated and are considered "spent out." Even though the money has not been paid out, the agency cannot enter into any more agreements to obligate funds or hire employees or contractors; it has spent out all its money, just as an individual's checkbook would be spent out if writing the rent or mortgage check drew the available balance down to zero the moment it was sent, notwithstanding the landlord or mortgage company might not cash it for a week or so. Approaching the end of the FY, an agency could be running a surplus in an account right to the last day of the FY, when it finally gets the contract it wants and signs it. The money is then obligated and, obviously, will be spent out in the next year.

Congress specifies the length of time that funds may be held, pending obligation in the annual appropriations legislation. Typically, Congress makes personnel and O&M funds available for obligation for one year, research and development funds for two years, procurement funds for three years and military construction funds for five years. Congress has created a special category for shipbuilding funds; they too are available for obligation for five years. Once the funds have been obligated, they will be spent out as progress payments on contracts occur or as paychecks are issued.

Unexpended Balances

A typical managerial measure used to track the obligation and outlay of funds within DoD and the federal government as a whole is unexpended balances. This refers to all funds that have been appropriated but have not yet been spent as outlays. Unexpended balances are made up of obligated and unobligated funds. Obligated funds mean a commitment has been made and the federal government will pay bills (make outlays) as soon as the work or service is performed or items are delivered. Unobligated funds indicate funds that have been appropriated, but where the government has made no commitment to put them to work; no people have been employed, no contracts entered into, no supplies ordered. The complexity of government with its purchase of capital goods and its support of intergovernmental spending mean that unobligated balances will never sink to zero. Good financial management means tracking these categories; after all, if the money is not outlayed in this FY, then money does not need to be set aside for it. In a deficit period, that is, most of the last 50 years, this would mean that the treasury would not have to issue billions of dollars worth of bonds and the taxpayer would not have to pay additional interest until the proper time. Thus tracking both unexpended balances and unobligated balances is a sensible thing to do.

An OMB study of the more than $1 trillion in unexpended balances in FY 2004 indicated that 70% would be obligated by the end of the year and 30% would not. We may note this amount is almost half the federal budget which may or may not be outlayed in the FY.

The OMB estimate is that 30% or $300 billion will not be obligated and hence cannot be outlayed. Federal funds (defense, Treasury, etc.) will

Table 8.3. Appropriation and Obligation Period

Appropriation	Service Abbrev.	Length of Obligation Period	Examples	Properties	Budget Activities Funded
Operations and Maintenance	O&MN, O&MMC, O&MA, O&MAF	1 year	Admin. expenses, labor charges, TAD travel for civilians and military	Used for daily operations and expenses, minor construction up to $500k	Operations forces, training and recruiting, administration and support
Military Personnel	MPN, RPN, MPMC, MPMCR	1 years	Officer and enlisted personnel salaries	Used for salaries, training, bonuses, PCS moves, allowances	Officer pay, enlisted pay, allowances, PCS travel, midshipment
Research, Development, Testing, and Engineering	RDT&EN	2 years	Expenses for developing new technology	Used for the development of new or improved capabilities until ready for operational use	Advanced technology, strategic programs, technology base, tactical programs
Other Procurement	OPN	3 years	Purchasing equipment or conducting modernization greater than $100k	Used to procure equipment not funded by operations and maintenance funding	Ships support equipment, ordinance equipment, electronic support equipment, spares and repair parts
Procurement Marine Corps	PMC	3 years	Purchasing equipment, weapons and munitions greater than $100k	Used to procure equipment not funded by operations and maintenance funding	Ammunition, vehicles, spares and repair parts
Aircraft Procurement	APN, APAF	3 years	Procuring 40 F/A-18s	Used for the acquisition of initial or additional aircraft and related equipment	Combat aircraft, trainer aircraft, aircraft spares and repair parts
Weapons Procurement	WPN	3 years	Procuring Tomahawk missiles	Used for the acquisition of initial or additional weapons	Missiles, torpedoes, ammunition, spares or repair parts

Category	Acronym	Duration	Scope	Use	Details
Military construction	MCON	5 years	Building facilities on a base or installation, acquiring land	Used for the construction, acquisition of installation of permanent public works facilities	Major construction (> $1.5 million), minor construction (up to $1.5m), planning, historical projects
Family Housing Operations	FHOPS	1 year	Maintenance of family quarters	Operations of quarters, leasing and maintenance	Operations, leasing, maintenance, interest payments, insurance premiums
Family Housing Construction	FHCON	5 years	Construction of family quarters	Construction of quarters and improvements to existing quarters	New construction, improvements and design
Shipbuilding and Conversion	SCN	5 years	Building of ships, submarines and other craft	Construction of new ships conversions of existing ships	FBM ships, amphibious ships, mine warfare ships, other ships
Base Realignment and Closure	BRAC	No year	Closure or realignment of shore infrastructure	One time, nonroutine operating and investment cost for closure or realignment	MCN, family housing, environmental, operations and maintenance, military personnel, homeowner's assistance.

Source: Hleba, 2002, p. 20

**Chart 1: Total 2004 End-of-Year Unexpended Balances
(S in Billions)**

Obligated - $761
70%

Unobligated - $331
30%

Chart 1 indicates that 70 percent of the unexpended balances will be obligated to pay for such things as grants awarded, supplies ordered and services received by the end of 2004. The other thirty percent will not be obligated at the end of 2004.

Source: Office of Management and Budget, 2003b, p. 2

Figure 8.2.   End of year unexpended balances: Obligated and unobligated.

make up 79% of the unobligated balances with the remainder coming from trust funds. Of the federal funds portion, 68% of the unobligated balances will be financial reserves and 20% capital investment (OMB, 2003b, Charts 1, 2 and 3). DoD differs from this somewhat. A study of DoD unobligated balances for 2002 reveals that procurement and military construction accounts for 46% of the total, as is shown below. Procurement, military construction, and research, development, training, and engineering (RDT&E) total 61% of DoD's unobligated balances. The largest single account is the Navy ship construction account, which had $5.6 billion in unobligated balances in 2002. This is an understandable outcome given our previous discussion of account spend-our rates. Figure 8.3 shows the historical trends for size and persistence of unexpended balances and unobligated balances in DoD.

Historically we can see the defense buildup of the 1980s that emphasized defense procurement. Unexpended balances are interesting to those in Treasury who must raise the money by taxes and bonds to cover the government's bills, but unobligated balances are (or should be) of intense interest to program managers and Congress: unobligated balances mean that for one reason or another, programs are not being executed as directed by law. Therefore close attention is (and should be) paid to unobligated balances. Funds not obligated could mean a normal profile, for

Table 8.4. Unobligated Balances in DoD for FY 2002 by Appropriation (Dollars in Billions)

Procurement and Milcon	$17	46.23%
Revolving Funds	9.2	25.02%
RDT&E	5.6	15.23%
Other	4.3	11.69%
Critical Financial Reserves	0.669	1.82%
Total	36.769	
(Dollars in Billions)		

Source: OMB 2003b, Balances of Budget Authority FY 2004.
Table 5: Federal funds end of year unobligated balances by agency and program
OMB. 2003b, pp. 33-34

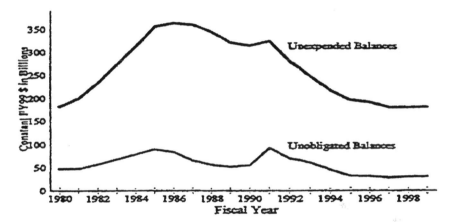

Source: Tyszkiewicz and Daggett, 1998, p. 10.

Figure 8.3. DoD unexpended and unobligated balances: 1980-2000.

example, in the shipbuilding account, but it could also mean problems have been encountered. The ethic within DoD is to obligate as close to 100% as possible, but the complex rhythms of budgetary life do not always make this possible. In addition, changes in inflation rates can impact unobligated fund balances; if inflation is lower than predicted, less money may be needed. The U.S. also spends money abroad for personnel and for procurement; if the U.S. dollar strengthens so that it can buy more for less, fewer dollars may be needed than were budgeted. As a practical matter, Congress also watches unobligated balances not only to see that programs are being executed, but also as a potential source of funds. For example it used unobligated funds to help provide offsets for supplementals in the late 1990s (Tyszkiewicz & Daggett. 1998, p. 10).

Historical patterns show a relatively high but consistent balance for unobligated funds, particularly given that most managers know that not obligating money within the FY is a cardinal sin, an automatic budget cut in the current year, a probable cut in the next year, and a potential indictment of other programs of that manager. In defense, the logic is that someone fought to get those funds into the Program Objective Memorandum (POM), the defense budget, the proposed appropriation bill, and through Congress, and now they are not being obligated. Reviewers in the Pentagon and Congress will want to know why not, especially when they might be able to divert some of these funds to other programs.

The DoD is not the only department with unobligated balances to track. What a 2002 study of the unobligated balances revealed is shown in Table 8.5.

Although DoD has the most dollars tied up in unobligated balances, it ranks far down the percentage ranking list (13 agencies or groups of agencies rank above it), mainly because defense puts great emphasis on obligating funds by the end of the year. Unhappily this also creates a "rush to spend it up" at the end of the FY which we will discuss later. Despite end of year spending patterns, unobligated balances are a fact of life in federal financial management and it is good that they are closely watched. Next we turn to an examination of the common events that transpire during a FY for budget execution.

FLOW OF FUNDS

Once the President has signed an appropriation bill into law and after the start of the FY, the Treasury Department issues an appropriation warrant to the OMB. This warrant establishes the amount of funds authorized to be withdrawn from Treasury accounts for each bill. Both the military construction appropriation bill and the main defense appropriation bill would result in such warrants. When a bill covers different agencies, OMB must apportion the correct amount to each agency covered by the bill, as for example to the Departments of State, Commerce, and Justice from the appropriation act covering all three departments.

Allotments

Each agency then uses allotments to delegate to subordinate components of the department (e.g., Army, Navy, and Air Force) the

Table 8.5. Agencies Ranked on Unobligated Expenditures

Agencies Ranked by Percent of Unobligated Expenditures 2002

2002 Actuals	Obligated	Unoblg.	Unexpend	Percent
OPM	$6,925	$27,720	$34,645	80.01%
Other Ind Agencies	$26,716	$56,519	$83,235	67.90%
Labor	$9,142	$14,160	$23,302	60.77%
Corps of Engineers	$1,071	$1,611	$2,682	60.07%
Treasury	$17,393	$24,502	$41,895	58.48%
Interior	$4,498	$3,853	$8,351	46.14%
Intl. Asst. Programs	$63,939	$41,516	$105,455	39.37%
Transportation	$59,709	$36,282	$95,991	37.80%
State	$4,743	$2,780	$7,523	36.95%
Agriculture	$19,604	$9,229	$28,833	32.01%
Homeland Security	$10,491	$4,498	$14,989	30.01%
HUD	$93,957	$37,356	$131,313	28.45%
Vet Affairs	$7,475	$2,840	$10,315	27.53%
Defense	$158,108	$39,128	$197,236	19.84%
Commerce	$4,341	$1,017	$5,358	18.98%
EPA	$11,480	$2,472	$13,952	17.72%
Energy	$10,453	$1,800	$12,253	14.69%
Education	$30,501	$4,822	$35,323	13.65%
Justice	$13,891	$2,081	$15,972	13.03%
Social Security	$50,038	$3,017	$53,055	5.69%
Health	$70,842	$2,443	$73,285	3.33%
Total	$675,317	$319,646	$994,963	32.13%

Source: Jones and McCaffery, 2003 compiled from OMB, 2003c, Exhibit 2:18.

authority to incur a specific amount of obligations. In DoD, this task is performed in the DoD comptroller's office under the oversight of the under secretary for defense (comptroller). The allotments go to the offices responsible for overseeing their execution, for example the military department comptroller's offices. Here the allotments are further divided to the relevant budget submitting offices and claimants, who may in turn further allocate the funds, until a portion of the DoD appropriation flows down to the lowest command level at which a comptroller is authorized to commit funds and initiate payments. For example, this chain could run from the DoD comptroller, to the Navy comptroller to the Pacific Fleet to the type command that controls air stations to the comptroller of an air station somewhere in the United States or overseas.

As dollars are allocated down the chain of command so is responsibility to meet the provisions of the Anti-Deficiency Act (31 U.S Code 1341 and 1517) and not overspend an appropriation. This responsibility may not be imposed all the way to the smallest commands; the office they report to may hold the responsibility. This office is then said to hold operating budget responsibility and the offices below it are guided by planning targets, such as operating targets, allowances, and expense limitations. Quite properly, the military departments have flexibility to set the appropriate level for the Anti-Deficiency Act responsibility in order to see that small budget holders are not subject to over-control or held to account for spending patterns not within their administrative control.

In the Department of the Navy, the Program Budget Accounting System (PBAS) is the mechanism through which funds are administered and controlled, eventually delivering fund authorization documents to the desktop computers of Navy offices. At the end of this process of allotment and allocation, commitments can be invited (bid invitations) obligations can be incurred (e.g., contracts let) and outlays can be paid when the work or service is completed or the equipment delivered. These allotment and allocation processes constitute the planning process for when funds will be spent, by quarter or month, by administrative level, and by category of expenditure. These plans are guided by a well-defined set of laws and rules.

Midyear Review

Just after the halfway point in the year, the budget office conducts a midyear review with its major claimants. This review monitors program and budget performance against the obligation and outlay phasing plans that were developed during the budget formulation process and formed the basis for the allotment request. The obligation rate is carefully measured against the time remaining. Funds may be taken from accounts that are under-executing and given to accounts that have run into some unexpected difficulty and need more money. Only urgent requirements are funded. Generally, these have to be newly emergent needs; after all, if they are old priorities they were considered and did not make it into the budget, thus they are of a lower priority than everything that is in the budget. Midyear review is about the time that reserves held by commands begin to be released to fund unfunded requirements. Holding reserves is a habit with commands and comptrollers; sometimes they admit to it up front, because it is only prudent to pull off a reserve to help solve unanticipated problems that arise. Sometimes they argue that they do not have a reserve and they mean it. What they do not say is that they know very well which accounts to go to if they need additional money. No major command comptroller could function well within DoD without such knowledge. Midyear review is also an opportunity to begin to set the scene with the departmental budget office for the summer budget process.

Fiscal Year Closeout

The end of the FY has great significance within DoD. Annual obligations must be fully obligated by September 30 of each year, a responsibility that DoD takes quite seriously, as will be discussed later. Funds not obligated are lost. Moreover, any underexecution may cause reductions in the following FY. At the very least, an underexecution will result in a budget cut or mark that the command will have to respond to in order to get the funds back. During the last 2 months of the year, an intense effort is undertaken to get control of unobligated funds and to apply them to unfunded requirements. This happens in almost every comptroller's office from the lowest field command to the Pentagon. Reserves are fully released and obligated. As the FY closes, various deadlines have to be observed about when purchases may be made and in what order of magnitude. Rules and procedures are in place to avoid the end-of-year rush to spend, but all they seem to do is put up additional barriers and hurdles and do not always prevent spending on lower priority items and areas. In fact, the end of year close out rush often results in less than optimal and sometimes outright wasteful spending. For this reason there is cause for DoD to request that Congress change appropriation law so that the one year obligation period for O&M and other 1 year appropriation accounts be extended to 2 years so that funds can be applied to highest priority needs without the pressure of end of year close out every FY (see more on this proposed reform in the last chapter). This proposal should not be confused with biennial budgeting, which would appropriate O&M for 2 years. Such a reform is not supported by the authors of this book.

To conclude our review of FY close out, we are compelled to observe that the new FY arrives on October 1, but it is often not accompanied by congressional passage of an appropriation bill for the new FY. This leads to the need for approval by Congress and the President of what is termed a continuing resolution appropriation (CRA) so that DoD can still operate and pay its bills until a regular appropriation is passed and signed into law by the President.

Continuing Resolution Appropriations

When no appropriation bill has been passed by the start of the new FY, DoD functions under a CRA. This provides funds to continue operating until an appropriation bill for the new FY is passed. Congress indicates in the CRA the rate of spending that DoD should attain during this period; usually it is set at the level of the lowest rate of the bill passed in either the House or Senate. It is Congress's choice, although the President can and has vetoed CRAs. Usually these are not controversial.

For DoD, the CRA means the pace of operations will be carried on as it was in the FY just concluded, but no new program starts will occur, no new purchases be made, and no new employees be hired, except, of course, if Congress has explicitly directed otherwise. For some programs and accounts, a CRA is not a problem; a civil servant in Washington or a Marine onboard ship in the Pacific will hardly notice the event. For someone hoping for a federal job, or a contractor waiting on a purchase or procurement obligation, the CRA could impose significant hardship and a CRA that lasts until late November or early December might severely affect the way the program is administered the rest of the year … and the financial viability of a small company waiting to execute a federal contract. In any case, CRAs are a fact of life for federal financial managers. Next we look at some individual accounts to surface some of the complexities that bedevil defense budget execution managers.

BUDGET EXECUTION PATTERNS

We now turn to an examination of patterns of budget execution for defense budget managers. In this regard we review the characteristics of some defense appropriation accounts.

Navy Military Personnel Account

Our first account is a military personnel account. This appropriation is guided by end strength billets granted the department in the department's authorization bill and then funded by the appropriation bill. It would be reasonable for the enlightened observer to assume that the budget appropriation for military personnel would be highly stable. A textbook description would indicate that Congress passes an appropriation bill and after the normal allocation and allotment processes, Navy fund managers know the amount they have for the year to pay for military personnel. However, the account structure itself is not simple. It supports the congressionally authorized number of personnel and includes funds for such items as basic pay, housing allowances, retirement pay, hazardous duty pay, basic allowance for subsistence, enlistment bonuses, and clothing allowances.

Table 8.6 shows the total military personnel for the Department of the Navy (DoN) MPN account from September 1999 through October 2000. The first column labeled "APPN YTD" contains the amount of money Congress appropriated to DoD to incur an obligation year to date. The second column labeled "OBLIG YTD" is the amount that DoD officials have obligated year to date in contracts that legally bind the government to a future expenditure, in this case recruiting personnel. The third column labeled "DISB YTD" is the total outlay from the Treasury for the FY in execution of the account's obligations. The fourth column labeled "% OBL Per MN" is the percentage of new

Table 8.6. Navy Personnel Account Profile (in Thousands)

Month	Appn. YTD	Oblig. YTD	Disb. YTD	% OBL Per MN	% Disb. Per MN
Sept.–99	16909133	16903230	15681715		
Oct.–99	16814377	1530961	1450599	8.76	8.30
Nov.–99	16831427	2889514	2825433	7.77	7.87
Dec.–99	17290760	4315097	4153254	8.16	7.60
Jan.–00	17310972	5712740	4875431	8.0	4.13
Feb.–00	17330725	7151770	6253722	8.24	7.89
Mar.–00	17362469	8603070	8381684	8.31	12.18
Apr.–00	17379878	10053077	9099385	8.29	4.11
May–00	17397810	11517176	10534485	8.38	8.21
Jun.–00	17418949	13012091	12699296	8.56	12.39
Jul.–00	17392367	14510662	13406883	8.58	4.05
Aug.–00	17420989	16102110	14948983	9.11	8.83
Sept.–00	17472900	17434859	17172535	7.63	12.73
Oct.–00	17728554	1669236	798025		
Total FY 2002				99.78	98.49

Source: Skarin, 2002, p. 12.

obligations incurred during the month compared to the total for the FY. The last column labeled "% DISB Per MN" is the percentage of actual money disbursed per month.

First, at the end of the year, 99.78% of the personnel appropriation is obligated and 98.49% is disbursed. Obligations never exceed the appropriation, not by month or in total. Disbursements do exceed the amount obligated by month, primarily in the end of quarter months, but disbursements never exceed the cumulative total obligated, notwithstanding some monthly variation. If no fluctuations were to occur in this account, the obligation rate per month should be 8.333% (100/12). In no month does this occur. Instead, obligations range from 9.11% in August 2000 to 7.63% in September 2000. Moreover, money actually disbursed or paid from the personnel appropriation ranged from 4.05% in July of 2000 to 12.73% in September. The appropriation itself is not stable; in no 2 months is it the same. Moreover, the appropriation increases from $16.8 billion in October of 1999 to $17.4 billion in September of 2000, all within the FY. Some of this turbulence was caused by a late appropriation bill.

DoD was covered by a CRA until October 25, 1999 when the FY 2000 appropriation bill was signed. For the month of October, the appropriation was less than the appropriation for FY 1999 (see September 1999). However, it also appears that the spending moved along at the CRA rate through November, but even at that, the personnel appropriation increased from December 1999 to September 2000. This appropriation is by far the most stable and most simple account Navy budget managers have to supervise and while its managers certainly conformed to legal norms of not obligating and disbursing more than the appropriation, the account exhibits a substantial degree of turbulence, most surprisingly in the size of the appropriation itself. In fact, the appropriation decreased between June and July.

These numbers appear to support an orderly scenario, yet no one month is the same as another and .22% of what is budgeted never becomes obligated (the appropriation for personnel has a 1 year obligation period). Budget managers do spend the appropriation up, but they do not overspend, thus they display and obey two key injunctions in the DoD budget world, (1) spend it up, but (2) do not overspend. It is hard to see how they could do much better than managing to 99.78% of the appropriation, still that leaves over $38 million appropriated but not spent. If they were able to transfer these funds to the procurement account, they might be able to buy another F/A-18, if they could get legislative approval. Still, the bottom line is that it is hard to see how the fund managers could have done better than 99.78%

Air Operations and Maintenance Account

The account designated "AG 1A" is a Navy Air O&M account. Individual line items in this account include mission and other flight operations, fleet air training, aircraft depot maintenance, and safety support. This account has been the subject of much budget debate because the large expense to maintain aircraft is weighed against the importance of maintaining pilot proficiency. In times of scarcity, funds are sometimes taken from this account to support higher priority items in the air operations world, like flying hours, in order to keep mission readiness high. Money is then paid back into the account when it becomes available. This account then should be marked by turbulence.

Table 8.7 displays the AG 1A account for FY 2000 and the transition months into and out of FY 2000, that is, September 1999 to close FY 1999 and October 2000 to start FY 2001. Differences from the MPN table previously described may be seen in the columns labeled "Revised," and "APPN vs. REV." The "Revised" column contains the dollars appropriated and includes any reprogramming, supplementals, and/or rescissions. "Reprogramming" occurs when funds are shifted from one program to another within the DoD budget. For example, Congress has authorized pay raises, but failed to provide all the dollars needed and DoD has reprogrammed funds from other programs to implement the raise. Recently Congress has given DoD authority to transfer $2 billion a year among accounts. DoD can transfer certain amounts below specified thresholds without notifying Congress. When the transfer exceeds a threshold, DoD must either notify Congress after the action or seek prior approval for the transfer from both the authorization and appropriation committees of each chamber. Just one negative Senator or Congressman can kill this kind of action. A "supplemental" is a budget request submitted separately to Congress and is generally related to unforeseen circumstances such as funding for disaster relief. A "rescission" accommodates changing priorities and occurs when BA is cancelled as approved by both the President and Congress. The cumulative revision column accumulates these changes as they impact this account. The month-by-month revision column indicates how much each month was revised over its preceding month. The last four columns tell the story of cumulative dollars obligated and disbursed by month and the percents of total dollars obligated and disbursed by month.

Unlike the MPN account, the Air Ops appropriation remains stable throughout the FY. The appropriation number never changes over the course of the FY. However, the revised level varies significantly, ranging from 12.12% more than the appropriation in December 1999 to 21.96% more in August of 2000 as the Department of the Navy (DoN) budget

Table 8.7. Navy Air Operations and Maintenance Account

Month	Appropriated	Revised	Cumulative Revision	Month by Month	Dollars Obligated	Dollars Disbursed	Cumulative Percent	
							Obligated	Disbursed
Sep.–1999	3741527	4163821	11.29%		4160307	3070814	99.92%	73.81%
Oct.–1999	3402490	3402490	0.00%	−18.28%	345635	167538	10.16%	48.47%
Nov.–1999	3402490	3895179	14.48%	14.48%	794859	367754	20.41%	46.27%
Dec.–1999	3402490	3814776	12.12%	−2.06%	1095921	696639	28.73%	63.57%
Jan.–2000	3402490	3922129	15.27%	2.81%	1498939	986682	38.22%	65.83%
Feb.–2000	3402490	4036298	18.63%	2.91%	1800733	1374413	44.61%	76.33%
Mar.–2000	3402490	4040249	18.74%	0.10%	2137559	1820672	52.91%	85.18%
Apr.–2000	3402490	4043529	18.84%	0.08%	2637064	2316700	65.22%	87.85%
May.–2000	3402490	4037861	18.67%	−0.14%	2905024	2771119	71.94%	95.39%
Jun.–2000	3402490	4064057	19.44%	0.65%	3160359	2664554	77.76%	84.31%
Jul.–2000	3402490	4151047	22.00%	2.14%	3625119	2945082	87.33%	81.24%
Aug.–2000	3402490	4149634	21.96%	−0.03%	3902857	3268183	94.05%	83.74%
Sep.–2000	3402490	4144233	21.80%	−0.13%	4136944	3635864	99.82%	87.89%
Oct.–2000	4275064	4263635	−0.27%	2.88%				

AG 1A Air Ops (In Thousands of Dollars)

Cumulative revision FY 1999	11.29%
Increase from FY 2000 to FY 2001	25.65%
Cumulative revision FY 2000	21.80%
Average of months	0.21%

Source: DoN FMB.

managers found ways to supplement the account so that the flying hour program would result in pilots and aircraft on carriers being at the appropriate readiness levels.

However the major theme of this account is of the turbulence in execution of this account. First, turbulence exists in the appropriation. Even as the account was being revised upwards during 1999 (by 11.29%), budget decision makers were further reducing the appropriation, from $3.7 billion in 1999 to $3.4 billion in 2000, notwithstanding that it actually executed at $4.16 billion in 1999. FY 2000 starts off with a cut below the base of FY 1999 and a cut of 18.28% from the program level that it had been executing at in September of 1999. It is no wonder that revisions in the account begin almost immediately. In November 1999, the second month of FY 2000, the account is revised upward by 14.48%, the largest revision of the year. Subsequently, every month shows a revision. In FY 1999, 100% (99.92%) of the appropriation was obligated, thus the turbulence did not prevent managers from making sure the money was obligated. Of the money obligated, almost 74% (73.81%) was disbursed during the FY. Since all money has a 5-year funds availability period, the rest of the disbursements could occur over the next 5 years. These numbers are within normal tolerances for an O&M account.

This finding is also true for FY 2000. Almost 100% of what was available was obligated (99.82%) and 87.89% of what was obligated was disbursed. This illustrates how DoD financial managers work to make sure they fully obligate their accounts no matter the turbulence. Nonetheless, this account is an interesting example of such turbulence. First, the account was cut by more than 18% going from FY 1999 to FY 2000. In the second month of the FY, the account was revised upward 14.48%, in reaction to the cut. The account manager can cope with this by pushing the cuts out into the third and fourth quarters and hoping that relief comes, as it did in this case. Then there are changes in every month of the year. Five of these are decreases from one month to the next and three are increases of more than 2%. Turbulence is a fact of life for this account manager and every month brought with it changes in the account. Toward the end of the year, the account shows some stability as end of year spending discipline takes hold, and very small changes are made in August and September. Taking the average of the 12 months in the FY indicates a change of .21% (average of the 12 month changes). This average disguises the real turbulence, however, as the extremes range from plus 14.48% to a negative 2.06%, and a negative 18% to open the year. Moreover, at the end of the year, the account manager has executed 21.8% more dollars than the enacted appropriation provided when it was passed.

The account fared better during the FY 2001 budget process as it was increased 25.65% over the FY 2000 appropriation, again perhaps as a reaction to the turbulence manifested in 2000. After all, the account manager can argue that the account was woefully underfunded. Naturally, a revision was made during the first month of the new FY, a slight negative adjustment (.27%). However, this was still 2.88% more than what the manager had to execute in the previous month. Thus notwithstanding that the appropriation was changed downward slightly, the account manager had a substantial increase with which to begin the year.

We may notice that in FY 2000, 99.82% of the revised appropriation was obligated (rule 1: spend it up), and obligations never exceeded appropriations as the account managers follow the two cardinal rules of defense budget execution: (1) spend it all, and (2) do not overspend.

Ship Construction: LPD-17—An Advanced Procurement Account

Our third account for study is in the procurement accounts or the ship construction and conversion account (SCN). These are longer-lived and hence more complex accounts to manage. Once an appropriation was made for a ship, the funds have a 5-year obligational availability period and are disbursed or outlayed as work is completed. This could add another 5 years that the account would have to be tracked due to the five-year expenditure availability period. The ship studied below is the first platform in the Navy's new generation of Amphibious Assault Ships, the LPD-17 SAN ANTONIO Class. This was approved by Congress on December 17, 1996. Due to design complications and construction delays, the completion date has been pushed back from September 2002 to an estimated date of November 2004 (Navsea News, 2002).

By focusing on a specific LPD-17 advanced procurement account designated 99/03 1611 in the Navy shipbuilding and conversion Navy accounts, the difficulties of planning and executing an unpredictable budget become clear. Table 8.8 lists appropriated, obligated, and disbursement numbers for selected dates from FY 1999 and into FY 2002 to provide a longer-term profile of this account. The column labeled "APPN vs. OBL" reveals the percentage of the appropriation that became obligated. The far right column labeled "OBL vs. DISB" is the percentage of the obligated amount that was spent.

What Table 8.8 reveals is that the LPD-17 program had significant funding in FY 1999, but delays resulted in obligation of only 56.06% of this money and disbursements, meaning pay for work that was completed were just as tardy (from the 1999 cumulative total line). Historical data indicate that this is on par for all SCN expenditures. On average, SCN accounts obligate 63% of their appropriation the first year and outlay

Table 8.8. LPD-17: A Ship Construction Account

Month	*Appn. FY*	*Obl. YTD*	*Disb. YTB*	*Appn.* *vs. Disb.*	*Obl.* *vs. Disb.*
Sep.–1999	636878	357023	23404	56.06	6.56
Oct.–1999	279855	1364	1345	0.49	98.61
Nov.–1999	279855	4904	3513	1.75	71.64
Dec.–1999	279855	10559	4570	3.77	43.28
Jan.–2000	275900	13808	3607	5.00	26.12
Feb.–2000	275900	22843	4690	8.28	20.53
Mar.–2000	275900	25309	10868	9.17	42.94
Apr.–2000	275900	28457	12421	10.31	43.65
May.–2000	275900	38121	15203	13.82	39.88
Jun.–2000	275900	46694	18737	16.92	40.13
Jul.–2000	275900	49506	21662	17.94	43.76
Aug.–2000	275900	50816	123616	18.42	46.48
Sep.–2000	275862	53038	26172	19.23	49.35
Oct.–2000	222824	920	2172	0.41	236.09
Dec.–2000	222824	9346	8826	4.19	94.44
Mar.–2001	222824	25527	19622	11.46	76.87
Jun.–2001	222824	30453	32112	13.67	105.45
Sept.–2001	222824	45096	55303	20.24	122.63
Dec.–2001	177727	7609	21108	4.28	277.41
Mar.–2002	177727	15774	46934	8.88	297.54
Jun.–2002	177727	97186	91043	27.11	156.26
Aug.–2002	177727	97925	107320	55.10	109.59

Source: Skarin, 2002, p. 17.

7.1% of it. At the end of 5 years, the legal obligation period, the SCN accounts obligate 100% of their money, but they have disbursed only 67.2% of it. Disbursements obligated in the 5th year still could be paying out in the 10th year of the program. Appropriations were cut by 56% for FY 2000, but delays persisted. Not until tangible progress began in the construction of the ship toward the end of the year did obligations and disbursements begin to approach appropriated levels.

Managing the O&M Account

What we found in our earlier discussion was that budget managers in defense must spend up the money they have, but they must not overspend. We examine how this operates in the O&M account in general

344 L. R. JONES and J. L. McCAFFERY

throughout DoD. The law requires that budget managers must obligate their money before the end of the FY or else they lose it. Table 8.9 shows the obligation rate of the defense department O&M account. This account is analogous to the supporting expense category in most other budgets; it funds everything from steaming hours to flying hours, from yellow writing tablets to yellow paint. It does not fund military personnel, weapon systems procurement or research and development.

This account is closely watched, for it is a major contributor to training and readiness and to the ability of DoD to go places and do things. It is also closely watched when new missions emerge or budget reductions are necessary. To assume a new, unbudgeted, mission in the national interest within a FY often means finding money in the O&M accounts and spending it, with the reimbursement coming later. When quick budget reductions are needed, the O&M account is often the first target, because it is usually obligated in one year and a dollar reduction means a dollar of saving in the particular year, while the ship procurement account may require $20 of reductions in program over the life of the item to get one dollar of reduction in the current year because ship construction is a multiple year account. In sum, the O&M account, which constituted 34%

Table 8.9. Average DoD O&M
Monthly Obligation Rates With High-Low Ranges (Percent) 1977-1990

Month	Average	High	Low
Oct.**	11.235	12.30	9.83
Nov.	8.018	9.73	7.03
Dec.	7.346	8.18	6.47
Jan.*	10.048	11.42	8.11
Feb.	7.165	8.35	6.10
Mar	7.223	7.96	5.65
Apr.*	9.083	10.61	8.30
May.	6.708	7.61	6.09
Jun.	6.726	7.49	5.89
Jul.*	8.778	10.57	7.43
Aug.	6.887	7.39	6.17
Sep.	10.616	12.05	9.78
Total	99.833		

Note: High = highest October and low = lowest October in 14 year period, ** = start of fiscal year; * = first month of quarter.

Source: McCaffery and Jones, 2004, p. 216.

of DoD BA in FY 2002 (DoD Greenbook, 2003c, p. 7) and has been the largest DoD appropriation since 1994 is a sensitive and important account that funds much of the daily business of DoD. For the same reason, it requires close and skillful management.

Mark Kozar studied the DoD O&M account over 14 FYs (FY 1977 through FY 1990) and found O&M account managers obligating over 99% of their funds before the appropriation expired at the end of the year (Kozar, 1993, p. 73). This is a remarkable performance by these fund managers, given legal constraints and penalties against overspending (Kozar & McCaffery, 1994). However, within this orderly pattern of commitment of resources, quarterly and monthly variations appear as do variations among the military departments.

In general, the *first and last* months of the FY are the highest spending months. October generally surges because of new contracts being let. The summer is low as year-end positioning takes place and then September is high as managers rush to spend their funds on what had been planned for and those new needs that have arisen. October shows the highest rate of commitment of funds and the first 2 months show a higher rate than any other 2 months of the year. September illustrates the end of year spending surge; it was the second highest individual month and the rate of change from the preceding month to it (53.6%) exceeded any other pair of months. The fact that August is the lowest month of the year helps create this relationship. A quarterly trend is also evident. The months that start quarters (January, April, and July) are the third, fourth, and fifth highest months, behind only October and September (October has a double bonus; it starts both the FY and a quarter).

The first month of any quarter is high and the next 2 months low. Thus, a substantial change occurs from the last month of a quarter to the first month of a new quarter. This would seem to indicate the existence of quarterly allotment patterns, or at least that new money has become available with the opening of each quarter. The quarterly pattern would seem to indicate that each quarter has pressing needs that are met in the first month of the quarter. Interestingly the surge does not take place from one FY to another; the change from September to October is only 5.6%. This is primarily due to the surge in September. More variation occurs within the year than from 1 year to the next for DoD as a whole.

While the military departments basically follow the DoD profile, some differences are evident. Table 8.10 presents this data. Not only are quarterly and monthly variations evident, but obligation rates also vary somewhat by military department, particularly in the first month of the FY. The Army department actually decreases from September to October, perhaps indicating that it fills many fall needs with September spending, while Air Force spending increases dramatically from September to October, per-

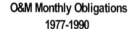

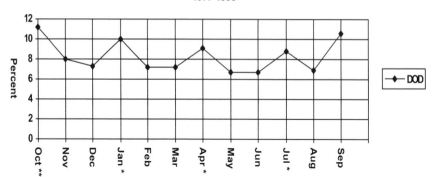

Notes: ** = Start of Fiscal Year; * = first month of quarter.

Source: Jones and McCaffery, 2004, p. 217.

Figure 8.4.   DoD monthly O&M obligation rates.

Table 8.10.   Military Department O&M Monthly Obligation Rates (Percent) 1977-1990

Month	DoD	Air Force	Army	Navy
Oct.**	11.235	14.192	9.188	11.69
Nov.	8.018	8.52	8.059	7.736
Dec.	7.346	6.815	7.968	7.212
Jan.*	10.048	10.526	9.117	10.814
Feb.	7.165	6.224	7.851	7.429
Mar.	7.223	6.251	7.979	7.166
Apr.*	9.083	9.399	8.209	9.69
May.	6.708	6.159	7.046	6.445
Jun.	6.726	5.881	7.141	6.49
Jul.*	8.778	9.324	8.113	8.525
Aug.	6.887	6.173	7.276	6.67
Sep.	10.616	10.662	12.473	9.319
Total	99.833	100.13	100.42	99.186

Note: Percentages exceed 100 due to rounding; ** = start of fiscal year; * = first month of quarter. Fourteen years of monthly obligation data provided by the Defense Finance and Accounting Service for 11 Department of Defense O&M accounts were studied for the fiscal years 1977 through 1990.

Source: McCaffery and Jones, 2004, p. 218

haps indicating that it really needs the new FY money. The average of these fluctuations makes it appear that little or no inter-departmental turbulence occurs around the close of one FY and the start of the next.

These exhibits clearly demonstrate the cyclical nature of those obligations. October, the first month of the FY is the highest. Then the first month of the next three quarters are high, but lower than October, with average obligation rates decreasing during the FY, with September, the last month of the FY, representing a significant surge, exhibiting an obligation rate only slightly lower than October. Two legal bases provide support for these patterns. The first may be found in the U.S. Code, Title 31 Section 1512, Apportionment and Reserves, that states that an appropriation available for obligation for a definite period should be apportioned to prevent obligation or expenditure at a rate that would indicate a necessity for a supplemental or deficiency appropriation for the period. An appropriation subject to apportionment may be apportioned by months, calendar quarters, operating seasons, or other time periods.

O&M appropriations are apportioned by calendar quarters by the Office of Management and Budget under the authority of Title 31 Section 1513. The apportionments, available on a cumulative basis unless reapportioned by OMB, are based on input from the military departments through the secretary of defense, as is the case for other department secretaries who receive information from their component agencies and bureaus, and so forth. According to Title 31 Section 1514, the department secretaries are then responsible for enacting regulations to administratively control and divide the apportionment. The system is designed to limit obligations to the amount apportioned, to fix responsibility for violations of the apportionments, and to provide a simple way to administratively divide the appropriation between commands.

Apportionments are an effective tool for preventing too rapid an obligation of total funds at any point in the FY. However, because funds are available on a cumulative basis, apportionments are better able to prevent obligation surges at the beginning of the FY than at the end of the FY (Kozar & McCaffery, 1994, p. 3). The data in the exhibits clearly shows an end of year spending surge. When compared to the actual average obligation rate of the last month of each of the first three quarters (7.098%), September's rate of 10.616% is 49.6% higher.

The second legal basis for these patterns may be found in Title 31 Section 1502 of the United States Code which states that the balance of an appropriation or fund limited for obligation to a definite period is available only for payment of expenses properly incurred during the period of availability, or to complete contracts properly made within that period of availability and obligated consistent with section 1501. A balance remaining in an appropriation account at the end of the period of availability must be returned to the general fund of the Treasury according to this section. This is reiterated annually in the general provisions of Appropriations Acts that typically state for DoD and other departments that, "no part of any appropriation contained in this Act shall remain available for obligation beyond the current FY, unless expressly so provided herein."

As a matter of practice, the budget execution box (limits) in which the budget fund administrator is forced to work is bounded by appropriation life, appropriation size, and the penalties associated with either over or underspending. The first quarter surge reflects the award of annual service contracts that must be in place for the entire FY but cannot be awarded until money is appropriated and allocated for a new FY. The second and third quarters each start with a "surge" as quarterly allocations are received in the first month. However, overall spending is tempered by the realization that funds must last for the entire FY. Some funding is held in reserve for unknown contingencies. For example, in the Navy the budget office makes an assessment of each O&M account and sets aside approximately 1% of the total program value to create a chief of naval operations (CNO) reserve. This is used to finance emergency situations that arise during the year (e.g., repairs to a ship that has run aground) and to finance unfunded requirements identified during the midyear review. The Office of the Secretary of Defense (OSD) usually withholds some funding from procurement accounts based on their estimate of the phasing of the programs execution. The OSD withhold may or may not be given back to the program, depending on how the phasing works out. The CNO reserve will be spent by the end of the year, usually, but not always, in the fourth quarter. It will be spent on unforeseen contingencies or, more likely, to cover requirements deferred from earlier in the year. In either case, the money *is* spent.

When the O&M appropriation expires at the end of the FY, any unobligated funds, with few exceptions, are lost. This gives managers a very strong incentive to spend all available funds and explains the 99% obligation rate and the September surge. The obligation rate data indicate that DoD managers, as those in other departments, are highly skilled at obligating funds before the end of the FY and are able to commit substantial funds quickly in September, despite numerous rules and provisions designed to forestall this process, for example, to prevent abuses in commitments of funds. Federal fund managers must find valid current FY needs to legally justify obligating the funds before the remaining obligational authority expires.

Undoubtedly, some of these September purchases help meet needs in the next FY and thus by supporting purchases other than current FY needs, funds can be freed up in the next FY which is, in September, less than 30 days away. Appropriation law makes it illegal to obligate or spend current year money to meet next year's needs. The money must be obligated to meet needs that arise in the current year. The actual outlay of funds may happen in the following year. For example, in 1984 the Government Accountability Office (GAO) found that DoD industrial funds illegally carried O&M funds over to the next FY. Reporting to the chairman of the House Appropriations Committee, GAO reported that the six DoD industrial fund activities carried over about $35.7 million from FY 1982 to 1983 through the improper use of industrial funds, extending the life of 1-year appropriations that would have otherwise expired. The primary causes for the improper carryover of funds were the lack of a legitimate current need for the good or service and the failure of the industrial activity to start the work before the end of the FY. These actions violated Title 31 Section 1502 and the general provisions of the DoD Appropriation Acts (GAO, 1984).

Managers of O&M accounts know they are managing a multiple year stream of resources. Replacing inventory or beginning a ship or aircraft overhaul are examples of obligations the manager may take at the end of the current FY to meet fiscal burdens that he certainly would have had to meet in the next FY. The O&M account is not a neat and orderly world. Prices and inflation rates change; commodity prices fluctuate; operating tempos change from what was anticipated and fund managers must adjust to these changes. This means that they pay the bills they must when they must, like service contracts that come due at the first of the year and in the first month of each quarter, and then hold back a little on other items that can be postponed until later in the year.

This pattern is understandable from the point of view of budget managers. The equipment purchase category constitutes a source of flexibility. If a piece of equipment scheduled for replacement is still operational, the unit may be willing to delay replacing it until closer to the end of the FY, then if no higher priority need appears, it may purchase the item or items previously budgeted. By accepting uncertainty and delaying certain purchases, the financial manager provides some slack that may help reduce the impact of unpredictable events upon him. Toward the end of one FY, budget managers are also positioning themselves for the next FY by examining actions they can take in this year that will help them meet the burdens of the next.

BUDGET EXECUTION PROCESSES AND ISSUES

Transfers Between Appropriation Accounts

Congress has officially recognized the need for agency discretion in budget execution since the growth of federal expenditures after World War II. Jones and Bixler (1992, pp. 58-60) found that the 1956 Defense Appropriations bill expressed congressional policy on reprogramming that still remains in force. The report attached to the bill stated that unforeseen changes in operating conditions and circumstances could always occur in the interval between building and presenting the budget and executing it. These might include changes in operating conditions, revisions in price estimates, wage rate adjustments and so on. Consequently, rigid adherence to the original budget might prevent the effective accomplishment of the planned program. The report cautions that it is not the intention of Congress to give the military departments and services unrestrained freedom to reprogram or shift funds from one category or purpose to another without prior notification and consent from the committee.

Since 1959, Congress has emphasized prior approval before implementation of certain transfers, certain threshold requirements, and periodic reporting on issues of special interest to Congress. In 1974, Congress added a restriction that prohibited DoD from transferring funds to restore budget items specifically denied by Congress.

Also of considerable importance to the authority and responsibility for budget preparation and submission, the 1974 Budget and Impoundment Control Act and related authorization legislation made budgeting in DoD a civilian function under the department secretariats and the secretary of defense, removing authority to formulate budgets for transmission to Congress from the service chiefs. The influence over resource allocation for the service chiefs was left to the preparation and submission of the 2 year POM and the POM preparation and approval cycle. Technically, nothing is supposed to be budgeted in the military department budgets unless it has first been approved in the POM. However, in practice this is not always the case due to changes made by budget submitting offices, Navy Budget Office or FMB (the department budget offices), the service secretaries, and numerous other players in the budget process, most notably the secretary of defense and his staff and, especially, Congress.

Each year the DoD appropriations act contains language that gives the secretary of defense authority to transfer funds, with the approval of OMB, between appropriations or funds in the current FY. Congress must be notified. Dollar limits are set each year in the Appropriation Act; in recent years, this has been $2 billion per year. Sometimes additional amounts are made available for transfer for specific purposes (Tyszkiewicz & Dagget, 1998, p. 48). In general, transfers must be to higher priority items where costs have grown beyond the budget estimate (e.g., fuel costs) and must not be to an item explicitly denied by Congress. Technically transfers are shifts from one program to another in a different account, while reprogramming is a shift from one program to another in the same account. It is useful to remember that DoD TOA involves BA created in past FYs to be spent in the current year, so transfers and reprogrammings may also include changes in the FY identification.

Reprogramming should be the easier action, since it is an adjustment within an account as opposed to between accounts, but both involve transfers to programs and programs have advocates. If those advocates are in Congress, a seemingly low-key reprogramming may run afoul of Congressional interest, say transfer from pier maintenance in a specific city to a general fuel account all within the same program, but with a different geographical—and Congressional—consequence. Normally this transfer would be invisible. Thus, for both reprogrammings and transfers the rules generally involve dollar thresholds, no new starts, avoidance or very careful handling of Congressional interest items and Congressional notification.

Reprogramming Within Appropriation Accounts

While transfers of money from one appropriation account to another within a FY requires what is in essence an appropriation amendment by Congress, reprogramming of funds within appropriation accounts is an every year occurrence and normal business in the budget execution world.

Jones and Bixler (1992) classified the DoD reprogramming actions into four types.

1. Prior Approval Reprogramming: This occurs when DoD wants to increase the quantity of a purchase regardless of the dollar amount, for example, increases in the number of missiles or aircraft purchased.

2. Congressional Notification Reprogramming: DoD notifies Congress when proposed reprogramming surpasses certain dollar thresholds or when it is for new programs or line items that may result in significant follow-on costs.

3. Internal Reprogramming: This includes changes within or between appropriation accounts but does not involve changes in

programmatic use approved by Congress during the annual budget process.

4. Below Threshold Reprogramming: Reprogramming which occurs
    below the threshold set by Congress. This is coordinated by the
    military departments and a   semi-annual report is provided to
    Congress (DoN Budget Manual p. 61).

Congressional interest can lead to complex guidance. In 1997, Congress imposed more restrictive guidelines upon DoD reprogramming.
This meant for the DoN that the prior notification threshold was lowered
from $20 million to $15 million for ship, aircraft, or intermediate maintenance changes and $15 million was set as the threshold above which the
Department would have to get approval to move money out of flying
hours, ship operating tempo and real property maintenance categories.
These thresholds vary over the years, sometimes as a product of the trust
relationship between a department and the Congress; in the 1980s they
were set at $10 million and advanced to $20 million in the 1990s after a
period of harmonious relations between the department and Congress.
Reprogramming has become a regular and anticipatable event in the
2000s. In part this has resulted and has been necessitated to fight the war
on terrorism and to deal with the fact that money appropriated for DoD
since 9/11/2001 has been provided through both regular and supplemental appropriations.

In the mid-1990s, the rules and thresholds for reprogramming were
changed to protect certain accounts from being used as "bill-payers" to
cover shortages in other accounts, for example, to avoid "raids" on the
Navy Flight Hour Program (FHP). These rules were passed as expressions
of congressional intent to control reprogramming from certain accounts
so that, for example, the DoN would not move money out of these
accounts to pay other bills without giving Congress a chance to approve
the move prior to its execution. In practice, this has meant any DoN
reprogramming change above threshold first has to be proposed by a
command comptroller, then approved at the FMB and Navy secretariat
level, then approved at the DoD level, then by OMB, and finally by the
House and Senate appropriations and authorization committees. If some
aspect of the reprogramming touch on intelligence matters, then the
House and Senate committees on intelligence also have to approve the
move. All it takes to block such a change is a no vote from one committee.
Within the department, creating these more restrictive thresholds for
prior approval meant that internal control systems had to be more closely
monitored. For example, when two or three units are spending flying
hour money and managing it to meet changing demands, their changes

each may be below the threshold, but above the threshold when summed together.

This problem was resolved by the Navy in 2007 to some extent by placing authority for all FHP accounts for the Pacific and Atlantic Fleets under one commander and comptroller (commander naval Air Forces). The change was part of the overall Navy initiative in adopting an "enterprise structure" so that the naval air enterprise (NAE) is under the command of one admiral. Other similar enterprises have been formed for additional war fighting communities of the Navy, for example, surface and submarine enterprises. Also notable is that under this structure all enterprises as well as Navy fleet commands (e.g., PACFLT, LANTFLT) and Navy system commands (e.g., NAVAIR, NAVSEA, SPAWAR) have the mission of providing services to the war fighter commands, and are subordinate to the admiral designated as the commander of each of the respective enterprises.

In summary, some reprogrammings require specific written approval of the secretary of defense or his designee. Others require prior approval or timely notification of one or more congressional committees and still others may be undertaken with Congress to be notified later in a semiannual report.

The complexity of this process may make it seem as if a large share of the defense budget is reprogrammed. This is not the case. Jones and Bixler (1992, p. 60) note that a relatively low percentage of the defense budget is reprogrammed annually. From 1980 to 1990, the percent change in outlays reprogrammed averaged less than 1% of the defense budget (.8%), ranging from a high of 1.2% in 1988 to .5% in 1981 and 1986. Jones and Bixler speculated this was because the process is unwieldy due to the multiple levels of review and coordination and because congressional staffers seize upon reprogramming as an important micromanagement tool, hence the DoD which they studied, and other federal government departments and agencies avoid using reprogramming as a policy vehicle. The requirement for monthly, quarterly, and semiannual reporting of below threshold reprogramming also provides a check on inappropriate use of reprogramming of funds.

Below Threshold Reprogramming

Nonetheless, under congressional thresholds, the agency has a great deal of flexibility in moving small amounts or money around to meet emergencies and to manage programs efficiently. These are generally called "below threshold reprogramming" and do not require prior submission to Congress or to the secretary of defense; they may be made at the discretion of the military department.

For example, annually the report that accompanies the DoD Appropriations Act contains language that grants to the secretary of defense authority, with the OMB, to move funds between appropriations or funds in the current FY, upon determination that such action is necessary and in the national interest. These reallocations are based on "unforeseen military requirements" and for "higher priority items than originally" appropriated. In no case are they to be for items specifically denied by Congress in the budget process. A dollar amount of reprogramming authority is usually set as a "not to exceed" target for the FY. Once these reprogrammings have been made, the secretary of defense must notify Congress of all transfers (DoN, 2002d, chapter 3).

In general, the guiding principle of below threshold reprogramming involves staying true to legislative intent by not using (or creating and using) temporary surpluses to create long-term obligations. For example, suppose an agency has budgeted for a substantial capital outlay item (central office remodeling). Suddenly an emergency occurs in safety inspection, one that cannot be satisfied by overtime, and more personnel are needed. The agency determines it can get through this year and maybe next, without the capital outlay item, thus it asks to move money from capital equipment to personnel accounts and subaccounts (SAGs).

How should the budget analyst look at this request? What the agency is asking to do is to convert a one-time outlay into an every year outlay. If the request is approved, the new position or positions will go into the personnel base and have to be funded in subsequent years, thus creating a substantial continuing cost greater than the cost in this year and one that will continue to grow as the person (or persons) is given merit increases and promotions. Budget analysts call this "the camel's nose" (Wildavsky, 1964), a type of expenditure that they delight in finding, exposing, and denying, if they can. The camel's nose strategy is intended to gain approval for moving a trivial amount of money to set a precedent for later movement of much larger sums. It may be that the proposed reprogramming has great political appeal to the service secretaries or chiefs, or to members of Congress in some cases (i.e., health and safety needs are often treated more kindly by DoD and Congress than administrative system's improvements); if so it might be unstoppable, although the budget analyst would ask why this could not be put into the next year's budget.

In this example the proposing command risks losing the money for remodeling for a long period of time. After all, the reprogramming means that the remodeling was not really that high a priority to the command and it will be some time before it can again present such proposals to reviewers with the expectation that it will be funded.

Congressional Special Interest Items

Not all budget adjustments are the result of internal DoD and military department (MILDEP) actions and not all budget increases are beneficial. Congress frequently adds "special interest" items to the military department budgets. These may be imposed through special language in the appropriation bill that results in fences ceilings or floors. A ceiling is a dollar limit above which money may be not spent; a floor is a lower limit which specifies an amount that must be spent on the specified item or service (more could be spent, but nothing less than the floor amount). A fence simply encloses or earmarks the money and says that it may not be spent for anything else. To the uninitiated, these additions often appear to be a real boon for the department. For example, for a number of years Congress (guided by former Georgia Senator Sam Nunn and Representative Newt Gingrich) added money for the Air Force to purchase additional C-130 transport aircraft despite the fact that the Air Force did not request the aircraft and had no requirement for them within their current inventory.

In 1997, $142.2 million was authorized for the procurement of three additional C-130J aircraft to the Air Force budget request for four. The Senate authorization committee report language acknowledged that the Air Force had no requirement. Congress did not, however, add $142.2 million to the Defense appropriation bill to pay for the additional aircraft. Instead, DoD was directed to find the money in its current appropriation. This meant that one or more programs that DoD had requested and had a bona fide need for were either not funded or under funded. Additionally, these aircraft have a logistics tail, including support equipment, maintenance personnel, fuel and oil, and flight crews. None of these items was funded, which meant that even more programs would be under funded. The service life of each of the aircraft is 20 to 30 years. Thus, the $142 million "gift" in 1997 will result in budget perturbations within DoD for the life of those aircraft.

Anti-Deficiency Act Control Points

One last point of interest is the ability of Congress, the GAO, the DoD inspector general (IG) and military department IGs, and the President's OMB for that matter, to structure appropriation control points for the financial management units within all of DoD. Budgets are passed at different levels of detail, but they are almost always more detailed in terms of control language written into law or in accompanying reports to appropriation and authorization legislation than the final numbers that appear in appropriations and authorization bills. Congress typically insets all kinds of ceilings, floors, fences and other requirements into appropriations and appropriation committee reports that are intended to convey congressional intent with respect to how money shall and shall not be spent by DoD and the military departments (MILDEPS) (for example by naming a specific item to be bought, replaced, or repaired). What is assumed by Congress is that the budget will be executed exactly in the way it was approved by Congress, not the way it was submitted by DoD and the President to Congress at the beginning of the annual congressional budget process.

For example, if the budget number was arrived at by computing the cost of five new employees and their supporting expenses, then what is assumed is that the agency will use the appropriated money to execute that plan, minus whatever amounts reviewers have pared off. What is required is that the central DoD comptroller's office report back any changes in the agency's request made in its passage by Congress. As it administers its program, the agency has some latitude to set what the barrier points are for its subdivisions.

For example, central authorities may only require that the agency control the totals for personnel and supporting expenses. But the agency itself may choose to require subordinate managers to report in more detail, for example in terms of line items that go to make up its major expense categories, for example, personnel, travel, computer supplies, computer maintenance, office supplies, utility costs, rent, and capital outlay. This provides managers with the necessary information to build budgets by providing them with cost data and leaves them with a template to execute the budget according to the way it was built. Because the agency's control is set at the total number, it then has flexibility to shift money between line items and between subprograms, while still keeping within the total dollar amount for personnel and supporting expenses.

In the federal government, departments may also choose at what level they wish to set responsibility for antideficiency violations. The higher the level, the more flexibility subordinate administrators in that chain of command have to move money around. Also, when an administrator knows a unit is in for a difficult year, he may set the antideficiency responsibility one level higher in the organization, thus giving him the ability to move money into that unit from others in case of a shortfall and exempting the unit from a legal charge of overspending the appropriation. The control is still on within the agency, so the appropriation number will not be overspent, but some relief is offered the lower level manager.

Conversely, an administrator may wish to punish or force a careful spending pattern on an administrator; he can do this by pressing the antideficiency responsibility down in the organization, by placing it on the administrator he wishes to control. Any shortfalls then become legal matters and the hope is that the threat of legal sanctions will create a more prudent manager. If the manger does not have control over a turbulent environment, then this strategy will not work. That legal frameworks do not always supply the anticipated behavior may be found in the federal government's constant battle to fend off a "spend it up" mentality in the fourth quarter, indeed the last month, of the FY. Later we discuss some of the ramifications of the fourth quarter phenomenon.

Congressional Control and the War on Terrorism Supplemental Appropriations

Over time, since the end of the Cold War and into the 2000s, DoD and the MILDEPS have experiences a considerable loss of flexibility to control internal allocation of funding of their base budgets as Congress has perceived the need to impose greater control and oversight (backed by extensive auditing) over DoD and its component departments. However, to some extent this trend has been temporarily reversed de facto as a result of the inception of the war on terrorism and the extensive funding provided since 2001 to DoD by Congress through "war fighting" supplementary appropriations. The availability of extensive supplemental funding has in fact reduced the ability of Congress to control internal reallocation of money to fight wars in various theaters. However, the overall control intended by the separation of regular base appropriations from war fighting supplementals has brought a great deal more complexity to budget execution and accounting within the DoD. Fundamentally, supplementals are intended to be used only to pay for the costs of war, yet the actual costs of war separate from the base budget are virtually impossible to identify with a high degree of accuracy due to a number of reasons not the least of which is the absence of accounting systems and trained personnel to identify and properly classify (with accounting code) costs in the field. Nonetheless, the MILDEPS and DoD have dutifully and to the best of their ability reported the costs of war to Congress as required in justification of supplemental appropriations. Whether programs or parts of programs should have been funded in the base rather than through supplementals and the implications of such practices will likely only be sorted out when Congress ceases to provide DoD with war fighting supplemental appropriations. It is reasonable to state that the comptroller community in DoD does not look forward to this point of reckoning, not because of anything done deliberately (although this point may be called into question is some specific instances), but that sorting out the differences between war and non-war money spent in the past and proposed to be spent in the future will take a great deal of time and effort in reestablishing a clear definition of what constitutes the base for MILDEP and DoD budgets.

RULES, CULTURE, AND CONSEQUENCES

The analysis presented thus far in this chapter has been somewhat technical because budget execution is of this nature—highly complex, complicated, and always changing. While we have described some the complexities of defense budget execution, we have not yet done justice in depicting the real life challenges that DoD financial managers face routinely. These challenges have resulted in some informal modes of behavior that are characteristic of budgetary negotiation and competition. Budget execution involves some invocation of informal "rules of the road" that are facts of budgetary life within DoD and elsewhere in the federal government. We discuss some of these informal rules and practices in the following section.

Rule 1: Spend it all (Spend it or Lose it)

In their introduction to the book *Reinventing Government*, authors Osborne and Gaebler (1993) state that the federal budget system encourages managers to waste money:

> If they don't spend their entire budget by the end of the FY, three things happen: they lose the money they have saved; they get less next year; and the budget director scolds them for requesting too much last year. Hence the time honored rush to spend all funds by the end of the FY. (p. 3)

This outcome results from the combination of a law and a cultural imperative. The law says that O&M appropriations are available for obligation only for one year. The imperative results from a culture that implicitly implies that good managers spend all the money available to them at all times. As a result, defense budget managers are faced with an incentive to spend as much as they can and to execute their TOA to the level of 100% annually. They believe they must obligate their money fully before the end of the FY or else lose the justification to ask for this funding in following years. Not only do they face this pressure, but they also tend to believe that having money left over is regarded as bad practice by their superiors in the chain of command. They also believe their superiors will punish them in some way for not obligating all of their budget, and apocryphal tales circulate through the comptroller community about military comptrollers who have been marked down on their fitness reports because they failed to obligate all the money in their budget. No one seems to know anyone to whom this happened, but all swear they know someone to whom it could have happened. Above all, they do not want it to be them.

Thus, air commands may fly more hours toward the end of the FY than they would have without budget pressure, even on the last day of the FY. Marine commands schedule longer exercises and invest more in preparing for fall exercises. Supply officers order more inventory than they think is necessary, ships get a load of new equipment at the end of the year, roads in military facilities get paved when they do not really need repaving, new office furniture and carpeting is ordered, and so on. A defense comptroller in at a second echelon command remarked,

> This is a fact of life. Every year I have been here I have received calls as late as 5 PM on the last day of the FY asking if I could execute some extra money. I keep an "unfunded" list of things I can commit money to and I just go down the list to spend additional funds. It happens every year.

Another noted:

> Every year, sometimes as early as July and nearly always by early September, our Wing, Groups and Squadrons commence to restrict flights, cut back on flight hours and threaten to "ground" their aircraft completely if "funds" are not made available. Then at the 11th hour (Sept. 29/30) a windfall of money is made available and the Wings/Groups/Squadrons flail themselves and gnash their teeth all the while chanting the same tune, "If we could have had this money earlier we could have spent it, but it's to late now." As comical as it may sound, I know for a fact this has happened, every year, for the last 14 years (I'm sure much longer than that but I've only been privy to it the past 14 years).
>
> The sad part is the whole "the sky is falling" evolution could be avoided if a few select personnel (accounting/fiscal/budget officers) would (1) make exercise and detachment cost projections on historical data (number of personnel times number of days, etc.) NOT on estimations from personnel without a monetary background; (2) recoup incomplete contracting dollars (the contracts will be invalid as of October 1).
>
> Here is an example; As the Fiscal Officer I am to support three squadrons going to "Scorpion Wind." I know 300 Marines are going and the evolution takes two weeks. Hypothetically let's say it costs $100 a day for each Marine (transportation, food, etc.). 300 Marines x $100/day = $30,000/day x 14 days = $420,000 for the exercise. I know from previous exercises all the Marines will not be staying the entire two weeks; 50 Marines will only stay one week (rotating personnel back and forth through the squadron). 50 x $100/day x 7 days = $35,000. I now have a $35,000 "rainy day fund." Now if I have 6 exercises per year with $35,000 excess per exercise I now have a

$210,000 rainy day fund to support the MAG Commander's "Oh, we really need to do this" ideas. I can make the money appear and the MAG Commander believes he has a monetary magician working for him.

Another way to recoup funds is to reconcile contracts. On the average I had contracts worth in excess of $3,000,000. These were annual contracts and expire on 30 September. By aggressively reconciling the contracts we would recoup about $250,000 a year. If left to expire those funds would expire and we would lose them.

The end of year "spend it all" behavior does not occur only in the federal government, but other governments may treat things differently. For example, the authors participated in a performance audit on a state government tax unit whose leader was said to be the best tax administrator in the country. He was certainly regarded as such by everyone in the state government hierarchy. We knew he ran a tight ship because one of his clerks kept a record of the supplies handed out, down to the last yellow pencil. Later, in the course of the audit, we compared his budget and budget execution patterns and found that he consistently executed about 95-96% of his budget and let the rest go back to the general fund. We were surprised because this did not fit with the notion of a tight ship. Later, we asked about this and he told us that he tried to do the best job he could without going overbudget. He was not much concerned with being some percentage points under budget: "Personnel is a big component of my outfit," he said,

> and you can not make that come out even. You can keep it from going over, but you can not make it come out even. Things happen to people. I spend all I have scheduled for capital equipment and I have my people see that we do not go over in supporting expense areas. I have too much to do to worry about coming out even.

The culture of the federal government is such that federal and defense budget managers are barred from this kind of thinking for fear that under-execution will be perceived as "slack" in the current budget and will threaten success in the next budget, someone else will get their share and their boss will get angry with them and punish them for underexecuting their budget.

It may be argued that spending it all is good practice if commands that do it spend end of year money on their highest priority unfunded needs. It may be argued that this practice is what Congress intends when it passes an appropriation. Moreover, it can be argued in national defense that the Planning, Programming, Budgeting and Execution System (PPBES) involves a complex process to meet the threat with appropriate capabilities, and that making this happen depends on every budget holder executing 100% of its budget to achieve this result, with each part of the process summing to the whole. In our view, these arguments give too much credit to the vision of Congress and PPBES planners in knowing how much spending each year will produce the desired results and too little credit to the individual budget administrators who are able to bring about satisfactory results at a lower cost than foreseen nine to 18 months earlier. Be that as it may, the current culture includes the incentive that spending it all in the term allowed and executing to 100% regardless of how it is done, as long as it is legal (and this is a major caveat), is the best way to perform the task of end of year close-out. Does this practice get the taxpayer or Congress the best "bang for the buck" in defense spending? We doubt that it does and we suggest in the last chapter of this book that appropriation law be amended so that the one year obligation period for one year appropriations such as O&M be extended from 1 to 2 years to reduce the need to execute O&M to 100% each year.

Rule 2: Do Not Overspend

Overspending funds may result in a violation of the Anti-Deficiency Act and appropriation law, depending on what level in the budget chain of command it happens. If an appropriation account is overspent at a third echelon command, this error may be caught and corrected at a higher level in the budgetary chain of command, up to the MILDEP budget office, for example, FMB. The Anti-Deficiency Act, the common name for appropriation law contained in U.S. Code Title 31 Sections 1341, 1349, 1350, 1512-14, and 1517-19, prohibits budget managers from obligating more funds than are appropriated. However, an ambitious financial manager who did everything right as a leader, but overspent his budget slightly could get himself in serious legal trouble, enough to end his or her military or civilian career, and in some cases criminal prosecution leading to fines and time in prison. Few budget officers wish to risk their careers over a single instance of spending within or between budgetary accounts. However, accounting systems in DoD are complex and often inadequate so that time lags in recording transactions (and accurately) are such that due to uncertainty in guarding against overspending, budget managers may underspend by unexpectedly large amounts. As one funds manager at a small command put it,

> You never want to execute any account to 100%. This would look fishy and arouse the suspicion of auditors. Also, doing so reduces your ability to go back into the account in the next FY and fix any problems that have resulted from accounting system errors and the end-of -year spend-out rush.

The Anti-Deficiency Act also prevents managers from obligating funds before an appropriation and an accompanying authorization have been enacted by Congress and signed off on by the President. This is true both for regular and continuing appropriations provided in a Continuing Resolution. Title 31 Section 1341 includes limits on the expenditure and obligation of funds and states that an officer or employee of the United States Government may not make or authorize an expenditure or obligation exceeding the amount available in an appropriation for the expenditure or obligation, or involve the government in a contract or obligation for the payment of money before an appropriation is made and authorized by law.

The primary section of the U.S. Code providing what is termed the "Anti-Deficiency Act" is 31 U.S. Code Section 1517(a). One interpretation of what this code section tells budget officers is as follows:

> You can't spend what you don't have. You can't overcommit, overobligate, or overexpend in an appropriation or any subdivision of the appropriation...this is a real time violation, and is not something that is monitored at the end of the month or quarter. (Candreva & Jones, 2005, p. 21)

Budget execution is highly contingent on responding to emerging crises and events. Sometimes things happen in the budget execution year that are beyond anyone's control. For example, a part costing $7,900 was ordered. The part was not in stock and was backordered. While awaiting delivery, a temporary repair was made using spares. The idea was to pay back the spare when the requisitioned part arrived. Unfortunately, the replacement part did not arrive for three years, and when it did arrive, the unit delivery price had risen to $57,000. Sufficient unobligated funding was not available to cover the difference from the original year appropriation in which the part had been ordered, resulting in a violation of the Anti-Deficiency Act because of over-spending—which resulted from the lack of an adequate ordering and cost tracking system.

In another example, various items of equipment were purchased with money from a specific account -- as per requirements and normal procedure. However, these items could not function independently. When assembled with other items purchased or added to an existing system, they created a complete and functioning system. The aggregate cost of all the equipment items that had either created or become part of the system exceeded the expense/investment criteria in place during the FY in which the funding was appropriated. Initially, a Sec. 1301 violation was incurred because the items were purchased from the wrong appropriation account. Then, a Section 1517 violation resulted because of insufficient funding in the "correct" appropriation account to pay for the items that had been received.

Budget execution behavior intended not to execute to 100% is the result of the incentive for financial managers not to break the law. However, in some cases overspending occurs without the direct intent to do it. Thus, many Anti-Deficiency Act violation investigations do not end up with anyone punished in terms of their careers or job status. However, such investigations are required to explain how and why the violation occurred and almost always include recommendations for preventing the error from happening in the future. Still, when systems failure (e.g., accounting systems) is the cause, the problems sometimes are not fixed. The best example of this is where an Anti-Deficiency Act violation occurs as a result of a coding error by a GS-3 accounting employee where the way to fix the systematic problem is to (a) improve the clarity of the rules for transaction coding and how they are written, (b) provide better training to accounting employees. Such training, typically delivered on the job, may not be done systematically so as to eliminate coding errors, and money for such employee training may be the first area cut from a command's budget when budget reduction is mandated from above.

Rule 3: Spend it on the Right Stuff

DoD financial managers also have to give great attention to the "Color of Money" rule. Section 31 U.S. Code 1301(a) requires that funds appropriated by Congress be used only for the purposes delineated by the appropriation. For example, funds for O&M may not be used for investment, thus money intended to pay for utilities or contracts for trash removal may not be used to procure weapons. In addition, there are limits placed on appropriation categories above which obligations cannot be made, for example, O&M dollars cannot be spent for a local area network system or to buy an expensive piece of medical equipment (e.g., an magnetic resonance imaging or MRI scanner) for a Navy hospital when the expense would be greater than $100,000 when this is the limit for O&M (this ceiling has varied in amount over time but it is used to differentiate operating from capital expenses). Obligations for minor capital items below the threshold are permitted, but above the threshold a procurement account must be used. Likewise, money from the investment accounts cannot by used for operating expenses.

Managing within the color of money constraints is relatively easy for budget managers who have only one type of appropriation to spend, for example, O&M or procurement, but many budget managers spend from several appropriation accounts. Their job is much more complex.

Rule 4: Keep it Legal

The higher in the budgetary hierarchy the larger the budget, and the more appropriations accounts that must be managed, each with their different obligational rules, availabilities and limits on funding discretion, congressional special interest items and reprogramming thresholds; also it is more likely that accounts will be obligated and administered over multiple year periods. Cheney (2002, p. 29) found that managers at higher levels in the budgetary chain of command committed more Anti-Deficiency Act violations. The Naval Air Systems Command, for example, was one of these; with nearly $16 billion (FY 2003) in appropriated dollars, it administered funds from six different appropriations, including O&M with its one year obligational limit. The others accounts including Navy aircraft procurement (APN) and Navy weapons procurement (WPN) all had multiyear obligational periods.

With all this complexity, it is worth noting how few violations of these rules occur. Cheney studied 60 violation investigations over the course of a decade and estimated the average occurrence at about 6 per year. It is obvious that ADA law is thoroughly a part of the DoD financial administrative culture; perhaps because penalties for those who break this law can be substantial. The penalties are described in sections 1349 and 1350 of the Anti-Deficiency Act, Title 31 and range from administrative discipline to criminal prosecution that can include suspension without pay, removal from office, fines, and imprisonment for 2 or more years. The same penalties apply to officials who authorize exceeding apportionments, allotments, or operating budgets as to those who actually commit the errors. Penalties associated with a willful act are more severe than those associated with a violation caused by ignorance or negligence.

Typically, a willful act results from an attempt to circumvent statutory law through "creative" budgeting and accounting practices involving a violation of Title 31, Section 1301 that restricts the expenditure of funds to the purposes for which they are appropriated. In the Navy, many military officers rotate through budget and financial management billets as subspecialty assignments secondary to their warfare specialty. Their stay in such posts typically lasts 2 to 3 years unless they choose to pursue a career in the financial management arena. Whether these officers are educated and prepared sufficiently so that they are appropriately careful about avoiding a career-ending mistake over a budget or financial matter unrelated to their warfare specialty depends to a considerable extent on the training they have received, which varies considerably from one officer to another. When competition for promotion is fierce, even one bad mark on a fitness report in a budget or financial management job can foreclose future promotion options. This tends to make military financial managers more conservative and focused on the rules but, as noted, the rules come close to being contradictory when they indicate the need to spend all the money available, but to not overspend it, and always spend it on the right things relative to the account from which the money is obligated. Military financial managers who must walk a fine line do so cautiously. In this regard, most military officers serving in comptroller billets depend greatly on their FM career civilian staff to make sure they keep it legal.

Rule 5: Do Not Become Confused by Complexity

History teaches that one error typically leads to another, just as a violation of one law often lead to violation of others, like a bank robber breaking the speed limit racing away from his latest crime. Many Anti-Deficiency Act violations (1517a) occur as a result of color of money errors (1301a). For example, let us use the example of a budget holder who obligates and spends money to purchase items with the wrong color of money, e.g., spending procurement money on maintenance. When the mistake is made but later discovered within the command by comptroller personnel, the accounts may be adjusted so that the color of money rule is not violated (by paying for the expense out of maintenance and restoring the procurement money to the procurement account), thus "fixing" execution during or post execution to make it right. However, now the financial manager and staff may find that the maintenance account is short and has been over-expended, thus a color of money error becomes an Anti-Deficiency Act violation. Also, the restored procurement account may now have a surplus in it, not a legal problem, but one that may get him in trouble with his boss, the budget analyst, and maybe, Congress.

In another example, O&M funds may be spent on minor construction, but only up to a limit of $500,000, and it is illegal to circumvent this limit by funding a project in stages of less than $500,000 (e.g., three $200,000 contracts). When an FM official is caught doing this, appropriate disciplinary action is enforced and the whole project ($600,000) must be obligated and paid for from military construction funds, including the costs of planning and design (perhaps another ($60,000 to $100,000). If the MILCON account does not have enough funds to cover the project, then an Anti-Deficiency Act violation has occurred and, if this is discovered in audit, sanctions may (should) be imposed.

Some recent violations of the Anti-Deficiency Act include a financial manager who used O&M funds in excess of the statutory minor construction limit while making improvements in a waste storage facility. Another example is the financial manager who obligated for a lease to pay for service to be delivered in the next FY out of current year money and committed money in a contract before the appropriation was available. In another instance a financial manager allocated more money than was available and let his subordinate commands commit more money than they should have obligated. In this instance, some subordinate commands committed less than the total made available to them, but the total at the superior command level was over the amount of funding available for outlay.

Cheney's (2002, pp. 24-25) study of Anti-Deficiency Act investigations in the DoN over a 10-year period found that 87% were section 1517 violations where obligations were greater than the amount apportioned or allotted to the command while 13% were section 1341 violations where the obligation was greater than the actual appropriation. Cheney analyzed these violations in terms of purpose, time, and amount and found that violations occurred because the money was used for the wrong purpose in 53% of the cases (color of money error), because the amount appropriated was exceeded (31% of the violations) and because the time period for the appropriation had been exceeded (16% of the violations).

Cheney concluded that the violations of "amount" occurred because of poor accounting practices where commands failed to post obligations or expenditures in a timely manner, leading financial managers to believe that they had more money available to obligate than they actually had, and thus they over obligated. Cheney found that the "time" violations occurred because commands unknowingly created liabilities in advance of appropriations as a result of unawareness of the funding complexity built into service contracts, or because of communication errors in the chain of command, including failure to provide timely information between administrative units. Cheney suggested that the "purpose" violations occurred because managers were confused about the purposes stipulated for appropriated funds and how they could or could not be used, and about the thresholds for specific appropriations. Consequently, some financial managers tended to use O&M dollars when they should have used other procurement, Navy (OPN) money, and they obligated O&M money above threshold ceilings. In some cases it appeared that financial managers also had trouble distinguishing between limits on the purposes and limits on obligation and expense in subaccounts within an appropriation account category (e.g., payroll, supplies, travel), including the O&M account and investment accounts (e.g., computers, hardware, additions to computer networks in the OPN account). Consequently, commands exceeded the thresholds for expense by subaccounts within the O&M appropriation and in some cases also made the error of spending money from an investment account illegally, for example, from OPN or military construction funds for O&M account purposes.

In the budget execution process financial managers are given dollar thresholds and other guidance to insure that Anti-Deficiency Act violations do not occur. For example, they are instructed that equipment costing less than $100,000 (where this is the mandated threshold) may be purchased out of the O&M account; equipment over that threshold is supposed to be charged to an investment account. This control follows from the logic that investments have future benefits and must be charged to specific procurement subaccounts within the procurement or military construction accounts.

In the military construction accounts, thresholds differ. For example, minor construction items below $750,000 (as of 2004—but this threshold is subject to change annually as noted) could be charged to O&M while items over that limit had to be obligated in the investment account. However, the rules sometimes are complicated to interpret in individual applications. Cheney (2002) illustrated this as follows,

> For example, if a command purchases a computer terminal that will be connected to an already existing LAN system, that computer terminal must be purchased using investment money even although its unit cost is below the investment/expense threshold because it is an addition to an end item or existing system whose collective value exceeds the investment/expense threshold. (pp. 24-25)

Cheney observed cases where commands tried to circumvent the threshold restrictions by purchasing "parts of buildings" whose value was less than the existing threshold at that time, but Anti-Deficiency Act investigation revealed that the sum of the parts of the completed building exceeded the threshold. Buying part of a building is an attempt to circumvent appropriation law, and the commanders and their staffs were charged with violating the Anti-Deficiency Act. Purchases of computers and electronic equipment also were frequent problems as commands procured multiple items individually at a cost less than the existing threshold, but when the items were joined together as a system, then their total costs exceeded the threshold.

Cheney's study concluded that the O&M account was used improperly in 65% of the violations as financial managers charged the O&M account when they should have used the OPN account. As we have noted, violations of the Anti-Deficiency Act are subject to a variety of punishment ranging from disciplinary letters to dismissal. In only two of the 62 cases in Cheney's study were administrators found to have knowingly and willfully committed a violation. In the other 60 cases it was judged that the administrator did not know which account to use or did not have an accurate appreciation for how much money was available for obligation, or that the money had not arrived at the command at the time the obligation was made and thus was not available for obligation (unauthorized prior obligation). How this can happen may be better understood by examination of the complexity of the accounts and the rules and guidelines that govern them in many commands throughout DoD. As a hypothesis we advance the proposition that where complexity increases, the risk of and the actual number of violations increases.

The Navy and the other military services have within their organizational structure a number of systems commands that have domain over the maintenance, construction, and equipping of ships, aircraft, other weapons platforms (including C4I and space systems) and the operational hardware the military requires to meet its mission. Although this is an oversimplification of the organization and functions of systems commands to an extent, generally speaking in the Navy for example the war fighting commands operate the weapons systems, platforms, and equipment while the fleet commands and the systems commands, including the Naval Sea Systems Command (NAVSEA), the Naval Air Systems Command (NAVAIR) and the Naval Space and Warfare Systems Command (SPAWAR), provide the ships, aircraft, and C4I technology that the war fighters use to deter threat or to operate under conditions of war. The systems commands are charged with procuring and maintaining the fleets of ships and wings of aircraft so that the Navy has modern and capable systems to deploy. These system commands are charged with fielding ships and aircraft through the procurement and construction process into deployment, and performing or supervising modifications of platforms and systems throughout their service lives. The collective budget of the system commands can be more than one-third of the Navy's total budget, depending on whether it is peace or war time. All of their operations are complex in terms of financing, but some are inherently more complex than others. For example, NAVSEA managed more than 1,400 foreign military sales contracts a year worth about $16.7 billion, involving 80 countries in FY 2001 (Cheney, 29). While some commands have only a few appropriations, these systems commands have to manage multiple appropriation accounts. For example, the NAVAIR comptroller will expend money from at least six different appropriations: Aircraft Procurement; Weapons Procurement; Research, Development, Test, and Evaluation (RDT&E); O&M; other procurement; and other. (We do not count accounting for military personnel because it is so stable and may not be accounted for at the systems command level; civilian personnel are part of the O&M account.) Moreover, most of these investment accounts may be obligated over periods longer than 1 year. Thus, money in an OPN account may be obligated over three years and spent (outlayed) over 5 years after the initial 3 year obligation period.

Cheney found that 38% of these system command ADA violations resulted from mistakes made in managing funding and accounts during the multiyear obligation period. While the complexity of such a task as noted above seems almost overwhelming, the bottom line is that only 62 cases of Anti-Deficiency Act violation occurred during the eleven year period studied and only two were judged to be willful. This is not a bad record given the billions of dollars appropriated and expended and the hundreds of thousands of transactions made during budget execution over a multiple year period of time.

OTHER FACTORS INFLUENCING BUDGET EXECUTION

The daily work process of a Navy comptroller, or comptroller in any of the military services, is complicated by a variety of factors that range from the imposition of "reserve funds" or "withholds" or "taxes" (the terminology varies but the practice is the same) imposed from above in the financial management chain of command that decrease the portion of the appropriated budget commands receive regardless of what they have justified in their budget requests, to late appropriations and continuing resolutions, congressional earmarks, spending ceilings and floors ("fenced" money) that decrease flexibility and demand monitoring of spending, accounts, performance, results, and eventual reporting on how all of these constraints have been accommodated so that compliance with controls is provided. And to this add having to deal with a continuous flow of data calls from above and coping with less than perfect and current accounting data and out-of-date computers and information systems, and one gets some idea of how challenging is the tasking of the military comptroller and his or her staff.

Withholding Funds for Flexibility

Initially it is necessary to remind ourselves that the money appropriated to DoD and the MILDEPS by Congress annually must be requested not only in the original annual DoD budget submission to Congress, but also after appropriations have been enacted. Once Congress passes and the President signs an appropriation then it is required that DoD and the MILDEPS request authority to spend what has actually been appropriated from OMB, which then has to approve this apportionment request and signal the U.S. Treasury that DoD and its components may begin to obligate and spend up to appropriation limits, subject to congressional controls written into both appropriation and authorization legislation each year, plus the myriad of controls imposed by the Anti-Deficiency Act and DoD's own administrative rules, regulations and guidance. Consequently we understand that a number of constraints on spending are engraved into the process once Congress and the President have completed their approval of the annual defense appropriations. This constrains all financial managers in DoD to full compliance with the directives and the intent of Congress, as implemented and enforced by the Executive branch and by DoD itself. In all of this we may distinguish statutory law (that passed by Congress) from administrative law (that imposed by the executive). The former has the full force of the U.S. government to support enforcement. The latter is enforced subject to the will and capability of the executive to do so.

At nearly every level of budget control, financial managers have an incentive and therefore a tendency to withhold and set aside a "little off the top" of the funding they have authority to administer to accommodate either known or unanticipated contingencies. Legal authority to establish reserves rests in U.S.C. Title 31 Section 1512, Apportionments and Reserves. Section 1512 states that in apportioning or reapportioning an appropriation, a reserve may be established only (a) to provide for contingencies; (b) to achieve savings made possible through or by changes in requirements or greater efficiency of operations; (c) as specifically provided by law. A reserve established under this subsection may be changed as necessary to carry out the scope and objectives of the appropriation concerned. This rule allows high level military department comptrollers considerable latitude to withhold money in response to orders to do so from Pentagon level or major command level military commanders or civilian officials.

These reserves or "withholds" range, for example, from those imposed by Congress such as the 2% reduction applied to all DoN procurement and research and development programs to pay for a shortfall in the Navy Working Capital Fund, to a 2% withhold by the DoN to all O&M accounts to fund a special initiative intended to save money in future FYs, to a fleet commanding officer keeping a small percentage of operating funds in reserve until midyear or a month before the end of the FY to fund emergent and previously unfunded needs. While each of these types of actions may be ordered early in the FY, each requires an adjustment of the level of spending which had been justified and planned for in the budget planning phase of the submission and review process.

Kozar and McCaffery (1994) found varying practices in the three military departments in the management of withheld funds. For example, until the early 1990s, it was customary for the chief of naval operations to hold back a 2% reserve at the beginning of the year; the Army chief of staff had a similar contingency policy, holding back about .5% of O&M

funds or approximately $100 million. Differing from Army and Navy practices, the Air Force chief of staff typically held back no reserve funds in the O&M account during this period. Normally these reserve funds would eventually be apportioned out to subordinate commands based on new needs identified and justified during the midyear budget review. At the same time, commands that were under executing their budget might well have money taken from them to be redistributed to others that were perceived as more needy; or they might be told to pick up the pace of execution (p. 10). However, as of the mid-1990s in the Navy, the CNO reserve, as it was known, disappeared and greater attention to execution was placed on what are termed midyear reviews where resources can be reallocated within and from one command to another based on a review and prioritization of "unfunded lists" submitted by each command holding 1517 authority (and therefore requiring a comptroller). Also, in addition to midyear review, an upper echelon command may call for a review of unfunded lists throughout the execution year in accompaniment of execution reports requiring justification of under or over execution of money relative to previously submitted and approved execution plans and schedules.

Because budgeting for contingencies is not allowed officially in Appropriations Law (one exception to this rule was the waiver provided by congressional appropriators to DoD in the mid-1990s) and budget submissions are carefully examined to eliminate "excess" (poorly justified) funding proposals, a withhold, by definition, means that the money received at the lower levels is less than that anticipated as a result of what was requested and then what was appropriated by Congress. Additionally, if a "contingency" does not materialize, money that had been set aside will likely be released late in the FY, requiring additional justification to the already revised (and re-revised) execution plan for the current FY.

Timeliness

In that the critical action that initiates budget execution (the passage by Congress and signature of appropriations by the President) rarely occurs on time before the start of a new FY, it is not surprising to find that the follow-on steps in execution can be less than timely and not according to formally adopted schedule. All appropriation bills have passed on time in some years including 1948, 1976, 1988, 1994, and 1996. Additionally, the defense appropriation bill has been on time in 2000, 2003, 2004, and 2006. (Note that when bills are sometimes late and sometimes on time, DoD budget managers have to be conversant with two different sets of rules and know which to apply.) When appropriations are late, adjustments must be made to the plans upon which budgets have been based. In the case of a CRA, Congress typically specifies not only the level of spending, but also imposes other restrictions, for example, no spending on new program starts. Spending levels are normally limited to prior year rates or the lower of the House or Senate versions of the appropriations bill, assuming that a conference between the House and Senate has not yet taken place.

In the acquisition world, for a program that is new, reverting to a spending profile that matches that of the previous year may require contractors to lay off employees, postpone tests and DoD to delay new or additional contract awards. Over the course of a FY, a new program typically will have hired additional people, started testing and increased the rate of spending in other areas. If the program must revert to a pre-expansion level of spending because of a CRA, inefficiencies, additional startup costs, and other costs will be the inevitable result, requiring program restructuring and perhaps additional funding.

Whenever funding is not received when it is anticipated, budget execution is affected. Even a slightly delayed appropriation may have large consequences for a project. One DoD manager said:

> If the weather during the winter and spring precludes work on an outside project, funding and subsequent contract delay in the first quarter of the FY may force some projects to be delayed for considerably longer periods of time than one might have predicted, given the length of time covered by short term funding legislation,

He also commented that late appropriations were not a problem for other accounts and programs he had to administer, unless an increase in program scope was planned. His comments reflect the essence of the budget execution dilemma: under a CRA some things roll on as planned, others do not.

Specific Categories: Ceilings, Floors, and Fences

Money for DoD is appropriated in different appropriations, and in specific categories identified within appropriations and accompanying report language, and also in language written into reports that accompany authorization legislation, each imposing specific rules and limitations. The categories of appropriations restrict the manner in which a budget can be executed. While all dollars are the same, restrictions within single appropriations such as those for WPN, OPN and especially MILCON make the execution of these dollars so different that they might as well be of different colors.

For example, in the defense department, the O&M appropriation has a 1-year life and finances the cost of ongoing operations (e.g., base operations, civilian personnel salaries, maintenance of property, training, etc.). Aircraft procurement has a 3-year life and finances the procurement of aircraft and related supporting programs. Military construction has a 5-year life and finances the purchase of land and construction of facilities. The restrictions on each of these appropriations create legal boundaries that put the budget manager in a "box" with walls that are difficult to breach. The ability to move money from one account to another is normally beyond the control of lower level funds administrators. Should one account or subaccount have too little money and another too much, typically it is not a simple matter to shift significant amounts of money from one to the other, because reprogramming rules always apply. Shifting funds can only be accomplished within the restrictions imposed by the rules, practices and process controls governing reprogramming, transfers, use of supplemental appropriations and inevitably glacial waiver approval processes, first within the DoD and the executive branch and then, for some types of flexibility, within Congress.

Politics and Budget Reduction

First we need to reiterate the things we have explained in our previous chapter on the politics of congressional budgeting are a preamble to all we observe here. Since the post-Cold War drawdown in defense spending followed by the burst in spending in the post-9/11 era, budget reduction in the military departments and services by specific appropriation account often has been executed by withholds of various types, commonly referred to by comptrollers as "taxes" to fund a plethora of emergent initiatives (e.g., IT-21) and persistent requirements for cutting specific appropriation accounts, especially O&M, to increase funding in accounts that pay for recapitalization of aging and worn out weapons platforms and systems. Over the past 5 years or so the Navy and other services have grown accustomed to the practice of up-front budget cuts imposed from above before or concomitant with the start of the FY, either by Congress, DoD or the MILDEPS themselves, to fund recapitalization and a range of other measures, particularly where "savings" are estimated to be achieved through implementation of processes such as Lean Six Sigma or organizational initiatives such as sea enterprise and the naval enterprise model. In practice comptrollers have little choice but to accept and then find ways to accommodate such reductions to the spending plans they have formulated previously. Thus these "savings in advance" are an everyday fact of life in budget execution, but one that confounds comptroller efforts to accurately forecast spending, to identify which requirements can be met and not and to what extent, and to specify what degree budget plans can be executed effectively without "program breakage."

Additionally, the DoD comptroller and the military departments and service budget and comptroller communities have to deal with congressional changes to the defense budget as they occur in each and every annual review and approval of program authorizations and appropriations. At times Congress adds new weapons systems and equipment (and sometimes requirements) to programs and appropriations that were not requested by DoD (new items), increases funding for existing programs (plus-ups), and makes cuts ("marks") to existing defense program and appropriation requests that force consequent changes to DoD and MILDEP budget execution planning. For example, if Congress adds three C-130 aircraft to the President's budget request for the defense department, the C-130 program manager has to make changes to adjust for the additional aircraft in everything from the quantity of training materials to the amount of support equipment required. Additionally, some other portion of the budget must be reduced by comptrollers and budget staff to pay for the additional aircraft support not included with the Congressional "gift," causing perturbations to programs unrelated to the C-130 program. This is the double-edged sword of special item additions or "plus-ups" in congressional language. Not only does such action by Congress rearrange the spending and policy priorities within DoD, it creates new financial burdens for support systems and services to bear that are not funded in the budget for the particular FY in which they occur and therefore must be taken from the base of support for existing programs; it also forces changes in future budgetary planning and POM preparation. Typically, these kinds of changes by Congress require redevelopment of a part of the budget for the following year or years and a recasting of programming in the POMs prepared by the MILDEPS. This is expensive in terms of staff time and effort but it is another fact of life in budget execution.

Flexibility

Flexibility in the budget execution process means the ability to make adjustments within the resources that have been allocated. As noted with respect to discretion, the higher the level of budgetary management in the organization, generally the greater the degree of flexibility that is available to financial managers. However, at the third and fourth echelon command and field levels, dollars for maintenance and repair are typically not tied to specific buildings or projects. The local manager can decide to use dollars to fix a leaking roof or paint a building. If money is appropriated to construct a new building, the local manager has essentially no option other than to build that building, even if repairing another building would be more cost-effective. Thus, the degree of flexibility depends on the amount of money available as well as the specificity with which it is authorized and appropriated. Even "common sense" decisions are sometimes restricted. For example, recently a program manager for a missile program negotiated a contract that would have allowed procurement of eight more missiles for the same amount of money. However, he could not sign that contract without prior congressional approval because the missiles were line item appropriated with a specific quantity specified. Going to Congress to request permission to make changes (buy eight more missiles for the same dollar amount) entailed the risk that members or their staffs would either make changes in the program or reduce the appropriation for the program. Thus, while it seems to make sense to buy the additional assets under such conditions, no change was sought for fear that some members of Congress might cost out the contract on the basis of eight additional missiles and then reduce the dollar amount for the contract on the basis that the original number of missiles was enough and the original appropriation was overpriced ... and that they had other places to put the money they had just found.

Management Information Systems

Budget comptrollers in the federal government are often at the mercy of their management information systems, particularly accounting systems. They struggle to provide specifics on funds available, funds obligated and funds expended, as well as for preparation of what often seems like an endless stream of special reports, some of which having very short deadlines, for example, in a matter of hours or 1 or 2 days.

Manual ledger sheets are long gone (except perhaps in small or field combat offices where personnel serving as bookkeepers maintain manual records, sometimes parallel to digital records to ensure accuracy) with their labor-intensive and inflexible routines, but automated systems with their speed and versatility also have their problems. As with any system, the quality of the input directly affects the quality of the output. If account or sub-account category definitions are not accurate or understandable, and if staffs are not trained or motivated to record data at the source of transactions accurately, entry errors are inevitable. Budget officers may not know how to simplify and clarify data entry instructions. Typically, the budget comptroller has little or no control in the short range (the longer term solution is to obtain funding and provide time for employee training) over inputs and insufficient resources for training accounting staff, many of which are employed at very low salary and benefit compensation levels. Data input may be made in distant field activities or in a contractor's facility, or worse, not at all. Some accounting systems inevitably are not compatible with the systems of other organizations or units. This in turn means manual intervention, with all its attendant risks of error, becomes necessary to compile data "creatively" drawing information from different systems.

In 1993, the DoD had 270 financial management information systems. It also had $24 billion of unmatched disbursements (Perry, 1996, p. 107). That is, bills had been paid but who had been paid and out of what accounts they had been paid could not be determined. Contractors were not demanding payments so it was logical to assume that they had been paid. Yet, how they had been paid and how much they had been paid was a mystery. Errors ranged from simple numerical transpositions (not surprising when strings of manually entered numbers and letters are 16 characters long or longer); to funds paid out of the wrong appropriations or the wrong years. As of 2007, many of these same problems persist, aggravated by the flood of funding through supplemental appropriations for the costs of war, the virtual impossibility of capturing the costs of war in real time, and the difficulties encountered in recording transactions accurately in the war environment in which many activities are performed by a combination of military and DoD labor and contractor labor. Budget execution documentation and records at field command levels often do not match or fit with those estimated at the headquarters levels.

Executing budgets without knowing precisely the status of funding and accounts causes significant stress even for the most experienced budget comptrollers, and many of the personnel who are performing the roles of accounting and budget administration in war zones lack the training to do so adequately. Fear of Anti-Deficiency Act violations and damaging audits is endemic in the budget offices of the federal government, but it is hardly a concern in the war environment. This typifies the irony of trying to comply with the huge body of restrictions that govern budget execution while attempting to execute budgets to greatest efficiency in an environment that is characterized by a high level of uncertainty.

How Much Control is Enough? Considering the Consequences of Excessive Control

One final question must be asked in any review of federal and defense budget execution. Why are funds so strictly controlled in the federal government and DoD? The general answer is absence of trust between the

Congress and the executive branch, and a legitimate desire to prevent fraud, waste and abuse of taxpayer money. However, as has been observed elsewhere (Jones & Bixler, 1992), the tendency of Congress is to control executive agencies related to achievement of political objectives, coupled with the fact that a number of members of Congress and their staffs have been or are lawyers. Lawyers prefer highly specific rules versus less specificity, or the use of other means of control, for example, incentives that economists would favor.

In the executive branch, control in departments and agencies (and the tendency to overcontrol) may be explained by a preference to avoid (a) actions that might be sanctioned by Congress or its legal arm, the GAO, or (b) punishment that would result from violation of overly complex permanent federal Appropriation Law, and annual appropriation controls erected by Congress and OMB. The federal government appears, relative to other levels of government, to "manage to audit" to an excessive degree, irrespective of evidence that the costs of control often exceed the benefits (Jones & Thompson, 1985, 1999).

CONCLUSIONS

While budget preparation is accomplished using a relatively rationally organized planning process, budget execution is a management process requiring a considerable degree of cash management expertise to respond to changing and uncertain conditions. Budget preparation involves planning for policy accomplishment, while budget execution involves managing the budget plan and actual appropriations in policy implementation. Bernard Pitsvada (1983) explained that budget execution is, "that phase of the budget cycle in which agencies actually obligate or commit funds in pursuit of accomplishing programmatic goals" (p. 87). Following plans made in the budget preparation cycle, employees or contractors are engaged, materials and supplies purchased, contracts let and capital equipment purchased.

The ordinary routines of budget execution are usually carried out far from the spotlight of media and public attention and the daily political crises that draw so much attention in the environment of the Pentagon and Capitol Hill. Budget decision making often involves major decisions committing government to noble purposes, such as sending a man to the moon, vanquishing diseases like polio, cancer, and AIDS, securing a life free from poverty and diseases through the social security program, and fighting wars in highly hostile circumstances. Budget execution decisions attempt to carry out the policy and programmatic promises developed and funded in budget authorization and appropriation processes within the law and the fiscal and other constraints imposed from within and from outside government. Budget managers do have discretion and their decisions are important, but few if any execution decisions change the course of history. However, these decisions make history in that they enable action.

Congress plays a role in budget execution both in terms of the restrictions imposed *ex ante* in the budget process and in oversight of budget execution. In his classic text on budgeting, Jesse Burkhead (1959) explained that budget execution is largely an executive responsibility (p. 340). However, Congress intervenes in execution to modify decisions it has previously made, to influence administrative actions, and to interpose independent checks on specific transactions (p. 341). Pitsvada (1983) added that, in general, Congress intercedes with specific constraints only if an agency has "performed in a manner that displeases Congress" (p. 86). He warned that the paradox of budget execution involves agencies that believe they must have more flexibility to meet changing needs, and Congress that believes unless it exercises meaningful control over budget execution, it is not exercising its most vital constitutional fiscal power, the power or the purse.

The tension over the correct amount of control versus flexibility in budget execution remains a fact of life that will not change. Some observers have suggested that too much control, and controls of the wrong kind, seriously impair the achievement of program efficiency (Jones & Thompson, 1994, pp. 155-193). Undoubtedly this is so. Still, given the growth in size of the defense budget, small percentage mistakes can result in large dollar errors, and the necessity for controls to ensure fiscal legality, managerial propriety and sufficient transparency to satisfy the procedural demands of governing under a democracy, have powerful practical and symbolic value. Where managers might prefer more flexibility, the Congress insists that public monies be safeguarded, even if this results in more control than might be necessary to meet policy ends efficiently. We also have to remember that to some extent Congress is more concerned with where and how money is spent than it is interested in establishing incentives that stimulate efficiency.

On its part DoD and many federal government budget administrators have developed ways to cope with the uncertainty they face in budget execution. These are not patterns of optimization, but accommodation, involving best estimate approximations that will provide for suboptimal solutions and allow for continuing corrections over time, as opposed to rationalistic, optimized solutions based on perfect information, rational decision mechanisms, and a once and for all solution.

Many budget preparation and decision process participants assume that the budget will be executed as planned once approved. They also assume that this is a relatively simple task compared to preparing and passing the budget. However, seasoned budget administrators know that this is not as simple as it seems. While many events do unroll as planned, budget administrators spend, "a substantial portion of their time ... rescuing carefully laid plans from unforeseen events, emergencies and nonforecastable contingencies" (McCaffery & Jones, 2001, p. 150).

Many of these events may involve relatively small amounts of money, but require the expenditure of disproportionately large amounts of time and effort to resolve. At the end of the year, in the aggregate and on average, budget execution may appear to have been a matter of uninteresting routine constrained by obscure and arcane rules and procedures and dominated by financial control mechanisms, clever cash management, and accounting creativity. It is unlikely to appear so uneventful to budget officers and staff or to the managers charged with carrying out programs. However, the saving grace to this situation is that most budget administrators are confident in their ability to solve many types of budgetary problems during the execution phase of budgeting, that is, to "fix it in execution." Those who execute budgets over a number of years have seen most of the types of problems they face before, and they generally know what must be done to resolve them again. Moreover, if familiar routines do not quite apply, they know where to go for help when they need to come up with new solutions, i.e., what rules, regulations, and people to consult (McCaffery & Jones, 2001, p. 150).

While it is productive to be optimistic that "the job can be done" in budget execution management, it is also necessary to understand that this is difficult business and errors will occur. When mistakes are made involving the management of defense budget accounts which result in large dollar discrepancies, and when external observers including auditors ask how this could happen, the key demands placed on comptrollers and budget managers are to clarify what happened and why, so as to preserve some level of perceived trustworthiness and reliability, while instituting measures to prevent them from happening again.

CHAPTER 9

BUDGET PROCESS PARTICIPANTS

The Pentagon

INTRODUCTION

When historians conclude that the secretary of defense (SECDEF) generally functions as a weak chairman of the board and that the Department of Defense (DoD) is so large and complex as to be "unmanageable," this perhaps *understates* the actual circumstance. This observation does not have the same impact as saying that the SECDEF is confronted with three massive military departments (MILDEPS) whose attitude at any point in time may be characterized as competitive as well as cooperative. While other decision processes in the Pentagon also reveal competition, none is perhaps so well situated as the budget process to capture the efforts of MILDEPS to get along with each other while, at the same time, getting a little more in resources than the others from the SECDEF. Senior leaders in the Pentagon know that budgets are "everything," that is, that without sufficient funding little can be done programmatically in the appropriate way or at the required level of intensity. Lack of budget success in a particular year not only means waiting for next year; it also may set crucial departmental

Budgeting, Financial Management, and Acquisition Reform in
The U.S. Department Of Defense, pp. 381–429
Copyright © 2008 by Information Age Publishing

programs back 5 to 10 years. As a result, senior leaders engage and negotiate at length over budget issues.

In this chapter, we describe how those who formulate and negotiate budgets in the Pentagon go about this task. First, we delve into the budgetary strategies employed by the MILDEPS and the behavioral patterns of budget officials and analysts. Second, we describe how the budget offices of the MILDEPS are organized and how they function. Third, we analyze the budget processes of the three MILDEPS and the U.S. Marine Corps. In part of the chapter, we view budgeting in detail from the perspective of the Department of the Navy (DoN) and the Marine Corps to understand by example many of the complexities of deciding upon and representing military department resource requests in the DoD budget. Finally, we briefly examine the budget functions of the office of the SECDEF.

MILITARY DEPARTMENT BUDGET STRATEGY AND BEHAVIOR

Within the DoD, the budget submitting offices for claimants including the major commands and the systems commands joust for dollars with budget reviewers in central comptroller offices of the MILDEPS and the Office of the Secretary of Defense (OSD). Competition for resources is intense and players earn reputations for their proficiency in the resource generation process. In terms of competitive strategy, entire MILDEPS become characterized in terms of their approaches. For example, the MILDEPS occasionally have been labeled by DoD level budget insiders as the "dumb, the defiant and the devious." None of these adjectives is complimentary but, as we will note, there are reasons to explain these characterizations.

Typically, the Army is cited as the "dumb" because of an observed practice of unwillingness to submit full budgets to OSD. A member of the OSD comptroller staff lamented, "We tell the (the Army) what we want but they can't produce it" (SECDEF, 1990). However, based upon our research, a more likely explanation is strategic in nature. The Army has been known to fail to submit requested budget data to OSD. Consequently, at times OSD is forced to "invent" the data for the Army in the DoD budget. This strategy permits the Army to deny authenticity and support for parts of the DoD budget when questioned in Congress. This is simply playing what our parents used to tell us was "dumb like a fox."

On the other hand, the Navy has been cast as the "defiant." In interviews with OSD comptroller staff we were told, "The Navy always has to do it their own way. We ask them to do something and they flat refuse ... period. This is always their first reaction. Then they [the Navy] give us what they want to do to comply with our request. This happens so much [that] we have come to expect it" (SECDEF, 1990). What is accomplished by this approach we may ask? Interviews with Navy comptroller staff verified that the OSD view was correct. No one in the Navy comptroller office even denied this strategy. We were told,

> What they [OSD] ask for often is wrong because they don't understand the Navy or our budget. We give them the best-scrubbed and most accurate budget of any service because we put our budget through a more thorough review process than the others [services]. Our numbers we can trust, but not theirs [OSD]. (Secretary of the Navy [SECNAV], 1990)

In this case, when questioned about what is in the DoD budget before Congress, the Navy can offer up its own numbers to supplement or replace what OSD has provided when it so wishes.

Both the Army and Navy strategies as described here are typical in budgeting. They fall into the class of tactics referred to as the "end-run" (Wildavsky, 1964). Both approaches allow the respective MILDEPS/ services to support the DoD and the President's budget, as they must once the SECDEF has submitted his budget to the President's Office of Management and Budget (OMB), while providing opportunity for deviating from it before Congress. This leaves us with the third characterization, the "devious." This approach apparently is used often enough that it is expected by OSD budget analysts.

> The Air Force always puts on the best show for us, but especially for Congress. They are the "high tech" people. They put up all kinds of graphics and "smoke and mirrors" and everybody in the room is "wowed." They [the United States Air Force] get what they want, then when they leave we all ask, "What did they say?" You can't figure it out. Their numbers are full of holes but the members and staffs [of congressional committees] don't care. The Air Force has their own strategy—dazzle them, get approval—and then do what you want. (SECDEF, 1990)

The obvious advantage of the Air Force approach is first to win at budget competition and, second, to gain maximum flexibility in managing the funds they receive. While this flexibility has been abused in some notable cases (e.g., cost overruns and funds management nightmares with the B-1 bomber program), it probably works more often than not to give the United States Air Force what it wants. Otherwise, why would the approach persist? Our point with respect to strategy is that the MILDEPS have many different ways to try to get their share of funding in the annual budget wars with the OSD and Congress.

BEHAVIORAL STRATEGIES OF CLAIMANTS AND ANALYSTS

In the complex world of budgeting, players often adopt strategies that simplify their perceptual costs of participation, increase their chances of winning or decrease the probability of losing. DoN comptrollers and budget analysts have shared their perspectives on the budget process with us over the last decade. A fleet comptroller explained, "Everything in (Navy) financial management is by the book. If you miss a time deadline, you are no longer a player; you have given up control of your destiny." Since analysts up the line are looking for dollars to cut, by not submitting something on time, an agency invites a cut when it misses a deadline. The rule then is always to meet deadlines, with the best that can be done, even if it is not perfect. Second, this comptroller emphasized, "You must give the desired product. If you are asked to give five issues with their dollar offsets, and you provided 109, 104 will be thrown out. If the remaining 5 do not have offsets, they go too." Again, the reason is that everyone is looking for money for other programs and will use any pretext to find it. In this instance this is also a work management issue for the claimant budget analyst; time is always short so one should not waste time on things that will be thrown out when that time could be better used working up other issues.

Finally, this comptroller said that the process lent itself to some simplifying assumptions:

> I assume that you [program sponsors in budget submitting offices] know your program and you will tell me when I make a mistake and you will do it within twenty-four hours ... 90% of the time people are not prepared to do this. If no appeal is made and made quickly, then the mark [cut] was a good mark. Our analysts think the proof of a good mark is when the claimant doesn't cut anything vital to its core function in response. If they don't cut flying or steaming hours in reaction to my mark, then it must have been a good mark.

One comptroller said he asked all his analysts to graph their accounts so that they could see the trend line in the recent past and the near future. Then the low point on the graph is taken as the starting point for a mark: "Whatever is the lowest point for your program in the past is a valid base, since you have survived, unless you can explain why it is too low." This comment is not applicable to the investment accounts, just to the operations and maintenance (O&M) categories where readiness is the main perspective.

One Navy Budget Office (FMB) budget analyst explained why some uniform personnel did not look forward to a tour of duty in the Pentagon:

Washington is different from the fleet. I just finished my executive officer (XO) tour. I had 300 people, all I had to say was "jump to it," and things would get done. Here it takes 16 chops (different signatures or clearances) on every memo I send. (Aboard the ship) I could really crank out the paperwork, sign things by direction of the captain. Here I have responsibility for lots of money, but no authority.... Without authority fighting battles takes more time; persuasion is your most important tool, but you do not have time to fight everything.

This officer was functioning as a budget analyst in the FMB and his superiors and peers all thought he was doing an outstanding job. He continued,

Peons like me do the work ... this hearing (a fleet budget presentation) was just for show. All the details get worked out between the analysts, people like me. We work all the technical things, then it goes to OSD and becomes a little political and then to Congress and it really gets political; people cut or change things because they do not like them, not on technical grounds.

He added,

When political decisions are made, weenies like me have to get $50 million out of a program. Sometimes we only have a day, or a couple of hours to do it in, so we make the cut and say, "Tell us what impact this will have on your program." You have to be prepared to be the bad guy here (in the budget office).

Working analysts from the lieutenant commander to captain rank in the Pentagon would often downplay their role in the resource allocation process; they see themselves as the "worker bees" implementing decisions that are made "above my pay grade." They would acknowledge that they had input to how the decision was shaped and how it was implemented; after all, that was the proper role of the worker bee.

THE PROGRAM VERSUS THE BUDGET

POM Glitches

Budget analysts typically have found problems with the program objectives memorandum (POM) and programming process. Some of these objections have been addressed in the 2001-2003 reforms that have merged the programming and budget analysis processes. For example, one budget analyst remarked,

> Programming is trade off time. The FYDP [future years defense plan] has a lot more program in it than we could possibly pay for. So everyone is looking to cut someone else's program. All the way into Congress, someone is looking to steal your money and put it into a better program.

Another added,

> Some programs are beautifully justified in the POM, but fall apart in translation into the budget. For example, suppose the money for military construction (MILCON) does not make it into the MILCON appropriation; now you can not build the building and the O&M analyst takes maintenance and operating dollars out of your activity budget and tells you to try again next year.

Another observed, referring to the early 1990s.

> The POM always bought back what you lost in the previous budget process. The POM directive was to meet every requirement by putting something in every cup. On average, the POM may have been underfunded by 20 cents on the dollar,

He added that he did not think this would work in an era of scarce resources and suggested that leaders ought to make vertical cuts in programs as soon as possible, rather than horizontal cuts across the board, so that whatever slack resulted from the vertical cuts could be put to use to prepare for the lean years. While this sounded like sensible advice, it did not seem to win many converts. Leaders preferred to keep programs alive across the board at diminished levels, like a golfer refusing to give up his short iron game because even at a low level he could keep the skill set alive and resurrect it quickly if it were needed.

One rookie claimant analyst was offered a deal his first year in Washington where he would get more money in the out-years (BY + 4 and 5) of a POM for his program if he was willing to allow his program to be zeroed out in the middle year. This seems to be the equivalent of sending someone for a "left-handed monkey-wrench." A program that is zeroed out is cut; it is no longer in the POM and now must compete with all those other programs that did not make it into the POM previously to get back in. Chances of success are not good. Furthermore, everyone knows that the out-years never unroll as promised and that everyone is promised to get well in the out-years. It is a cheap promise to make since even the most credulous program manager (PM) knows that years 4 to 5 beyond the budget year are so remote from the current year that they are almost imponderable. Thus, experienced players will take almost anything in a current year as opposed to a rich promise in the out-years. The out-years could unroll as promised, but the chances are remote. To some extent, players are willing to judge the sophistication of other players by how they respond to claims about the out-years; seasoned players tend to express a bit of well-mannered disbelief. Our "rookie" friend refused the deal and kept his program alive.

Budget Process Glitches

If analysts are cynical about some aspects of the POM, they also recognize there are glitches in the budget process. A retired director of the FMB said, "It's the purpose of green eye shade types (budget analysts, comptrollers) to find loose money for things the CNO wants to do." To counteract this, he advised claimants to, "have your numbers accurate and with good justification ... so you do not come back empty-handed." He added,

> When you talk to your boss about cost figures, do not wing an answer. Instead say, "last time I checked it was X and as soon as the meeting is over I'll get back to you right away if that is wrong. Nothing is worse than having a reputation for 'bum dope.' "

Another FMB director advised claimants not, "to put non-starters in your budget. You will lose the money and people will suspect the rest of your budget." A nonstarter is a budget item that has insufficient support. A claimant must "ask around" about this kind of idea to see who will support it. For example, a big discretionary program enhancement in a lean year is a nonstarter. He warned that some items always seem to be nonstarters, for example, staff travel or things that are nice, but not critical. When money is lost for nonstarters, it is lost to another claimant and may threaten the rest of that program in the budget base. At the DoN level, it could be money lost to another military department, thus increasing their budget base in the near future. The civilian world may allow for the expression of nice to do ideas in the budget process as consciousness raising items, but military budgeteers are death on this tactic. To some extent, the arena for this kind of behavior is in the domain of the authorization committees when they debate service roles and missions, force structure, nuclear weapons policy, and treaty provisions, controversial items that usually lead to intense discussion and late authorization bills without really threatening funding.

In the investment accounts, finding the correct cost estimate is a problem. One FMB leader noted, "Everyone has a number. The contractor, the PM, the Navy Budget Office, OSD ... analysts will take the lowest number, so you have to be sure you can explain the difference between your number and the lowest number." Another analyst added, "You find out what numbers other participants in the decision process have by calling around, by 'working the system' by 'dropping in for a cup of coffee. " A retired FMB director commented that analysts should, "Know what the boss is thinking … try to answer questions from his perspective." He advised preparing for the budget hearing by calling the officer who would brief the Admiral before the hearing, the pre-briefer. At a flag officer review,

> Someone is going to tell him what he is going to hear. Make sure you talk to the pre-briefer. You might say, "Can I help you get ready for your briefing of the boss?" Then compare numbers and explain them. Try to avoid surprises.

One retired senior admiral explained,

> Nobody likes surprises, particularly when they cost more money. The field activity is often the first to see it … to recognize program growth. They should take it up the chain of command. Unexpected program growth suddenly dropped on senior managers is very hard for them to handle.

Since they have trouble accommodating it, their first response is to deny it and cut it. He also warned, "Little changes can mean a lot." He noted that a small change in some electronic gear for a helicopter was "a minor upgrade, but it eventually caused some very expensive program changes, in training, repair manuals, even the airframe. This small change drove a huge change in program cost." Moreover, he warned that priorities change all the time and that when a claimant loses a program or item in his budget, it could be because of new technology elsewhere, or maybe something in the black budget has made the need for it obsolete. Additionally, treaties can introduce turbulence into operating budgets by changing roles and missions and their support costs, so claimants have to be prepared for a turbulent environment.

Even after the appropriation act is passed, claimants have to be careful. One analyst warned,

> If you manage resources, you should know where your money appears, O&M, OPN, MILCON, because they all have different legal requirements. You should know if there are special requirements with it, say an Environmental Impact statement, of if Congress has written special provisions in the Conference report to guide how you spend it.

Analysts and claimants cope in this environment through prior training, by successive tours of duty in the budget world of increasing responsibility, by seeking budget or resource allocation tours in the Pentagon, the center of all the action, and by cultivating good interpersonal skills that allow them to communicate effectively up and down the chain of command and laterally with other players who impact their budget process, for example, by taking money they might have won, or by losing money through unsophisticated budget behavior. They cope by being flexible and reacting to fact of life occurrences, by, "making lemonade out of lemons." To learn how this is done, and under what forms of organization it takes place, we examine the military department budget offices.

ORGANIZATION AND DUTIES OF
MILITARY DEPARTMENT BUDGET OFFICES

The Department of the Navy Budget Office

The FMB, officially titled fiscal management and budget, is the central budget office of the DoN. It is responsible for the preparing both the Navy and Marine Corps budgets. The goal of FMB is to fuse the strategic demands and requirements of both services with the strategic plans and guidance of the SECDEF to produce a single SECNAV budget estimate submission (BES) to the OSD, represented by the OSD comptroller.

The FMB is part of the Office of the Assistant Secretary of the Navy, Financial Management and Comptroller (ASN/FM&C) who functions as the Navy comptroller. The ASN/FM (assistant secretary of the Navy, financial management) is a civilian presidential appointee. The Navy comptroller was established initially in accordance with the provisions of Title IV of the National Security Act Amendments of 1949 that formalized control of DoD under the SECDEF. The mission assigned to the Navy office of the comptroller is to implement principles, policies, procedures and systems to ensure the effective control over all financial matters within the DoN. The Navy comptroller is required by Congress through legislation enacted in the 1970s to prepare the budget. Congress required budgeting to be a civilian function in DoD so that the military department budgets are issued under the authority of the military department secretaries rather than the military chiefs of staff (e.g., the chief of naval operations—the CNO—in the Navy). The comptroller has broad budgetary responsibilities performed almost exclusively by the FMB. While a variety of information and support units are located in this office, the two main analytical divisions responsible for the budget are operations, which handles personnel and their supporting expenses, and investment and procurement, which handles ship, aircraft and weapons procurement. The leader of the FMB is called the director and is typically a rear admiral. This position is unique in that the director (N-82) reports directly to both the CNO and the SECNAV (i.e., this position is "double-hatted"). The professional staff is composed of approximately one-half military officers who rotate through a tour of duty as a budget analyst and one-half civilians who are career civil servants. The civilian executive staff of FMB are senior executive service (SES) employees with special employment status that makes them accountable not only to the secretaries of the Navy and defense, but also to Congress. (One year the House Appropriations Committee was displeased with the DoN FMB SES people so it cut them out of the appropriation bill; the Senate restored them in conference work; being visible to Congress is sometimes a mixed blessing.)

The Navy budget guidance manual states, "The budget functions of the Comptroller of the Navy occur during all phases of the budget cycle, including formulation, presentation, and execution" (SECNAV, 2002f). Some of the duties performed by the budget office include:

1. Establishment of the general principles, policies and procedures that control the preparation, presentation and administration of the DoN budget.

2. Establishment of the appropriation structure for preparation and justification of the budget.

3. Supervision of the analysis and review of DoN budget estimates, and submission and negotiation of the budget with the SECDEF, the OMB, and Congress.

4. Supervision of any reprogramming of funds by DoD or Congress.

5. Provision of information as principle point of contact for outside agencies and other military department budget offices in all DoN budgetary matters.

The Navy is the only military department with two services (Navy and Marine Corps), each of which prepares separate budgets that must be melded into one. FMB has developed a unique structure to facilitate this process. As noted, the FMB performs duties for the comptroller (ASN/FM&C) and the chief of naval operations as the fiscal management division (N82). Although the FMB and N82 offices are one in the same, the accountability structure for the two positions differs. As part of the CNO operating Navy structure, the office is designated as N82 and has delegated budget preparation and execution responsibilities, although this duty is subordinate to the authority of the SECNAV. For the comptroller, FMB has responsibilities to oversee preparation and execution of the Navy and Marine Corps unified budget. Historically, this office has functioned in managing "blue dollars," that is, those that support the Navy only, "green dollars" that support only the Marine Corps, and blue-green dollars—Navy dollars to support combined Navy/Marine Corps functions. as noted subsequently, the Marines have their own budget and comptrollership function, somewhat analogous to the large Navy fleet claimants.

The FMB organization consists of six divisions: appropriations matters office (FMBE); operations division (FMB1); investment and development division (FMB2); program/budget coordination division (FMB3); business and civilian resources division (FMB4); budget and procedures division (FMB5). Each division plays an essential and specific role in the development and implementation of the DoN budget. The following is a brief synopsis of the responsibilities of each division as specified in the budget guidance manual.

FMBE: Liaison Responsibilities

This division is responsible for maintaining liaison with the congressional appropriations committees, the office of legislative affairs, and the congressional liaison offices of the secretary of defense, secretary of the Army, and secretary of the Air Force for functions related to congressional hearings and congressional staff matters for all oversight committees. The FMBE coordinates all matters related to DoN participation in hearings before the House and Senate Appropriations Committees and keeps FMB and all its fellow divisions advised on the current status of congressional action and appropriation requests. It provides the schedule of committee hearings, arranges for DoN witnesses, coordinates the review of transcripts of hearings, coordinates responses to committee questions, arranges briefings for members of the committees, their staffs, and the members of the professional staffs, and provides any other coordination activities associated with these committees.

FMB1: Military Personnel (MILPERS) and O&M

The FMB1 is responsible for reviewing, recommending, and revising estimates for the military personnel (active and reserve forces) and operation and maintenance (active and reserve) appropriations, and other funds of the Navy and Marine Corps. It looks for proper pricing of PM initiatives and their translation into the budget. It ensures that the primary readiness accounts are properly funded, those that fund ship and aircraft tempo and depot maintenance, adequate funding of base support and appropriate funding of pay and allowances. It develops and uses operational cost models for programs such as the flying hour or steaming hour programs, uses average cost rates for personnel costs and checks the O&M accounts for the reasonableness of their estimates. The office is also responsible for assisting in the justification of estimates before OSD/OMB and the Congress, the continual review of program execution, and the recommending of adjustment to allocations, when required. For the OSD/

OMB review, the FMB1 analysts act as the primary DoN contact with the OSD/OMB staff analysts. They are responsible for publishing schedules of hearings, attending hearings, and coordinating and clearing all responses to requests for additional information. They also prepare or review reclamas to program budget decisions (PBD) and issues for the major budget issues (MBI) meeting. Finally, they are responsible for ensuring the accuracy of the OSD decision recording system and for updating the DoN tracking system. For the congressional review, the FMB1 analysts are responsible for preparing or clearing budget material provided to Congress in support of military personnel and operation and maintenance appropriations. This may include budget justification material, statements, transcripts of hearings, answers to questions, backup or point papers, and appeals to authorization and appropriation reports. Representatives from FMB1 may attend hearings as backup or supporting witnesses.

FMB2: Investment and Development

FMB2 is responsible for reviewing, recommending, and revising estimates for the investment and development appropriations, including procurement, research and development, construction, family housing, and base closure and realignment. This office is also responsible for assisting in the justification of estimates before OSD/OMB and Congress, the continual review of program execution, the recommending of adjustments to allocations when needed, and the reporting of selected acquisition costs and data to the Congress. The focus is on most likely cost, including proper pricing and pricing consistent with the pricing of previous years. It looks for realistic phasing of projects to see that contract dates and delivery dates are consistent and it examines development and procurement milestones to see that they are properly phased. The office looks for a reasonable funding profile and an absence spikes or dips. They also check for prior year execution performance to see that contracts were executed on schedule and that obligations met obligation rate targets.

For example, in a ship construction project where only 10% of the appropriation was to be obligated in the prior year, it would check to see if all of the 10% were obligated. If only 7% were obligated, this could mean the program had a problem and imperil funding for the budget year. Critical indicators for these accounts involve time-phasing of production schedules, rates of production, lead-time, slippage in production schedules, and congressional approval for production. For the investment account, the watchwords are most likely cost, realistically phased programs, with a reasonable funding profile, based on good execution performance.

For the OSD/OMB review, the FMB2 analysts act as the primary DoN contact with the OSD/OMB staff analysts. They are responsible for publishing schedules of hearings, attending hearings, and coordinating and clearing all responses to requests for additional information. They also prepare or review reclamas to PBDs and issues for the MBI meeting. Issues which cannot be solved between analysts or supervisors of analysts may rise to MBI status and be resolved by SECDEF. In any year 5 to 10 issues may do this; usually the military department has to bring a suggested solution with it, for example, fund X by cutting Y, both within its own jurisdiction. It is not seen as fair to argue fund Navy X by cutting Army Y, although if SECDEF wanted to do this it would be seen as okay by Navy. MBI generate lots of paperwork, sometimes filling several three-inch binders for each issue. Finally, FMB2 is responsible for ensuring the accuracy of the OSD decision recording system and for updating the DoN tracking system.

For congressional review, the FMB2 analysts are responsible for preparing or clearing budget material provided to Congress in support of investment and development appropriations. This may include budget justification material, statements, transcripts of hearings, answers to questions, backup or point papers, and appeals to authorization and appropriation reports. Representatives from FMB2 may attend hearings as backup or supporting witnesses.

FMB3: Budget Guidance and Procedural Coordination

This division is responsible for the preparation of DoN budget guidance and procedures; control and coordination of budget submissions; coordination of reclamas to SECDEF PBDs; preparation and/or clearance of all program and financing schedules included in the budget; coordination of DoN participation in appeals to congressional action; development and operation of ADP systems in support of the budget formulation process at the DoN headquarters level; administration of financial control systems and procedures for the apportionment, allocation of funds and the reprogramming process; and preparation of fund authorization documents for the appropriations under its cognizance.

This office also reviews and makes recommendations on departmental budget issues and appraises the effectiveness of budget systems. Additionally, FMB3 prepares reports for the DoN on the status of the OSD/OMB review, coordinates the DoN participation in the MBI meetings, monitors the OSD automated decision recording system, and operates the DoN system for recording all decisions. FMB3 also coordinates the preparation of DoN input into the President's budget, preparation of financial and other summary budget documents, and preparation of the budget officer's statement. This office is responsible for preparing budget material provided to Congress in support of advisory and assistance services. FMB3 prepares bill digests and congressional action tracking system tables pertaining to the DoN budget at each stage of authorization and appropriation committee action on the President's budget. It is also responsible for reviewing any issues that may arise during congressional review concerning appropriation responsibility and all proposed legislative changes that affect the budget.

FMB4: Working Capital Funds

FMB4 is responsible for reviewing, recommending, and revising estimates for the Navy working capital fund (NWCF) and civilian personnel for inclusion in the budget and the justification of these estimates to OSD/OMB and the Congress. This office also reviews and validates funding estimates in working capital fund activity budgets to ensure proper balance between NWCF "providers" and DoN appropriated fund "customers." For the OSD/OMB review, the FMB4 analysts act as the primary DoN contact with the OSD/OMB staff analysts. They are responsible for publishing schedules of hearings, attending hearings, and coordinating and clearing all responses to requests for additional information. They also prepare or review reclamas to PBDs and issues for the MBI meeting. Finally, they are responsible for ensuring the accuracy of the OSD decision recording system.

For congressional review, the FMB4 analysts are responsible for preparing or clearing budget material provided to Congress in support of NWCF activities and civilian personnel accounts. This may include budget justification material, statements, transcripts of hearings, answers to questions, backup or point papers, and appeals to authorization and appropriation reports. Representatives from FMB4 may attend hearings as backup or supporting witnesses.

FMB5: Budget Policy and Guidance

FMB5 is responsible for the development, coordination, and issuance of DoN budget and funding policy and procedural guidance for all DoN appropriations, funds, and organizations. This includes the promulgation of DoN policy guidance required in the development of the budget; review and appraisal of budget policy and procedures and their implementation within the DoN; development of improvements in organizational responsibilities and interfaces related to budgeting and funding; continuous appraisal of adequacy and effectiveness of financial management systems to ensure conformance with budget policy; resolution or adjudication of audit and inspection findings involving budget policy and procedures matters; analysis of implications of audit findings for financial management policy of the department; review of budgetary policy impact of legislative proposals; identification and clarification of congressional direction concerning DoN budget policy and procedures; and development of functional standards for and review of comptroller organizations. This division adjudicates such issues as when government funds may be used to provide cake for the U.S. Vice-President when he visits a ship to give a speech, for example, it is legal when he awards a medal and praises the ship; it is not legal when he awards a medal and gives a campaign speech. The role of FMB5 is so important that often it is referred to as the FMB.

Department of the Air Force Budget Office

The Air Force budget office was established to aid the assistant secretary of the Air Force (financial management and comptroller—SAF/FM) in formulation of the Air Force BES (Department of the Air Force, 2002a). Its goal is to obtain funding to support the Air Force mission by translating program requirements into approved budget estimates. The Air Force budget office is composed of five directorates, including the directorate of budget investment (FMBI), the directorate of budget and appropriation liaison (FMBL), the directorate of budget management and execution (FMBM), the directorate of budget operations (FMBO), and the directorate of budget programs (FMBP). The Air Force budget office depends on these directorates to help in the development and execution of the budget estimate. The following is a description of each directorate's mission

FMBI: Estimation and Execution

The FMBI directorate develops the budget estimate and tracks financial execution of aircraft, missile, munitions, and other procurement. It also aids in the formulation and execution of the research, development, test and evaluation (RDT&E), military construction (MILCON), military family housing (MFH), base realignment and closure (BRAC) and security assistance activities accounts. The directorate is organized into five divisions, namely military construction, program support, security assistance, missiles, munitions, space and other procurement, and aircraft and technology.

FMBL: Congressional Liaison

The FMBL is the Air Force liaison to the congressional budget and appropriations committees and the Congressional Budget Office. Its job is to develop and implement strategies to ensure Congress is aware of pertinent Air Force budget positions and issues. In addition, it is responsible for monitoring congressional activity and keeping Air Force officials informed of actions taken on the Air Force budget. Furthermore, it acts as the Air Force point of contact for House Appropriations Committee surveys and investigations.

FMBM: Policy and Procedures

The FMBM directorate establishes financial policy and procedures. It provides oversight of Air Force defense business operating fund activities and manages the Air Force's financial data systems, prior year financial adjustments and processes for appropriation distribution to subordinate activities. The FMBM reviews and validates all Air Force requests of the Department of the Treasury, and the schedule of apportionment and reapportionment to the office of secretary of defense (comptroller).

FMBO: MILPERS and O&M

The FMBO directorate is the Air Force focal point for all matters pertaining to planning, formulating, integrating, defending, and executing of the Air Force's operations and maintenance and military personnel appropriation budgets that support approved programs and mission priorities.

FMBP: PPBES and Budget Coordination

The FMBP directorate integrates the Air Force budget within the Planning, Programming, Budgeting and Execution System (PPBES). It also coordinates the Air Force actions for the BES and budget review process leading to the President's budget submission. The FMBP manages the Air Force database for the Force and Financial Plan and all fiscal control adjustments. In addition, it acts as the principal advisor to the assistant secretary of the Air Force for financial management and comptroller, and the deputy assistant secretary for budget on total force comptroller and budget issues between the Air Force, Air Force Reserves and Air National Guard. These function are core to Air Force budgeting.

Department of the Army Budget Office

The Army budget office is the lead agency in the Department of the Army (DoA) responsible for the development and defense of the Army budget. It directly assists the assistant secretary of the Army (financial management and comptroller (ASA (FM&C)) by providing a link between the Secretary of the Army and the organizations that make up the DoA during the budget process. The deputy assistant secretary of the Army (DASA) for the budget is the head of the Army budget office. The Army budget office is composed of four directorates, including the management and control directorate (BUC), the operations and support directorate (BUO), the investment directorate (BUI), and the directorate for business resources (BUR).

In addition to these units, the Army budget office also has a congressional budget liaison office to assist in the handling of budget issues that occur on the congressional level. Each of the directorates that make up the Army budget office has explicit tasks, and it is also implicit that each directorate must seek to ensure that the budget estimate presented by the DoA to the SECDEF speaks for every organization in the department from the lowest to the highest. The following is a brief synopsis of the responsibilities of each directorate taken from the ASA (FM&C) organization and functions manual.

BUC: Budget Formulation and Execution Guidance

The BUC directorate is responsible for Army budget formulation and justification processes, issuing Army-wide budget formulation and execution guidance. In addition the BUC analyzes the impacts of changes to the Army's budget during the formulation, justification, and execution phases of the DoA budget process. The BUC directorate is organized into three divisions, for budget formulation; budget execution, policy, and funds control; and budget integration and evaluation.

BUO: MILPERS and O&M

The BUO directorate is responsible for formulating, presenting, defending, and managing the execution of the operation and maintenance, Army and military personnel, Army appropriations. The directorate coordinates budgeting of these appropriations from program development completion through budget execution completion. Also, this directorate participates in the program development process by membership on functional panels that interface with programs previously given resources in the budget cycle or being executed by the field. In addition, the BUO serves as the focal point for the major Army commands (MACOMs) to interface with DoA headquarters on operating budget issues. The BUO directorate is comprised of three divisions, the current operations division; the military personnel division; and the operating forces division.

BUI: RDT&E, Procurement and MILCON

The BUI directorate is responsible for financial management operations, budgeting, and execution for the Army's procurement appropriations; research, development, test, and evaluation, Army appropriation; military construction, family housing, and chemical agents and munitions destruction, Army appropriations; and for the defense department's homeowners assistance program. The director serves as an assistant secretary of the Army (financial management and comptroller) (ASA (FM&C)) representative to the Army system acquisition review committee. This directorate is the primary office for interfacing with the office of the under secretary of defense (comptroller) on investment in military construction and multiyear appropriation matters. The directorate is organized into four divisions: weapons systems; acquisition and integration; facilities; other procurement, Army.

BUR: Working Capital Funds, FMS, IT and PPBES

This directorate is responsible for formulating, presenting, and defending the Army working capital fund (AWCF), foreign military sales (FMS), and information technology systems budget (ITSB) to OSD, OMB, and Congress. It develops and issues Army policy for business resources and manages the interface between the Army, other military services, DoD, and other non-DoD government agencies. The BUR advises the DASA for the budget DASA (B) and assistant secretary of the Army (financial management and comptroller) (ASA (FM&C)) on issues relating to all other working capital funds and serves as the focal point for all aspects of the PPBES for the Army's working capital fund, foreign military sales, and information technology budget. The BUR directorate is comprised of four divisions: supply management division; depot maintenance/ordnance/information services division; business integration division; and the special business activities division.

In conclusion we may observe that each of the MILDEPS have organized their budget offices differently, but commonalities exist in terms of functions performed and duties. All of these budget offices serve a "gatekeeper" function, as do budget offices in virtually all public organizations. To get funding and to use it, all parts of the organizations served must go through the budget offices to get their resources. All budget offices must organize to formulate, analyze, propose and execute budgets within department guidelines and according to established procedure. All budget offices have to defend their proposed budgets to OSD and Congress. Finally, all DoD budget offices have to be accountable for executing budgets and budget policy according to appropriation law so that, above all, what they do is legal and auditable.

Now that we have some understanding of how the military department budget offices are organized under their respective department secretariats, we turn to an analysis of the military department budget processes.

MILITARY DEPARTMENT BUDGET PROCESSES

Although the goals of budgeting in each of the military department are the same, the methods they use differ, as indicated in the following descriptions of the military department budget processes.

Department of the Navy Budget Process

To develop a budget, the DoN depends on a decentralized budget formulation process driven by bottom-up responses to top-down controls. In the DoN the budget process portion of the PPBS consists of four phases, including office of budget (FMB) review; OSD and the OMB. review; congressional review; and appropriation enactment and execution.

Navy Budget Office Review

During the first phase of the DoN budget process, budget submitting offices (BSO) submit budget estimates for the organization they represent to the FMB. BSOs are also known and referred to as major claimants. In the DoN, about 24 major claimants submit budgets, including the Atlantic and Pacific Fleets, the bureau of medicine, the bureau of naval personnel, the chief of naval education and training, the military sealift command, the naval air systems command, the commander-in-chief, reserve forces (NAVRESFOR) and so on. The commandant of the Marine Corps (CMC) is also included as a major claimant.

The submissions by the major claimants are developed from the lowest level budget estimates usually referred to as cost center estimates. These cost centers are at the lowest tier in the DoN financial chain of command, for example, air stations in the Pacific Fleet. They submit budget estimates to the activity comptrollers who in turn review, revise, and combine the cost center estimates into an activity budget. For example, the type commander for air operations for the Pacific Fleet will assemble the budget for all air stations (and all the aircraft carriers too). This new consolidated budget is then forwarded to the major claimant who will ensure its reasonableness. The major claimants will then consolidate the activity budgets into one major claimant budget submission. As is discussed later, the Pacific Fleet claimant will submit a budget that consolidates air, sea, and undersea operations and their supporting expenses. This is then submitted to the assistant secretary of the Navy (financial management and comptroller) (ASN (FM&C)) through the FMB. The Marine Corps has its own internal planning, progrmming, and budgeting process and also submits its budget to the assistant secretary of the Navy (ASN (FM&C)). Organizing the process like this allows the civilian comptroller position to adjudicate the needs of the Navy and the Marine Corps outside the military chain of command.

The process begins in late winter, when the budget call is issued and claimants prepare their budgets. These are based on control numbers from the previous year's budget and new local needs. During this process, claimants also include items from the POM if it is concluded in time; if not their budget is adjusted to include POM items during the summer in the FMB. In June, these budgets are then submitted to the FMB where the analysts review the package, including the official submission, summaries, and backup data. Analysts ask questions of the claimants and dialogues ensue, both over the phone and face to face as claimants try to sell budgets and analysts try to keep them within fiscal guidelines. Some questions may be sent in written form to the claimant, asking for a written statement in support or clarification of a position in the original submission.

Upon receiving the budget submissions from the major claimants, the FMB will conduct a review of the estimates. If FMB officials think that certain estimates submitted need to be revised, they will issue a mark. A mark is basically an alert to the major claimant that their BESs will be altered. At this point, a major claimant is permitted to submit an appeal (reclama) stating its position. It is only during this phase that the major claimants within the DoN are provided with an opportunity to state their objectives and priorities for resources in the context of an executable budget. Beginning in June, major claimant budget hearings are held and issues are discussed. Subsequent to the hearings, "marks" are distributed to the claimants in the name of the comptroller of the Navy. These marks are normally cuts, made as adjustments, corrections, or denials of and to the requests for funds made by the claimants. A reason is provided for each mark to the claimant; usually the reasons have to do with incorrect pricing, a challenge to program executability during the fiscal year, hence a reduction of funds, or timing concerns where it is not believed that the funds will be needed during the year for which they have been requested.

The next step then is for the claimant to reclama or appeal the mark. In some years, some claimants appeal all marks; in other years, only a portion the marks are appealed. In general, the FMB considers a mark a final decision, unless new evidence can be brought to bear which gives the analyst new grounds for analysis. The FMB position is that the original mark is a studied and rational decision and not capricious or random, thus to appeal a mark is to question the analyst's best judgement. Hence, to make that appeal and still save face, a claimant would have to bring in new information that the analyst did not have at his disposal when he or she made the original decision. As one FMB director said, "Reclama's are for money that is already lost. They have to be supported by new technical information." To get back 5 or 10% in reclama action then, is a good score, since the analyst fully intended that nothing be given back.

While claimants can hold back some information and appeal all marks, submitting new information at the later juncture, this ultimately would not be seen as "fair." Defense is a resource constrained arena and for one claimant to get more, someone else has to get less. A claimant could appeal many items 1 year, but in subsequent years, his credibility would suffer. The resolution of the reclama process occurs at the level of the director of the budget office, between the director, the claimants, and the analysts. Some issues may be of a high enough visibility that the claimant's community may appeal it to the secretary of Navy level. Every year sees some of these, but most are settled at the analyst levels. Both claimants and analysts know where most of the trouble spots are in the budgets and the claimants usually begin to work these issues by alerting their analysts to potential needs long before the budget is submitted to FMB.

Budget analysts within the FMB view initial budget estimates with a preconceived downward bias. They assume some amount may always be cut from the claimant's initial request, if for no other reason than the claimant has had to begin his budget so much earlier in the year that accurate costs on certain items or programs or economic trends were not available. Consequently, analysts believe that program officials tend to over-estimate total program costs in order to compensate for this greater uncertainty. This logic is certainly typical of all budget reviewers and FMB analysts are no different. A reclama review is the final step in this process and provides a forum for resolution of adjustments. Its purpose is to guard against arbitrary or incorrect adjustments made to BSO budgets. Normally reclamas are due on the fourth day after the mark and the review session will be held subsequently to get final resolution on that issue.

The results of this process are designed to produce a budget that is timed correctly, and accounts for all known delays and disruptions (i.e., do not ask for 12 months of funding, if only 10 can be executed); contains current pricing of cost estimates based on execution experience (i.e., latest estimate of steaming hours) and latest cost factors (i.e., on local labor rates); is executable and avoids any mismatches between budgeted resources and program requirements.

Budgeting is often referred to as an exercise in timing, pricing, and executability so that program requirements and their dollars of support are matched to the fiscal year. Historically, some evidence exits about how these factors have been used in Navy budget account management. Smart and Shumaker studied Navy procurement accounts and found that executability problems (27%), pricing problems (19%), prior year performance flaws (10%) and congressional action (22%) accounted for 78% of the marks analysts made in these accounts.

Marks studied marks to Navy O&M accounts from 1988 to 1989 and over the FYDP—6 years, from 1988 through 1994 (Marks, 1989). In a detailed appraisal of four large claimants, Marks counted both the number of changes and dollars affected by the change. He found that 75% of the FMB marks involved pricing changes. Timing changes drove 8.5% of the cuts and programming decisions accounted for the rest. Marks found that programming changes drove almost 47% of the dollar changes, even though they were fewer in number than the pricing changes. Marks also added that in the reclama phase FMB gave back almost all it cut, but not necessarily to those from whom it was cut. He also indicated that the OSD review did not have large dollar consequences, averaging about 1.36% for the accounts he studied, and ranging from a cut of 11% to some accounts to an increase of 6% in others. In general, Marks found that aggressive claimants fared better, both in the original submission and in the appeal phase of the budget process. However, it is good to remember that these are incomplete studies (they only covered some accounts for some years) and dated. The current Navy budget manual says that marks may be made for various reasons, including pricing, congressional interest, program slippages, and so on. At the end of the budget process, FMB works to construct a budget that has best pricing (most accurate, if not the lowest) to accomplish the mission, the best schedule, strong budget justification, clear dollar and manpower balance, timely execution plans, and a clear statement of funds needed during the fiscal year. The Navy budget guidance manual that indicates how budgets are to be prepared may be found online at http://dbweb.secnav.navy.mil/guidance/bgm.

In 2002, the budget process was changed with issue papers replacing the mark-reclama process. Proposed cuts were issued as issue papers on the FMB internet Web site, with e-mail notification to other offices or claimants affected by the issue. If the issue affected them, they could write a response to the issue paper supporting or opposing the FMB position. Once FMB is satisfied with the DoN budget estimates, they will then forward their version to the SECNAV for final approval. In 2003, the mark and reclama terms were dropped from the vocabulary of the Navy budget world. One budget officer said, "There are no marks anymore, just areas of interest and anyone can raise an issue for adjudication" (SECNAV, 2003). He suggested that the process was more collaborative, but still not an easy process. The jury is out on this. It may be a permanent change—or not.

The mark and reclama process had a long tradition in the Navy, a service given to cherishing its customs. Moreover, the reason budget offices are constituted is largely because no one else wants to raise the difficult little questions about inflation rates and workload, or botched procurement programs or badly sequenced ship repairs and exorbitant base repairs...all the little things that budget analysts get paid to raise questions about. Budget offices have been "bad guys" for a long time. Their job is to say "you can not do that because you do not have the money," or "Congress only gave us so much money for that," or "Congress did not give us any money for that." Budget offices enforce economizing discipline. In the next chapter, we describe the midyear review process. Students of the budget review process realize that something has to replace the mark—reclama process simply because that process embodies a function that must be carried out. It is not clear that the budget review process can be delegated to others through a web-based system of issues and responses.

Phase II: Negotiation with OSD/OMB

Once approved by the SECNAV, the budget estimates are then submitted to the OSD and the OMB. This initiates the second phase of the DoN budget process. OSD and OMB conduct a joint review of the DoN budget to ensure the reasonableness of the DoN budget estimate. During this phase, the PBD is developed. The PBD, like the marks issued by FMB in the first phase of the DoN budget process, is a list of requested adjustments. These requests are submitted to each service in response to their budget estimates. Initially a draft PBD is issued in order to allow each service to reclama. Some PBDs are very simple, for example a change to an inflation or currency exchange rate; others can involve complex weapon

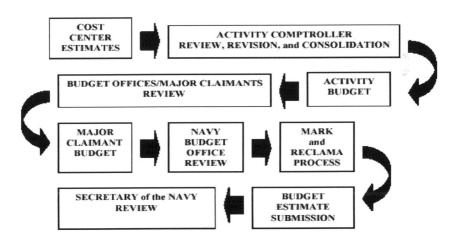

Source: Taylor, 2002, p. 36.

Figure 9.1.   Navy budget process from cost center to secretary of the Navy.

procurement acquisition strategies spreading out a buy over several years or considering going from one source to multiple sources. Once all the reclamas to these PBDs have been reviewed and changes have been made, the finalized PBD is developed and issued to the services for action.

After making the changes required by the PBD, each service will then resubmit its BES to the OSD and OMB in order to create the DoD budget. The DoD budget will subsequently become the President's budget and be submitted to Congress. This marks the end of the second phase in the DoN budget process.

Phase III: Negotiation With Congress

In the third phase of the DoN budget process, the President's budget is submitted to Congress for review. By the time the budget is ready for this congressional review, it has been reworked and revised many times to make it accurately reflect the DoN needs while staying within the DoD budgetary constraints. This is important since the DoD budget will now compete with other departments for funding. Appeals may be made to Congress when a bill that has passed one or the other chamber has a provision that is detrimental to the needs of the Navy. No appeal may be made if identical bills are passed in each chamber. However, when bills differ, the Navy may appeal, for example, to support the Senate's position against the position taken in the House. Appeals are directed to the conference committee charged with resolving issues where the two versions of the bill differ.

Generally, appeals must stay at or under the higher number. Some analysts believe that successful appeals start by staying a little under the higher number. Conventional wisdom says that DoN does not appeal all differences because it does not want to wear out its welcome and that it knows the bottom line at the White House so it must be judicious about appealing in order not to exceed that White House number. There are also cases where DoN would like to appeal, but DoD (SECDEF) prefers that it not appeal. Usually the DoN must support the position closest to that in the President's budget, but can negotiate with the DoD comptroller to diverge from this position. DoN and the other services exercise this appeal process because the two houses of Congress often treat issues differently as a result of differing philosophies or constituency issues and the outcome may be injurious to Navy programs, for example, in reducing a multi-year procurement program to where the contractor can not perform to bid pricing specifications.

All things considered, the legislative process is even more complex than it seems. One would think that the conference committee proceedings would be a limited to legislators, but the administration can and does appeal bills even at this late stage. Moreover, sometimes how issues are finally settled in the conference committee has ramifications for the ongoing budget process in DoD. For example, in 2003, DoN needed to know the outcome of several issues in the FY 2004 appropriation bill then under consideration, before it could finalize related issues in the FY 2005 President's budget, being built in the Pentagon during the same time. No matter what DoD did, if it acted before the conference committee's actions on the FY 2004 appropriation bill, it would be offering up some programs for cuts or others for deeper cuts in the FY 2005 budget or even cutting programs itself that Congress might not have cut.

This kind of dependency is particularly annoying when the appropriation bill is late, which is the usual case. When this happens, DoD delays its decisions and makes them in the November-December period, after the final appropriation bill has passed. Usually, but not always, these issues are few in number and concern weapon systems purchases, either the absolute number or changing the sequence of purchases over the next several years. This is why these decisions can be held in abeyance; they affect the future more than the present or budget year.

Phase IV: Execution

The fourth and final phase of the DoN budget process marks the enactment of appropriations by Congress in the DoD appropriations bill. This bill, once signed by the President, allows the DoN to incur obligations and to make payments out of the Treasury. A detailed description and analysis of the Navy budget execution process is provided in chapter 9.

The above section emphasizes the similarities between military department budget processes. However, the DoN is a study in contrast to the Army and Air Force in that it has a somewhat more complex budget process in part because it budgets for two services, the Navy and the Marine Corps. The Marine Corps has its own financial management office under the commandant that prepares, examines, and executes the U.S. Marine Corps (USMC) budget. The USMC FM (civilian comptroller and director of the budget) has status comparable to the Navy ASN/FM. Because fiscal guidance is issued by OSD to the military department budget offices through their secretariats rather than to the military services, additional computations must be conducted within the DoN to divide Navy from Marine Corps funding. The assistant secretary of the Navy, comptroller (civilian presidential appointee) plays an important role in adjudication between Navy and the Marine Corps budget officials in close coordination with the USMC FM. In the section below we analyze aspects of the Marine Corps budget organization, budget process, and budget accounts with emphasis on the interdependence between Navy and Marine Corps accounts and management.

U.S. Marine Corps—Navy Budget Process

When determining how to divide money between the Navy and the Marine Corps, the DoN depends on a procedure known as the "blue-green split." This procedure is based on a formula established by the Navy and the Marine Corps in a letter of agreement more than 25 years ago. Although the division of funds is consistent, dollar amounts are not fixed and can be altered significantly in favor of one service or the other by the SECNAV in different fiscal years. On average, the Navy has allocated roughly 86% of its annual appropriated funding to the Navy and 14% to the Marine Corps (Williams, 2000, pp. 84-85).

Marine Corps Appropriations

The DoN funds that are spent by or on behalf of the Marine Corps are concentrated in two appropriations clusters; the first is termed the "green" appropriations and the second the "blue in support of green" appropriations. The green appropriation consists of dollars controlled directly by the commandant of the Marine Corps. These include military personnel, Marine Corps; reserve personnel, Marine Corps; operations and maintenance, Marine Corps; operations and maintenance, Marine Corps reserve; and procurement, Marine Corps.

Control is shared between the Navy and Marine Corps in the second group of appropriations, the blue in support of green. these include military construction; military construction, reserve; family housing; research, development, testing and evaluation; and procurement of ammunition. These accounts are composed of funds provided by the Navy that provide "direct" and "indirect" support for the Marine Corps. The direct support funds are provided directly from the Navy budget. This category provides the money required to procure, operate, and maintain Marine Corps aircraft. The indirect support part of the "blue in support of green appropriations" is comprised of funds that the Navy would have to spend even if the Marine Corps did not exist. The blue in support of green appropriation budgets amphibious ships and their equipment, naval surface fire support, corpsmen, and chaplains. Figure 9.2 demonstrates the division of funds in the DoN.

Green Appropriations

Once the blue-green split is completed, the Marine Corps can begin to build their POM, which in turn leads to the budget. The first thing Marine Corps financial officials do, once their amount of the DoN TOA (Total Obligational Authority) is determined, is to "pay the bills" associated with the green appropriation. Funds must first be set aside for resources that have already been committed by the Marine Corps in

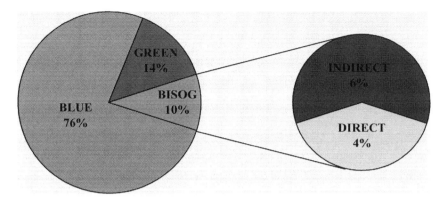

Source: Taylor, 2002, p. 50.

Figure 9.2. Blue-green and blue-in-support of green dollars.

previous years. These funds are referred to as the "core." The "core' is simply the summation of the previous funding decisions that the Marine Corps does not want or need to revisit, what non-DoD analysts would call the base (Burlingham, 2001, pp. 60-64). Its purpose is to help the Marine Corps to:

- Fence entitlements,
- Maintain programmatic stability for well defined, executing programs,
- Recognize the cost of doing business,
- Establish a programmatic baseline, and
- Create a discretionary portion of program.

The core is developed from the minimum requirements of each account that makes up the green appropriation. In short, this category includes all the "must fund" fixed costs of the command (USMC, 2002). Budget analysis proceeds by examining the basic, minimum requirements for the core appropriation accounts in order to set core dollar number.

Green Dollar Core Review

In the military personnel and reserve personnel accounts, analysts determine the cost of maintaining the Marine Corps at its authorized end strength. In addition, these analysts price bonus plans, accession phasing, and all of the pieces that must fit together to ensure that the bills are covered. Since the Marine Corps is a people intensive service, it is no surprise that this account consumes the majority of the funds in the core. What funds should be set aside in O&M, for both active duty and reserve accounts is the most difficult to determine. The O&M account has been called an accountant's nightmare, because so many different people spend these funds in so many different ways (Williams, 2000). If there are any disagreements over the funds that make up the budget core, it is safe to assume that they will occur within this account.

In the procurement and procurement of ammunition accounts, analysts determine minimum requirements based on two procedures. in procurement, budget analysts meet minimum requirements by continuing to fund obligated programs. Analysts for the procurement of ammunition account determine minimum costs by looking at training and combat requirements. The training requirements are consistent from one year to the next, while the combat requirements are simply replenished when used.

In the research, development, testing, and evaluation account items that the Marine Corps intends to buy later, and those which they already started to invest in are funded. At the same time, the analysts for this account assess the minimum requirements to fund the science and technology facet of the Marine Corps. These dollars will be used to support the Marine Corps war fighting lab and various advanced concept exploration programs.

The military construction and family housing accounts factor in both the current inventory and the resources needed to begin construction on the most urgent requirements of the Marine Corps when determining minimum required amounts. This account is the smallest one in the core.

Once the minimum requirements of the green appropriation are identified, the core is set. the Marine Corps will only use this money to fund the requirements determined in the core-setting process. The funds that remain are referred to as the discretionary funds and are available to the commandant to satisfy all of the demands of the operating forces and supporting establishment of the Marine Corps.

Green Dollar Discretionary Funds

The discretionary funds of the Marine Corps are identified during the planning phase of their planning, programming, and budgeting system. These funds are only a small portion of the resources received during the blue-green split. In the POM for FY 2002-FY2007 for example, about 5% of Marine Corps funds were designated as discretionary funds—about $5 billion over the 5-year period.

The programming phase of the Marine Corps is a period of internal competition. Whereas the planning phase determines the discretionary fund amount, the programming phase decides to whom the fund should go. Requests for the discretionary funds are called "initiatives." In the

POM for FY 2002-FY 2007, about 525 "initiatives" were submitted. In order to fund all of these, the Marine Corps would have had to spend over $17 billion more then they had estimated. As a result, over three fourths of the 525 initiatives were not funded. The goal of the Marine Corps, as it is with the other three services, is to get the most for its money. In order to do this, Marine Corps leaders must ensure that the initiatives receiving the funds are the most beneficial. The core is not questioned, since the resources it funds have already been found to be beneficial in previous year reviews. The challenge is selecting from the numerous initiatives submitted, the ones that will most benefit the Marine Corps in the future and funding those with very limited resources.

Just as the other PPBES processes function, the Marine Corps uses a sophisticated committee review system to sift and winnow out the best initiatives. Initiatives are grouped into categories and compete against each other in the POM process. Winners then are funded in the budget process to the extent funds are available. In the POM process, initiatives are reviewed first by program evaluation groups (PEGs). These committees are composed of lieutenant colonels, majors, and civilian equivalents who are tasked with conducting the initial evaluation of the initiatives. For an initiative to receive resources in a POM, it must first compete successfully within its own PEG, for example, investment, manpower, military construction, and so on. The PEGs are not fiscally constrained. It is their job to hear briefings on selected initiatives that represent different Marine Corps missions or sponsors, judge priorities and relative benefit among the selected initiatives, and consider any objections. Each PEG prioritizes the initiatives in terms of its benefit to the overall mission of the Marine Corps, rather than by cost. Their ranking lists are then forwarded to the POM working groups (PWGs) for benefit/cost analysis and re-ranking.

The POM PWG process begins with the consolidation of the PEG lists into a single benefit-only list to create a merged list from the different PEGs so that all initiatives may be rated against each other. The PWG then refines the list by taking the benefit value of each individual initiative and dividing it by its cost. This will readjust the order of the list by presenting one based on both benefit and cost rather then just cost. In addition to this refinement, the PWG makes further adjustments based on the professional knowledge, judgment, and experience of the lieutenant colonels, majors, and civilians that make up the group. Once each of the initiatives is properly ranked the PWG initiates a process called "order to buy." In the order to buy process, the PWG begins at the top of the list of newly ranked initiatives and starts "spending" the discretionary funds. This process continues until all discretionary funds are spent.

The results of the PWG are submitted to the program review group (PRG), a committee of the most senior officers in the Marine Corps, where they are combined with the core for final assessment. The PRG objective is to assess the war fighting capabilities, verify compliance with guidance, resolve intermediate issues, and make corresponding program adjustments. Once completed, the PRG will then form a single Marine Corps POM including both the fixed core and the discretionary initiatives, which is forwarded to the commandant along with any major issues that need to be resolved. The commandant may make final adjustments to this package.

When determining which initiative to support, members of both the PEGs and the PWG use various criteria. Although each analyst may have his own specific standards of determination, as a whole, they tend to look for the same things. The fist thing they look for when determining validity of an initiative is whether its sponsors have provided a concise, specific statement of fiscal need, based on sound funding estimates. They tend to favor initiatives that identify trade-offs, offsets, and overlaps, and avoid blanket claims, slogans, and buzzwords. They are more likely to support initiatives that define their programs in simple terms and clearly explain the impact on the Marine Corps. In addition, they tend to reject claims of cost savings that cannot be identified by activity, amount, or year. The more quantifiable the supporting information, the more likely the initiative will be highly ranked (Taylor, 2002, p. 59).

Department of the Air Force Budget Process

The Air Force budget office is the lead agent in the budgeting phase of the PPBES. The Air Force budget process is described in "The Planning, Programming, and Budgeting System and the Air Force Corporate Structure Primer" (Department of the Air Force, 2002b). The budget office is the key player in this part of the PPBES process. Its objective is to formulate, execute, and control the allocation and use of resources based on requirements identified during the planning and programming phases of PPBES. In the Department of the Air Force the budget process consists of three phases, namely (1) investment budget and operational budget review, (2) BES, (3) budget review.

Phase I: Scrub the Base

The investment budget review (IBR) of the first phase begins with the review and evaluation of the execution and performance of programs funded with investment dollars within the major commands and system centers. It is during this phase that analysts from the Air Force budget office determine the expected obligation and execution rates of each program. Their goal is to identify and adjust obligation and execution problems. If the Air Force does not identify and adjust these problems, the OSD will usually do so in the budget review. However, if the OSD adjusts a program, the savings do not automatically belong to the Air Force.

Based on the findings of the IBR, Air Force budget office analysts propose specific adjustments to specific investment accounts in selected programs. These proposals are then forwarded to the investment budget review committee (IBRC), which is chaired by the director of the directorate of budget investments (FMB-I). The IBRC will review each proposed adjustment in order to determine which will be sent to the Air Force board (AFB), which is chaired by the director of the Air Force budget office. This board reviews the IBR recommendations and decides which to keep, adjust, or delete. These results are then submitted to the Air Force chief of staff (CSAF) and the secretary of the Air Force (SECAF) for final approval.

The second aspect of the first phase of the Air Force budget process is referred to as the operational budget review. It progresses similarly to the IBR except that it focuses on the O&M accounts. The operational budget review group (OBRG), chaired by the director of the directorate of budget operations (FMB-O), reviews the proposals and briefs them to the Air Force board. As in the IBR, this board evaluates the proposals and submits their findings to the CSAF and the SECAF for consensus.

The main objective of the investment and operational budget reviews is to prevent the OSD from adjusting the Air Force TOA. By correcting funding issues within the department, the Air Force is able to make changes that will result in net savings. The recommendations developed during these reviews, once approved by the Air Force chief of staff and the SECAF will then be used in the development of the Air Force budget estimate.

Phase II: Adjust Scrubbed Base to POM

The beginning of the Air Force budget estimate development marks the commencement of the second phase in Air Force budgeting. It is during this phase that, the approved recommendations of the IBR and OBR are merged with the guidance provided in the program decision memorandum by the Air Force budget office in an attempt to readjust the POM. The BES is developed from the newly adjusted POM.

The BES is like a bill and it is the job of the budgeters and programmers within the Department of the Air Force to find offsets to lessen its cost. Once the BES is determined, it is briefed by the AFB to the Air Force council (AFC), CSAF, and SECAF. Once approved the BES is submitted to the OSD thus concluding the second phase of Air Force budgeting process.

Phase III: Negotiate With OSD

Once OSD has received the Air Force budget estimate, the third and final phase of the Air Force budgeting process begins. During this phase, the OSD and OMB conduct a joint budget review of the Air Force BES. It is their objective to identify more cost effective pricing or programming alternatives. These alternatives are presented in PBD memorandums (PBD). Initially the OSD budget analysts will prepare PBD drafts in order to alert service representatives of the impending marks (cuts) certain accounts will receive. It is imperative that Air Force representatives be proactive and involved in the hearings during this process. Often OSD budget analysts will not write a draft PBD (make a cut) if the Air Force can explain away the analyst's concerns. If however a disagreement emerges, then the Air Force has an opportunity to challenge a draft PBD in the form of comments or reclama (appeal).

Once a reclama is initiated, it is the job of specific representatives, appointed by the Air Force budget office, to defend the programs involved in the reclama. This entire process takes only a few days, but its effects can be lasting. After this process has been completed, the director of the Air Force budget office signs a memorandum of either acceptance or rebuttal, in part or completely, to the OSD comptroller. If a rebuttal is initiated, the program in question is elevated to the status of a MBI and is negotiated between the Air Force chief of staff, the SECAF and OSD, represented by the defense review board.

Once all marks and reclamas are final, the Air Force PBD is used to adjust the department BES that is delivered to the OSD for action. The Air Force BES will then be merged with the BESs of the other services into the DoD budget. This final action will conclude the Department of the Air Force budget process. The Air Force budget process is more centralized than that of the two other MILDEPS.

Department of the Army Budget Process

The Army budget process is managed by the assistant secretary of the Army (FM&C) through the DASA budget. It is the deputy assistant secretary who takes charge in the budgeting and execution phase of the Army's PPBES through the Army budget office. The Army budget office divides the budget process into three parts, namely formulation, justification, and execution.

Budget Formulation

The formulation phase of the budgeting process begins with the development and approval of the Army BES. It is during this phase that the first 2 years of the programs in the POM are converted into the department BES. The Army budget office supervises the entire formulation process. MACOMs and installations, such as airfields, barracks, camps, depots, and other facilities, aid in the development of the BES by providing a budget request by means of their command budget estimates (CBEs). In the DoA, about 16 major Army commands, ranging from Europe, to the Pacific and Korea, and including such commands as the corps of engineers, medical, traffic management, and special operations submit command budget estimates.

These estimates developed by the installations are summarized in the MACOM current budget estimate, which is next reviewed by the program Budget Committee and the Army resource board. Then it is merged with the program revisions submitted by the director, program analysis, and evaluation who works directly for the Army chief of staff. Next, it is sent to the secretary of the Army and the chief of staff, Army by the ASA (FM&C) for final approval. Decisions are made on the budget at all these stages. Once approved, it is the job of the Army budget office to forward the DoA BES to the OSD and OMB.

OSD and OMB review the submission to ensure reasonableness and Army estimates are either approved or adjusted. Usually disagreements with adjustments are handled at the lowest level; however, if the dispute persists they are labeled MBI and forwarded to the defense review board and Joint Chiefs of Staff for examination. Once the adjustments have been finalized, the results are combined with the other services in order to form the PBD document, which is then sent to the deputy secretary of defense for final approval. The PBD list the required changes that each service must make to their budget estimates. After these changes have been made the services budget estimates can then be included into the DoD budget and eventually the President's budget that marks the end of the DoA budget estimate formulation phase.

Justification in Congress

The second phase of the DoA budget process, budget justification, initially begins after the President's budget is submitted to Congress for review. The House Budget Committee and Senate Budget Committee begin the congressional review. Their goal is to ensure that the President's budget is within the discretionary spending caps of the Omnibus Budget Reconciliation Act of 1990 as updated. During this period that the SECDEF and representatives of each service within the DoD testify before the House national security committee, Senate Armed Services Committees, and appropriation committees in order to justify to Congress the DoD budget estimates. To support the SECDEF and the DoA representatives, the Army budget office provides detailed budget justification books to the authorizing and appropriations committees as well as any other assistance that might be needed to prevent a congressional adjustment to the DoD budget estimate.

Execution

The third and final phase of the DoA budget process is execution. Here it is the job of the ASA (FM&C) to supervise and direct the financial execution of the funds appropriated by Congress. By utilizing the command budget estimates submitted by the major Army commands and installations for guidance the Army budget office acting on behalf of the ASA (FM&C) distributes all funds approved by the budget and monitors their execution during the budget year. This picture of the Army budget process is drawn from "Budget Office of the Deputy Assistant Secretary of the Army" [http://www.asafm.army.mil/budget/budget.asp] August 2002. Notice that the Army description emphasizes Congress more than the Air Force does.

DoD BUDGET ACCOUNTS: WHAT DO THE DOLLARS BUY?

Previous sections described and analyzed budget office organization and budget processes, that is, who does budgeting and how the process works. Now we examine military department budget accounts using a case study of the DoN budget to explain how the funds structure is organized and what is purchase by account, as well as some of the issues that analysts face in making budget decisions. The DoN has a total of five major appropriation accounts that fund both Navy and Marine Corps activities. These appropriations include O&M; military personnel (MILPERS); procurement; research, development, testing, and evaluation (RDT&E); and Military Construction (MILCON). The majority of the funds go to O&M, military personnel, and procurement as can be seen below. In the following section, we discuss what is in each of these accounts and how they are treated in the budget process.

Military Personnel

The DoN MILPERS appropriation is divided into two categories, active and reserve. The active MILPERS is composed of military personnel, Navy (MPN) and military personnel, Marine Corps (MPMC). The reserve MILPERS consists of reserve personnel, Navy (RPN), and reserve personnel, Marine Corps (RPMC).

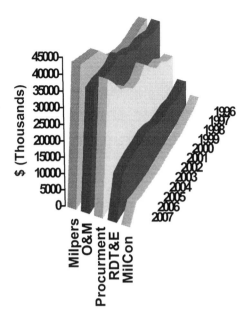

Source: Taylor, 2002, p. 40.

Figure 9.3. DoN appropriated funds 1996-2007.

The budget activities funded by the MPN and MPMC include pay and allowances of officers, pay and allowances of enlisted, pay and allowances of midshipman, subsistence of enlisted personnel, permanent change of station travel, and other military personnel costs. The budget activities funded by the RPN and RPMC appropriations include unit and individual training and other training and support. The following exhibit (Table 9.1) shows each MILPERS appropriation and its budget activities.

When determining the MILPERS budget estimate, analyst use an average cost basis. The numbers of people promoted, departing, arriving, and already serving are all factors that affect the level of the MILPERS budget estimate. The most important factor is the determination of the average cost rates utilized by the Navy and Marine Corps analysts in the formulation of their MILPERS budget estimates. Although the rates for military pay and allowances are established by law, the average rate of base pay must be established from these estimates. The base pay average is the largest contributor to the MILPERS account.

Table 9.1.   Components of the Navy Military Personnel Account

MPN	MPMC
• Pay allowances officers	• Pay allowances officers
• Pay allowances enlisted	• Pay allowances enlisted
• Pay allowances midshipmen	• Pay allowances midshipmen
• Subsistence of enlisted personnel	• Subsistence of enlisted personnel
• Permanent change of station travel	• Permanent change of station travel
• Other military personnel costs	• Other military personnel costs
RPN	*RPMC*
• Unit and individual training	• Unit and individual training
• Other training and support	• Other training and support

Source: Taylor, 2002, p. 43.

Other estimates that affect the level of the MILPERS account include: the number of promotions expected, the number of personnel gains or losses, and the longevity raises which will accrue during the budget year. Utilizing the above estimates, an average cost rate is computed for each paygrade from the lowest recruit to the highest officer. The same procedure is followed to establish the average cost rate for basic allowances for quarters. Turnover rates are very important for analysts, because turnover rates affect the cost of allowances for changes in duty station, reenlistment bonuses, and clothing for new recruits, and separations. The personnel accounts are not as complex as the other accounts in that they proceed from known factors like specified salary and bonus costs and as well as the number of authorized personnel and use historical averages to develop average costs for other factors; however, it is a large job that takes a lot of brute computational power.

The MPN account may not be as complicated as others, but its issues are important and subtle. In a volunteer force, the innate capability of personnel is of great importance. Over the last decade, CNO guidance to MPN managers and others in the resource allocation process has stressed that people are of paramount importance. MPN managers have recruiting accession and standards goals as well as retention targets. The MPN account has support for recruiters, bonuses to stimulate reenlistment and retention and to fill out communities where expertise is always in demand, as well as support for such things as advertising and change of station travel. All of these matters can be important. For example, when change-of-station dollars get short, moves are often pushed into the next fiscal year. What this does is extend time on station for both officers and enlisted personnel often for another three to four months. This can mean extended sea duty, with a resulting decline in re-enlistment, billets ashore gapped or left vacant for months, and a slowed career progression, directly affecting morale as well as again threatening retention and force structure goals.

Some of the costs in this account are historically based on standards; others are driven inversely by the economy. When the economy is good, recruiting tends to decline and thus more money for recruiters and advertising is needed. Basic salary is also a concern. While the employment cost index, a standard measure of wage cost in the economy, increased by 6.4% from 1989 to 2000, enlisted pay for the same period was increased 2.8% and officer pay only 1.3%. Personnel managers know that you get what you pay for, no matter what is said about making personnel a high priority. The pay and pension reform in FY 2000 budget were a significant step forward, but like many issues, their effect will only be felt in the long run. In the meantime, analysts look to historical averages, pay structure, the health of the economy, competing labor markets (for pilots, for example), and decide on dollars for advertising, recruiter activity and bonuses. Information technology also appears to be a large unmet bill in the MPN account because the systems that deal with personnel include many unrelated systems, with many requiring frequent manual data entry. Adopting modern corporate business systems practices could help manage this account.

Operation and Maintenance

The DoN O&M appropriation, like the MILPERS appropriation is divided into active and reserve. The active category includes operation and maintenance, Navy (OMN) and operation and maintenance, Marine Corps (OMMC). The reserve category includes operation and maintenance, Navy reserve (OMNR) and operation and maintenance, Marine Corps reserve (OMMCR). The budget activities funded by the OMN and the OMNR appropriation include operating forces, mobilization, training and recruiting, and administration and service wide support. The budget activities funded by the OMMC appropriation are similar to the OMN and OMNR appropriation, but do not include mobilization. The OMMCR appropriation funds operating forces and administration and service wide support.

The O&M account is very diverse and includes everything from civilian pay to fuel, paper, and ammunition-all the items, which the DoN consumes as an operating entity and most of the expenses it incurs to keep operating. It is easier to state what is not in the account then what is in it.

Table 9.2 shows the budget activities under which all the items of the O&M account fall.

The O&M account is the most inclusive of the five major appropriations in the DoN. It includes funding for items such as the flying hour program, costs for operating the fleets, civilian salaries, base maintenance, ammunition, yellow tablets and pencils, other administrative expenses, and temporary active duty travel for both military and civilians, all the items that the Navy consumes as an operating entity and most of the expenses it incurs to keep operating. O&M budget requests are determined either by formula or by a historical cost average. Formulas determine such budget categories as the cost of steaming hours for the Atlantic Fleet. Historical costing procedures determine budget amounts by making an estimate based on cost already incurred, such as the average cost for temporary employees in the Washington D.C. area or office supplies or base maintenance over the last 3 years.

When determining the reasonableness of these estimated amounts, analysts depend on experience, work measurement standards, cost accounting information, employment trends, price level changes, and prior budget execution performance. Selecting the best measurement technique however depends on which program is being estimated. For example, cost and work measurement data are best used when examining ship and aircraft overhauls, fleet operations, flight observations, medical care, supply distribution, and real property maintenance. The flying hour program and ship steaming hours are examples of formula driven categories. Due to its variety, no standard methodology can be described for review of this account.

Table 9.2.   Components of the Navy O&M Account

OMN	*OMMC*
• Operating forces	• Operating forces
• Mobilization	• Training and recruiting
• Training and recruiting	• Administration and service wide support
• Administration/service wide support	
OMNR	*OMMCR*
• Operating forces	• Operating forces
• Mobilization	• Administration and service wide support
• Training and recruiting	
• Administration/service wide support	

Source: Taylor, 2002, p. 44.

Procurement

The DoN procurement appropriation encompasses a number of appropriations. Aircraft procurement, Navy (APN), Weapons Procurement, Navy (WPN), shipbuilding and conversion, Navy (SCN), other procurement, Navy (OPN) procurement, Marine Corps (PMC), and procurement of ammunition, Navy and Marine Corps (PANMC) are all included under the Procurement appropriation as is shown below.

The planning of pricing and milestone schedules in the acquisition cycle of DoN programs is dependent on accurate procurement appropriation funding. Production schedule, inventory requirements, spare part philosophies, and lead-time are all taken into account when determining how much of the budget should be allotted to this appropriation. It is, however, the determination of an accurate cost per unit estimates that is most important to analysts. These cost per unit estimates are determined in two ways, depending on whether the item is newly acquired or already in development. The cost estimates for existing items is a historical cost supplied by the accounting system, while cost estimation for new items is developed using engineering cost estimates.

In addition, cost per unit estimates for newly acquired items also use factors such as amount of inventory on hand, projected consumption rate, requirement for spare parts, status of research, development, testing and evaluation programs, production time schedules, slippage of production schedules, required lead time, mobilization base and approval for production to aid in the determination of the most accurate cost per unit. These procurement accounts are a particularly interesting part of the Navy budget in that they represent the purchase of capital goods items, whose mission is often unique to warfare (but not always-search and research and communications hardware have obvious civilian uses) where DoD specifies an end use and a private corporation attempts to estimate what it would cost to build that item and how much profit can be made. The account is also interesting because defense corporations have large amounts of money at stake; loss of a big contract as an old contract phases out could threaten the very existence of a corporation. Congressmen and Senators in whose jurisdictions these corporations are located have an obvious stake in these events. Whether new systems are being considered or whether the question is to buy a few more aircraft or ships, each session of Congress sees a lot of lobbying focused on the armed services and appropriations committees to ensure that DoD keeps buying weapon systems.

Budget folklore has it that members of Congress are particularly interested in these committees because they can use them to get defense spending for their states and districts. Members of Congress from states rich in defense spending have to get on these committees to protect the investment in their state in its human and physical capital

Table 9.3.   Components of the Procurement Account

APN	WPN
• Combat aircraft	• Ballistic missiles
• Airlift aircraft	• Other missiles
• Trainer aircraft	• Other weapons
• Other aircraft	• Torpedoes and related equipment
• Modification of aircraft	• Ammunition
• Aircraft spare and repair parts	• Spares and repair kits
• Aircraft support equipment and facilities	

OPN	PMC
• Ships support equipment	• Ammunition
• Communication and electronics equipment	• Weapons and combat vehicles
• Aviation support equipment	• Guided missiles and equipment
• Ordnance support equipment	• Comm. and electronic equipment
• Civil engineering support equipment	• Support vehicles
• Supply support equipment	• Engineer and other equipment
• personnel and command support equipment	• Spares and repair parts
• Spares and repair parts	

SCN	PANMC
• Fleet ballistic missile ships	• • Ammunition, Navy
• Other warships	• • Ammunition, Marine Corps
• Amphibious ships	
• Mine warfare and patrol ships	
• Auxiliaries, craft and prior year program costs	

Source: Taylor, 2002, p. 46.

base, and to ensure that they get their appropriate share of defense dollars.

The mechanics of the weapons account are also interesting. First, the budget for a weapon system is developed by a PM who is a DoD employee, usually a uniformed expert in the military system being purchased. It is his job to oversee it, although few PMs ever oversee a program from inception through RDT&E to full production through the close out of the production line, a process that could take 10 to 15 years or longer. Thus, the PM enters into a program, somewhere in midstream, and takes over from someone else. His job is to guide the program over the hurdles that will occur "on his watch"—like new requirements, or unexpected glitches, or fiscal slowdowns. He works closely with the corporation which produces the system and when it comes time to budget, he and his business manager sit down with their corporate counterparts and estimate what the cost will be to produce x number of missiles for the next budget year or years. It is a cooperative, yet adversarial relationship over the cost to produce hardware where no one knows precisely how much the item will cost until it is built.

The central budget office analyst exercises a review function, just as he or she would were the ship or aircraft or weapons accounts the budget for temporary labor to paint the walls, something which can be understood more easily and for which a multitude of real world reference points exist for comparison. Analysis in these accounts often proceeds from a series of assumptions about what happens to cost when experience building a system begins to accumulate. The central budget analyst will expect to see unit costs in a program decrease under certain circumstances. Some of these include:

1. As experience accumulates building a weapon—this is called the learning curve.

2. As quantity increases: this is a cost versus quantity relationship that notes efficiencies of scale should lead to decreased unit costs

3. With repetition: repeated construction of the same system should drive down the unit cost of the system. This is the learning curve in another guise, but it is often broken when new requirements are put into the systems and modifications made.

4. With dual sourcing: competition between or among contract bidders is expected to drive down the price that a corporation can ask, hence apparently driving down cost. How far down is often a source of contention between the PM and the central budget office analyst.

Costs may also change when a corporation decides to take less in the way of profit from a program for reasons it may not readily disclose, for example, to keep its plant fully occupied, to maintain core skills while it waits on the outcome of another contract, or perhaps because it knows that once the buy has started it can build back its profit margins as change orders come in to modify the weapons system over its lifetime. This is likely to be proprietary information that government analysts may be able to estimate some years later, but it is not likely to be available during the annual budget process.

These attempts to guess which way cost will go and how large the increment of cost change will be are important because these weapons are bought with appropriated funds that have to be requested from Congress. Since dollars are always in short supply, a weapon system that can be made to appear to cost less looks better than one that costs more, if their capabilities are approximately equal. It is no wonder that weapon systems incline toward a low-ball cost strategy when the contractor is anxious to get the contract, elected representatives are anxious to have him get it, PMs are focused on fielding a good weapon system and improving their chances for promotion, and budget analysts at several levels are interested in seeing unit costs decline.

Another phenomenon experienced observers detect in the weapons procurement area is often described as the bathtub curve. It describes the attempt by a corporation to keep its cash flow, and hopefully profits, up as it is transitioning out of an old program and production lines are shutting down, while the new weapons contract that will replace it is just beginning to phase in. This results in a gap which occurs in the corporate profit stream when one system is phased down and another phased in. Lobbying efforts are often focused on the size of the gap, in an attempt to decrease it, and keep the cash flow stream to the corporation going, by extending the construction from the current weapon system or beginning a new system before the old production line has closed down, thus advancing the new profit stream before the old stream has dried up.

The task for analysts in these accounts is quite complex. An investment review checklist suggests that the basic question for these accounts was to ask if it made sense to program the item in question at the requested quantity. Analysts were urged to review the document that provided the basis for the requirement, find the inventory objective and how it was derived, and review any munitions, spare parts or overhaul or rework schedules. Then they could move to a budget scrub, concentrating on such questions as: is the item budgeted to most likely cost? Is the program realistically phased? Is the total program funding profile reasonable? and Is the current program being executed on schedule? For program pricing help, they might request a unit cost track between fiscal years to find any fluctuations, request identification of nonrecurring costs to make sure they do not show up in future years, verify a learning curve was used in program pricing, ensure that only OSD/OMB inflation rates were used for cost escalation, and request a contract status report to analyze it for cost and schedule deviations compared to the budget request.

In terms of program phrasing, the analysts might ask if the production build-up is too rapid, review the equipment delivery schedule to ensure that the funded delivery period does not exceed 12 months and examine the factors limiting production ramp-up such as test equipment, personnel, and raw materials. In the funding profile, the analyst is cautioned to watch for and avoid funding spikes as well as programs where the profile is too stretched out to be effective, investigate the possibility of multiyear contracting and acquisition improvement program initiatives like competition and economic production rates.

Investment account analysts have access to multiple budget schedules that allowed them to cross check programs, numbers, and dollars. They can look to see if the out-year profile made sense, what system was suggested for replacement, how it would interface with other systems and if so, how it was funded. For example, if an item were related to military construction, was the military construction item funded? Certain basic themes persist in these questions, none more important than ascertaining if the future profile is reasonable given the past record. "What is the basis for the estimate?" and "how does it compare to the last negotiated cost?" followed by "what are the elements of change from the last negotiated cost?" are questions that are asked by those who are responsible for budgeting these accounts. However, each type of account has its own intricacies, from airframes, to engines, to ships, to missiles, to ammunition, and so on.

Research, Development, Testing, and Evaluation

The DoN research, development, testing, and evaluation appropriation includes basic research, applied research, advanced technology development, demonstration and validation, engineering and manufacturing. Table 9.4 shows these budget activities more clearly.

The budget estimate for the RDT&E appropriation tends to be fixed across the DoD, despite some annual fluctuations depending on variations in budget climate. Insiders say that when the account is larger than 10% of total budget, the risk is that not all projects started can be put in the field. When the account is below 10% the risk is of limited innovation, that not enough new projects are being started. Additionally insiders know that some systems will have teething problems while others may never develop as expected. Running this account at a bare minimum ignores these fact of life situations.

Budget insiders believe that historical logic suggests that an investment of under 10% in this account indicates that the DoN is not investing in enough weapons development to keep up with potential competitors in the long run while investments of over 10% of the [DoN] budget raises concern about the ability of the organization to successfully man, deploy, and maintain the range of weaponry under development. Although this metric may have been based on the cold war world, it tends to persist in the conversation of knowledgeable budget insiders.

Table 9.4.   Components of
the RDT&E Account

RDTEN
• Basic research
• Applied research
• Advanced technology development
• Demonstration and validation
• Engineering and manufacturing development
• RDTE management support
• Operational systems development

Source: Taylor, 2002, p. 48.

An examination of the past computed from Table 6.8 of the DoD comptroller's green book (Figure 9.4), indicates that total DoD RDT&E has been about 12.6% from FY 1981 through 2003, lower that than during the Reagan build-up—which was primarily a procurement build-up—and higher than that during the down years of the 1990s as DoD fought to protect its technological edge. With the procurement holiday of the 1990s, aging systems (as described in chapter 11), and a new foe, DoD plans to allocate 15.3% of its budget each year on RDT&E over the course of the FYDP. History would indicate that this is a somewhat optimistic figure.

Whatever the truth to the 10% rule, analysts for the account have some set routines. They annually review each program's financial balance in order to determine the status of unobligated and unexpended balances. Unexpended and overobligated funds are automatic warnings to analyst that a program needs to be reviewed. For example, a program which is underobligating and has high balances may have run into substantive trouble (underobligating because contractor can not make some part of system work and thus can meet milestones and thus not get paid) and ought to be reviewed more deeply before any additional money is committed. Conversely, a program that is in danger of overobligation may also be in trouble since it is costing too much and also ought to be reviewed. Perhaps the original cost estimates were wrong and the engineering cannot be done for the price that was estimated.

The budget office also reviews the account for balance with the other accounts, for total level of expenditure, for areas in which further research ought to be pursued or curtailed, and the availability of scientific personnel and research facilities. Analysts may look for unrealistic growth planned over a prior year, no recognition of schedule delays, unobtainable

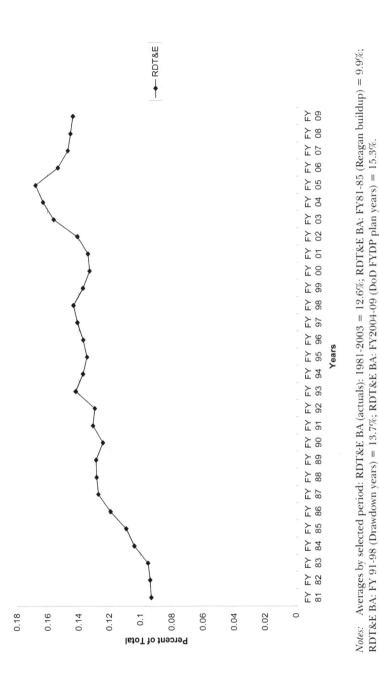

Notes: Averages by selected period: RDT&E BA (actuals): 1981-2003 = 12.6%; RDT&E BA: FY81-85 (Reagan buildup) = 9.9%; RDT&E BA: FY 91-98 (Drawdown years) = 13.7%; RDT&E BA: FY2004-09 (DoD FYDP plan years) = 15.3%.

Source: DoD, 2003c.

Figure 9.4. RDT&E history (RDT&E as a percent of total DoD Budget Authority, average by fiscal year).

milestones, and failure to consider changes in limiting factors like personnel or test equipment, and proposed changes in technical scope. Analysts also have to be sensitive to congressional action that may restrict progress or funding, pending submission of data to Congress. Analysts also look for duplication of effort where other services or government agencies are developing a similar item or where other programs satisfy a similar requirement and could lead to a cut against this program. Analysts examine the budget estimate to ensure that only those funds needed to carry the program over the 12-month budget period are requested. They may examine contractor and service data to see when the work will be performed and they may examine outlay data to see that the program is on schedule thus far. Schedules and milestones will be examined to make sure the program is still on track and no slippage from the approved plan is evident.

THE BUDGET FUNCTION IN THE OFFICE OF THE SECRETARY OF DEFENSE

The budget functions of the OSD are performed by staff of the deputy secretary of defense, comptroller. The title of the DoD comptroller has changed over time. Presently it is principal deputy under secretary of defense (comptroller), and deputy under secretary of defense for financial management reform. All of the tasks associated with budgeting are managed under the auspices of the DoD comptroller, not to mention some of the tasks related to programming not elaborated here. Briefly, the DoD comptroller budget related responsibilities include issuance of guidelines for budget preparation and review (including preparation of exhibits), review of military department budget submissions, preparation of budget marks to these budgets and responses to military department reclamas, issuance of PBDs, compilation of military department and defense agency budgets into the budget of the SECDEF and the President's budget. Preparation and issuance of the official DoD budget document, the national defense budget estimates or "Green Book" for each fiscal year is an important responsibility of the DoD comptroller staff.

When the defense budget is reviewed by the committees of Congress, DoD comptroller staff support testimony by various defense department officials including the DoD comptroller, review proposed and enacted legislation for fiscal impact, and provide a myriad of reports required and requested by members of Congress and their staffs. They also conduct various special studies on behalf of the SECDEF and develop estimates of savings projected to result from implementation, for example, management efficiency initiatives.

As explained in previous chapters, within the PPBES process the development of the DoD budget begins with the dissemination by the SECDEF of the defense fiscal guidance that specifies the "top-line" budget targets for each military department and agency. This comprises roughly the amount of money targeted as not to be exceeded when the MILDEPS set their respective resource requirements. Once the fiscal constraints are issued, the DoA, Navy, and the Air Force may begin formulating their budgets.

In budget execution, the DoD comptroller and staff have responsibility for submitting the military department and defense agency apportionment requests to the OMB, allocating regular and supplemental appropriations into appropriate DoD accounts, and internal allotment of the budget to what DoD analysts refer to as the military components (MILDEPS and services). DoD comptroller responsibilities also include establishing and distributing spending targets and procedures when Congress makes regular and Continuing Resolution Appropriations, monitoring spending patterns across DoD accounts, monitoring and reporting to the SECDEF and other DoD officials on spending trends and patterns. In budget formulation and execution, DoD (C) officials review the control of accounts and spending activity to comply with the law and congressional requirements relative to their authority to do so, responding to congressional budget initiatives including those originating in the Congressional Budget Office, and from committee and subcommittee staffs. They submit DoD reprogramming requests and monitor the reprogramming process, submission of requests and distribution of supplemental appropriations. They maintain coordination of end of year accounting and reconciliation to appropriations requirements, responding to internal (inspectors general) and external (e.g., GAO) audit findings and a number of other post-spending year tasks. Among these other tasks is monitoring the working capital funds operated and managed by the MILDEPS, and provision of all types of information to the SECDEF, and to other DoD officials including data to be issued to the public and news media through press releases. This is only a synopsis of tasks performed by DoD comptroller staff.

The DoD budget process cycle revolves around the schedule established by the DoD comptroller. The Comptroller also coordinates with the other DoD level secretariats (e.g., the under secretary of defense for acquisition, technology and logistics) to provide information essential for administration of the responsibilities of these offices. This is especially important for the acquisition budgeting process that operates semi-autonomously as described later in this book. This summary of the budget responsibilities of the DoD office of the comptroller is illustrative and not comprehensive. More details on the roles of the DoD comptroller and of OSD staff are provided in the first four chapters of this book.

CONCLUSIONS

In this chapter we have demonstrated the complexity of the budget processes that operate within the MILDEPS and in the office of secretary of defense in the Pentagon. We have reviewed how each department budget office is organized and functions within its budget processes to produce submissions to the OSD. We have explained that the PPBES process has undergone some recent modifications. Since 2003, a combined POM and budget review performed by the military department budget and programming offices and the comptroller staff of OSD have attempted to determine the desired force structure and fiscal estimates.

In this chapter we have not explored in detail what happens once the military department budget submissions to OSD are reviewed and approved by SECDEF and staff. The budget process as it operates outside of DoD is detailed in earlier chapters. As noted, OSD does not have a separate budget office that operates in the same ways as those of the MILDEPS. The staff of the OSD office of the comptroller performs the function of budgeting and analysis for the SECDEF. In analysis of the military department budget processes we analyzed the PBD and reclama cycle that eventually leads to translation of the DoD budget into the formats required by the President's OMB, and then the appropriation format employed to by Congress. We have traced the origin of the Pentagon budget process beginning with preparation of future year estimates shown in the FYDP, and the Quadrennial Defense Review. We have noted that these data are not incorporated directly into budgets because they are generally formulated prior to the beginning of the budget process. We have explained how great care is taken to make the budget year estimate precise and accurate.

Each of the MILDEPS maintains a central budget office to coordinate, facilitate, and oversee this process. It is the job of these offices to review the budget for conformance with the directives issued by OSD and DoD comptroller at the start of the budget process and through any subsequent changes to those directives. Ultimately, the military department budget offices must produce budgets on behalf of their secretaries and military chiefs that integrate the guidance set forth by the SECDEF with the requests of the organizations within their respective departments. As we have shown in this chapter, the budget offices and the budget processes of each MILDEPS and OSD are designed to accomplish a very complex task of rationing and coordination. We have shown that each military department and the budget officials and analysts that work in them employ their own unique strategies to obtain and protect their budgets and programs.

In the next chapter we examine how budgets are prepared and executed outside of the Pentagon, in the field at the major force command and subordinate levels where "the rubber meets the road," that is, where spending occurs and assets are consumed in execution, and where much of the data originates that support military department and DoD budgets.

CHAPTER 10

BUDGET PROCESS PARTICIPANTS

The Commands

INTRODUCTION

With his headquarters within walking distance of the Pearl Harbor Memorial, the commander of the Navy Pacific Fleet (CPF) is designated as one of 24 Navy budget submitting offices (BSO). CPF is a claimant in the budget process and builds and presents a budget for itself and its subordinate commands and presents that budget to the Navy Budget Office (FMB). This budget will later be presented to Department of Defense (DoD) as part of the Navy budget and then journey to Congress as part of the President's budget (PRESBUD). When the resulting appropriation bill has been signed, the FMB will pass CPF back his share of the budget to execute. This chapter will describe participation by CPF in the Department of the Navy and Department of Defense Planning, Programming and Budgeting Execution System (PPBES) by describing its processes and relationships, both internal and external. Although many of the issues faced by CPF may vary from the issues faced by other claimants, both in the Navy, in the other military departments, the processes and procedures to plan,

Budgeting, Financial Management, and Acquisition Reform in
The U.S. Department Of Defense, pp. 431–480
Copyright © 2008 by Information Age Publishing

431

program, develop, and execute the budget reflect great overall commonality with other major claimants.

The mission statement of CPF states:

> The mission of the U.S. Pacific Fleet [PACFLT] is to support the U.S. Pacific Commands (PACOM) theater strategy, and to provide interoperable, trained, and combat-ready naval forces to PACOM and other U.S. unified commanders. This mission reflects changes since 1986, when the U.S. Congress passed the Goldwater-Nichols Act of 1986 to engender more cooperation and "jointness" between the armed services. PACFLT's role has transitioned from that of war fighter to that of force provider, sustainer and trainer for the unified commanders. The net effect of this change is that the operational chains of command are now shorter and more direct, while PACFLT and other force providers are able to focus on maintaining readiness. (Commander, Pacific Fleet, 2001)

The Goldwater-Nichols Military Reform Act of 1986, reorganized or revolutionizes the way the military does business (Hadley, 1988, p. 17). Goldwater-Nichols empowered regional joint commanders to exercise operational control over all forces in his region of the world. In the Pacific region, the Joint Pacific Command (PACOM) has operational authority over combatant commands of assigned forces through commanders of service components (e.g., the Navy), subordinate unified commands, and joint task forces. Operationally, CPF is the naval force provider for PACOM. Administratively, as an echelon two commander, CPF reports directly to the chief of naval operations, the Navy echelon one commander, Thus, CPF reports administratively to the chief of naval operations, and operationally to the U.S. Pacific Command.

In addition to CPF, other service component commanders reporting to PACOM include U.S. Army Pacific, Marine Forces Pacific, and U.S. Pacific Air Forces. As the naval forces component commander in the Pacific region, CPF is the world's largest naval command. The CPF area of responsibility (AOR) mirrors PACOM and includes the Pacific Ocean, a significant portion of the Indian Ocean, and about half of the continental United States. The CPF AOR extends from near the African coast on its west side to Oklahoma on its east side as depicted in Figure 10.1. The AOR covers more than 50% of the world's surface; approximately 105 million square miles, and 16 time zones and includes 56% of the world's population. It includes 43 countries, and 10 U.S. territories. It also includes or touches on the world's six largest armed forces: (1) Peoples Republic of China, (2) United States, (3) Russia, (4) India, (5) North Korea, (6) South Korea (Pacific Command, 2002, p. 1).

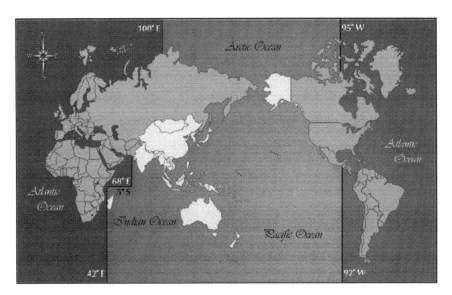

Source: Pacific Command, 2002.

Figure 10.1.   Pacific Fleet command area.

Commander Pacific Forces Resources

To perform its mission as force provider in its AOR that includes many of the most militarily significant regions of the world, CPF requires an enormous amount of resources. Human resources include 196,000 active duty military personnel, 13,000 reserve personnel, and 30,000 civilian personnel. Its physical infrastructure consists of 20 major installations, 15 minor installations, 191 ships, and 1434 aircraft that are distributed among subordinate type commands (TYCOMS), shore commands, or other commands, stretching from Oklahoma to Diego Garcia in the Indian Ocean (Pacific Command, 2002).

While CPF generally administers using money from the OMN (operation and maintenance, Navy) account, in fact money is drawn from a variety of appropriation accounts to support a multiplicity of operations. Its total budget exceeds $13 billion, but it is the operations and maintenance (O&M) budget that maintains operations and fleet readiness and takes the most managing. This is the focus of the chapter. For example, the CPF estimate of the fiscal year (FY) 2002 OMN account totaled $7,565,000,000. Including price and program growth, the OMN account estimate for FY 2003 was $7,477,000,000. These are baseline figures and do not include congressional supplementals

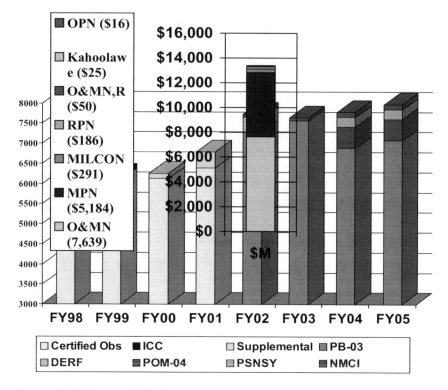

Source: Pacific Command, 2002.

Figure 10.2.   Pacific Command Resources for FY 1998-2005.

received due to costs associated with the war on terrorism and other funding shortfalls identified in the CPF midyear review (nor reductions due to congressional adjustments). The amount of supplemental funding received for FY 2002 operations was an additional $290,000,000.00 (Pacific Fleet, 2002c).

Not only is the CPF AOR very large, but it is also very volatile. Consequently CPF is continuously involved in contingency operations, most of which end up generating supplemental appropriations. In recent years, CPF has provided forces to Operations Southern and Northern Watch in Iraq, Noble Eagle in Afghanistan and Enduring Freedom that expands the war on terrorism worldwide. CPF also provide forces to support U.S. policies throughout the region in areas such as East Timor and the Philippines and is involved in homeland security as a component commander who reports to the U.S. Northern Command for operations regarding homeland security. Much of the supplemental money provided to CPF is the result of the costs incurred supporting these contingency operations.

SUBORDINATE COMMANDS

The CPF claimancy includes five Navy operational commanders, four type commanders (TYCOMs—echelon three commands), and six Navy regional commanders. CPF acts as the BSO for all activities contained within these organizations. These organizations include the operational commanders for seventh fleet, third fleet, the maritime defense zone, task force 12 (undersea warfare), and task force 14 (missile submarines). The type commanders include: air (COMNAVAIRPAC), surface (COMNAV-SURFPAC), submarines (COMNAVSUBPAC), and fleet marines. The regional commanders naval forces include the Mariana, Japan, Korea, Pearl Harbor, San Diego, and Seattle. In addition to these regions, the CPF claimancy also includes small naval forces in Singapore and Diego Garcia, British Indian Ocean territories. Each "sub-claimant" has its own comptroller responsible for providing programming estimates and budget submissions to CPF, their BSO.

COMMANDER PACIFIC FLEET INTERNAL BUDGETING AND PROGRAMMING ORGANIZATION

The CPF internal programming and budgeting structure was recently changed to reflect that of the Department of the Navy (DoN). Prior to the change, CPF programming and budgeting functions were both performed under the direction of the CPF comptroller. Recently however, CPF reorganized and divided these functions between the comptroller, a Navy captain, who leads the comptrollership functions and a civilian GS-15 who leads the programming side of the equation.

Budgeting

The budgeting functions are directed by a comptroller who is responsible for the budget preparation and execution process at CPF. The comptroller also supervises the fleet accountants who maintain official records of obligations and expenditures throughout the fleet. The comptroller does not perform programming functions, but develops the budget for the fleet based on programming requirements as directed by SECNAV (secretary of the Navy) instruction 7000.27 that states:

> The commanding officer or head of an activity that receives allocations or suballocations of funds subject to the Anti-Deficiency Act (31 U.S.C Section 1341 or 31 U.S.C Section 1517) shall have a qualified comptroller who reports directly to the commanding officer. (Secretary of the Navy, 2002b)

The responsibilities of the comptroller are described as follows:

> The comptroller shall ... have overall responsibility for budget formulation, budget execution, financial management, managerial accounting program analysis, and performance measurement. (Secretary of the Navy, 2002a)

The comptroller directs various budget and accounting divisions. Among these is the budget division that prepares the budget submission for the DoN budget. This submission is reviewed in the FMB, integrated with the rest of the Navy budget and passed on to the office of secretary of defense for review, revision, and approval and later becomes the PRESBUD. This office is responsible for apportionment for the upcoming execution year and reconciles the executed budget at the end of the year by certifying the obligations made against it. In order to perform these functions, the division employs approximately 13 budget analysts who prepare OMN budgets that are divided into five major areas, including air operations, ship operations, combat support, ship maintenance, and BOS.

As each FY begins, the new budget is executed at CPF by the budget execution division. This function is performed by a Navy commander. Under the guidance of the Comptroller, he is responsible for the execution of current budget year control numbers. His department has five budget analysts who report directly to him. Analysts in this division are responsible for monitoring the obligation of funds at subordinate commands. Additionally, they coordinate with the budget analysts and program managers to effectively coordinate funding of emergent issues, execution adjustments, conduct midyear review, request for supplemental funding based on unfunded requirements, and end of FY collection of unobligated money for reallocation across the command, termed the "end year sweep."

Programming

While SECNAV instruction 7000.27 directs that the comptroller for activities that receive allocations or suballocations report directly to the commanding officer (PACFLT), no corresponding requirement exists for the head of programming, consequently the programming function reports to the commanding officer through the deputy commander, a two star admiral. The programming function sets the fleet requirements that the budget must resource on an annual basis. These fleet requirements are set by the office of N8 for warfare requirements and assessments and N43 for fleet maintenance which includes aviation, surface ship, and submarine maintenance and N46, who is responsible for programming for shore activities. Not surprisingly, this arrangement of program managers all designated by an "N" and a numeric code is termed the "N code" structure and when asked where a requirement came from people say "It was set by the N codes," and by a specific code.

These "N code" program managers maintain close contact with program sponsors in Washington, D.C. as well as points of contact at subordinate commands within the CPF claimancy such as the type commanders including AIRPAC and SURFPAC (type commands) for fleet warfare and maintenance requirements and regional headquarters for base operating support (BOS). They function to coordinate fleet resource assessments and requirements that are used to develop program inputs to resource sponsors in Washington, D.C. While primarily active in the programming phase of PPBES, these program managers maintain contact throughout the fleet and with their resource sponsors in Washington and serve as resident experts within CPF for emergent budget and execution issues.

RESOURCE MANAGEMENT COORDINATION AT CPF

Although CPF employs budget analysts in two departments within the comptroller organization whose missions are different (budgeting and execution) and develops warfare requirements, assessments and programs within different N codes, programming, budgeting and execution are not conducted in a vacuum within CPF. Although budget analysts are familiar with the inputs, marks (cuts), and budget decisions that helped to formulate budgets, they will rely on the program managers who performed the assessments and developed requirements for their areas of the fleet program and budget when conducting analysis on subordinate command budget submissions. Similarly, program managers must be aware of the issues that arose during the development, submission, and execution of prior year budgets when conducting their assessments and developing their requirements.

Changes to the PPBES process in 2001 and 2003 within DoD have reinforced the necessity for programmers and budgeters to strengthen their working relationships. As one of the Navy BSOs, CPF planners, programmers, and budgeteers participate in all phases of the PPBES. In the planning phase, CPF participates in the integrated warfare architectures (IWAR) analysis; in programming CPF requirements officers develop fleet resource requirements and program managers prepare capabilities plan (CP) inputs based on those requirements; in the budgeting phase, CPF provides control numbers to activities and solicits their budget inputs based on these control numbers. They then consolidate fleet budget estimates and provide required supporting exhibits to the FMB. These are used to develop the Navy budget estimate submissions (BES) sent upward in the financial management chain to the FMB in the Pentagon. CPF monitors and provides feedback, and adjusts its budget submissions throughout DoN and Office of the Secretary of Defense (OSD). It is then up to CPF to execute that budget in coordination with its activities and the FMB.

It must be kept in mind that while these budgets are developed, the PRESBUD for the upcoming FY is being refined, and the current FYs budget is being executed, thus as explained in chapter 3, there are actually three budget processes performed in any given year. For example, the Navy will execute the FY 2003 budget at the same time it is preparing the FY 2005 PRESBUD. Meanwhile Congress is debating and enacting the FY 2004 budget and also may be enacting a defense supplemental bill. Ripple effects result from what happens in the execution year, from how Congress treats the DoD appropriation bill and any supplemental. The ripples influence both budget and program decisions for subsequent years. At the BSO level, programmers, budget, and execution personnel are all actively engaged in all three budget processes and a decision on an item or program in 1 year usually has consequences for the item in the other 2 years.

PPBES IMPLEMENTATION AT PACIFIC FLEET HEADQUARTERS

As a major Navy BSO and second echelon command, CPF participates in every phase of the PPBES cycle. In the planning phase, it participates by developing IWAR and in conducting IWAR analysis in integrated process teams (IPT). The results are incorporated in the chief of naval operations program assessment memorandum (CPAM) that is the foundation for developing the Navy program.

During the programming phase, CPF develops fleet resource requirements and provides monetary programming inputs to resource sponsors in the development of the Navy program. In the budgeting phase, CPF develops control numbers for its subclaimants based on tentative POM/ budget control numbers, consolidates subclaimant budget submissions, and provides FMB with the BES. CPF then monitors the DoN program/ budget review process at both the DoN and OSD levels ensuring any CPF claimancy issues are addressed and resolved. It can do this because it is a major BSO; a type command such as AIRPAC might follow its own issues through OSD review, but it could not actively engage with OSD; it has to rely upon CPF or the FMB to complete this task. CPF revises its budget throughout the process until it is submitted as input to the PRESBUD.

CPF is also a major player in budget execution. It develops execution control numbers and plans for activities, monitors funding obligation rates, conducts a midyear review, oversees the end of the execution year "sweep-up" of funds, and allocates supplementary funding. The following section will analyzes CPF integration into the Navy PPBES cycle according to a number of dimensions including planning, integrated warfare architectures, programming, and budgeting.

Planning

CPF provides planning inputs including IWAR focus area recommendations and IWAR focus area analysis to the Office of the Chief of Naval Operations (CNO OPNAV N81) via commander, United States Fleet Forces Command (USPAC or CFFC). CFFC (commander fleet forces command) was established during CNO realignment in 2002 and operates in coordination with the commander in chief, U.S. Atlantic Fleet (LANTFLT/CLF). CFFC is responsible for overall coordination, establishment, and implementation of integrated requirements and policies for manning, equipping and training Atlantic and Pacific Fleet units during inter-deployment training cycle (FMB Presentation, April 2002, Slide 12). CFFC consolidates inputs from both CPF and CLF before forwarding them to N81. CFFC also coordinates fleet inputs during the programming phase of the PPBES cycle.

Integrated Warfare Architectures

Primary participation by CPF in the planning phase of PPBES is via the IWAR process. In May, CPF receives a data call from the CNO staff (OPNAV N81 via CFFC) requesting priorities for IWAR analysis for the current year. CPF develops areas for consideration in the IWAR process based on CNO guidance, Theater concerns, and previous IWAR (Overby, 2002). CPF develops their priorities, focusing on areas where they feel either capabilities are nonexistent or inadequate to meet threats, or resources are misaligned causing incomplete capabilities. After CPF forwards its desired focus areas to CFFC for consolidation, N81 reviews and develops Navy IWAR focus areas. In 2002, CPF developed proposed focus areas in four of the five IWAR war-fighting areas: air dominance, sea dominance, power projection, and information superiority and sensors and six of the seven IWAR support areas: infrastructure, readiness, sustainability, training, manpower/personnel, training/education, and force structure.

In June, N81 determines yearly Navy IWAR focus areas based on responses to its data call. CPF will then participate in IWAR analysis via the IPT. The IPT is comprised of program managers and requirements officers from COMPACFLT, COMLANTFLT, COMUSNAVEUR, and the OPNAV staff. Most IWAR interaction from this point is Web-enabled allowing rapid interaction among IWAR focus area points of contact on the CPF, CLF, and NAVEUR (Europe) staffs. IWAR analysis continues through October. CPF, N80 coordinates CPF interaction during the IWAR analysis phase. This concludes when the focus area study results are briefed to resource sponsors, assistant secretaries of the Navy, SECNAV, CNO/Vice CNO, and fleet commanders in October. Following IWAR briefs, CPAM are developed and consolidated to become the summary CPAM briefed to fleet commanders in February of the following year. This CPAM then becomes the Navy input to the PPBES programming phase.

Programming

Programming at CPF is mostly concerned with O&M accounts. The mechanism for providing program inputs to OPNAV is via CPs. These were previously known as baseline assessment memoranda (BAM). In October 2003, OPNAV N801E distributed the first CP serial that changed program assessments from BAM to CP. A CP is an identification and critical evaluation of a baseline valid requirement for specific programs and estimates the funding necessary to achieve 100% of the valid requirements. For example, CP inputs are developed for CPF for the flying hour program. The CP will provide "an assessment of the flying hours required to meet stipulated readiness goals and the resultant flying hour support requirement necessary to support those hours" (OPNAV, 2002b).

The CPF initial formal interaction with the CP process is the draft CP procedures memorandum serial that they receive from OPNAV N80 via CFFC in September. In 2002, CPF received a draft BAM serial in September for review. The draft was a rough copy of the BAM guidance for program review (PR) 05. CPF makes corrections to the draft as it determines necessary and submits its corrected copy back to CFFC.

Adjustments are made to the draft copy and the actual guidance is then distributed. From 2002 forward, guidance will address the CP. The CP procedures memorandum specifies assessments that will be conducted; assigns assessment sponsors who will oversee the assessments; provides guidance for conducting the assessments; and delineates responsibilities for producing the CP. For PR 2005 sponsor responsibilities included:

- Assessment sponsors: Assessment sponsors should prepare the assessments listed in enclosures (1) and (2) and deliver them to OPNAV (N81) no later than 31 January 2003. Use the FY 2004 Navy BES resource allocation display as the resource baseline.
- Resource sponsors (N-41): Work closely with assessment sponsors for CP development. Satisfy CP requirements wherever possible when developing PR 2005 SPPs (sponsor program proposals). Clearly document compliance and explain areas where the assessed requirement has not been met.
- Assessment division (N81): N81 will participate in the development of CP to ensure that each requirement is valid based on analytically sound methodology
- Claimants: Participate in IPT and work groups as requested by assessment sponsors to support the respective assessments
- Fleets: CFFC is the fleet requirements integrator and the conduit of Fleet requirements to OPNAV. As such, fleet (PAC/LANT FLT/ NAVEUR/Lead TYCOMs) and NAVSEA, where applicable (i.e., as operator of government-owned shipyards) input will be coordinated, consolidated, and submitted via CFFC CP points of contact to assessment sponsors. The CP procedures memorandum (OPNAV, 2002a) also details factors to consider when preparing CPs for submission to N81. The procedures memorandum does not tell assessment sponsors how to conduct their assessments, but it does request that assessment sponsors provide a detailed description of the methodology that was used to conduct the assessment. This means that there is no standard method for conducting the CP process and different programs develop program resource requirements using differing methodologies. PR 2005 CP and assessment sponsors are detailed in Table 10.1.

Based on requirements developed for its claimancy, CPF program managers provide assessment sponsors with the dollar amount their programs cost to meet 100% of the valid requirements. CP prepared by the assessment sponsors provide N81 with the cost to fund 100% of the validated requirement being assessed across the Navy. Later, based on resources actually available, programs will be funded at a percentage of the 100% requirement. Actual program/budget funding level target numbers are provided by N80 in its program guidance.

Table 10.1.   CPF Topics and Assessment Sponsors

PR-05 CP Topics	Sponsor
Total force management	N1
Ashore readiness	N4
Contingency engineering (naval construction force	N4
Maritime readiness/sustainment	N4
Aviation readiness/sustainment	N4
Conventional ordnance readiness/sustainment	N4
Antiterrorism/force protection	N3/N5
Fleet readiness	N4
NMCI	N6
Individual training and education	N1

Source: OPNAV, 2002b.

For PR 2005, assessment sponsors were required to submit their CP to OPNAV N81 no later than January 31, 2003. Based on the CP, IWAR, and CPAM, N80 develops its programming guidance. Programming guidance provides preliminary advice to resource sponsors for developing their SPP. If N80 guidance is promulgated prior to the DoD level defense planning guidance (DPG), N80 will revise its guidance based on the DPG. CPF monitors program development as the resource sponsors adjust funding to meet the program. CPF program managers decide if changes being made to funding within resource sponsors are major enough to attempt to justify adding money back in during the programming phase.

Prior to 2001, when the POM was almost complete, CPF would participate in the "end game." End game was a small window of opportunity before the POM was completed at the Navy Pentagon level ("locked") where CPF could address the most important major programming issues that could not be resolved in their favor. The commander would either transmit a message or even contact the CNO directly regarding his last major issues. According to the CPF Comptroller, the general message sent was, "Based upon the SPPs, what is in the programming data base at this point in time—we cannot live without these programming changes. This does not meet our requirements" (Morris, 2002).

The process prior to 2001 was sequential and well understood. After the end game, the POM was locked and budget control numbers were developed by FMB and issued to BSOs. The 2002 PPBES cycle that developed the POM for FY 2004 and budgets for FY 2004 and FY 2005 became a pooled process as the budget was prepared and the program was revised concurrently. In actual implementation at the time when writing was completed on this book, the new PPBES process was not implemented satisfactorily from the view of the military departments and services. For more on implementation of the modified PPBES system in practice, see the conclusions to this chapter.

Budgeting

Throughout the budget year, the CPF budget staff simultaneously coordinates three budgets, the Department of the Navy (DoN, OSD, and PRESBUD) as well as assisting with the execution of the current year budget. The process will be described by developing a timeline starting with the POM/PR budget submission.

Budget Guidance

The DoN POM/PR budget process at CPF begins with receipt of the budget guidance memorandum (BGM), serial 1 from FMB. BGM serial 1 outlines the budget calendar, required budget exhibit submissions, and pricing factors for inflation and foreign currency exchange rates. CPF uses FMB guidance as a foundation for preparing their guidance to their activities, adding any claimancy specific guidance. As FMB issues further serials to its budget guidance, CPF will also update their guidance as necessary if the updates affect their activities. For example, CPF budget guidance 02-01A (first update) provided updated inflation rates for activities to apply in the submission of their budget exhibits.

Budget Preparation

As the BSO for its claimancy, CPF receives control numbers from FMB based on the T-POM (2001 and subsequent), and develops and distributes control numbers for its activities (regions, TYCOMs, numbered Fleets), that the activities use to build their portions of the overall CPF budget upon.

CPF develops controls based on T-POM (program controls developed by FMB) and in coordination with their in-house requirements, program managers, and activity level budget and programmer inputs. Controls are based on budget models for certain major programs such as surface ship operations and the flying hour program (FHP), and on historical data and obligation rates for other major programs such as BOS and are also adjusted for pricing and known new requirement changes.

Additionally, CNO goals are established for funding ship operations, the FHP, ship maintenance and base operations as a percentage of the 100% valid requirement identified in the programming phase. Funding for large portions of FHP and ship operations are "fenced" (protected) and CPF must solicit FMB approval before attempting to modify funding levels for those programs. Other items such as civilian personnel and utilities (within BOS), although not fenced are not discretionary and must be funded at 100% of requirement. CPF must determine funding levels for those programs that are discretionary and are neither fenced nor require 100% funding. The sustainability, restoration, and maintenance (SRM) account is an example of a discretionary account.

Budget Review

CPF budget analysts review budget exhibits provided by activities with their programmers and requirements staff as necessary to ensure that activities are preparing budgets based on valid requirements. Although analysts are intimately familiar with their programs, the program managers have been working directly with their fleet counterparts and have developed points of contact throughout the fleet during the planning and programming phases of the POM or PR for which the budget is being developed. As activities submit budget exhibits, analysts pay particular attention to items that show a marked increase or "spike" over previous years' funding levels. They also ensure that exhibits are fully justified by requirements developed throughout the planning phase and that exhibits are detailed enough to support the resources requested (Shishido, 2002). When spikes occur or the exhibits are not detailed enough, analysts examine circumstances which may have led to the spike to determine if the increase is justified or request more detail from activities.

Once fleet inputs are received, analyzed, and revised based on analyst review, CPF combines the inputs and builds its budget submission in late May. Budget submissions will balance to controls given by FMB and are submitted using pricing factors, exchange rates, and exhibits required by the budget guidance. Once submitted, the request undergoes the concurrent program/budget review and issue/comment process.

Issue/Comment Phase

In June, FMB analysts review the budget submissions and generate issue papers. Issue papers are posted to the Navy Headquarters Budgeting System Web site and BSOs are notified by FMB that there is an issue that affects them that requires comment. CPF may either comment on the issue in an attempt to restore funding, concur with the issue, or simply choose not to comment on the issue. If CPF does not comment, the issue is resolved as FMB chooses. If comments are generated in an attempt to restore or justify resources, the issue and comments will undergo FMB analyst, department head, and division director reviews. Issues that remain unresolved at lower levels are addressed at Program Budget Coordination Group reviews that take place in July and August for major program/budget issues. Prior to 2001, this was the "mark/reclamma" phase of the DoN budget review. As DoN reviews progress and issues are resolved, CPF updates its budget exhibits for submission to OSD. The OSD budget submission incorporates any program and budget issues that affected CPF funding and programs during the DoN review.

Budget Sweep-Up and Certified Obligations

From the end of August through October, CPF budget analysts work in conjunction with the execution department to help conduct the "end of year sweep-up" of funds and to certify the obligations against the execution budget. The end of year sweep-up is the process of obligating execution year funds before the end of the FY on September 30. Analysts coordinate with both CPF execution analysts and reporting activities to ensure that either (1) an activity can execute the remainder of its operating budget prior to the end of the FY, (2) an activity will not run out of operating funds prior to the end of the FY, or (3) have excess operating funds at the end of the FY. If an activity has unobligated money at year-end that cannot be executed, these dollars will be transferred to other activities that are in danger of running out of operating funding prior to the end of the FY.

Certifying obligations is the process of comparing the budget to the actual obligations that were executed during the execution year. Any variances are reported to FMB for analysis and possible adjustments to future years' programs and budgets.

OSD Review and PBDs

In September, OSD programmers and analysts review service component program and budget submissions and release PBDs (program budget decisions) that are "marks" against the DoN budget. CPF monitors the PBDs to ensure that they prepare reclammas when necessary in an attempt to restore funding. Some PBDs are simply informational and address such issues as working capital rates or foreign currency adjustments. Although changes in these rates may negatively affect CPF funding levels, they are "fact of life" adjustments that must be incorporated into their budget. The PBD process lasts until late November to early December. At the same time, DoD receives undistributed congressional marks that act the same way as a DoN issue or mark based on pricing issues, specific increases and general provisions in the language. If Congress eliminates or realigns funding however, CPF has no right of reclama, but must absorb the loss.

The President's Budget and Midyear Review

In December, after PBDs have been adjudicated, CPF is issued new control numbers to be used in developing their PRESBUD submission. After OSD and OMB (Office of Management and Budget) review of the component service budget submissions is finalized, the President submits his budget to Congress on the first Monday of February. Over the next 9 months, the PRESBUD will be closely scrutinized and serve as the basis for the congressional authorization and appropriations legislation.

In March, after their PRESBUD submission, CPF receives guidance for conducting its midyear review of the execution year appropriation. Midyear review guidance provides specific guidance to major claimants for exhibit preparation and other submissions such as current unfunded requirements. The guidance also highlights the scarcity of funds available to solve problems identified at midyear review and directs claimants to "craft their submissions of unfunded requirements to reflect only those that are most critical to mission accomplishment" (OPNAV, 2002a). When the midyear review is complete, the budget cycle begins again for the next POM/PR year with the tentative POM numbers full a full review cycle or the POM Review process for the short cycle.

PPBS has been used in DoD for over 40 years to plan, program, prepare, and execute budgets, but it is constantly under change (see conclusions and chapter 4). The system is in a continuous evolution of innovation and change. The year 2002 was particularly interesting for PPBES participants. Disturbed by the length of time that system took to produce decisions, Secretary of Defense Donald Rumsfeld issued a memorandum in 2001 that directed a concurrent program/budget process. This was a major change for the system. In addition, BAM were replaced by CPs. Then there are also routine budget issues that occur from year to year and emergent or "pop-up" issues that occur throughout the budget year. This chapter will discuss some of the major routine and emergent issues at CPF as well as the new concurrent program/budget process from the CPF perspective. We begin with routine issues.

BUDGET SYSTEM ISSUES

Players Must Anticipate Change

The PPBES process is cyclical with different elements taking place throughout each year. Even though changes were made to the process during 2001 and 2002, inputs although extremely intricate, are still reiterative in nature from year to year and unless major changes occur, do not change significantly. Even with changes occurring, in many ways planners, programmers and budgeters are on "auto-pilot" and wait for guidance from the Pentagon to provide specific guidance and deadlines for required inputs. They deal with complexity by anticipating what will need to be done, often starting before formal guidance has been received. This allows them to produce a quality product under tight deadlines, or even when deadlines seemingly are in conflict. For example, the POM is supposed to be done before the budget, but historically for CPF the POM has been completed after their budget has been sent to the FMB. CPF budget staff anticipates the POM decisions and rely on the central budget office to make changes when a POM decision contradicts what they anticipated.

There is also anticipatory behavior undertaken to meet all the different guidances that routinely are issued to direct and coordinate the PPBES process. During any given year, the Navy issues guidance that affects every phase of the PPBES cycle. During the planning phase, OPNAV N81 issues a call for IWAR focus areas based on theater concerns, CNO guidance, and previous IWAR and develops specific IWAR to be analyzed. During the programming phase, the Navy issues BAM or CP guidance for developing baseline program requirements that will be used to build the Navy's program. The budgeting phase of PPBES begins with the FMB budget guidance and is adjusted based on follow-on serials.

The Navy issues guidance for virtually every stage of the PPBES process. However, due to the reiterative nature of the process and the tendency of participants to anticipate what will be needed, much of the work addressed in the guidance has either been done prior to guidance being issued or has been started and is in progress. Participants do this because they count on the products they are expected to produce not changing dramatically from one year to the next and they know if they wait until the official guidances there will not be enough time to meet the deadlines or to meet the deadlines and still produce a product that will allow them to compete advantageously for programs and funds. They know, for example, that a program that is "under-justified" or seems carelessly written begs for a cut. They also know that massive data calls drive out analysis and that it is best to anticipate and start early. They also know that the planning, programming and budget input requirements simply do not change dramatically enough from year to year to wait for guidance to be issued. Even in 2002, with the change from BAM to CP, CPF programmers already were engaged in developing the data required to provide the assessment sponsors with CPF program guidance.

Participants deal with the pressure to complete complex tasks under limited time boundaries by anticipating. This was a common theme throughout discussions with CPF planning, programming and budgeting staff. According to the CPF assistant fleet programmer, "This is typical—

you get the directive after most of the work [on the BAM/CP submission] has been done" (Catton, 2002).

In fact, for the PR 2005 assessment, CPF had not received final BAM guidance as of late October and when guidance arrived, it directed and provided guidance on the preparation of the CP instead of the BAM. At the top of the hierarchy in Washington, those who write guidance do not purposely design in short time cycles, but in effect the guidance goes through drafts and must be signed off in various offices and then they trickle down to the fleet, examined and interpreted as they go. When they finally reach the fleet analyst level, the analyst is usually caught with a short time deadline. Thus, they count on things staying about the same even if their titles change and they begin data collection and analysis processes as they did in the past.

The views expressed by the CPF assistant fleet programmer were reinforced by both the CPF IWAR coordinator and budget department Head. Referencing the IWAR data call, CPF IWAR coordinator stated, "It's standard, so we start generating information prior to the data call. Its on semi auto-pilot" (Pacific Fleet, 2002b). With regard to the FMB budget guidance, the CPF budget head stated,

> You don't wait for it, you're getting prepared. And we know how they have [the regions and type commands] spent their money in the last few years. They're not going to make any major changes unless a region goes totally BOS contracted or something. Other than that, we know where they're going to spend their money. We could do their budget for them. (Pacific Fleet, 2002b)

Readiness Trumps Support Accounts

In its OMN budget, CPF supports both readiness programs and support programs. While both these programs seem equally discretionary, readiness is seen as such a high priority that it is almost a mandatory account. The importance of the readiness accounts is supported by directives from CNO and from Congress requiring their submission. When readiness accounts are short, the support accounts end up as a "billpayer." No matter how fully readiness seems funded in the current budget, operating tempo always seems higher than budgeted and funds for maintaining readiness always seem short. The support accounts help pay the bill. While this issue can be anticipated, it is not easy to resolve.

Readiness accounts are those accounts that actually support the warfighters when conducting interdeployment training cycle training and deployed operations in support of PACOM operations and include the FHP, ship operations, and ship maintenance. Support programs such as

BOS provide resources for regions to fund base operations, and support for operating units. BOS consists of funding for shore activities that support ship, aviation, combat operations, and weapons support forces. Base support includes port and airfield operations; operation of utility systems; public work services; base administration; supply operations; and base services such as transportation; environmental and hazardous waste management; security; personnel support functions; bachelor quarters operations; morale welfare and recreation operations; and disability compensation (Pacific Fleet, 2002c).

While Navy programming develops requirements and resource allocations for both types of programs based on 100% valid fleet requirements, a major portion of program funding for readiness accounts is protected by congressionally imposed restrictions. Money cannot be removed from them in excess of $15,000,000 (Navy-wide) without congressional approval. Consequently, it may be said that these accounts are not discretionary. Major readiness accounts at CPF include the FHP and ship operations accounts. In addition, the CNO provides goals for the percentage of the valid fleet requirement to be funded for these accounts. For the FY 2003 budget, goals for programs within the FHP ranged from 89% for TACAIR (tactical air forces) to 92% for fleet readiness squadrons. These goals were met in the FY 2003 FHP budget (DoN, 2002c, pp. 2-12). For FY 2003, CPF execution controls equaled 94.1% of requirement for the FHP and 95.2% of requirement for surface ship operations.

These accounts are developed using metrics based on operating characteristics of the various platforms within them. As an example, OPNAV develops the operational plan 20 (OP-20), the primary FHP budget exhibit and directive. To develop the OP-20, OPNAV works closely with major claimants such as CPF and coordinates with the TYCOM, commander U.S. Naval Air Forces Pacific (CNAP/AIRPAC) (Navy Budget Highlights, 2002).

CNAP assists in developing the OP-20 by providing FHP cost inputs to N78 via its flying hour cost reports (FHCR) that consolidate FHP costs provided by squadrons and air stations on a monthly basis throughout the year via their budget (operating targets) budget OPTAR [operating/operational target] report (BOR). Factors reported in this report include the number of and type, model, and series of aircraft assigned, funding obligation totals, flight hours flown for the month, and the total gallons and type of fuel consumed for the month and FY to date. These data from CNAP are input by OPNAV into its Flying Hour Projection System (FHPS). This projection system relates annual budgeted flying hours to forecasted flying hour costs and produces future year projections (Phillips, 2001). Based on readiness, training, operational capability requirements, available resources, and programming guidance requirements,

CNAP distributes FHP funding among the various T/M/S commanders. Since this is money is dedicated to squadrons operating types of aircraft, the money is said to be fenced, to be used by that or those squadrons only. However, a portion of the moneys within the FHP is not fenced. The flying hours other (FO) account provides funding for temporary duty, training under instruction, support equipment, and so forth. It is developed by averaging the previous 3 years' budgets and is not based on metrics, as are other FHP accounts. Essentially, it is a support account within the FHP and may itself be raided for a higher priority use.

As with the FO account, many support programs such as BOS have no model for building major portions of their budget submissions, but rely on previous budget funding and execution levels (Catton, 2002). Because there are no congressional restrictions and because they are not directly supporting readiness (buying fuel, spent on maintenance of aircraft, etc.), and have no established CNO funding levels, these accounts are inherently underfunded and become "bill-payers" for emergent unfunded requirements. Senior Navy leadership is aware of this issue. Two IWAR focus areas addressed BOS funding for 2002, Shore Infrastructure Recapitalization and BOS Readiness Metrics Review. FO also becomes a bill-payer account, because even though it is a part of the total FHP, it is outside the congressional fence. Fleet comptrollers tend to see these accounts as "free-money" for meeting emergent funding issues during budget execution. In this contest, formulae trump averages. To put it another way, a formula that relies on a rolling 3-year historical average is not as powerful as a formula that collects projected costs to fly an aircraft. These projected costs combine historical costs, engineering projections, and contracted prices for consumables such as fuel, and parts.

The bill-payer issue is a matter of concern throughout all Navy fleets. While BOS is a major bill-payer, CNAP is concerned about funding being re-programmed from FO by CPF to support emergent requirements (Phillips, 2001). As an example, when the Navy started to convert to the common access card (new ID card) FMB funded the conversion entirely from BOS. CPF was concerned that this would have too big of an impact on BOS and "taxed" accounts across the board to fund the conversion. FO, as a discretionary account within the FHP paid its fair-share of the tax (Scott, 2002). These taxes leave the TYCOM, and every other activity that was taxed with some other unfunded needs within their budgets.

The assistant fleet programmer, the head of fleet budgeting, the BOS budget analyst, and the aviation budget analyst all identified the facilities SRM account as a prime bill-payer. These comments are supported by CPF execution control levels for FY 2003 and midyear review requests and subsequent supplemental funding for FY 2002. Compared to the FHP and ship operations funding levels of 94.1% and 94.2% of requirement respectively, SRM was only funded at 54.8% of the CPF requirement (Pacific Fleet, 2002c). Table 10.2 displays CPF FY 2003 execution controls by account, for example, readiness and support, and Table 10.3 shows CPF midyear review priorities and request for supplemental funding for FY 2002 and Table 10.4 is the CPF actual supplemental funding received for FY 2002.

There is a discrepancy between CPF priorities and the supplemental funding received. CPF priorities listed BOS and SRM, which was funded at a fraction of requirement for FY 2003, as part of their first priority and as their overall second and third priorities. However, CPF received no additional funding for these bill payer programs except to fund additional force protection requirements within the region because of the September 11, 2001 terrorist attacks.

While these accounts may be safely underfunded in the short run, in the long run inadequate funding will lead to deteriorating infrastructure including runways, hangars, and piers that support operations and may eventually have a negative impact on overall fleet readiness and result in increased costs to upgrade more severely degraded facilities.

Concurrent Program and Budget Review is New

As noted, in 2002 the PPBES process was changed to run program and budget review concurrently. Prior to August 2001, the military departments and services developed and submitted their POMs to OSD for review in May. The services would then start to build their BES based on the POM. In 2002, military departments and services were required to submit both their BES and their POM to OSD simultaneously in late August. Because of this change, the Navy budget submissions were developed based on tentative POM (T-POM) control numbers issued in late May. We view this change from the operator perspective with CPF.

Table 10.2. CPF Execution Controls for Selected Accounts

	FY 2003 Control	FY 2003 Requirement	FY 2003 Shortfall	Percent Funded
Air Ops.	2,119	2,252	−133	94.1%
Ship Ops.	1,231	1,292	−61	89.1%
OBOS	1,144	1,281	−137	89.1%
SRM	556	1,015	−459	54.8%

Source: Pacific Fleet, 2002c.

Table 10.3.   CPF FY 2002 Prioritized Midyear Review Submission

O&MN Priority #	Issue Title		Unfunded Amount ($000)
	Prioritized List of Critical Unfunded Requirements		
1	Cost of war		307,238
	Force protection	11,759	
	SRM program	6,665	
	BOS program	10,600	
	Ship maintenance	214,237	
	Ship operations	53,000	
	Combat support	10,977	
2	SRM baseline program		55,000
3			13,371
4	Ship maintenance baseline program		44,346
5	Combat support baseline program		1,064
6	PREPO		4,954
Total			425,975

Source: Pacific Fleet, 2002c.

Table 10.4.   CPF DERF and Supplemental Funding for FY 2002

Categories	DERF Funds	Supplemental
DERF and Supplemental Funding		
Enhanced Force Protection	**36**	–
Antiterrorism (AT) force protection (FP) task force findings	14	–
Fund base operations for AT/FP/force protection modernization	22	–
Increased Worldwide Posture	**327**	**250**
Increase in flying hours	17	143
Increase in steaming days	108	48
Combat support force operations	24	3
Communications	5	2
Ship and AC maintenance	173	54
Initial Crisis Response	**7**	–
1. NCW	7	–
Total	**370**	**250**

Source: Pacific Fleet, 2002c.

A CPF programming official explained the advantages of the change as follows:

> It eliminates unnecessary duplication of effort. Prior to the change, the POM would be finished in May. Then, budgets would be prepared for OSD review. Emerging issues could cause services to change the program while developing the budget…. Gives the services longer to finish the POM while incorporating emergent budget issues. Services can re-visit the program based on budget issues…. Prior to the change, OPNAV N80 would finish the program and then it was "out of their hands…. The new process leads to more cooperation between programmers and budgeters…. I think it [the new process] gives claimants more input into the program…. Take Information technology—if many claimants have issues with funding, it can become a major issue. Now you can revisit the POM; before you couldn't. Claimants can also say they can't execute the program as funded by controls. (Allen, 2002)

When the process ran consecutively, coordination was more difficult. If the POM was not completed until after CPF finished its budget, how could the POM guide or control the budget? Neither N-80 nor anyone else could "enforce" POM numbers. Budget staff believed the POM process was flawed, that it operated on a "fair share" principle where priorities were not realistic. They were not confident of the POM process that seemed to "give everybody something." Budget staff had to "fix" the POM in the budget, at least for the budget year. This had consequences for future POM planning that programmers often objected to, because they did not support having the budget drive the POM, especially because it happened virtually every year after 1990 when budget staff were forced to make cuts and corrections irrespective of the POM. For these reasons and others, friction arose between the budget and POM communities. The concurrent POM-budget process may ameliorate some of this tension. However, by 2003 this had not occurred.

At CPF, comments by the deputy comptroller were reinforced by the head of the budget department and the comptroller. Both agreed that the change could provide more coordination between programmers and budgeters. According to the CPF comptroller,

> Secretary Rumsfeld has been talking about transformation and new ways of doing things. He's saying we can't continue to think of things the way we always have. To me, this is transformation applied to resource allocation, programming, budgeting, and requirements determination. They're (OSD) looking to streamline things and make them more efficient, to eliminate redundancies, and to ask questions just once instead of over two different processes…. When you have a concurrent process, you're forced to work together. (Morris, 2002)

The budget department head echoed these comments:

> I think the intent was to streamline the process so that there's not so much flux. Before, once the program locked, you had to wait a whole cycle or try to fix the program in the budget. What we're trying to do instead of trying to fix it (the program) in the budget is to make the program executable in the programming stage and only have to concern ourselves with pricing and pop-up issues in the budget ... so there also were not as many required exhibits. (Catton, 2002)

While there was agreement among CPF staff on the reasoning behind the change to a concurrent program/budget process, there was some disagreement about the new process. Observers noted there was very little direction given as to how the new process was to be implemented. The only guidance provided initially was a one-page memo from the secretary of the defense (SECDEF). The CPF comptroller indicated that the change to a process that had been conducted in much the same way for many years caused anxiety among personnel within the CPF programming and budgeting organization, but he, "did not want to be too quick to jump to conclusions" (Morris, 2002). Comments from other staff members were not as encouraging. One analyst described the process as "chaos." As late as late June, 2002, after their original POM 2004 budget had been submitted to FMB based on draft POM control numbers, there were still questions among CPF staff on how the new process would work.

In the end, CPF was able to work through issues related to the concurrent program/budget and submitted their DoN FY 2003 budget on-time based on both programming and budgeting changes that occurred as a result of issue papers, comments and reviews at the FMB, N80 and other levels.

It is useful to reflect on how these relationships change because of the change in the PPBES process. According to Thompson (1967), and Nadler and Tushman (1988), there are three types of interdependence in complex organizations:

- Pooled interdependence
- Sequential interdependence
- Reciprocal interdependence

Pooled interdependence occurs when separate units operate independently but are part of the same organization and share certain scarce resources. An example is a bank with several branches. Individual branches function independently of each other but share certain resources of the main corporate entity such as advertising or marketing. The branch banks do not depend on each other for their functioning.

Sequential interdependence occurs when a unit or task downstream of another depends on the prior unit's output or task completion. Sequential interdependence demands a greater degree of coordination than pooled functions. The work of one unit can be affected by upstream units. Coordination must exist to ensure that workflows remain constant. An example of sequential interdependence would be an oil company. First, oil must be extracted from the ground, then it must be refined into different products, then it is shipped to customers. One task cannot be completed prior to the previous tasks and coordination between tasks must exist to ensure that workflow remains constant. Resources must be expended to maintain such coordination. This type of dependence characterized the pre-2001 separate POM-budget review process.

Under reciprocal interdependence, work groups must work continuously with other units in the production of common products. Reciprocal interdependence imposes substantial problem solving requirements between units because no single unit can accomplish its task without the active contribution of other units. The new concurrent POM-budget review process is an example of reciprocal interdependence. As tasks become more interdependent, the amount of coordination and communication between tasks increases. Reciprocal interdependence represents the highest degree of interdependence and therefore the highest degree of required communication and coordination between units.

As noted, prior to the 2001 POM-budget review change, the system operated during the preparation phase of PPBES with a high degree of sequential interdependence between programming and budgeting where each successive process (programming then budgeting) was dependent on the one prior to it. Once the input was received, downstream tasks were not supposed to have an effect on the output of the previous phase of action, that is, budget changes were not supposed to cause changes in the existing POM. However, this was not an accurate description of what actually happened as we have demonstrated. The 2001 change to establish a concurrent program/budget review process recognized that in reality there is a high degree of reciprocal interdependence between the two functions. Actions taken during the budgeting phase of PPBES have significant impact on the upstream process of programming.

Personnel at CPF long recognized the reciprocal nature of planning and budgeting. Had the threat and budgetary environment been stable, perhaps no PPBES changes would have been necessary. However, neither condition held. The threat environment was not stable and the budgetary environment fluctuated during the 1990s and with the advent of the war on terrorism as operating tempo remained high in response to contingencies. As a result, there was friction throughout the system, most significantly at the top where the old PPB system was not producing the outputs desired by Secretary Rumsfeld quickly enough. The problems with PPBS that irritated Secretary Rumsfeld had been identified by critics for decades but it took a determined SECDEF to make changes (Jones & Bixler, 1992, p. 32; Puritano, 1981).

In conclusion, whatever the structure of the formal system, a tremendous amount of informal coordination and communication is necessary to make programming and budgeting work effectively at all levels of DoD. Much of this is informal contact; participants allude to "common-sense" guidelines that they develop as they work in the system. These contacts take place day-in and day-out as a part of the work between planning, programming, and budgeting personnel at CPF, and between CPF and their counterparts in the activities that report to them and in the Pentagon.

Experienced participants believe that those who do well in resource decision making are highly skilled at working in formal relationships. While PPBES is a highly formalized process, the day-to-day interaction between players in the process is anything but formal. Analysts at various levels of command coordinate with each other on a daily basis, not just when a required submission is due. They develop a deep understanding of issues that affect both subordinate and senior personnel in the process. This helps them understand which office to engage and with whom they should speak when an issue arises.

Related to informal networks is the fact that different tactics for programming and budgeting in field commands are required when resources are limited. These tactics allow analysts to attempt to maximize their resources without drawing unnecessary attention to the particular account or line item in question. Analysts also learn where to look for resources that players at other echelons of command may be trying to hide. Participants learn tricks of the trade to avoiding the appearance of large spikes in funding, to account carefully for increases in accounts scrutinized closely in prior years, or burying resource requirements within accounts that are general in nature to make them less easy to differentiate. These tactics are the product of experience and cultivation of informal knowledge of the process.

MIDYEAR BUDGET REVIEW

In the section that follows, we show how midyear budget review in the Navy at PACFLT was used to help build the budget base and prepare for the next DoN/DoD and presidential budget submission. As this was happening, Congress was in the middle of its debate on the budget resolution, and it was too early to tell how the appropriation bill would be treated. The midyear review is basically a process of adjusting current year budget execution to fact of life occurrences, changes in operating tempo, emergent equipment and repair costs, and other changes in the pace of program and spending that were not anticipated when the budget was submitted. Midyear review usually involves finding a little bit of money in one account to help another account or claimant and so forth. Claimants are expected to solve their own problems, but sometimes the comptroller has held money back or knows where another claimant is underexecuting and has some money that may be reallocated. Midyear review also is used to educate claimants about major issues they will face in the summer budget review process. Retrenchment is an always an issue looming on the horizon. Moreover, at the operating level, resources are always scarce compared to the tasks expected. At midyear, we find budget analysts and claimants at work figuring out what is happening in the current year to their budgets while also working on the budget request that will be submitted to the FMB later in the summer.

Much of defense budgeting is unfamiliar to budget practitioners, filled as it is with acronyms, unfamiliar chains of command, and relatively sophisticated decision techniques, particularly in the procurement accounts. One tends to forget that DoD also has a large operating budget (O&M) for supplies and services and that these operating budgets basically are built and reviewed like those Wildavsky (1964, 1988) described in which he captured the essence of the dialogue that takes place between those who budget and those who review budgets (the "spenders" and the "cutters"). His observation that it takes more than numbers to be successful in the budget game is still true everywhere, including defense; his identification of confidence, competence, and clientele as pervasive themes in the budget process is still persuasive, and his identification of contingent strategies, for example, making a profit or workload, as a means of articulating need in the budget dialogue, is still eminently useful. Wildavsky (1964) noted, "several informants put it in almost identical words, 'It's not what's in your estimates, but how good a politician you are that matters' " (pp. 64-65). Being a good politician according to Wildavsky requires essentially three things: cultivation of an active clientele, the development of confidence among other governmental officials and skill in following strategies that exploit one's opportunities to the maximum. Doing good work is viewed as part of being a good politician (Wildavsky, 1964, pp. 64-65). Wildavsky's focus may have been primarily at the intersection of administrative and legislative worlds, but it also works within the administrative world where complex decisions are taken based on expert knowledge and trust. The snapshot of claimant budget review that follows reveals the components of Wildavsky's analysis at work.

Background

What follows is an analysis of midyear budget review hearings that took place at the comptroller's level in the CPF. The analysis is based on the 1990 midyear review (McCaffery, 1994) and updated with supplementary interviews performed in 2002 and related materials. Our experience indicates that the budget review issues discussed are so fundamental as to be timeless, just as is Wildavsky's (1964) portrayal of the interaction between legislators and executive budget makers. For purposes of clarity, we have updated organization names when necessary and FY sequences in the examples.

The purpose of these midyear review hearings is to review the current budget and establish good control numbers for the budget currently in Congress, since these will serve as the target numbers in the budget base. The hearings also provide a first cut at the following year's budget, which will be built in June at CPF as claimants discuss current shortfalls and future needs. This scenario is acted out in anticipation of increasing resource scarcity. By May, Congress has begun its debate on defense funding and defense budget additions or reductions beyond what the President had suggested. As a result of discussion over how big these add-ons or cuts might be, expectations at the CPF comptroller level vary, but the outcome is that most expect future years will not be as well-funded as the current year, given the need for Navy recapitalization at the expense of O&M and other accounts, plus the effects of continuing high operating forces optempo in the war on terrorism. Thus, to CPF the future always looks lean and the comptroller intends for his budget analyst group to pass this message down to subordinate claimants. Hence, the tone of the review is to make sure the "fat" is out of subordinate command budgets and that they are prepared to present tight and accurate budgets for years to come.

The CPF comptroller also intends to see that claimants build executable budgets. Since control numbers are handed down from the fleet level to type commands, historically the tendency in some claimant commands was to execute to the control numbers, but not to manage their budgets by, for example, transferring between accounts when one account had a surplus and another a shortage. Part of the comptroller's job is to use the budget as a management tool in business operations while adhering to the budget as approved by reviewers. There are always operational plans for claimant guidance. The comptroller seeks a closer connection of the budget to claimant operations on behalf of his two-star admiral commander, CINCPACFLT. Moreover, in the era of rapid change, another purpose of the midyear budget hearings is to introduce claimant commands to the idea that they have flexibility to propose reprogramming of funds to highest priority. Historically, CPF comptrollers have sometimes held reserves back during the FY so that they could 'bail out' commands that had problems. However, since 1990 CPF comptrollers have not held substantial or in many cases any reserves. Instead, subordinate commands have been told no reserves are available and that they ought to plan for and fix their own problems. This is another part of the process of making commands develop an executable budget. After all, if their budgets were executable, they would not need to ask the comptroller for help, barring unanticipated emergencies.

In summary, the purposes of midyear budget hearings are to ensure that the claimants:

- Produce a budget that is an executable fiscal plan; thus, the question is often asked, "can you execute this budget, or are you just giving us back our control numbers."

- Justify changes, and that the justifications are adequate. Dialogue occurs over how long a justification should be. The answer is that two words are too short, but that more than a paragraph (e.g., 50 words) is too long and that "you have to get their attention in the first couple of lines."

- Understand that the program (work plan) drives the budget numbers and that the budget review is not just an exercise in manipulating lots of little numbers established by the comptroller.

- understand that resources are scarce and that they are expected to find and solve their own budget problems, and that the framework they create in the budget process, including the way resources are spread, sets the format for following year processes and profiles. If the budget is just numbers prepared exclusively for the CPF budget office, then this is a waste of time.

Typically, midyear budget hearings last several days. Each claimant makes an opening statement and then the CPF analysts examine and question their areas in each type command claimant budget. Hearings often open with a statement from the CPF chief budget analyst about the importance of using the budget as a management tool—an executable business plan—and proceed from there, lasting perhaps 2 or 3 hours and involving the claimant comptroller, a deputy, and the CPF chief budget analysts. The CPF comptroller, a captain's billet, often does not attend hearings. The deputy comptroller, a civilian, runs the hearings and the issues raised are to be surfaced and solved at the analyst and type command representative level, if possible. If not, then the comptroller becomes involved.

The concept of role is widely recognized in the literature on budgeting. Just as in Miles Law, "where you stand depends on where you sit," claimants are expected to take certain postures and budget analysts others. Participants also have expectations about how each other should behave: these are called sent or perceived roles. The claimants often show great surprise that the comptroller staff expects them to turn in a budget that is a plan different from their control numbers. After all, the CPF comptroller had given them the control numbers as part of the budget execution process at the start of the year and the claimants think that fleet should be pleased to get them back. For the claimants, this is a violation of their perceived role. It also violates their perception of how the comptroller is supposed to operate. Moreover, the CPF comptroller staff informs claimants to make sure they have an executable budget even if they have to reallocate some of the control numbers. This is an expectation that claimants have to meet, a new sent role. This makes their job more complex. No type command wants to "fudge" numbers to meet fleet demands, because once they buy into a distortion they are responsible for justifying it in current and future budget submissions. Most type commands are willing to move money to highest priorities—the problem often is that fleet and TYCOM priorities differ and such differences must be negotiated between the two sides. Sometimes this leads to using FMB as an arbitrator, but usually fleet and TYCOM officials would prefer to work these matters out themselves without involving FMB. However, TYCOM have been known to go around fleet to FMB at times to win an argument. The fleet comptroller and staff loath this when it happens.

With every midyear review, there are new complexities combined with old and familiar problems. For example, certain issues keep coming up during budget review, including:

- technical considerations, such as average salary fluctuations; inconsistent entries on different forms such as one form identifying a reduction of 18 people while another indicates 11, and incorrect entries such as civilian positions entered on the wrong line.

- Analytical problems that elicit questions such as: "Is this a one time or continuing expense?" "How can you increase support when the number of ships is decreasing?" "If you can live with a low number in FY 2002 and FY 2004, why should there be a high number in FY2003?" "Why have you exceeded your flight hour program other (FO) budget for TAD (travel)?"

- A series of policy guidances that developed out of the give and take of the budget hearing such as building an accurate work plan; justifying changes because FMB will find inconsistencies; say the right thing ... not "defer spare parts, but transfer to reserve budget"; show your budget decreases; remember that numbers tell two stories: ask yourself what someone could make of these numbers: if FY 2003 is up, then you have to justify it, justify why you got by in FY 2002 and cannot get by now. Otherwise the increase in FY 2003 may be seen as "just fat."

An underlying theme in many hearings concerns time; time was always short and claimants have to make their arguments short and precise. Thus they are warned by fleet analysts, "Analysts have one month to look at the budget ... one nice paragraph is what they look for ... analysts do not have the time to look at attachments ... so hit them right up front." "Look for linked categories—consumables is directly proportional to parts. The two should be linked together (move up or down together)." "Avoid bad words; travel is a bad word ... FMB does not like to see an increase in travel, so we need to see justification.... This really runs up a red flag" Each of the claimants has special problems to consider.

To understand the nature of roles and the detail of budget review, we examine a typical midyear review for three type command claimants below, using a set of recent midyear review hearings at the Pacific Fleet Command Office of the Comptroller as examples. Quotations in this section are not attributed to meet an agreement with CPF officials that in observing and documenting the budget review we would maintain anonymity for participants.

The Submarine Claimant (SUBPAC)

SUBPAC budget strategy was clear right at the start. "We are different," they said. This was a suggestion that the whole budget system ought to treat them differently, or perhaps that the budget system was inappropriate to their concerns. "Operating tempo (optempo) has no meaning for us, because when we tie up submarines we have to pay utility bills which we do not pay when we are underway" (since electricity is generated by the nuclear reactor). A budget strategy to reduce the operating costs of the fleet is to decrease optempo by tying up ships, which decreases steaming time, which decreases the amount of fuel used and hence fuel costs as well as other associated costs. SUBPAC also is different in that it is funded out of the budget activity for strategic forces and many in the Navy believed that it was these ballistic missile submarines that persuaded the USSR to keep the peace during the years of the Cold War.

The CPF lead analyst was equally clear in his opening remarks. He noted that SUBPAC was underexecuting its budget for spare parts. The analyst warned, "It is hard to go in and ask for $10 million when you are executing at $4 million." The message here was twofold: first, it told the SUBPAC comptroller and analysts that it was just like anyone else when it came to the potential for typical budget errors and, thus, the comptroller's role in budget review was legitimate and, secondly, that there was no additional money for underexecutors, another typical budget message.

SUBPAC argued that they were underexecuting only because of a lag in credits from the accounting system when spare parts were returned to inventory. This turned into a complex argument, but the budget role had been set in the context of being told that there was no additional money for people who underexecute budgets. Thus, SUBPAC was told it was not unique. A series of technical criticisms reenforced this message.

- Firefighting/personal safety: "You can't cut these; you know these are two of your highest priorities ... so make sure you do not submit high priorities as cuts."
- Be alert to one time costs in a specific year: specifically: "Make sure you take it (one time costs) out the next year."
- Be realistic: include "no nice to do's, only must do projects."
- Watch for "blips" in a trend line: "You have to give us a reason for that blip." SUBPAC argued the blip was a control number that had been passed through to them, traceable up the chain of command to FMB. "You gave us that control number" CPF responded by saying, "Well, then you have to tell us what you are not going to do the next year, or what you did this year that you didn't do in last year. You have to explain that blip. Besides, a control number is a starting point, not an ending point."
- When price adjustments are made, make sure they get put in the right line: "We lose money that way" Price adjustments are usually funded, so any data entry error with them can mean that something gets cut that would otherwise be quickly funded.
- Make sure changes show up in narrative justification: "All changes have to be justified."
- Explain illogical patterns, for example, "You show funding going down, but you show you are going to be underway more hours. You'll have to explain that." The response was, "We had our operators actually count days underway," and it might be illogical, but it was true because a better indicator had been used. The rejoinder

was that they ought not to change performance criteria unless they had a good reason, but if they changed them, then they, "have to put it in the budget narrative, so that people do not jump on us." This was a warning that CPF work was reviewed by others, and hence helped legitimize their role as well as justify any "hard line" they might seem to be taking.

- Two schedules do not crosswalk: "This is maybe a typo (typographical error) or a rounding error, but the numbers for program and pricing growth on these two forms do not agree, and they should. It is not a lot, but it does get into important areas of the budget, like utilities. Since you are the only command in this budget account, this will stick out like a sore thumb." The problem here was that there were two different numbers on two different exhibits for the same item and the same number should have appeared on each. If the discrepancy goes to the budget review stage, a budget analyst will automatically pick the lower number and cut the higher to make it conform to the lower, irrespective of whether it is the correct number. The analyst may even cut below the lower since this kind of error seems to indicate either the budgeting unit does not know what it is doing, or is treating the budget system with disrespect.

- Failure to justify loses dollars: "Per Diem average costs jumped by $10, but there is no justification for it. First chop (review), you lose the $10. So if it is in there, you must justify it." Here the claimant had simply failed to justify a change. This is an easy cut for reviewers.

- One program had a fluctuating profile over five FYs: CPF remarked, "This is a bad profile. Can you *execute* at this level? What is the absolute minimum you have to have to execute in this program?" The response was that this was a commands and staff category, and that they were not very popular, but that this money was for communications. "These are the guys who communicate with the guys who shoot the missiles. JCS still has to talk to these guys." By linking this seemingly obscure support category to the main mission, communicating orders to fire missiles, SUBPAC basically made the category a mandatory expense, although they would still have to write the narrative justification to explain what it was for. We notice also that CPF was concerned about SUBPAC ability to carry out the program; line comptrollers do more than just cut dollars from budgets. They have to be concerned about the ability of the claimant to execute programs.

As the hearing proceeded, SUBPAC became irritated by the level of detail of review and at one point said, "I want to come when AIRPAC comes to see if they have to justify $5000 to $10,000 items. Why do we have to justify something that is .5% of our budget?" Since the CPF comptroller was a pilot by trade, what SUBPAC was hinting was that CPF might go easy on the claimant who controlled aircraft. Community rivalries are persistent, but the CPF lead analyst again chose not to fight on these grounds and supplied an answer that made sense in the budget context, saying, "We often get questions about $2000 items from FMB." Another said, "we may get questions on it from FMB, but if we are pre-armed, we won't have to pass the questions on to you." Another added, "This budget line is by itself—things stick out." Thus, the "it's so small" argument was detoured, indicating questions over small items were a part of normal budget routine, that PACFLT was asked such questions by the FMB, that knowing the answer would mean they would not have to bother SUBPAC and that since the account was by itself, it was sure to draw a question. Hence, the fact that SUBPAC was different worked against it here. After several more technical errors were discussed where associated budget categories moved in different directions or where numbers which should have been the same from different schedules were not the same, the comptroller said,

> The day is over when you have a budget which is just a document or file. You have to stand behind it and execute it. You have to look at it to see that it is properly balanced—if steaming hours go up, fuel should go up—and that it is justifiable the way it is shown.

Thus, SUBPAC was urged to treat their budget as a working document and a business plan that they were actually going to execute.

In various parts of the hearing, budget analysts followed through on this philosophy. One cautioned *against* a large cut in travel, "Is that a realistic reduction in travel? Maybe you ought to take a look at this, because we will cut it more if we have to cut travel." The probing to see what was behind a change remained steady and usually ended with a warning to justify changes; claimants also learned that decreases had to be justified, and justified with more than just a few words. How much justification was required was more problematical, "For a decrease of $200k, four words are not enough." But then where there was no justification at all was even worse: "What program are you going to decrease (next year) that you had (this year). You did not justify the decrease from one year to the next, so I will go back to the current year and take the money away, too." This is a way of working from the future into the present, penalizing the claimant in the current year for not writing justification in his budget for the future year.

Review went from the very detailed to the more general. At the detail end, the typical errors appeared—forms that did not crosswalk, numbers that should have been the same, annualization percentages incorrectly done, one time programs continued. These were sometimes introduced apologetically by the budget reviewers, "This looks nit-picky, but it is really not...it gets me into other issues. Your OP-32 (another budget schedule) shows decreases in all other items, but fuel goes up.... Why is that?"

Another issue had to do with support of computer war-gaming exercises. The program and budget exhibits reported different numbers. The CPF analyst suggested that just saying the number of exercises would be reduced was not good enough: "need to say what percentage of the contract we would be supporting if x dollars was taken away ... to try to stop questions from FMB."

The claimant responded, "War games do not track to dollars ... it is man hours provided that make costs fluctuate ... I have to be careful about providing that kind of data in budgets because it is 'business sensitive' information." Here the claimant alerted CPF to the possibility that full provision of information might be illegal, because it would allow another contractor to gain inside information; if the contractor knew what the full cost of the civilian budget page was going to be, he could fix his bid to be just under the civilian cost and gain an advantage in the contracting process.

The budget analyst replied: "We can not have war games provided go up and down and funds not fluctuate ... FMB will ask." The emphasis was on trend analysis—looking at a line over time—and the expectation was that program variations ought to lead to funding variations. These were two fundamental concepts that the budget analysts always used, looked for, and asked questions about. The statement that FMB would ask was both a legitimization of the question asked by CPF, and the truth. FMB would ask.

Command Pacific Fleet

CPF itself had a small claimancy of supporting units for some fleet operations. The opening statement stressed the harsh fiscal climate and then emphasized that what the budget office wanted was a budget as balanced as it could be between resources and program needs and that there would be no money for unfunded requirements, so the claimant was advised to, "reprogram if you have left a high priority unfunded out of the budget." The decision rules were easily identified:

- Do not just give us back the control numbers; make it a budget you can execute

- Look for level functions where the dollars vary and ask for explanations of the variations

- Find the actor who benefits; if it is not CPF, exclude it from the budget or charge them back in some way, but get it off your budget.

- Explain all program changes; we do not need to get embarrassed by FMB because we (CPF) do not know the changes and they do.

- Drop a program out which has a history of being unfunded by higher levels: "are we putting the money in and losing it every year? If it is still not supported in the next budget, then we should take it out."

- Make sure promises are kept and that claimants realize they made a promise: "When they pound on my door on 1 Oct., saying 'you didn't recognize my deficiency,' I want to be able to say 'we saw you, you had some good consolidations, they made sense, now go do them.' "

- Beware of crossing trend lines: "How can you increase by three people each year, but have a decrease of $80,000? In a few more years they will be free"

- Check average salary for the unit: "There is $300k that sticks out like a sore thumb ... that is $64k apiece, that is too high an average salary for this unit."

- Make sure your anchor number is correct: "I checked this year's number for this year against last year's number for this year. What you show as last year's number is incorrect. If you start right, then the changes will be right."

- Do what you say: "You told me [in the narrative] the change was in vehicle rental support, so I checked it and did not find it ... that will make analysts ask questions about the program."

- Use the right numbers: "Your program and budget forms have different numbers for the pay raise. Which is right? You have to get annualizations from the old budget form and then do price growth on the new budget form."

- Extrapolate into the budget year: "There is a growth of $2 million between FY 2002 and 2003. What were those dollars for ... because they come in the budget and stay."

- Challenge the unusual number: "FY 2000 and FY 2002 look comparable; FY 2001 looks out of place; it is $50k more than FY 2000. Which is the real base? You have to convince me that FY 2001 is the real number." The claimant responded, "We are in deep trouble … [In this account]." The budget analyst quickly interjected, "Looking at the numbers, I can not tell whether you are in deep trouble in FY 2002 or just fat in 2001"

- Weed out the unfunded requirements that won't sell: "CPF is decreasing ships, but you are not decreasing funding and your unfunded requirements are not decreasing"

- Let the operators make the decisions, but make them stick with them: "Operators have to decide how much training time is enough; financial guys can not make this decision."

- A good scrub [tight budget] prevents end runs: "If we can say this is a hard core program (no nice to do's, just must do's), then we can turn back the $1 call by the operators to the admiral that gets the budget decision reversed."

- Link changes to the PBD (program budget decisions made at the secretary of defense level) but explain the PBD: "When a program budget decision is made that changes one of your accounts, tell us the logic of the PBD, don't just tell us the PBD number … unless it is a crosscutting PBD that is obviously understandable, a 4% inflation adjustment for fuel for example."

- Build the detail level program first, and then aggregate at the program level; this was actually an instruction about which form to do first, about how to start the budget construction process.

General discussion at the end of this hearing focused on the claimant's difficulty in finding historical records to explain changes in budget items and about how much justification was enough. The claimant asked how much narrative should be provided, "We know that AIRPAC has pages and pages," said the claimant. This comment indicated that the claimants talked to each other and strategized about issues before the hearings. The budget analyst replied,

> Well, pages won't get it. What you need is the meat … just one paragraph (holding thumb and forefinger apart) … we don't have time to read pages and pages. If you don't get us [our attention] in the first two lines, it is gone [the money is cut].

At the conclusion of this hearing the analysts told the claimant they had done a good job on their basic budget document, because, "Everyone else had disconnects." The claimant responded by saying they had created a budget team that actually moved into a separate room away from the phones in the afternoons and did nothing but the budget—that they "worked hard on it," but there were difficult questions to answer, basic questions that even senior people could not answer. These basic questions would persist as funding decreased and the organization attempted to match the decline in resources to the cost of the Fleet program structure.

Naval Pacific Air Command (AIRPAC)

The CPF budget director's introduction to the AIRPAC hearing was much the same as to the others; it emphasized that there was no midyear money for distribution so it would be AIRPAC responsibility to take care of its "hard core" unfunded requirements on its own. Moreover, AIRPAC also was warned that the problem was going to get worse, "We know now that we are going to have a serious problem." There had been a contingency reserve pulled out by FMB when the control numbers were passed out at the beginning of the year [in January]: "not only are we not going to get that back, but we may also lose an additional amount [3 times the reserve]." Thus the stage was set for the hearing.

AIRPAC is responsible for supporting the fuel, maintenance, and related costs for all Navy aircraft carrier and land-based air wings in the Pacific Fleet. The basic AIRPAC mission is manage the Navy FHP to support pilots and aircraft readiness to fly off carriers to meet threat contingencies and training requirements, and to support the carrier readiness work-up cycle for training and sustaining the skills of pilots. There is an extensive training task for each carrier deployment cycle. AIRPAC supports the aircraft when they are deployed and it also must provide for equipping planes. Support for AIRPAC weapons systems—aircraft [fixed wing and rotary]—by type, model, and series is defined through the programming and budgeting process, and OMN and OMMC money to support its mission is provided to AIRPAC through the OP-20 budget issuance from OPNAV. The major task of the AIRPAC Comptroller and staff is to manage the annual cash flow of the FHP. The FHP pays for a diversity of accounts beyond fuel, including almost all of aircraft maintenance and a variety of other expenditures such as TAD and other spending from the FO account. AIRPAC comptroller staff also must submit and defend program and budget detail to fleet for all POM and FYs.

The AIRPAC budget process requires paying close attention to pricing, timing, and execution questions with precise dollar figures for the budget and current year. In addition to the comptroller's budget office and the fiscal management chain of command from the type command to fleet and FMB, there is also a resource sponsor chain of command. This includes the FHP manager—a Navy commander billet—at the type command level, N-41 at the fleet level—a civilian position (this assignment at fleet has fluctuated over time), and N-78—a Navy captain billet—at the OPNAV level. All of these parties participate in the programming and budget execution processes as sponsors for the war fighting and support resources. If the resource is an aircraft, sponsor interests include type of aircraft, what it should do (mission), how it should be configured to meet the mission, how and when it will be deployed. These issues also are of significant interest to the wing commander on behalf of the customer-war fighter community. FHP managers support a mix of aircraft for whatever mission is necessary and maintain and upgrade aircraft as funded to do so over time. The air wing commanders who use the assets are not responsible for long-term maintenance and management of assets.

It is necessary to understand a little about this management structure for this hearing, because if there is a frictional interface between the resource sponsor (who "owns" the aircraft) and the budget funds holder (who supports them), then these different perspectives must be reconciled. This is not unusual because the type command comptroller, staff and budget officer manage accounts and money for the current year while the air resource sponsor is responsible for configuring the shape of the air asset base over a longer term horizon. Furthermore, the management tool for the OPNAV sponsor is the OP-20, a budget report that provides money for the FHP to AIRPAC on an historical basis that is not accurate enough for cash management purposes (and is typically underfunded). Thus, for example, at times the fleet-level sponsor (and comptroller for that matter) and the type command comptroller and staff may disagree over numerous issues related to support funding and FHP management. Cash management in execution requires the type command comptroller and staff to react to immediate pressures of cost and cost reduction caused by budget, mission and asset changes, for example, deployment of more or fewer aircraft carriers and air wings in the execution year than planned in the POM budget (McCaffery & Jones, 2001, pp. 423-440; Phillips, 2001).

The point here is that resource sponsors control the input to the POM, which is the resource allocation program for all air assets, and operations, including the FHP that determines how many hours will be spent flying each year by type of aircraft. The budget, then, costs this out in precise detail. As noted, AIRPAC is responsible to two masters, one for the executing the FHP and one for executing the budget. At midyear review, AIRPAC had discussed the FHP with the OPNAV and fleet resource sponsors and had received certain guarantees from them about what would be in the POM and, hence, later in the budget. However, some of this detail was not in agreement with what CPF saw as the resource constraint under which AIRPAC had to operate.

The AIPPAC budget team chose not to make an opening statement, indicating they would prefer to get to the details. As the hearing developed, it became clear that the AIRPAC comptroller representative had put himself at a disadvantage and should have made an opening statement, saying perhaps that it had been overexecuting its control numbers, realized that it would be a bad year, but that its resource sponsor stood firm on several key points. Thus, AIRPAC could have argued it was between a rock (the resource sponsor) and a hard place (the fleet comptroller and eventually the FMB) and would appreciate all the help it could get. It did not make that argument.

The hearing opened with a program that was growing and would be fully operational in the following year. The CPF budget analyst asked; "Has anything been done to eliminate unnecessary expenses in this program?" He wanted an example to show that the program manager had "really looked at this to see what we can do without if we get cut. The justification has to speak to cost efficiencies made to keep costs in line." The AIRPAC response was that a multimillion dollar building had already been built and that the program was growing. CPF responded, "We need to highlight this to Washington so we can say, 'Hey, we do not have money to fund this program when it comes on line in 1992.'" AIRPAC argued that they had just received word from their program sponsor that resources were included in the POM that was due to be released the following week. The budget analysts were skeptical, noting that this sponsor did not have a lot of money, and may not have the resources to fix the program. AIRPAC argued it had been fixed. Since the POM was not out yet, nobody knew the truth and AIRPAC had to argue what it had been told in good faith. Reluctantly, CPF gave in, "Well, if your sponsor has fixed the POM, we may be all right." The AIRPAC response was brief, "I guarantee he has." However the budget analysts remained skeptical, saying, "we'll see."

The end game of the POM process is such that when final decisions are made, the AIRPAC request could have been sacrificed to a higher, and later developing priority, hence the "we'll see" response from fleet. This argument highlighted the problem of a growing program in a declining budget era. It also marked the first time in the hearings where the resource sponsor was introduced in direct opposition to the budget office and its perceived mission. AIRPAC did not give any ground in this issue because it thought it was right, and would be proved right by subsequent events. Moreover, at that time, the Admiral who headed the resource sponsor directorate outranked the Admiral who headed the FMB (N-80), so the AIRPAC comptroller believed that if push came to shove, the resource sponsor would win. The AIRPAC comptroller's analysts knew that budget review authority came down to them through the FMB under N-80, so they knew who outranked whom. Restructuring of POM-budget review in 2003 and thereafter to run concurrently should reduce this kind of dispute.

The next phase of the hearing concerned some typical budget issues including a 40% pay raise that could only have been a typo and a missing justification when a transfer was made from one program account to another. AIRPAC explained, "We show it on the reprogramming exhibit." The fleet analyst responded, "Every exhibit should stand by itself." This was followed by a typical search for "fat," for example, "You took money out of this account? Why did you have a surplus there?"

Then a proposal of greater substance was advanced by CPF, "What if you substituted more simulator hours for [real cockpit] flying to achieve your PMR [primary mission readiness where 2% traditionally has been satisfied by simulator time]?" How much could you save with a ten percent use of simulator time? This proposal has been pushed by FMB analysts and others for years. More modern, high tech simulators make this option even more attractive to budget analysts and other resource administrators, but not to operators. AIRPAC responded that simulators were no substitute for actual flying hours, a point that the fleet analyst listened to without agreeing. This question was perhaps a prelude to a budget battle in an upcoming year. The CPF analyst seemed to be alerting AIRPAC that the issue would be pressed further in the future. The key to future budget decisions in this area would be what was decided in the Navy POM.

AIRPAC was then admonished that even when CPF itself changed control numbers, it was up to AIRPAC to, "come back and tell us what you are going to do [how they would accommodate the cuts or changes ... or protest if they could not.] This brought the argument back to the opening dialogue:

AIRPAC: It's hard for us bean counters to do that... [forecast operational decisions which will be made in response to comptroller dollar cuts].

CPF: I assume you talked with your admiral about this budget?

AIRPAC: They reject the cut.

CPF: Well, find something else to cut. What you suggested [in the budget] is unacceptable up the line ... so find something else. Not everything is a gold watch [impossible to cut] ... be brave.

> AIRPAC: The admiral does not want to cut something [a program or vertical cut] and see that others haven't made the same cut [suffered equally].
> CPF: If he feels he is being treated inequitably, he can appeal. The flying hour program [the meat of this budget] has got to come down.

However, the FHP would not be discussed until later, and not in depth. The next series of activities illustrated the usual budget concerns, pricing incorrectly, inadequate justification, and so forth. Then, as now, errors in original data entry reverberated through the system. Once again a difference surfaced in this regard, rising out of a detail on one form.

> AIRPAC: Well, we have to get back to the control number, don't we?
> CPF: We want you to rebalance this between accounts so that you have an executable document ... because we are going to hold you to it."
> AIRPAC: We thought we had to come to the controls.
> CPF: Bottom line, you did ... but we all work hard on this and we want to know it stands for something.... What did you do in the past when you just budgeted back to control numbers? Did you make up your own business plan (internal plan)?"
> AIRPAC: Yes ... for example I make up my own work plans.... Then we would come to you where we had to do reprogramming.

This was followed by a discussion of the methodology for making air support plans and led into review of the FO account, where control numbers seemed to cause some problems.

> CPF: Looks like you are overexecuting travel [a perennial problem].
> AIRPAC: It is a control number problem.
> CPF: What is your real number? What is your real execution document plan number? How do you know you are doing good or bad? You do it backwards ... you should make a plan that will allow you to execute what you need to do and then make a budget ... If you had sat down at the beginning of the year and made a real financial plan, you would not be in the position of saying you are going to overexecute your travel by 50%.

AIRPAC was silent. The fleet budget analyst added, "I want you to put together the budget you think you can execute at the beginning of the year."

> AIRPAC: We have to come to our sponsor's flying hour schedule … your control numbers are not our only guidance."

In this context, guidance does not mean advice, it means targets. For AIRPAC, the flying hour program set targets (constraints to be worked out in execution) that it had to meet just as surely as it had to meet the budget execution numbers.

> CPF: Well, fight the flying hour battle right up front.
> AIRPAC: You cannot make a flying hour program for FY 2005 in FY 2002.
> CPF: Yes, you can … that does not mean it will not change 12 times.

AIRPAC did not respond to this, thus CPF tried another tactic.

> CPF: What would you do if you lost an additional $50 million?
> AIRPAC: Cut squadrons or stop flying.

For AIRPAC, this was the ultimate threat. Over the past decade or so, one of the favorite strategies of DoD budget reviewers has been to pick a large, even unreasonably large, number and suggest it as a hypothetical cut. This tactic was used extensively under the President George H. W. Bush administration under Defense Secretary Dick Cheney by the DoD deputy comptroller (Jones & Bixler, 1992, pp. 131-151). And in some cases, the hypothetical cuts became real cuts after the end of the Cold War. The Navy FMB also used this tactic in numerous budget drills in search of OMN cuts to reallocate dollars to procurement. If the response was that nothing particularly bad would happen, then the reviewer might actually proceed with the cut.

However, in this midyear review budget hearing, AIRPAC provided an answer that indicated important parts of the program would have to be cut—squadrons or flying hours—and therefore mission capability. The threat of a big cut was opposed with the threat of an unacceptable outcome. The result was a draw. AIRPAC has had to use this approach virtually every year since the early 1990s to some extent to defend its execution projections by specifying at what point flying would have to cease and aircraft would be parked if it did not receive budget relief. This threat became more credible when the CPF comptroller supported this approach and actually shut down air squadron flying in FY 1994. In the following budget, the FHP was increased. Sometimes, establishing a "drop dead" date for cutting mission performance appears to be the only way to enforce the point that budget execution estimates are real numbers and not just made up data for tactical effect.

At the conclusion of the hearing, the fleet budget analyst noted in a conciliatory manner:

> What we are trying to do is to make sure that anything that goes to FMB does not come back with a mark (cut) because we missed something ... we are trying to get a budget here that we can stand behind and support.

Explaining the level of detailed scrutiny, another analyst said, "FMB told us it was extremely important that our OP-32s (a key budget control report) be perfect." This was not likely to be persuasive to AIRPAC—their reaction was, in essence, "that's your problem."

Competition in budget execution is just as evident as in budget preparation—perhaps even more so because in execution all parties are working with real dollars rather than proposed dollars. In execution, "the rubber meets the road," as one AIRPAC analyst put it informally [to us] after the conclusion of the hearings.

> We fix things in execution that are broken in the POM and the budget and our thanks for trying to execute the budget [well] is to get grilled by fleet analysts that don't know our budget nearly as well as we do. It bugs us because it means they don't trust us to do our jobs. But what really "ticks me off" is when they call [by phone] directly to our analysts to question our numbers or justification without even alerting the comptroller or budget officer about what they want.

Why would a fleet budget analyst do this? To find a weakness that would justify a cut is one answer. Again, competition for control of dollars is evident throughout the budget process.

There was one issue that irritated fleet analysts particularly in this hearing. AIRPAC admitted in essence that it had to maintain two sets of accounts, one containing the CPF numbers and one with their real execution plan numbers. This issue needed to be rectified from the fleet perspective. However, AIRPAC saw no alternative if it was to execute effectively. At a more fundamental level, AIRPAC challenged the nature of the fleet and Navy budget process when it claimed to be unable to project a FHP accurately 3 years into the future. What AIRPAC indicated was that it would be unwise and a waste of time to do this in an era when steep funding increases or declines and dramatic changes in the threat environment were obviously going to occur. Also, AIRPAC indicated that it would be foolish to do so until it saw others "bite the bullet" too.

Because of these budgetary doctrinal differences, there was a strong undercurrent of potential rivalry and conflict in the AIRPAC hearing, different from the atmosphere in the other two hearings with their focus on wrong numbers and good faith. Basically, AIRPAC believed it was the best judge of how to execute the FHP while the CPF analysts questioned how this was done and at times wanted to pursue option other than those chosen by AIRPAC. As Wildavsky (1964) has noted, budgeteers in the public sector generally have adopted as an article of faith that when it is just a number in dispute, reasonable people can find ways to compromise, with civility, knowing that this year's losses may by made good next year, just as this year's wins may be eroded in following years. However it works out, the budget process teaches that reasonable people can reach compromises over numbers. As Wildavsky put it—it is just a matter of dollars (pp. 11-12, 60).

This hearing was fought largely on doctrinal grounds and was not a typical budget hearing. However, to the extent that a budget office can let a claimant retreat to a high ground above the budget fray, as SUBPAC did by claiming it was different or as AIRPAC did by claiming the resource sponsors plans for the budget trumped the budget office review process, the fleet budget reviewers believed they had lost control of the process. This is why the fleet budget analysts introduced questions that were clearly budget questions. It was a way of attempting to reassert their domain over the process. Thus, the little questions about salary buried in the middle of an obscure form really stood for the larger issue of who had authority over what decisions.

In summary, these midyear budget hearings were intensive training sessions in getting claimants to say what they needed, and to say what they meant. Time and again claimants referred to their clienteles within the DoN, but so did the budget reviewers; their client was the DoN budget office. Time and again claimants used contingent strategies, for example, fire safety training. Time and again reviewer and claimant became involved in sorting out the right numbers and the right forms from the multitudes of numbers and forms upon which they had to be entered. All of this is consistent with Wildavsky's (1964) description of budget negotiation and strategy (pp. 63-126).

Budgetary technology has changed since Wildavsky's classic original *Politics of the Budgetary Process* book was published in 1964. Most of the work for these hearings was based on electronic spreadsheets, with hardcopy sometimes piled half a foot high on the table. However, computers did not solve any problems other than basic manipulation of data, valuable enough in itself. All of the decision making surrounding the schedules and the review of decisions was a human process done by experts in their areas, some to propose a budget, others to review those proposals. Also illustrated time and again was that even experts do not know everything, because expertise is compartmentalized and because issues move and change over time and present themselves in different ways in different years. Moreover, even subject matter experts have to be coached at times to make their best case in the budget process.

In the last three chapters we have been concerned to a large extent with how people inside DoD make the budget process work. The title below "A Budget Claimant's Perspective" is a joint perspective on budgeting from the field claimant perspective, from a group of experienced officers who had held field claimant billets of one sort or another in DoD, for example, from those who put together and operate budgets. We gave them a couple of short paragraphs as a starting point and they edited it to what they saw as true about the budget process from their perspective during their times in budget and financial management billets.

A Budget Claimant's Perspective

The following is a statement from one officer with budget experience:

> In budget preparation, I pad because I expect to get cut, but I don't add 10% so they will cut 5 points and I will get an increase of 5 points. What I do is add projects or items to my budget that are complementary to my strict needs list. I think of it as needs, wants, and wishes. If someone else wants to fund one of my wishes as his or her need, that is fine with me, so long as they do not net it out against my needs. I make my best estimate of what I need and I can operate with less than that if I need to, but some things may not get done. I am ready to accept that what I see as a need, might not be a need in the greater scheme of things, but if CNO or the President or Congress do not want to fund my critical need, well, then they must be prepared to suffer the consequences. I believe that there are some needs so critical that everyone ought to agree to fund them. I will fight for these, but there comes a time when I have made the best case I can and I have to settle for the wisdom of people above my pay grade or the majority vote if it is a legislative issue. After all, government is not a solely owned proprietorship. Of course, I will make sure they understand what is at stake on the big issues and I will make the case again next year and, indeed, the first chance I get. I recognize that not all needs are critical and that it is generally a bad idea to fight tooth and nail for every single item.
>
> I recognize the budget is critical, but it is not the only critical event. I need to get what I want in the POM. If it is in the POM, momentum is on its side and all I have to do is tend the fire and it should carry through in the budget. If I neglect the POM or neglect to co-ordinate the POM/budget

process, I could lose my project/and money. But if I am a good steward, the POM should carry through into the budget. In multiyear projects I recognize that bad execution in the current year threatens what I will get in the budget year and may threaten what I get in the out years of the POM. I recognize that if I get dropped out of the POM, it is hard to get back in. I will seek compromises to avoid being zeroed out in the POM.

The budget is critical, but I know I have other opportunities to get money. First, there is the midyear review. If I have real needs, I may be able to get some money here. I have to be careful, because there are always problems at midyear and I could be asked to be a donor for someone worse off than me. Leading up to midyear review, I plan ahead and make sure that I am executing at 100%, because I know I will lose money that is underexecuted and it will cast suspicion on that account for the rest of the year.

Second, there is the end of year sweep. Someone up the chain of command always has money they have been holding in reserve for an emergency, so I am likely to have access to end of year money, but I have to have my list prepared ahead of time and be able to obligate it quickly (there is no time here for puzzling out what I might be able to spend it on). The closer to the international date line [IDL] I am, the more likely I am to get end of year dollars during the last week or even the last day of the FY. If I am far away from the IDL, and I can identify a surplus, I may give it up with the hope that I can trade it for goods in the new FY.

Third, it is also possible that I may get some money in the annual supplemental; usually this will be in repayment for money already spent and will not quite pay for what we really did. Still I can put it to good use.

I do not keep any reserves, but I know where I can find some money if I have to. I like to look at BOS accounts, e.g. pier maintenance ... even the flying hour program has a support program in it that can be used for unforeseen emergencies. I know where the Anti-Deficiency Act control points are, so I know when I may transfer money and I know what a go-to-jail offense is. I avoid the latter. Other people may keep reserves; while I like to solve my own problems, it does no harm to ask the comptroller up the line for help. Maybe he can find some money from me in someone else's reserve. After all, these are not my personal economic shortcomings I am solving, but rather bona fide Navy (DoD, AF, and DoA) financial needs.

I build my budgets based on what I need and then I enhance the justification by using whatever buzz word is hot that year. If quality of life is a current buzzword then that is what I use: I ask for more tanks, because more tanks help safeguard Marine lives and more live Marines means a better quality of life for everyone related to the Marine. This is far-fetched but I mention it so you can see how far this can be spun out. Transformation might lead me to ask for more yellow tablets so people can plan better. Enhanced war fighting capabilities might result in more dollars for staff travel accounts for war gaming. This approach is no guarantee of success, but it might help.

Some accounts are dogs and I know I will take hits there. I try to minimize the damage. I always expect to get cut here (staff travel) and when I do not,

I am pleasantly surprised. I try to fill the account in good years and then let it dwindle down in the bad years. I always ask for something because I need money in this miscellaneous account to carry out the mission. It is just not a high priority.

Before and during budget preparation I strategize with the boss about what to include, how to package the requests, and how to phrase them. By working together—my employees and I, my boss and I, my boss and his boss—we optimize our budget package. I look at my budget and ask myself, "How will the boss see this?, as good stuff that he can defend to others?, or as stuff that it would be nice to have but is not crucial?, or as me re-fighting battles I lost last year and probably rightfully so?" So I ask myself, do I have non-starters in here that I am doomed to lose and is the value of consciousness-raising worth the losing the money and maybe jeopardizing other parts of my request? I understand my boss can say no, but I want him to think seriously about what the consequence of a "no" is. I know there is not enough money to go around, but I want to make sure that I express my true needs. I also want my people to know that I went to bat for them in the budget process.

In budget execution, when the budgetary authority comes down, I immediately and regularly examine my program for executability. For example, if my budget request included 1,000 computers and service/support for 1,000 computers, it is possible that the number of computers was cut to 500 but the service/support line was not cut at all. Or, if a program is stretched out, I may find myself with an unexecutable excess in one account and a shortage in another. Such "asymmetries" provide the opportunities for horse-trading within statutory limits and within the friendlier confines of my own department.

Also, I know my appropriation Anti-Deficiency Act control points: if you are controlled on personnel, supplies and services, and capital outlay, you may not be able to transfer money among these accounts, but you will be able to transfer within personnel between bonuses, and other allowances for example, and within supporting expenses between postage, computers, oil, and ammo. If you are controlled on these, then you still may be able to transfer between tank ammo and rifle ammo. It all depends where the control authority rests.

I would like to think that numbers, logic, and clear evidence of need win out in the budget process, but I also realize that there is not enough money to go around and other programs may be valued more highly than mine at any particular point in time or in any specific decision process. They may get more funding than my programs, even if they can not clearly demonstrate the extent of their need or how much of the need their recommendation will satisfy. When this happens, I must ensure that I fulfill my fiduciary duty to represent my programs. I do not take these outcomes personally, and may even be happy for a brother officer who is the comptroller for a popular program and got more money with less effort.

CONCLUSIONS

We conclude our analysis of claimant budgeting with the following observations. These are based to some extent on observation of budget hearings and also on interviews over a series of years with participants in the budget process who work in comptroller offices, and also in some cases resource sponsor offices, at echelon one, two, three, and fourth level commands—in FMB, at Pacific Fleet, at type commands, and in field commands.

Incremental Routines Are Dominant

Although inputs (exhibits, analysis, etc.) are extremely complex, discussions with personnel in the budget process at CPF indicated that there is an incremental nature to the resource requirements and allocations from year to year. Most accounts change incrementally. Moreover, they are built and reviewed incrementally, meaning that comptrollers and budget analysts rely heavily on the account history and trend lines. Most of the players are seasoned veteran analysts. They have seen most of the issues before. They know what the critical issues are and what questions should be surfaced. They also understand where the permissible answer is likely to lie; that is to say, they know an unacceptable answer when they hear one and they recognize an answer that will play up the line, even if it is not as well-documented as it might be. They make good use of these skills and perceptions in the budget process.

Resources Are Limited and Justification is Critical

Programmers and budgeters are working with a limited pool of resources and must decide how to distribute available resources among competing priorities. In recent years, the amount of program planned in the POM has exceeded the amount of program bought in the budget. Put another way, the budget typically under-resources the plan. Within this over-all condition, another tension resides, as readiness accounts take priority over support accounts. This is seen in the CPF OMN account, where readiness-related funding took priority over support-related funding. At CPF, ship operations and the FHP were both funded at over 90% of their requirements for 2003, while sustainment, restoration, and maintenance for facilities was only funded to 54% of its requirements.

Analysts at CPF and elsewhere explained the difficulty they faced in justifying to the FMB the funding of support accounts versus readiness accounts. They indicated that metrics for developing budgets within support accounts are either nonexistent or inadequate to provide justification of increased funding within a resource-limited environment at the expense of readiness. For example, the best metric available for developing the FO account projection has been to average the funding for the previous three years even though, in the words of the CPF budget department head, "you're taking an average of three years that were also underfunded." Type command analysts indicate exasperation with the inability to project FO accurately and are in search of a model to perform this task.

The Process Is Undergoing Transformation

Changes to the PPBES process, transforming it from a system designed to accommodate a sequential task flow to one that accommodates a reciprocal task flow, have increased the communication and coordination demands between programmers and budgeters. Prior to the change, the program was locked and the budget was built based on the POM numbers. However, comptrollers in many instances were faced with "trying to fix the program in the budget" and would actually change the program approved by the CNO and Secretary of the Navy while trying to make the program "executable." Some indicated that this meant that budgeteers were remaking decisions originally set by programmers for war fighters. Some believed this was inappropriate, and budgeteers generally agreed with this view. To the extent that it happened, budgeteers believed they were forced to make decisions by budget deadlines that either they were not prepared to make or that should have be made at higher level in the operational and/or financial management chain of command. The revised PPBES concurrent programming and budget review process may help solve this problem, but this is an issue to be evaluated as reform moves forward.

CHAPTER 11

FINANCIAL MANAGEMENT AND DEFENSE BUSINESS PROCESSES

INTRODUCTION

Financial management in the Department of Defense (DoD) is performed as an integral part of a diversity of managerial functions including budgeting and the full Planning, Programming, Budgeting and Execution System (PPBES) process, comptroller, accounting and reporting, auditing (financial, management, and performance), treasury functions (from revenue generation to investment), operations and maintenance, capital asset planning, acquisition and procurement, logistics management including inventory management and control, contracting, personnel and human resource management, risk analysis and management, program assessment, evaluation and policy analysis, and retirement account management and fiduciary stewardship over such accounts. While this list does not include all of the functions in which financial management plays an important role, it is sufficient to indicate the breadth of involvement of financial managers in virtually everything done in and by DoD. Further, while defense financial management carries out many of the roles and responsibilities under the same rules and roughly in the same manner as

Budgeting, Financial Management, and Acquisition Reform in
The U.S. Department Of Defense, pp. 481–525

these functions are performed in other departments within the executive branch of the federal government, financial management in DoD is dissimilar to other federal departmental counterparts in some significant ways. First, organizationally DoD is huge relative to most of the other federal departments whether one measures this by dollars executed annually, employees, numbers of contracts let, the range and amount of capital assets required to perform its mission or other metrics. Second, DoD operates all over the world every day of the year. DoD business is similar in this respect to large multinational corporations in the private sector. In this regard, DoD does business with literally more than 100 nations a day and operates in this manner continuously. In addition, DoD does not just spend money; it also generates revenue through programs such as foreign military sales of military hardware and provision of security and financial assistance to allied nations in areas related to national defense. As noted, in this respect DoD performs in part a treasury function that is not present at all or not present in other federal departments in size and scale that matches the level of DoD activity.

It is axiomatic to observe that to do anything programmatically in the federal or any other government you need money. And everywhere that money is authorized for expenditure, spent, and accounted for financial management knowledge, skills and abilities are required to perform the functions and to ensure that fiscal law and administrative rules governing fiscal policy are abided by and enforced. The public and the legislative and executive branches of government rely on financial managers to be the guardians of public fiscal affairs to see that financial functions are performed to the letter and intent of the law and to provide the means for assuring accountability in spending of public money. Financial management in DoD is, as a consequence, a very important function, in fact one that is crucial to the achievement of national security and defense missions and objectives.

In this chapter we assess the many challenges that confront financial managers in the DoD. While readers may perceive that in some instances we concentrate too much on problems and "bad news" our purpose in delving onto the problems that confront DoD financial management is intended to be entirely constructive. We believe that no problems can be resolved until they are clearly identified and communicated to those responsible for their resolution. It is in this spirit of the need for continuous improvement that we investigate many of the problems faced by DoD financial managers and provide analysis of some alternatives for addressing pressing issues and problems related to DoD financial operations.

CHALLENGES TO DEFENSE FINANCIAL AND BUSINESS PROCESS MANAGEMENT

Appearing before a House of Representatives subcommittee of the Committee on Government Reform the Comptroller General of the United States David Walker explained that the:

> The Department of Defense and the military forces that it is responsible for are the best in the world. We are an A on effectiveness, as it relates to fighting and winning armed conflicts, when those forces have to be brought to bear. At the same point in time, the Department of Defense is a D-plus, as it relates to economy and efficiency. In fact, the Department of Defense has 6 of the 22 high-risk areas evident within its confines. It experiences challenges with regard to human capital, which is a government wide challenge, information technology, especially in the computer security area, which is a government wide challenge; they've had other information technology challenges, serious financial management challenges, face a number of excess infrastructure challenges; the acquisitions process is fundamentally broken, the contracts process has got problems, and logistics as well. (Walker, 2001)

These challenges have led to the following problems cited at the March 7, 2001 hearing of the House National Security, Veterans Affairs, and International Relations Subcommittee of the Government Reform Committee:

- Odd could not match $22 billion worth of expenditures to the items purchased.
- The Navy had no financial information on $7.8 billion of inventory for its ships, and that it wrote off as lost over $3 billion worth of in transit inventory
- In May 2000, the General Accountability Office (GAO) found that the Odd has nearly $37 billion of unnecessary equipment.
- In March 2000, the Odd inspector general reported that of 6.9 trillion in Pentagon accounting entries, 2.3 trillion were not supported by evidence to determine their validity.
- No major part of the Odd has ever been able to pass an independent audit (Congressman Dennis Kucinich, 2001).
- While DoD maintained 500 bases in over 137 countries, "each (service) has done its own thing; each has created its own system. And therefore, you have a number of independent, free standing information systems that are not integrated, that don't talk to each other" (Walker, 2001).

- Each service has its own coding structure.... We had the one example where I believe there were 66 characters to enter one transaction and those characters and the way they would be set up would differ by department. There is a lot of rivalry between the departments, and a cultural issue that each entity develops its own systems (Steinhoff, 2001).

On this last point, GAO official Jeffrey Steinhoff indicated what a dramatic problem this cultural imperative to have each military department develop its own systems created:

So, you end up ... with ... 22 major systems that can't communicate; they weren't designed to work together. You have 80 percent of the basic financial information coming from non-financial systems. Many of those are non-financial systems. And that information was really being derived to provide the degree of control you want. So, you've got a lack of standardization, you've got a very complex environment, you have these 22 major systems, and you have a host of other systems that are feeding into those. In FY99 the Defense Finance Accounting Service processed $157 billion of payment transactions. And $51 billion of those, or one in three dollars, was an adjustment to a previous transaction. You would find no business in the world that would be entering one third of their transactions as adjustments. (Steinhoff, 2001)

A quick look back to 1994 finds a similar litany: DoD overpaid contractors $753 million in the first 6 months of 1993 and lost an estimated $2.3 million in interest from contractors who held on to the money too long; Army payroll systems overpaid some $7.8 million, including 6 ghost soldiers and 76 deserters; and $19 billion was disbursed, but not matched to contact obligations (Bowsher, 1994).

In its 2007 high risk update, GAO cited the following programs as shown in Table 11.1.

GAO's report notes that substantial progress has been made in all areas. Some of this can be seen in Table 11.2.

Despite these improvements, DoD continues to lag. GAO says, "Furthermore, the Department of Defense continues to dominate the high-risk list. Specifically, DoD has eight of its own high-risk areas and shares responsibility for seven government wide high-risk areas" (GAO, 2007, p. 2). Two DoD areas went on the list in 1990; one in 1992; two in 1995, and so on. In 2005, DoD's approach to business transformation made the list. No DoD high risk area has ever gone off the list. DoD makes an A in war fighting prowess, but can it afford to limp along with a D or worse in ancillary and supporting services? Seventeen years of history since 1990 seems to say "yes," but is this a necessary outcome? The

Table 11.1. GAO 2007 High Risk Areas

Area	Year Designated High Risk
Medicare Program	1990
DoD Supply Chain Management	1990
DoD Weapon Systems Acquisition	1990
DoE Contract Management	1990
NASA Contract Management	1990
Enforcement of Tax Laws	1990
DoD Contract Management	1992
DoD Financial Management	1995
DoD Business Systems Modernization	1995
IRS Business Systems Modernization	1995
FAA Air Traffic Control Modernization	1995
Protecting the Federal Government's Infomation Systems and the Nation's Critical Infrastructures	1997
DoD Support Infrastructure Management	1997
Strategic Human Capital Management	2001
Medicaid Program	2003
Managing Federal Real Property	2003
Modernizing Federal Disability Programs	2003
Implementing and Transforming the Department of Homeland Security	2003
Pension Benefit Guaranty Corporation Single-Employer Pension Insurance Program	2003
Establishing Appropriate and Effective Information-Sharing Mechanisms to Improve Homeland Security	2005
DoD Approach to Business Transformation	2005
DoD Personnel Security Clearance Program	2005
Management of Interagency Contracting	2005
National Flood Insurance Program	2006
Financing the Nation's Transportation System	2007
Ensuring the Effective Protection of Technologies Critical to U.S. National Security Interests	2007
Transforming Federal Oversight of Food Safety	2007

Source: GAO (Jan. 2007) High Risk Update, GAO-07-310, p. 6.

Table 11.2.   Changes in High Risk Areas

	Number of Areas
Original high-risk list in 1990	14
High-risk areas added since 1990	33
High-risk areas removed since 1990	18
High-risk areas consolidated since 1990	2
High-risk list in 2007	27

Source: GAO (2007) High Risk Update, GAO-07-310, p. 4.

rest of this chapter opens a discussion of what there is about DoD which makes it both so hard to reform and a continuous source of concern.

FINANCIAL MANAGEMENT SYSTEM COMPLEXITY

The fact is that DoD is a very complex administrative entity. Sheer budget size does not begin to describe what happens in a fiscal year (FY). While progress is being made in improving financial and business systems and processes, DoD financial managers still have to deal with at least 32 different accounting systems, and the line of accounting that must be applied to introduce a transaction (e.g., a purchase) into the system and eventually pay them off consists of a fixed number of 11 coding elements and is 48 characters long. These digits identify the department, the FY, the appropriation, the administrative entity (Pacific Fleet), the object class (travel), an allotment control number, the identity of the activity executing the event, the transaction type, a property code, and a cost code (Hleba, 2001, p. 75). While the line of accounting is manipulated automatically through much of its life, it must be captured manually at least one time and errors occur here due to its complexity and the sheer volume of transactions processed in DoD finance and accounting systems. Moreover, the fact is that the line of accounting is not just entered once. Speaking of the contract payment system Comptroller General David Walker of the United States said:

> What we have here is a situation where a single transaction, which I think this illustrates (indicating chart), has to be entered multiple times, because rather than having an integrated information system, which is what most modern corporations would have now, where you enter the data once—it goes to a number of other subsystems—it has to be entered independently,

which, by definitions, means a lot more activity, a lot more opportunity for error. (Walker, 2001)

The result is that millions of transactions must be keyed and re-keyed into multiple systems. To illustrate the difficulty that DoD faces, Figure 11.1 shows the number of financial systems involved and their interrelationships for one business area—contract and vendor payments. Transactions must be recorded using a complex line of accounting that accumulates appropriation, budget, and management information that varies by military service and fund type and an error in any one character in such a line of code can delay payment processing or affect the reliability of data used to support management and budget decisions as well as introduce error into associated systems.

Since DoD financial management functions manage billions of dollars, even a low error rate will result in an embarrassingly high dollar figure of seemingly mislaid (or wrongly paid) dollars. In fact, DoD has been accused of some alarming financial management problems. A clean audit under the Chief Financial Officer Act (CFOA) may still be 3 to 6 years away and other problems seem plentiful no matter that they have been under attack for the last decade. In the hearing in 2002, Deputy

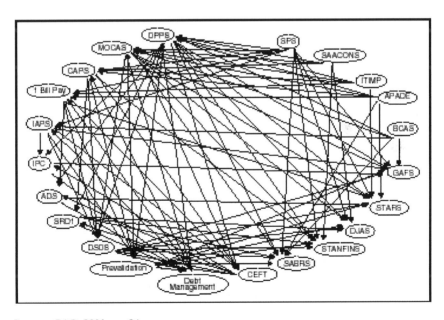

Source: GAO, 2001e, p. 34.

Figure 11.1.   DoD current system environment is complex.

Inspector General of DoD Robert Lieberman testified that he was cautiously optimistic about financial management reform, saying,

The department has taken a major step forward by finally accepting the premise that the financial management improvement effort needs to be treated as a program, with all the management controls that a very large program should have. Those include a master plan, well defined management accountability, full visibility in the budget, regular performance reporting and, resources permitting, robust audit coverage. (Lieberman, 2002)

> Representative Shays (2002) interrupted him, "You mean we haven't been doing that in the past? I mean even with the other plans, when we've had hearings, I thought all of those things were there?" Mr. Lieberman responded, "No, sir. I think the plans were consistently deficient in all of these aspects over the years. We always had a reasonable top level vision of where we wanted to go, but the details of the implementation were always lacking." (Lieberman, 2002)

GAO (2002e) noted that DoD had serious financial management problems and that they, "are pervasive, complex, long-standing and deeply rooted in virtually all business operations throughout the department" (pp. 1-4). GAO further noted that DoD loses track of assets, keeps unreliable budget and cost data, wastes billions of dollars on poor management of excessive inventory, makes billions of dollars of erroneous payments to contractors and purposely low-balls out year project costs (GAO, 2001f). Below we discuss some of the reasons these problems persist, reasons ranging from a failure to modernize, to existing regulations that impede reform, to the false allure of technology, to the lack of adequate controls, to the transient nature of DoD leadership, to the lack of incentives, to the reluctance of followers to follow, all of which lead to a cultural resistance to change.

Failure to Modernize

Review of financial management in DoD reveals a set of ironies. State of the art warplanes and satellite-guided precision weaponry are budgeted and accounted for with systems not too different from those of the Eisenhower era. It was not always this way. David R. Warren of the GAO stated:

> The business processes within the Department of Defense largely were developed ... in the 60s and 70s, and at that time, they were quite good systems, and based on modern business and practice at the time. However,

over time, they have evolved and have not modernized. So what you're faced with is what is often referred to as a brute force system. It gets the job done but in many respects, is very inefficient. (Warren, 2002, p. 15)

The failure of central guidance routines to initiate, stimulate, oversee and standardize modernization of DoD financial architecture left individual services and separate commands to create their own solutions to problems facing them. What resulted were literally thousands of nonstandard business systems. The current result is still a system where multiple systems perform similar tasks, the same data are stored in multiple systems and, despite the advent of modern times, much data entry is manual (Kutz, 2002a). In June 2002, DoD identified 1,127 different financial and nonfinancial feeder systems, in such areas as property management systems, inventory systems, budget formulation systems, acquisition systems, and personnel and payroll with approximately 3,500 interfaces providing information among them. Deputy Under Secretary of Defense for Financial Management Tina Jonas testified in June 2002:

> It is impossible to be accurate or timely with this type of business environ-
> ment. People wonder why we can't get a clean audit statement…. The fur-
> ther you get out from one of the core systems, accounting systems, the more
> likely it is that an error will have been made. (Jonas, 2002)

This multiplicity of systems has resulted in challenges that have not always been met. Training new personnel to operate old and arcane systems, each one different from another and different from accepted practices is time consuming and costly. Moreover, the profusion of systems has also increased the difficulty of accurately tracking costs throughout the department. For example, in 2002, the GAO conducted a study to investigate the accounting paths of two items.

The first item, a chemical-biological protective garment called JSList (joint service, lightweight, integrated suit technology), was selected because it was unique to DoD. The second item was a commercial computer obtained with a DoD purchase card. The results of this study reveal some of the consequences of DoD "brute force" accounting systems. For the chem-bio protective suit, GAO found that it took 128 processing steps to acquire, control inventory, and pay for the JSList chem-bio protective suit and 78% of these were manual. Not only was this process overly complicated, but also the high percentage of manual operations is an open invitation to error (Kutz, 2002b, p. 13). Gregory Kutz of the GAO stated, "The chem-bio suit inventory process was characterized by stove-piped, non-integrated systems with numerous costly error-prone manual processes. Of the 128 processing steps that we identified, 100 or about 78 percent were manual" (Kutz, 2002b, p. 7).

In contrast, the DoD purchase card process was impressive. Kutz (2002b) observed, "The system was mostly automated and provided the flexibility to acquire goods and services on the day needed" (p. 8). However, Kutz added that the Defense Finance and Accounting Service (DFAS) still received monthly credit card statements mainly by mail or fax and personnel were required to manually reenter each line of the purchase card statement because DFAS did not have the ability to accept the credit card data electronically.

While this seems a small problem, it is a costly one for there is a $17 per line processing fee per manual transaction. An analysis of the Navy monthly purchase card statement revealed 228 such transactions. This resulted in DFAS charging the Navy almost $3,900 in extra processing fees. As might be expected, modern business concerns do it differently. For example, both Wal-Mart and Sears make extensive use of electronic data transmissions within their internal systems and suppliers (Kutz, 2002b, p. 8).

DoD is not insensitive to these issues. Modernization is occurring with the department. For example, the Defense Contract Management Agency runs a system called Mechanization of Contract Administration Services (MOCAS). The MOCAS system is capable of processing contracts and receiving invoices electronically. The electronic processing charge is $20 cheaper per invoice than the manual rate. However, to receive the reduced electronic rate, both the contract and the invoice must be received electronically (Boutelle, 2002, p. 33). In the chem-bio suit example above, DFAS used MOCAS to pay financing and deliverable invoices, but DFAS only received 74% of the invoices electronically. Each of the remaining invoices in the 26% group resulted in a charge that was $20 an invoice higher than it needed to be.

Simple changes can add up in this environment. Representative Kucinich (2002b) cited an example where electronic processing instead of manual data entry could have saved a significant amount of money. On one purchase card statement, a sample of line items read:

- Vendor, Staples: amount of purchase $4.37, processing fee $17.13;
- Vendor, Culligan Water Conditioning: amount $5.50, processing fee $17.13;
- Vendor, Office Depot, amount $8.59, processing fee $17.13

Not only were the processing fees more than double the amount of the individual purchases, the entry charge would have been only $6.96 if sent electronically instead of $17.13 (Boutelle, 2002, p. 36).

Contract overpayment is another product of the complexity of the financial structure. Chairman of the Defense Financial Management Study Group Steven Friedman, stated in June 2002:

> These systems that grew up over decades, hundreds and hundreds of feeder systems, typically were at the service level or lower, were old. Roughly 80 percent of the systems were not in control of the DoD's central financial management. These feeder systems funneled information to DoD's central financial and accounting system. Over the years, standardization and compatibility had not been mandated; these really couldn't speak to each other. (p. 5)

The consequences of this failure are severe. GAO notes that for FYs 1994 through 1999, contractors returned over $1.2 billion of overpayments to DoD representing problem accounts that had been resolved. Still unresolved are problems that could amount to more than a trillion dollars. For example the DoD inspector general has said that there may be as much as $1.2 trillion in transactions that cannot accurately be accounted for by the DoD (Kutz, 2002, p. 14). These are the consequences of multiple antiquated systems with their unique and quasi-unique feeder systems. Perverse incentives also operate to the disadvantage of government. DoD has to pay interest if it does not pay promptly, but a contractor who is over-paid faces no such penalty. In effect, DoD has made him an interest free loan. Comptroller General David Walker (2001) noted,

> We have some perverse incentives. We have a situation where the government ends up having to pay penalties if it doesn't promptly pay. On the other hand, if the contractor has been overpaid, they do not have an affirmative responsibility to tell the government that they were overpaid. And, in addition to that, if they don't end up telling the government, and if they don't refund the money within a reasonable period of time, they're not charged any interest. And so, therefore, we have an unlevel playing field. (p. 7)

Walker noted that $900 million dollars was refunded to the government by contractors who were overpaid in the 2000. He said that this amount was lower than usual and that the situation was getting better. In 2001 current DoD financial management systems could not tell the total amount paid incorrectly, or if it had all been paid back, or how long it had been held without paying interest. With electronic funds transfer the overpayment dollars would benefit the recipients immediately. If the firm did no business with the government, it would be an obvious error and certainly should be repaid within a reasonable time before interest was charged, say 30 days since it was the government's error.

If the overpayment were a progress payment on an ongoing contract, the error might be harder to discover. Nonetheless the government has lost the use of that money and someone is getting a free loan. At 5% interest on $900 million, each month that goes by amounts to $3.75 million in lost interest on the interest free loan, plus an additional interest cost the public has to pay to fund it out of borrowing against the national debt, all due to inadequate systems. Then of course there is the $2 million a month DoD pays to contractors as a result of the Prompt Payment Act because it systems do not allow it to pay contractors promptly (Jonas, 2002). It is obvious that while progress is being made, problems still exist.

Leadership Failure

Another irony in DoD is the failure of leadership to bring DoD financial and management systems into the modern era. DoD top leadership has been committed to change for at least the last decade and since 1991 it has been spurred on by the imperatives of the CFOA, but the fact remains that DoD has failed to achieve a clean audit and fails tests of adequacy on many other financial dimensions as has been mentioned above.

For example, the Pentagon estimated in 2002, more than a decade after the CFOA, that gaining a clean audit opinion might still be 8 to 10 years away (Kucinich, 2002a, p. 3). Why do not these problems yield to committed leadership? Part of the answer is that while leadership may be committed, its tenure in the job is too short and this changeover slows down the process of reform. The sheer size of the task of fixing the DoD financial management requires long term commitment, but structural factors make that commitment hard to sustain. New administrations bring on their own team and the size of DoD means that it takes a relatively long time to get the team in place. Filling the top several spots may be done quickly, but much of the implementation depends on the second and third tier political appointments in the military departments (MILDEPS) and in DoD itself. These are the people who will actually see that reform gets developed and implemented and it takes awhile for them to get onboard, learn what is needed, and pull together as a team. Then, too, hardly is a team in place before turnover begins to occur.

Top DoD political appointees have averaged only a 1.7-year tenure and this hinders long term planning and follow through (Kutz, 2002, p. 25). Some suggest that the lack of a champion for long-term reform has contributed to the failure of previous initiatives, like Corporate Information Management (CIM) and Defense Business Operations Fund (DBOF), and even when there is a champion, he must remain in place awhile to make a difference.

Another irony is that a new appointee may not understand enough about DoD and its systems at the start to be able to move forward expeditiously with reform and by the time he does his time left in that position is too short to do much good. Another irony is that the really good people who could make major changes are often the first to leave, moving up in the hierarchy to better positions, because they have demonstrated competence and reliability. Optimistically, he or she may be able to do more good in the new position, but realistically his time in that position will be just as short and he faces another buy-in period. As Franklin Spinney (2002), a tactical air analyst for DoD, puts it, "We don't have the kind of corporate memory, so a lot of people come in and they go along with this stuff in the short term, they don't really get the big picture until they leave" (p. 29).

Leadership turnover in the political class also has a direct affect on the attitude of the long-term civil service employees. Speaking to the causes of the lack of continuity in reform efforts, Stephen Friedman (2002) said:

> If people there in the Defense Department believe this is a flavor of the month and that their bosses are going to be leaving in—whatever the actuarially measurable time is—a year and a half for senior people and then this will not be a continuing priority, you will not have the sustained effort. (p. 11)

Then, in addition, the temptation exists for top leadership to leave the problem to others to resolve. This is particularly harmful when it comes to changing the culture of the confederated Pentagon. David Walker (2001) recommends:

> We need to have a task team or some structure that is focused on resolving these six high-risk areas. It needs to be an integral part of their (DoD's) strategic planning; it needs to involve other people in the executive branch, including OMB [Office of Management and Budget] and other players that have a stake here; and it needs to provide periodic reporting back to the Congress to make sure that satisfactory progress is being made. You know, providing policy direction, quite frankly, just doesn't get the job done. I mean, we're talking about the need for sustained attention at the highest levels. It's very tough work, and you can't rely upon the individual service to do it—the job's just not going to get done if you do that. They have a role to play, but you've got to drive it from the top.

Another factor is the length of time military personnel fill these positions. The Air Force has career tracks for financial management personnel, but the Departments of the Army and Navy largely rotate officers through financial management positions every 2 to 3 years, in positions ranging from deputy comptroller at field commands, to budget analysts in the department budget offices in Washington, to the directors of those offices. They, like their civilian leadership, are not there long enough to decide on, implement, and sustain major reform. While this is a problem, it is not the critical one, for major financial management reform is not their job.

The U.S. commitment to civilian control of the military means that initiating and sustaining financial management reform is largely a task for top civilian political leaders, be that reform directed from outside DoD under the auspices of the CFOA or OMB, or from the inside in the desire by staff of a new administration to act more efficiently. Many "model" private sector companies realize how crucial leadership is in this process. Companies such as such as General Electric, Pfizer, and Boeing, all identify leadership as the most important factor in making cultural change and establishing effective financial management (Kutz, 2002, p. 5). The irony for DoD is that its top leaders have measured up to this task, but structural factors have meant that the amount of leadership actually provided has proved insufficient to the task.

Failure of Followers

One of the reasons that leadership is so important is that a cultural resistance to change is evident in DoD where many feel that not passing an audit or not making progress to fix an ineffective financial system will result in few consequences for them. Thus, followers feel they do not necessarily have to go where leaders want to take them on matters of reform. Some say there is a culture throughout DoD that supports the belief that DoD can virtually ignore changes with which it disagrees. DoD Inspector General Robert Lieberman (2002) spoke of the pessimism regarding implementing a new financial architecture for the department,

> The DoD might lack the discipline to stick to its blueprint. The DoD does not have a good track record for deploying large information systems that fully meet user expectations, conform to applicable standards, stay within budget estimate, and meet planned schedules. (p. 24)

GAO cited one of the reasons the CIM program from the late 80s failed was:

> After about eight years and $20 billion, that whole effort was abandoned. And one of the reasons that the GAO, General Accounting Office said it was abandoned was that there was resistance between the Department of Defense components and a lack of sustained commitment to the program. It said some military departments did not want to participate in this

corporate information management believing their financial management systems were superior to that which was being proposed by the CIM. (Tierney, 2002, p. 17)

The belief in the superiority of one's own system has contributed to the proliferation of stovepipe systems and service and even office-unique feeder systems that have allowed the services and DoD agencies to develop redundant and conflicting solutions to business needs. Speaking of the tendency of components within DoD to take standardized off-the-shelf software solutions and customize them, David Walker (2001) noted:

> We need to elevate a lot of these issues to the highest levels of the Department of Defense, such that there can be some discipline to assure that there's commonality, and that we're focusing on what there is a universal need for, and resisting attempts by these individual silos to basically take something that was intended to be standard and make it customized. To where, in effect, we have not accomplished the intended objective. (p. 47)

Another consequence of this reliance on the superiority of one's own service solution is that it creates a culture that changes systems that are fully effective. For example, when DoD attempted to improve the credit card payment program, GAO Director of Financial Management and Assurance Gregory Kutz (2002) stated:

> There is nothing wrong with the system; people simply are not following the controls in many cases that are in place. We're paying monthly credit card bills with nobody actually reviewing the bill, so we find ourselves paying for things the government shouldn't actually be paying for ... I think there's a lot of people that are probably trying to wait this out and hope that this too will pass. (p. 37)

Part of the reason employees will not follow is related to the perversity of the incentive system in DoD. The stereotype of the civil servant is as someone who does just enough to get by, is hard to fire, and can indulge in mute insubordination if irritated. While overdrawn, some of this is true and the path to change leads through a better incentive system. DoD financial management employees must be motivated to pursue change. Comparing managers in the DoD with those in the private sector, Friedman (2002) stated:

> If you looked at a manager's incentives, he hasn't gotten any material bonus for doing better work. It's hard to measure whether he's in fact, done better work. It's hard for him to discharge an employee that he considers to be incompetent, and then at the end when he looks at it, there aren't the

incentives to really stick your neck out and do anything other than manage
the budget. (p. 11)

Some believe that just as rewards are hard to come by, so are penalties.
They say that no penalty is imposed on DoD for failure to comply with
modern financial management practices because the defense function is
simply too important, and that imposing budgetary penalties on DoD
would harm national security. For example, Rep. Dennis Kucinich has
said:

> I do not believe the Department of Defense will fix this broken, unsustain-
> able system on its own. What motivation does it have? Despite its routinely
> dreadful performance, Congress almost never rejects a Pentagon request for
> more money. The time has come for Congress to treat the Department of
> Defense as the market treats any commercial enterprise. Just as investors
> withhold their supply of capital to a company that fails to meet its expecta-
> tions, Congress must refuse to supply additional funds to the Pentagon until
> its books are in order. If Congress keeps appropriating more and more
> money, despite these horrendous practices, what's the incentive for the Pen-
> tagon to reform? (Kucinich, 2002a, p. 3)

Representative Shays added, "We need defense so we keep operating.
But if we knew that we couldn't function unless we got our act together, I
think it (reform) would happen more quickly (Shays, 2002, p. 6).

The above speaks to organizational failure, but some observers believe
that more needs to be done to place personal responsibility on people
when things go wrong. For example, according to the GAO, the department lost 250,000 possibly defective bio-chem (joint, service, lightweight
integrated suit technology) suits, and could not track 1.2 million new
suits, many of which were discovered being auctioned on the Internet for
three dollars. Commenting about those in charge, Representative Janice
Schakowsky (D-IL) states, "Nothing happens to people. On the purchase
cards ... little to nothing happens to people who misuse them, so there
are absolutely no consequences. DoD keeps failing us and we keep passing
higher and higher budgets" (Schakowsky, 2002, p. 25). Commenting
about people in positions of responsibility who do not perform, Representative Tierney (2002) said: "What happens is they get promoted. It's not
that nothing happens to them. They get promoted by longevity being in
there" (p. 8). Whether or not this is factually true, it is a widespread perception, both within and without government and that alone hampers
improvement efforts.

Hundreds of thousands of DoD employees perform meritoriously, but
some do not. It is clear that those who do not have tainted perceptions of
all the thousands who do. Moreover, the task of reform is really a

leadership task; it is above the pay grade of most, if not almost all, DoD career employees. Still when changes are made, followers must follow the new rules and faithfully implement new procedures, and not complain that "it's not as good as my system," or "it's too complicated," or "that will never work." Just because it was "not invented here" does not give just cause for ignoring it.

Failure of Practice

DoD has been unable to borrow and implement best business practices. For example, large businesses like Wal-Mart and Sears have excellent inventory management practices that include standardization of data, little or no manual processing and systems that provide complete asset visibility. Wal-Mart requires all components and subsidiaries to operate within its framework and does not foster stovepipe system development. GAO found Wal-Mart and Sears had visibility over inventory at the corporate distribution center and retail store level. GAO noted that in contrast, DoD does not have visibility at the department, military service, or unit levels.

Integrated or interfaced systems and standardized data allowed both Sears and Wal-Mart to specifically identify inventory items. Wal-Mart headquarters staff was readily able to identify the number of 6.4-ounce tubes of brand name toothpaste that were available at a Fairfax, Virginia store. Other information was also available, such as daily sales volume (Kutz, 2002b, p. 8). A similar system does not exist at DoD and would be difficult to implement because of the complexity and size of the DoD system. In DoD, there are warehouses full of items, each with a different national stock number (NSN). During a GAO study, DFAS was handling 1.8 million unique items of inventory. By comparison, the typical Home Depot carries about 70,000 (Coyle, 2002, p. 23).

Experts know that in the modern business world reliable and timely information requires that systems interface with each other. For example, the Defense Supply Center has over 22,000 customers, with many having noncompatible systems. John Coyle (2002) notes, "The key to success in a company like Dell or Wal-Mart is they do have inventory visibility. They know where the inventory is up and down their supply chain" (p. 18). This problem is solvable, but the information architecture must be fixed before a reliable inventory management system can be attained.

The lack of a proven inventory control system has important consequences for DoD. GAO suggests that being unable to accurately track items directly contributes to a lack of readiness, a culture of waste, and high susceptibility to fraud. In 2001 GAO found that DoD could not

properly account for and report specifics on its weapons systems and support equipment (GAO, 2001f). For example, the Army did not know the
extent to which transport ship inventory had been lost or stolen, and the
Navy was unable to account for more than $3 billion worth of shipped
inventory, including some classified and sensitive items (Kucinich, 2002a,
pp. 22-23). These weaknesses in inventory control procedures caused
some Congressmen to worry that American war materiel may end up in
the hands of its enemies (Shays, 2002, p. 17).

GAO's David Walker (2001) warned:

> In the acquisitions area, DoD's practices are fundamentally inconsistent with
> commercial best practices. The result: billions wasted, significant delays,
> compromised performance standards.... We need to be following commer-
> cial best practices for acquisitions, unless there is a clear and compelling
> reason, from a national security standpoint, not to. Unfortunately, DoD all
> too frequently is motivated by—get the money, spend the money, hit the
> milestone that was set years ago, irrespective of the results of the testing,
> irrespective of whether or not one might question whether there should be a
> delay in order to further assess. This, as I said results in billions wasted and
> a lot of other adverse consequences. (p. 14)

The GAO JSList study provides insight into the scope of the problem
facing DoD. The Pentagon contract called for production of 4.4 million of
the two-piece bio-chem suits over a 14-year period for approximately
$100 each. GAO's first clue that something was wrong occurred when they
found 429 of the first 1.2 million suits had been auctioned on the Internet
for less than $3 each.(Kutz, 2002b, p. 9). GAO also found that some mili-
tary units kept no records on the number of suits they possessed in inventory. Others used dry erase boards to maintain their tally. When told of
these abuses, the program manager for JSList, Douglas Bryce (2002),
said, "I had no idea that these re-sales were occurring" (p. 3). While this
was a small percentage of the suits, the fact that it happened indicates a
failure in control systems. Congressional investigations also surfaced
problems with the JSList program. Many suits were discovered to be missing and needed to be removed from inventory, but as of June 2002, the
Pentagon was unable to find 250,000 of these defective suits (Kucinich,
2002b, p. 4). This happens because after the initial issuance of equipment
to commands, the Pentagon relinquishes responsibility to individual units
and tracking methods at the command level are dominated by unique
and largely manual systems. Program manager Douglas Bryce (2002)
observed, "They build an Excel spreadsheet or a Windows spreadsheet or
some spreadsheet to track themselves internally, and what that creates is
manual processes" (p. 39). The consequences to military personnel of
donning a defective bio-chem suit are obvious. Fortunately this did not

happen in the 2003 Iraq war, but DoD keeps inventory a long, long time and this is a problem which must be fixed.

The private sector provides good examples of how a standardized inventory and tracking system contributes to problem-solving. For example, in the early 1990s Johnson & Johnson had to recall Tylenol (Kutz, 2002b, p. 13). Their system enabled tracking all the way to the retailer's shelves, and resulted in swift action. DoD has demonstrated it has the capability to maintain tight inventory controls over priority items. GAO Director of the Defense Capabilities and Management team David Warren (2002) testified that sensitive items such as firearms are controlled in a much better manner than other items (p. 23). It seems that the JSList was not deemed a high enough priority when it was bought to result in tight inventory control. The lesson is that one can never know what will be crucial and thus a systems architecture must be designed that will allow for the treatment of all materials and supplies in the inventory systems as crucial. While this seems a huge task, all it really relies upon is the creation of a main system with standardized secondary systems good enough so that subordinate commands will not find creating command-unique feeder systems useful or necessary.

Inventory control is a significant task for DoD. According to DoD estimates, in 2002 it had about $200 billion dollars worth of inventory in various storage facilities. Moreover, not all of this is current and useful equipment. Rep. Schakowsky states (2002),

> DoD continually stores huge amounts of material and equipment that has no use. Additionally, the DoD process for tracking acquisitions and purchases is antiquated and seriously flawed. Oftentimes, the DoD can not find records of procurement, accounting, control and payment. (p. 5)

Good practice would include destroying, disposing, or surplusing inventory that is in this condition to at least save the costs of record-keeping and in the case of some items, preventing environmental damage as different ammunitions decay.

Another part of this problem involves accountability. Even the program manager may not have visibility of the total program. For example, Bryce (2002) testified about his procurement experience with JSList:

> There are twenty-four major steps to the process. Of those twenty-four steps, I have visibility of five that I can track through some type of system that I have access to or monitor or input to. That leaves nineteen that I do not. Those who have control over the other nineteen include various agencies within DoD, which could be DFAS, it could be DLA, Department of Defense.... Each one of those have processes and do things that I have very little visibility of as the program manager. (p. 39)

It is not a good sign when complexity defeats accountability, but this is what happens in many DoD situations due to stove-piped systems unique to different parts of DoD and decentralized and shared responsibilities. When the overarching information system is comprehensive, well-maintained and used by all participants, decentralized structures can rely upon centralized information for purposes of decision, review of decision, and accountability. When the information system is fragmented and corrupted, subordinate units create and maintain their own parallel systems.

To some extent DoD financial management is so difficult due to a host of environmental problems ranging from contract turbulence, perverse incentive systems, the lack of people with the correct training and congressional overcontrol.

Contract Turbulence

The turbulence of DoD contracts has caused inaccurate payments and led to potentially thousands of man-hours of financial reconciliation to correct errors. DoD data for FY 1999 showed that almost $1 of every $3 in contract payment transactions was for adjustments to previously recorded payments—$51 billion of adjustments out of $157 billion in transactions (GAO, 2002f, p. 3). GAO found that DoD contracts containing multiple fund citations and complex payment allocation terms were more likely to have payment errors because of the amount of manually entered data and the consequent opportunity for error in the manual processing.

In one example of a case reviewed for closed contract adjustments, GAO found 548 different accounting classification reference numbers (ACRN). The contract had been modified over 150 times and had received two complete contract reconciliations to correct payment problems, including one that produced 15,322 accounting adjustments. DoD said that it had further plans to complete a third reconciliation for this contract to correct about $3 million of illegal and otherwise improper closed account adjustments and it estimates that the reconciliation will take over 9,000 hours to complete (GAO, 2002f, p. 3). This is over 4 person-years of labor costs, but then benefits can range in the millions of dollars to say nothing of the benefit of getting it right.

Progress has been made by DoD in reducing closed account adjustments. These adjustments are held to be illegal when initial disbursements (1) occur after the appropriation being charged had already been canceled, (2) occur before the appropriation charged was enacted, or (3) were charged to the correct appropriation in the first place and later adjusted when no adjustment was necessary. Also included are adjustments not sufficiently documented to establish they were proper (GAO,

2002f, p. 5). For FY 2000, DoD reversed $592 million of $615 million illegal or otherwise improper closed account adjustments involving forty-five contracts. Thirty of those had additional accounting errors that required correction. Because of the complexity of the contracts and the time it takes to complete a reaudit, officials at DFAS estimate that it would take over 21,000 hours to correct the accounting for the 30 contracts (GAO, 2002f, p. 4).

In July 2001, GAO recommended implementing controls to increase management oversight and apply renewed vigor to the 1990 account closing law, prohibiting adjustments. In September 2001, DFAS upgraded the contract reconciliation system (CRS) to identify and prevent illegal adjustments. This measure is geared to stop disbursement charges until an appropriation has been enacted. A sample of FY 2001 closed appropriation account adjustments found $172 of $291 million (59%) were either illegal or otherwise improper; an improvement from 96% the previous year. Gregory Kutz of GAO noted, "Our review disclosed that CRS routinely processed billions of dollars of closed appropriations account adjustments without regard to the requirements of the 1990 account closing law" (GAO, 2002f, p. 5). Improvement has been significant. During the first 6 months of FY 2002, DoD reported making $200 million of closed account adjustments—including only $253,000 of illegal adjustments—which was 80% less than the $1 billion of reported closed account adjustments made during the same 6 months of FY 2001 (p. 3). Kutz concluded,

> The lack of fundamental controls and management oversight had fostered the idea among DoD contracting and accounting personnel that it was acceptable to maximize the use of available funds by adjusting the accounting records to use up the unspent funds in the closed accounts, regardless of the propriety of doing so. (p. 3)

There is little doubt that DoD financial managers believe that they are in a culture that demands they use up their money. Here this cultural imperative surfaces in the practice of using up unspent funds in closed accounts, legal or not.

Failure of Technology

Adopting modern technology has also failed to fix DoD's problems in financial management, not because it has not been tried, but because efforts to implement new technology have been undertaken without examining and reengineering the system before installing technological fixes. A bad system converted to a mechanical system even with the latest technology is still a bad system. John Coyle (2002) commented,

> You have to re-engineer, because if you throw technology at the problem, it doesn't solve the problem. Every company I've ever worked with that tried to throw technology at the problem have ended up costing themselves a lot of money. They have to start with the basic processes. (pp. 23-24)

In addition to restructuring the overall architecture, the piecemeal practice of disbursing IT money has not worked. Gregory Kutz (2002) stated,

> IT money is being shelled out all over the place within the department, and that is how you get the proliferation of systems and everybody building their own systems. One thing Congress could do, which has been done at a place like IRS, is to try to centralize that funding to get control over it. (p. 27)

He observed, "There are buckets of money all over the department that are being spent on IT improvements or upgrades that are not being controlled properly at this point" (Kutz, 2002a, p. 33.) Lawrence Lanzillotta (2002) cited a DoD study where it became apparent that trying to bring small IT fixes together, was not going to work and that there had to be an overarching architecture or a plan for people to follow (p. 15).

Another problem is that when current technology is recommended as part of the solution, it may be outdated by the time it is implemented because technology advances so rapidly and because of lags in the procurement process for these kinds of systems (Shays, 2002, p. 35). Moreover, even the advisory panels constituted to advise DoD may not be up to date on the latest technology when their members are selected from high visibility candidates who are divorced from the cutting edge of technology: as John Coyle (2002) explained, "The problem is that sometimes retirees like myself are appointed to those advisory groups and some of them aren't always up to date on the most modern technology" (p. 28).

Congressional Oversight and Information Demands and Effects

DoD is hampered in its reform efforts by rules and regulations that are time-consuming and restrictive, including cumbersome appropriation accounting requirements, detailed record keeping and reporting mandates, and obstacles to private sector partnering in areas that are inherently commercial. John Coyle (2002) concludes that government policy, "In effect, preclude some of the types of strategic acquisition practices that are going on in the private sector and allow a company like Dell to do the kind of things they do" (p. 29). To enact significant reform, DoD must overcome budget language and strict procurement regulations, but it must do so with processes that still leave Congress with appropriate oversight controls and a procurement process that guarantees protection from fraud and abuse and an efficient procurement cycle.

Failure of Incentives

Incentives to spend the money up at the end of the year and out of closed contracts have been discussed. Another instance of perverse incentives involves contracting for weapons systems. Both within and without, DoD planners and contractors know that less expensive systems stand a better chance of selection than more expensive systems, all things roughly equal. This gives rise to a low ball gaming strategy. This has been referred to by DoD insiders as the "political engineering process." One part of it includes bidding the lowest price with the intent to make it back later in contract changes as requirements are added back. Another part of it involves spreading the contract around geographically to create a wide constituent base so that many members of Congress will have a vested interest in supporting the project.

Low-ball cost estimates have severe ramifications throughout the defense industry. Franklin Spinney (2002) notes:

> Biased numbers hide the future consequences of current policy decisions, permitting too many programs to get stuffed into the out years of the long-range budget plan. This sets the stage for unaffordable budget bow waves, repeating costs—cycles of cost growth and procurement stretchout, decreasing rates of modernization in older weapons, shrinking forces, and continual pressure to bail out the self-destructing modernization program by robbing the readiness accounts. (pp. 10-11)

An example may be the found in the F/A 18 fighter program. This flagship airplane is the Navy's top carrier aircraft and to be without equal as an inexpensively produced aircraft. However, as Spinney (2002) says

> You'll see I have a box there that highlights pre-production cost estimates; they're way low. Costs on the F-18s went down. They just didn't go down as far as we said they would so we bought far fewer of them than we thought. And in fact the actual costs were twice as much as predicted. (p. 74)

Because of the misestimation of unit costs, the Navy only could afford to buy fewer aircraft than predicted and production rates became lower than anticipated. This resulted in a lower replacement rate, meaning not enough of them were purchased in a timely manner to replace the older equipment. Therefore, there was an increase in the average age of equipment, exponentially increasing operating costs. When this happens, the consequences are dramatic. Ultimately, DoD is left with a shrinking force structure and potentially degraded readiness.

A DoD study of the C-130 program revealed the costliness of the political engineering process. In testimony before Congress, Spinney (2002) observed that the C-130 was a very simple airplane, built in an underutilized factory in Georgia. What struck him about it was that fuselage sections, which could have been made in the same factory, were contracted out. When asked if it were cheaper to contract out, an assembly line worker said, "No way at all. We did this for political reasons" (p. 32). As a result, the cost increased exponentially over the years, from $11 million per plane in 1969 to between $41 and $42 million per plane in 1993 (adjusted for inflation), when the last C-130H was assembled, virtually identical to the first one (p. 32). At this hearing, a dialogue ensued between Rep. Kucinich and DoD's Spinney (2002) that outlined the parameters of the problem:

Rep. Kucinich: You have mentioned defense power games and front-loading and political engineering, and looking at that, do you really mean to suggest that defense planners and contractors misrepresent the latter year costs of these programs, and seek to spread subcontracts around the nation to ensure the survival of those weapons programs?

Spinney: I think it's very deliberate, yes, sir. I've talked to many contractors about this and of course they won't come up and testify that they do that, but they told me that they do it.

Rep. Kucinich: And that means Congress is part of it?

Mr. Spinney: Yes, sir. Congress is going along with it. I would like to —I had a conversation with one corporate vice president. He was an executive vice president of a major aerospace company, and I took him through the whole frontloading argument. It was part of a 5 hour lecture that I had and we gave to the entire staff of the company. And basically the bottom line, he says, "Look, we have to do this, because if we come clean, we won't get the contract because everybody else is doing it." And that's the dilemma. And the same thing exists inside the Pentagon. Because there's a

> constant competition for resources, you have different factions fighting with each other to try to do what they think is best. I'm not talking about malevolent behavior here, but they naturally try to win the competition, and so they tend to be overly optimistic. And the basic argument that you make when you do that is if I don't do this, I'm going to lose the battle (p. 30).

This problem is compounded because the same competition exists inside the Pentagon where there is constant competition for resources and different factions fight with each other to make programs look the most appealing for their own survival. Usually this means meeting all current requirements at less cost. Then once a program has been started, it attains a momentum that makes it hard to stop, even as the costs increase. Representative Kucinich observed, "The contention is that once the out years are reached and the true costs of production become evident, there is no longer the political will to cancel the program" (Kucinich, 2002a, p. 29).

To some extent, the cancellation of the Army Crusader artillery system in 2002 is a case in point. The Crusader grew to a size where it would not fit many requirements, but even though it was too big and very expensive, it was still very difficult for Secretary of Defense Rumsfeld to stop the program.

BARRIERS TO IMPROVEMENT OF
FINANCIAL AND BUSINESS PROCESSES

Some critics want DoD to adopt the best private sector practices—would that it were so simple. DoD dwarfs the largest civilian companies in size and personnel, and it is a unique business. Major corporations in the United States have undergone similar financial restructuring recently. These include Gillette, Cisco, and Hershey. They routinely took between 4 and 4 years to finish their transformation. Speaking of them, Deputy Under Secretary of Defense for Management Reform Lawrence Lanzillotta compared these changes with what confronts DoD, stating, "These are individual efforts that took three years. We are not happy with the complexity of the problem that we found (in DoD)" (Lanzillotta, 2002, p. 9).

DoD challenges are further complicated by the fact that as it looks for companies to emulate, they are also changing, thus DoD is chasing a moving target. For example, dramatic changes occurred in the business organization landscape in the 1990s, such as supply chain consolidation, globalization, increased governmental deregulation, dramatic changes in technology and increased access to information (Coyle, 2002, p. 17). DoD problems are further complicated by human capital issues. DoD may not possess enough people with the skill sets to operate on the cutting edges of technology and management: Stephen Friedman observed, "People were trained in many systems that we are trying to move away from, and that we need more advanced degree professionals, more people who are trained in business practices" (Friedman, 2002, p. 12). Part of the human capital problem is due to downsizing of the 1990s. Comptroller General David Walker (2001) explained that DoD:

> was downsized significantly in the 1990s, but it was done so in a way where there was not effective work force planning. And, as a result, over 50 percent of DoD's work force is over 50. And a significant percentage of its work force is going to be eligible to retire—the civilian work force within the next four years. There are major succession planning challenges as well as major challenges with regard to attracting and retaining skilled personnel, both in the uniform as well as the civilian area.

Our analysis thus far has made it clear that there are high stakes involved in financial management reform, ranging from lawful operation and suppression of fraud, waste and misuse, to the freeing up of billions of dollars from inefficient systems. However, one of the major hurdles has to do with a culture that is rooted historically in confederation of strong, independent, semiautonomous MILDEPS which support their war fighters and have created financial and nonfinancial systems to provide that support. This culture deems that the war fighters are critical and the supporting functions need to do whatever is necessary to support them. This eventually led to a multiplicity of systems with attendant problems.

Table 11.3 captures the raw data from the work of the process action teams in devising the financial architecture for DoD. The exhibit divides the effort into four domains, accounting, human resources, logistics, acquisition, and budgeting. It then shows the structural consideration (what must be done), the enabler (how it will be done), the impact and the change hurdle, thus centralized DoD accounting practices will be enabled by executive sponsorship, have the impact of reducing the number reconciliations, unmatched disbursements, and so forth, and have to overcome the change hurdle of loss of control at local levels. The exhibit is a scorecard of what must be done, what it will take to do it, what the desired outcome may be and what will have to be overcome. Of the 30 bullets in the hurdles column, about half (14) are focused on issues of control, eight on skill changes needed and four on law or policy changes. While additional funds are mentioned or implicitly needed here or there, money is seldom mentioned as a problem. Changing the culture is the problem. We may notice in the enabler column how often "standardization or standardized" is used. It is clear here that improvement comes from increased central

Table 11.3. Organizational Readiness Summary of Raw Data

Domain	Structural Consideration	Enablers	Impacts	Change Hurdles
Finance, Accounting Operations and Financial Management	• Centralized DoD accounting procedures, policies, processes and codes.	• Executive sponsorship to enforce changes with a governance structure to enforce new processes and systems.	• Reduce number of reconciliations, unmatched disbursements, and interfund billing.	• Loss of control at lower levels. Lose ability to influence own accounting structures.
	• Costs captured in a consistent and standardized manner across DoD.	• Centralized education and training across DoD.	• Reduce duplication and improve efficiency to free up resources.	• Lack of cost accounting skills in DoD.
	• Change organizational behaviors to dissuade data interpretation.	• Institute standardized cost accounting models and methods with standard data capture and entry approach.	• Easier to generate certifiable financial reports. Easier to measure progress using standard benchmarks.	• Resistance to cost accountability in the field, linked to budget disincentives for saving money.
	• New role and function to support a leading industry practice—extending credit.	• Standardized data templates and accounting codes supported by clear business rules and integrated systems.	• Greater accountability for DoD costs.	• Consolidation of training functions and funding.
	• New approach for collecting debts.	• Legislative/administrative relief to enable new activities such as spot credit.	• Data changes at the source. This will eliminate data corrections throughout accounting process.	• Significant culture change to not adjust data. New skills required.
	• New rules for applying cash and allowing small write-offs.	• Strategic collections strategy/governance to support strategic collections.	• Facilitate greater competition by vendors/suppliers.	• Obtaining legislative or administrative relief.

Table continues on next page.

Table 11.3. Continued

Domain	Structural Consideration	Enablers	Impacts	Change Hurdles
Finance, Accounting Operations and Financial Management	• Enterprise wide data strategy and management across DoD.	• Change in government wide budget disincentives for collecting debts and late fees.	• Proactive collection of outstanding debts ($4 billion outstanding) and late fees.	• Obtaining changes to federal budget policy to allow collection fees to be kept by collecting agencies. Retain Budget Authority within source.
		• Business rules that allow thresholds for write-offs instead of zero balances.	• Reduced transactions, costs and payment mismatches.	• Acceptance of new business rules.
		• Clear and visible sponsorship by the secretary of defense (SECDEF). Standardized business rules.	• More timely, accurate, reliable, auditable and detailed data available to support better decision making and comply with CFO Act.	• Clean audit statements not viewed as valuable.
		• Communication and governance supporting new strategy, policies, and business rules.	• Significantly reduce or eliminate data gaps, data calls, and interpretation of disparate data.	• War-fighter acceptance.
		• Training and incentives for cooperation, integration, and governance.	• Accurate data can be pulled electronically in real time. Certification of data may be obtained electronically.	• Lack of clarity exists around data ownership. Numerous systems exist which are not integrated.
			• Improved traceability and delivery of financial management reports.	• Large archives need reconciled. • Funds for training.

Human Resources Management	• DoD-wide unified Human Resource guidance, regulations, and policies.	• Standardized systems and data that allow self-service. • Unified training and training requirements. • Governance model in place to support.	• Streamlined systems and processes. • Integrated personnel data from recruitment to retirement - one employee profile accessible via one source for all in DoD. • Improve capability to match employee skills to org needs.	• Infrastructure funding. • Cultural resistance to reduced data control and sharing information. • Legacy data on antiquated systems. Skills needed to maintain are scarce. • Diverse training requirements and approaches exist.
Logistics	• Integrated, centralized, and standardized buying requirements, buying power, logistics planning, capabilities, and guidance across DoD. • Use of integrated strategic sourcing and contract management. • DoD-wide balanced logistics performance targets and measures.	• Expand scope, role and responsibilities of Joint Logistics Board (JLB) to drive leading practices of a DoD-wide logistical view. • Use outsourcing when more cost efficient. • Use integrated data and the FMEA architecture to facilitate data sharing, capacity and demand management.	• Increased logistics consistency and reliability. Better alignment of logistics requirements and activities with capacity utilization to get maximum throughput across DoD. • Streamlining of support organizations. • Centralized decision making.	• Cultural resistance to giving up power and sharing information. • Alignment of scorecards, cascading objectives and measures. • Some training will remain unique, for example, war-fighter needs.

Table continues on next page.

Table 11.3. Continued

Domain	Structural Consideration	Enablers	Impacts	Change Hurdles
Logistics	• Consolidate logistics training DoD-wide.	• Standardized data fed into Joint Logistics Board can enable logistics scorecards.	• Scorecards enabling greater accountability.	• Regulations prevent some of the most cost effective solutions, i.e., outsourcing and Just In Time inventory for moving and storing goods and materials.
	• Common real property guidance, policies, regulations, procedures and training, and environmental and space management.	• Real Property Center of Excellence Program.	• Understand and address common training needs. Share requirements with human resource management.	• Cultural resistance to relinquishing control / ownership of real property information.
		• Centralized Real Property inventory database with standardized inventory information that supports requests for information and 'what if' analysis.	• Access to Real Property records with consistent, comparable, and accurate data.	• Alignment of real property inventory requirements and other information standards.
		• Unique Identification codes	• Current inventory valuation.	
			• Increased accountability, stewardship of government owned assets and improved planning and budget estimates.	

		• Potential to redeploy / share facilities using analysis linked to mission impacts. • Reduced real property costs.	• Acceptance of loss of control.
Acquisition and Procurement	• Consolidation of DoD-wide buying needs and power via use of shared strategic sourcing and contracts. • Many FM, procurement & acquisition systems across DoD. • Behaviors reflect multiple reviews, reconciliations, & pro-rations of data between FM, acquisition, procurement, and contract management.	• Early stakeholder involvement to minimize control resistance. • Integrated acquisition, procurement and FM life-cycle business rules.	• Better prices, reduced costs and overhead, and reduced delivery time. Improved tracking and control of funds. Facilitate auditable financial statements. • Facilitate DoD-wide process controls. Greater visibility into buyer and seller transactions. • Automated and integrated systems that can facilitate auditable financial statements and provide timely business information to support better decision making. • More efficient disbursements.	• Acquiring people with credit management skills (new role). • Change in law or policy, or both. • Interdepartment funds not linked. • Large backlog of existing contracts. Dealing with archived data.

Table continues on next page.

Table 11.3. Continued

Domain	Structural Consideration	Enablers	Impacts	Change Hurdles
Strategic Planning and Budgeting	• Link budget and performance using metrics. Many different priorities across DoD derived from different policies.	• Strong executive leadership and sponsorship of changes. Incentives to reinforce change behaviors.	• Better information and guidance to develop DoD budgets to reflect current Administration priorities.	• Significant change to existing processes.
	• Inability to define standardized capabilities and metrics.	• Standardized performance metrics, budget and correlation with resource allocations.	• Optimize DoD resources.	• Culture changes that represent loss of power and control over money —loss of historical autonomy within the services.
	• Nonstandardized policies. No DoD-wide perspective.	• Change in incentives to support budget savings.	• Budget will reflect the mission.	
	• Budget approval will be linked to DoD requirements and not to service "shares."	• Legislative or administrative relief.	• Increased DoD credibility with financial stakeholders.	
	• Change Quadrennial Defense Review timing requirements—limits effectiveness.		• Balanced and executable budget tied to strategic goals that reflect actual execution.	

Source: DoD Response to P.L. 107-314, National Defense Authorization Act for FY 2003, Transition Plan, Annex G, Organizational Readiness Assessment, Table 2.1, Organizational Readiness Summary of Raw Data, p. 13, from the DoD Business Management Modernization Program, May 2003 at www.dod.mil/comptroller/bmmp.

control at the DoD level. This was also clear to DoD process action teams (PATs). The report says:

> Wide scale organization resistance is expected as the power base shifts and control of power, processes, systems and information shifts from Service and Agency to a consolidated, integrated and standardized enterprise-wide environment.

This particular finding was common to all the PATs because in the current environment power, processes, standards, and technologies have all been unique to each service and agency. The scope of changes represent losses of power, money and information that will affect many DoD levels, from civilian executives and combatant commanders down to field units, budget planners and accountants. The autonomy of these organizations dates back many years and reflects decision making, Budget Authority, and buying conventions within each of the services and agencies. "This represents a significant change to the DoD culture" (Business Modernization Management Program [BMMP], Annex G, 2003, p. 14)

It is clear from the "impacts" column that the stakes are high; it is not clear that they are high enough to overcome the culture whose history is based in decentralization.

RECENT DEVELOPMENTS IN
DoD FINANCIAL AND BUSINESS MANAGEMENT REFORM

Continuing Business Process Reform

Any assessment of recent changes in DoD financial management must include mention of the evolution of the Business Modernization Management Program (BMMP), established in 2003 under former Defense Secretary Rumsfeld, into the Business Transformation Agency (BTA) in 2005, and the BTA's role in setting policy and standards for the MILDEPS to employ in their individual financial management and related information technology (IT) system reform initiatives. BMMP was created once it was recognized that many of what appeared to be DoD financial problems actually were the result of broader management deficiencies. The predecessor of BMMP was the Financial Modernization Management Program (FMMB), established in 2001 at the deputy level within the DoD office of the comptroller. Notably, the BTA reports to the under secretary for AT&L and not the DoD comptroller.

To some degree the FMMB and its head as a deputy comptroller in DoD comptroller's office was established to handle problems such as the fact that as of 2001 there were more than 4000 "feeder systems" through-out DoD, all of which provided the same, similar or different types of information to be used within DoD and the MILDEPS for accounting, budgeting, reporting and other financial management functions. The existence of so many separate and quasi-independent systems was per-ceived as somewhat of a "mess" by Congress, the GAO and DoD leader-ship. In addition, DoD had been unable to comply with the requirements of the CFOA of 1990 and this was a source of embarrassment for DoD leadership. It was believed that, correctly in our view, the presence of so many unlinked systems contributed to the impossibility of DoD receiving a clean audit opinion from the Office of the Inspector General.

To its credit, the leadership of BMMP expanded the understanding of the problem as something that resulted from poor and badly integrated management systems and not just bad financial management systems. BMMP thus shifted the emphasis of reform from mere compliance with the CFOA to establishing an architecture and uniform standards for IT and related systems development, and gained the authority to approve, direct and oversee the development of all new IT systems in DoD.

BTA, whose combined government and contractor staff had grown to approximately 350 employees by mid-2007, is responsible for setting pol-icies and plans for (a) the business enterprise architecture which was introduced earlier in this chapter as something new, which it is, but is now in its fourth iteration, (b) the enterprise transition plan, and (c) the finan-cial improvement and audit readiness plan. Further, there is now a DoD requirement for joint chief information officer (CIO) and CFO (the DoD comptroller) approval over development and modification of any and all financial management systems and a separate acquisition executive has been established for oversight of such systems. Some of these changes were DoD-initiated, some were made in response to congressional initia-tives and direction, and most have been designed to address the IT and related systems issues and incentives to implement reform in an orches-trated and coordinated manner and to overcome resistance or efforts to undermine reform. How successfully BMA is in achieving these objectives remains to be seen. BMA also is working to establish performance criteria, metrics and measurement methods based on the balanced score card approach so much heralded in the private sector, for assessing the imple-mentation of new systems throughout DoD (Candreva, 2007).

A major issue that DoD FM IT decision makers have faced since the 1990s is whether to develop a single master financial management (FM) system and to impose it top-down from DoD, or to invest in the design and improvement of smaller IT systems used or to be used in the MILDEPS. The answer to this dilemma appears to be that both top-down (system-wide architecture and standards) and bottom-up initiatives are being pursued at the same time. In our view this is inevitable and probably the best way to achieve the overall goal of having superior FM IT systems throughout the organization. However, the cost of achieving this goal is high and funding for these initiatives is scarce during time of war and recapitalization of war fighting platforms, systems and equipment which has a higher priority within DoD.

Defense Financial Accounting Service (DFAS) Success

Another area of successful reform to mention is the continued evolution of the DFAS. DFAS has continued to expand the scope of its business has so that the agency now provides services to as much as 30%-50% over the numbers 5 years earlier. Further, DFAS has continued to consolidate its operations, for example, they have reduced their operations from over 300 sites to only 15 and have a goal of only five sites by 2009. In addition, where 5 years ago GAO complained about interest penalties for late contractor payments, the latest GAO audit indicates that DoD now pays invoices too quickly. The accuracy and timeliness of invoice payment has increased markedly due to successful IT improvements (Web-based invoicing, wide-area workflow, etc.) which have largely resolved some of the contract payment issues we noted earlier in this chapter.

Observing recent accomplishments and activities of the DFAS demonstrates how progress is being made in improving some very significant financial and business processes.

In FY 2006, DFAS:

- Paid 145.3 million pay transactions (5.9 million people)
- Made 7 million travel payments
- Paid 13.8 million commercial invoices
- Posted 57 million general ledger transactions
- Managed military retirement and health benefits funds ($255 billion)
- Made an average of $424 billion in disbursements to pay recipients
- Managed $20.9 billion in foreign military sales (reimbursed by foreign governments)
- Accounted for 878 active DoD appropriations

DFAS has fifteen operating sites while they all do accounting and financial services, what each does varies a little from the other. This may be seen in the below in the materials provided online by DFAS Charleston and DFAS Columbus.

DFAS Charleston provides the following services to the Department of the Navy:

- 3,048 financial/accounting reports
- 37,000 invoices
- 5,000 travel vouchers

The site also pays 164,000 payroll accounts worldwide with bi-weekly gross earnings of more than $250 million. The payroll office works three shifts a day to pay the worldwide base of DoD civilian employees.

In FY06, DFAS Columbus was responsible for:

- Accounting support and financial statements, analyses, and reports to customers.

- Cash reconciliation and the processing of interfund and reimbursable billings, including those from foreign governments and international agencies.

- Agency-level financial reports and statements required by the CFOA of 1990.

- The clean audit opinion of three defense agencies.

- Contractor and vendor payments of more than $246 billion annually.

- Travel pay operations' processing of more than 135,000 temporary duty (TDY) and permanent duty travel (PDT) claims for defense agency, Army, Army materiel command (AMC), and Navy (working capital fund and revolving fund activity) customers in the amount of $126.5 million.

- Over 7,800 W-2s prepared for calendar year 2006, pertaining to taxable entitlements paid in conjunction with PDT moves.

- The payment of 5.7 million disbursements, exceeding $189.1 billion, including DFAS Cleveland checks that are printed by DFAS Columbus. Over 99% of the Columbus disbursements are processed electronically.

- The processing of over 191,000 collections, totaling more than $20.6 billion.

DFAS was established January 15, 1991 to improve DoD financial management by consolidating and standardizing DoD finance and accounting procedures, operations, and systems. In 1992 DFAS took control of 338 DoD finance and accounting offices operated by the military services and defense agencies. It has steadily consolidated and centralized these offices into 5 centers and 20 operating locations, reducing the number of personnel from 31,000 in 1992 to 20,000 in 1998 to 16,000 in 2007. Finance and accounting systems have been reduced from 324 in 1991 to 109 in 1998 to 65 in 2005 with plans to go to 9 finance and 23 accounting systems (Candreva, 2005, p. 113; Hleba, 2001, pp. 73-74).

The 324 systems that DFAS took over in 1992 used multiple nonstandard practices and procedures. All such legacy systems were developed to meet the unique demands of each military component as it interpreted high level policy in support of fund control of appropriated funds, budget execution laws and rules, and reporting requirements. However, all these systems were a little bit different because they were designed by different people in different parts of DoD and the situation was worsened as many smaller nonstandardized accounting systems grew that fed data into primary DoD systems. These feeder systems varied from automated applications to spreadsheets to records kept on paper. The result was that DoD finance and accounting practices were not compliant with federal accounting and financial management requirements and could not pass an independent audit test.

While DoD has been responsive to the CFOA of 1990 that set much of this modernization in motion, DoD and its major components still cannot pass an independent audit 12 years later, nor is a remedy expected soon. Contract administration continues to be a significant problem, particularly with the use of interagency contracts, DFAS provides evidence of what can happen when expert civilian senior executive service employees remain in charge of operations and management for an extended period of time and are allowed to manage what is an essentially civilian function without using military labor—only 250 of the DFAS workforce of 12,000 are uniformed service members.

Debate Over Establishing a Chief Management Officer in DoD

Since this chapter focuses to a considerable extent on broader business improvement along with financial management reform, it is worth noting the prolonged debate over the establishment and the role of a chief management officer for DoD. GAO advanced this proposal assertively over a period of several years, but DoD formally responded in mid-2007 that the deputy secretary of defense, in fact, presently performs the role GAO has envisioned for a DoD CMO. However, the Senate Authorization Subcommittee on national defense put language in its version of the 2008 authorization bill to formally create the new position of under secretary of defense (USD) (management) which would serve directly under the secretary of defense, as does the USD presently, and above the undersecretaries. Whether this amendment will survive full committee, floor and conference committee review and will be enacted (and not be vetoed by the President) also remains to be seen.

Finally, it should be noted that the individual military departments and services continue to pursue their own approaches to organizational and system (including IT system reform, e.g., the NAVY's continued investment in an ERP-type financial and management information system. Another major DoD management initiative, apparently successful thus far, is performance-based logistics and RFID (radio frequency identification device) technology. Implementation an enterprise management system and process is a further reform in progress.

CONCLUSIONS

While financial and business process management may appear unexciting, its failures have significant consequences. Speaking of the planning and accounting processes in DoD Franklyn Spinney (2002) stated:

> The historical books cannot pass the routine audits required by law and planning data systematically misrepresents the future consequences of current decisions. The double breakdown in these information links makes it impossible for decision makers to assemble the information needed to synthesize a coherent defense plan that is both accountable to the American people and responsive to the changing threats, opportunities and constraints, of an uncertain world. (p. 81)

DoD financial management problems are both long-standing and persistent. Opening yet another hearing on DoD, on the June 4, 2002, Chairman Dennis Kucinich (2002a) said in his opening statement:

> Today marks the eighth hearing this subcommittee has held relating to the Department of Defense's financial management or mismanagement problems during the 107th Congress. The Subcommittee on Government Efficiency, Financial Management and International Relations chaired by Representative Horn, has also held hearings on the subject. The House Armed Services Committee, too, has heard testimony about the Pentagon's accounting troubles. So has the Senate Armed Services Committee and other senate panels, but we needn't stop there. Since the Chief Financial Officers Act of 1990 which established basic financial reporting requirements for federal agencies took effect, this subcommittee has held dozens of hearings on the Defense Department's financial mismanagement difficulties. In all these sessions, no matter who has testified, the comptroller general, the inspector general, the chairman of independent

commissions such as Mr. Friedman who is with us today, the message has been constant that the Defense Department's financial mismanagement situation or management situation is in a shambles. That no major part of the defense department has ever passed a test of an independent audit. That the Pentagon cannot properly account for trillions of dollars in transactions.... That no less than six Pentagon functions, more than any other government agency are at high risk of waste, fraud and abuse, and show little prospect for improvement. That the Department of Defense writes off as lost, tens of billions of dollars worth of in-transit inventory and that it stores billions worth of spare parts it doesn't need. That it will take nothing less than a complete transformation in culture at the Defense Department and a full commitment at the highest levels of leadership at the Pentagon to fix the situation. (p. 3)

Ultimately, these seemingly prosaic difficulties in financial management lead to an underprepared defense establishment bought at a higher cost than necessary. In this chapter we have attempted to understand what this means and why reform is so difficult. Testifying before Congress in 1994, then DoD Comptroller John Hamre (1994) noted that DoD had paid contractors $1.3 billion more that it should have in 1993, had paid 1100 personnel after they had left the Army, including some deserters, and could not match $19 billion in disbursements to acquisition contracts. Hamre said the department had troubled systems, but that he would not blame his predecessor. The problem, he said, was actually much older than that.

I do not want to lay these charges at the feet of my predecessor, for he too inherited this flawed system. Indeed, our deep-seated weaknesses stretch back to the founding of the Republic. In 1775, the Continental Congress appointed James Warren to be the first Paymaster General. He was in effect the first Comptroller for the Department of Defense, my predecessor. After 6 months in the job, he wrote to the Continental Congress saying he could not do his job properly because of the flaws in the financial management systems he inherited. He complained that each of the 13 colonies insisted on its own payroll system and they were not standardized. The overall system was open to abuse. Frequently individuals would sign up for the militia for one colony to receive the sign-up bonus, only to desert and join another militia to receive its bonus. Pay was not standardized. Uniforms were not uniform. It was chaos. When Dr. Perry asked me to undertake a thorough assessment of our financial management systems, I reported back to him that we have actually made tremendous progress in the past 200 years. We have added 37 states to the Union and only 5 additional payroll systems. (p. 6)

Hamre (1994) went on to explain that when DoD was established in 1947, it retained the existing military departments (Army, Air Force, and

Navy) with their vertical chain-of-command mode of operations. Hamre observed that this vertical chain-of-command organization was essential for success on the battlefield, but it had distinct consequences for peacetime operations:

> Management systems, including financial ones, were geared to report information up through these vertical channels. When computers came along and every organization sought to automate its processes, these organizations were not compelled to emphasize horizontal connections across organizations of like functions, such as pay or contracting. Instead, computers were used to automate formerly manual procedures. Financial management systems were designed within the chain of command to support the commander of that operation. (p. 7)

This is the explanation for the development of vertically separate systems, so picturesquely described as stovepiped. Hamre (1994) explained that this process did not stop. While the business of defense demanded integration, the business of defense financial management continued to fragment: "As the Department of Defense matured, certain activities—such as contract management—were made common across the Department. But, this process of standardization really produced yet additional collections of vertically-oriented chain-of-command organizations" (p. 8).

All sorts of bad results flow from this problem. When the DFAS was created in 1991 there were 66 major finance systems and 161 major accounting systems. These numbers have been reduced substantially, but the original stovepipe tendencies still exist and in practice are reflected in the operations of the department, in particular in the unique and disparate feeder systems that pour information into the main accounting and finance systems. While experts differ, the accepted count by the BMMP in May 2003 was 2274 (shown in Table 11.4). We may note that the human resource domain has the most systems, but that the Logistics domain systems are the most expensive on average. The number of systems changes as DoD goes about mapping out its architecture.

DoD continues to confront and whittle away at the problems it faces, but their durability makes one wonder about the prospects for reform. Bad data, overly complex and redundant systems, unique feeder systems that may or may not be reliable, outdated business practices, use it or lose it rules, a focus on audit to the detriment of providing useful management data, DoD continues to face these problems. Deeply troubling is the idea that budgets and appropriation systems were the primary driver for the vast majority of DoD financial system users, but that these uses and systems were incompatible with good management

**Table 11.4. DoD Systems Inventory, April 30, 2003:
Number of Systems by Domain and Dollar Cost**

Domain	*Number*	*Cost*	*FY 2003 Av. Cost*
Acquisition and Procurement	143	$135,281	$946
Finance, Accounting, and Financial Management	542	$726,080	$1,340
Human Resource Management	665	$1,524,809	$2,293
Installations and Environment	128	$229,899	$1,796
Logistics	565	$2,044,490	$3,619
Strategic, Planning, and Budgeting	210	$215,405	$1,026
Technical Infrastructure	21	$37,728	$1,797
Total	2274	$4,913,692	$2,161

Source: Business Management Modernization Program (BMMP), 2003. "Organizational Readiness Assessment, Table 2.1," and "Organizational Readiness Summary of Raw Data," May, www.dod.mil/comptroller/bmmp

information systems, generally accepted accounting practices as required by the CFOA and preclude the use of commercial, off the shelf software. Some users felt they even did not suit good budget systems (Friedman, 2002) although others disagreed, "They were designed to do appropriations and congressional reporting requirements ... and they did that very well" (Lanzillotta, 2003, p. 4). Still, the allure of reform is there, not only to conform to legal requirements, but because reform of financial management systems could free up between $15 to $30 billion a year that is estimated to be wasted by not adopting modern business systems (Friedman, 2002; Platts, 2003). This money would be better spent purchasing badly needed force structure assets.

Reformers are serious, there can be no doubt about that, but the foregoing seems a recipe for failure. Moreover, DoD culture is strong and significant change is slow. Gregory Kutz (2003) of GAO has noted that reform has been difficult in DoD up to now because of, "the lack of sustained top level leadership and accountability, cultural resistance to change, including service parochialism, lack of results oriented performance measures and inadequate incentives for change" (p. 5). What all of this seems to say is that an "A" grade for military operations but a "D" for financial management modernization is better than vice versa. Nonetheless, financial management reform in DoD soldiers on, a triumph of hope over history.

Some Positive Results

Speaking for DoD, Deputy Under Secretary of Defense for Management Reform Lawrence Lanzillotta (2003) offered encouragement and described how reform will come about:

> In the past, each major DoD organization was allowed to design and manage its own systems without having to integrate it into a DoD wide architecture. This created a stovepipe support structure that was inefficient, unresponsive to leaders' needs.... We plan to use a DoD wide architecture to describe standard business and financial rules, employ a DoD wide oversight process directed by senior leaders to implement the architecture and to guide spending, refine and extend the architecture to create a seamless connection between it and other federal and DoD transformation initiatives and in the near term, address the critical financial problems, notably financial reporting. (p. 9)

Lanzillotta (2003) continued by describing how DoD financial management had been divided into seven domains including logistics, acquisition, procurement, each studied and led by experts to re-engineer the processes of that domain across the department,

> Domain leaders will implement the architecture by re-engineering business processes and develop a system solution that is consistent with the enterprise architecture. In this way, leaders who are expert in each domain will reengineer the way the department does business and formulate business improvements. (p. 9)

While Lanzillotta (2003) was optimistic about outcomes, members of the committee alluded to the long track record of inconsistent performance and stovepiped systems and suggested that when it came to changing feeder systems that belong uniquely to the different military departments, progress would not be as swift or the outcome as sure as changing those systems owned by DoD at the central level. Lanzillotta indicated that the budget would be the disciplining force, but acknowledged that it would be a long haul, if for no other reason than this was a ten-year program embedded in an environment where technology changed every 18 months. Still he said,

> I never believed ... that we'd just have a bad opinion and then one day in the future we would just have a good opinion. It was going to be an incremental approach. We would get things like liabilities taken care of.... We'd get more Defense agencies. And before too long, we'd hope to have the service organization pop and then I think the rest of it will come. (p. 9)

Much of what Congress focuses on in oversight of DoD financial management is reported with great pessimism, but the fact is that progress has been made. In 2003, Secretary Rumsfeld continued to fight the battle for transformation on the financial management front. He gave reform a top priority and created a top-level leadership team. In April, Team IBM, a consortium of six contractors, reported its financial management architecture for DoD. Director of the Defense Auditing Service Paul Granetto (2003) said,

> On April 30, 2003, the Business Management Modernization Program delivered the initial Business Enterprise Architecture, which is currently in the implementation phase. The Architecture is essentially a blueprint describing the Department's future financial management systems and processes. (p. 12)

GAO finds fault here and there with the structure, but this is an ongoing effort. Granetto added that DoD sees the process of reform as a program, just like any other major DoD program and this must take into account, "all of the appropriate controls required of a very large program. Those include a master plan, well-defined management accountability, full visibility in the budget, regular performance reporting, and comprehensive audit coverage" (p. 22).

Here and there within DoD, substantial progress has been made. For proof of this, we look to the achievements of DFAS. The Director of DFAS, Thomas Bloom (2003), first described the scope of DFAS initiatives:

> DFAS is the world's largest finance and accounting operation. In Fiscal Year 2002, the DFAS team paid 5.7 million people. We processed 11.2 million invoices from contractors, recorded 124 million accounting transactions and disbursed $346.6 billion. We paid 7.3 million travel vouchers, managed more than $176 billion in military retirement trust funds, and accounted for more than $12.5 billion in foreign military sales. We are responsible for 267 active DoD appropriations. (p. 27)

Of the world's largest finance and accounting operation, Bloom (2003) then said,

> I am proud of DFAS success in reducing costs to the taxpayers. In Fiscal Year 2002, we reduced our costs to DoD customers by more than $144 million from Fiscal Year 2001. We are forecasting another $108 million reduction this fiscal year (FY 2003). (p. 28)

He noted that DFAS had reduced its operating costs notwithstanding that its workload had been increasing. Bloom (2003) gave three examples. First, military and civilian pay services had Web-enabled customer access to pay account information through its "myPay" Web site. Bloom noted that approximately 1.7 million customers were using this system and the number increased daily, with some signing on from Iraq. Second, Bloom observed that, in the last 12 months, commercial pay services lowered by 30% the amount of interest paid per million dollars by decreasing the number of overaged invoices to the lowest level in DFAS history, from 9.03% in April 2001 to 4.1% in January 2003. Projections for FY 2003 indicated a savings of approximately $3.5 million in interest payments compared to FY 2001. Third, Bloom said that in FY 2002, accounting services achieved a 99.96% timely delivery rate for departmental accounting reports and reduced the average number of days to produce the reports from 14 to 13 days (Bloom, 2003). Bloom also noted that there were other improvements:

> We reduced problem disbursements by 90 percent from the 1998 baseline. We achieved our third straight DFAS clean audit in Fiscal Year 2002, and we enabled the Defense Contract Audit Agency, the Defense Commissary Agency, Military and Retired Trust Fund and the Defense Threat Reduction Agency to achieve unqualified opinions on their consolidated financial statements.... Because of our increased efficiencies. DoD spends less than one half of one percent of its budget on our services. This is a 20 percent decrease from Fiscal Year 1999 to Fiscal Year 2002. We expect that trend to continue. (p. 6)

He also noted that DFAS success had been recognized outside DoD.

> When we competed our civilian payroll system and operations with the private sector, no one chose to bid against us. More recently, the Office of Personnel Management selected DFAS as one of four agencies to provide payroll services across the Executive Branch. Of the four agencies selected, our payroll operation and system unit costs are the lowest. (p. 7)

Bloom observed that DFAS had also made strides in data automation. He noted that the audited financial statement (DDRS-AFS) enabled rapid data collection from numerous sources and transformed it into financial statements that automated and improved the timeliness of departmental reporting. He explained,

> DDRS transformed a manual process into a web-based solution that promotes standardized processes and report generation from a single DoD database.... The DDRS budgetary module provides the Fiscal Year and Appropriation Level reporting required by the U.S. Treasury and DoD. It is the vital link between the DoD installation level accounting systems and the financial statements. To date, DDRS has completely transformed the department's financial statement process. (Bloom, 2003)

In 2004, DFAS continued to improve its record. The director's annual report on the state of DFAS said:

> The quality of DFAS products and services also improved this past year. We reduced the time to deliver quarterly accounting reports from 45 to 21 days and the amount of time for annual reports from 80 to 45 days. The team lowered the amount of over aged Unmatched Disbursements from $134 million in FY 2003 to $23 million in FY 2004, decreased the amount of interest paid per million disbursed from $160 in FY 2003 to $138 in FY 2004, and expanded the MyPay customer base to 2.9 million people. (Gaddy, Zack. Director. State of DFAS in FY2004. www.dfas.mil)

The performance of DFAS has not gone unnoticed. A senior DoD official commented,

> They have indeed continued the trend you describe ... the facts and figures on the scope of their business are up as much as 30%-50% over the numbers you cite and their consolidation ... they have reduced from over 300 sites to only 15 and are still falling to a goal of only five sites by 2009. Where five years ago, GAO complained of interest penalties for late contractor payments, the latest audit said we now pay invoices too quickly!... DFAS is an interesting case study of what can happen when you leave expert SES's in charge for an extended period of time.

These are major accomplishments. They arise from persistent efforts by committed leaders. With the size and number of challenges facing DoD financial management, there is no other way than to persist and endure. Critical to the progress of financial management reform is how long the current management group stays intact and in place in DoD, and the extent to which a future group of leaders will adopt these approaches. On balance, it appears that reform of DoD financial management will continue, but it also appears that this will be a moving target, replete with harsh criticism on a regular basis, and that whatever is done will lead to cries for more and better systems, information, and leadership. This is good because, in the end, without criticism and identification of system weaknesses, there is no incentive to improve. And, as we have indicated in this chapter, DoD has made significant strides forward in improving its financial and business process management over the past 5 years.

CHAPTER 12

BUDGETING AND MANAGING WEAPONS ACQUISITION

INTRODUCTION

Members of the legislative and executive branches, including the Department of Defense (DoD), the military departments and services, defense agencies, program managers, and special industry interest groups all are involved in budgeting for the acquisition of defense weapons. In the acquisition process, specific attention must be given to the actions and responses of program managers to their external political and budgetary environment. Program managers are the first and last lines of defense for their programs in the political process. However, dealing with the complexities of acquisition budgeting requires the attention of all players in the process. A number of issues seem to persist over time to confront acquisition budgeting. Some of the more important of these issues are addressed in the following sections of this chapter.

REACHING CONSENSUS ON HOW MUCH TO SPEND ON DEFENSE ACQUISITION

As detailed in previous chapters, the President's budget is submitted to Congress each February. Congress regards the President's budget as a

Budgeting, Financial Management, and Acquisition Reform in
The U.S. Department Of Defense, pp. 527–563
Copyright © 2008 by Information Age Publishing

statement of executive priorities to address the national interest and constituent needs. The budget is debated and modified before the final reconciliation is done to enable spending to begin for each new fiscal year
(FY). The FY 2003 budget proposed an addition of $48 billion for
national defense. This represented a 13% increase in constant dollars—
the greatest one year increase since the defense build-up of the 1980s
(Department of the Navy (DoN), 2002a). Defense outlays were projected
to approach 3.5% of gross domestic product in FY 2003, the highest levels
since 1995. In contrast, during the mid-1980s defense spending averaged
nearly 6% of GDP. Funding for acquisition was increased by 13.2% in the
FY 2003 budget and subsequent increases averaged over 10% for FY 2004
through FY 2006. Below is a graphic which captures the current dollar
growth in the procurement account from FY 2003 through FY 2013 as
forecast by DoD in the President's FY 2008 budget. The years after 2007
are estimates and could change dramatically depending on outcomes in
Iraq and/or the elections of 2008.

The President's FY 2003 budget projected a much slower growth in
Budget Authority for defense—for an average annual rate of 3.2%
through 2012 (DoD, 2003c), but by the FY 2008 budget this projection

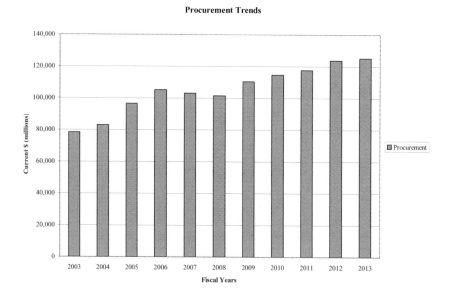

Source: Authors, 2007, from Table 6.8: Department of Defense Budget Authority by Title,
p. 155, DoD Greenbook, March (in current dollars in millions by fiscal year).

Figure 12.1.   DoD procurement trends, 2003-2013.

increased slightly as the procurement account was forecast to increase an average of almost 5% from FY 2003 through FY 2013. Nonetheless it is not clear that these increases will solve DoD problems with aging weapon systems and the need for recapitalization (e.g., buying ships, aircraft, and tanks on a planned replacement schedule to modernize the inventory). As presented earlier, FY 2008 and FY 2009 budget negotiations illustrated that pressures to recapitalize weapons systems had not abated. These issues and their resolution demonstrated the inevitable tension in funding national defense between how much is enough and how much can be afforded. As noted, the perception of threat and is crucial and always open to interpretation. Further, constituent demands for defense spending, and particularly for acquisition dollars, typically is well-represented— new acquisition programs often mean new jobs and increased revenues to companies in the defense industry.

There is no magic number for spending for weapons asset acquisition. For perspective on the relative size of acquisition/procurement spending as a proportion of total DoD spending, Table 12.1 shows DoD appropriations by major account. Procurement represented approximately 22% of the total while O&M was 36% in FY 2008. Generally speaking, the Pentagon wants as much money as would seem commensurate with meeting the threat and realistic program development and execution. However, defense cannot have it all and Congress often articulates and debates the trade-offs that must be made between "guns and butter." Some insiders believe the defense acquisition budget process is best characterized as a reaction to contingencies. The budget is expected to respond to contingencies in the external threat environment, and the changing political priorities of the President and Congress. Further, the change in the annual defense budget is in many ways less important than the trend represented in the President's projected requests for future years and Congressional Budget Resolution out-year spending targets.

Table 12.1.   DoD Appropriations by Account FY 2005 to FY 2008

	FY 2005	*FY 2006*	*FY 2007*	*FY 2008*
Military personnel	121,279	128,483	118,740	118,920
Operation and maintenance	179,215	213,532	191,598	165,344
Procurement	96,614	105,371	103,211	101,679
RDT&E	68,825	72,855	75,684	75,117
Revolving and management funds	7,880	4,754	2,241	2,454
DoD bill	473,813	524,995	491,474	463,514

Source: Authors, 2007, based on DoD Greenbook FY 2008.

The debate about how much spending is enough in effect results in a de facto multiyear budgeting approach for defense and acquisition, subject to annual revision. Budgeting for the rest of the discretionary budget generally focuses more on annual appropriations. In this respect, national defense budgeting is different. The stakes are higher when budget decisions over annual spending in the tens or hundreds of billions of dollars are accumulated over a multiyear period. Only the overall federal deficit or surplus and spending for entitlement programs command the type of attention given to defense and acquisition budgets. Because the amount of spending is of such volume and magnitude, reaching consensus on how much to spend, and what to spend it on, is always difficult to negotiate in both the executive and legislative branches. Absence of consensus tends to result in longer-term swings upward and downward, which makes acquisition of military assets harder to plan, budget, and execute (Jones & Bixler, 1992, p. 9).

SUSTAINING THE INDUSTRIAL BASE FOR DEFENSE ASSET PRODUCTION

The President's FY 2003 to FY 2009 and out-year defense acquisition spending plans all have been intended to assist in recovery from defense spending cuts at the end of the Cold War and during the 1990s. These requests and projections have drawn concern from industry groups. Among those most vocal were defense industry executives, including former Assistant Navy Secretary John Douglass, President of the Aerospace Industries Association; retired Air Force Lt. General Larry D. Farrell, President of the National Defense Industries Association; and former Congressman Dave McCurdy, President of the Electronic Industries. For example, Douglass remarked about ship construction funding, "You can't maintain a defense base on five ships a year" (Douglass, 2002). Defense industry representative distress was caused by a projected build rate of only five ships per year—not enough in their view. They said that the nation needed a viable defense industry to be prepared for a time of war and cited the reduced number of makers of major defense platforms as a cause for concern.

To combat the decline in shipbuilding and to maintain a force of 310 ships, the Navy acquisition plan includes the construction of an average of nine ships per year. However, getting the money in the DoD budget for construction at this rate has not been successful. After two DDG-51 destroyer ships were requested in the FY 2003 President's budget, the shipbuilding association asked Congress to add $935 million to the budget to procure a third DDG-51 in FY 2003 and fund advance acquisition for a third DDG-51 in FY 2004. This increase would, "move the Navy closer" (Farrell, 2002) to the requirement that it procure four DDG-51s a year, the rate needed to sustain a fleet of 116 destroyers and other surface combatant ships according to the National Defense Industries Association. In response to industry lobbying, a third DDG 51 was added by Congress to the FY 2003 budget.

Consolidations, mergers, and bankruptcies have reduced the number of major weapons systems contractors. Contractors making Navy surface ships shrank from eight to three from 1990 to 2000, as did the number of companies producing fixed-wing military aircraft (Ahearn, 2002). As of 2003, there still were two makers of submarines, although this was only because Northrop won a bidding contest with GD Corporation. Rotorcraft makers, such as helicopters and the V-22 Osprey tilt-rotor airplane, declined from four to three, while makers of strategic missiles shrank from three to two during this period (Ahearn, 2002). The number of companies filling contracts in the undersea warfare area fell by two-thirds, from 15 to 5, while producers of torpedoes slipped from 3 to 2 (Farrell, 2002).

Farrell (2002) estimated that, "shipyards are operating at 50 percent of capacity," which is inefficient, and costs the Navy, "hundreds of millions of dollars annually" compared to costs of operating shipyards at higher, more efficient output levels. Rep. Norm Dicks (D-Wash.), a member of the House Defense Appropriations Subcommittee, charged that the Bush administration budget fell short of what several studies showed was needed to make up for years of underfunding acquisition. The Congressional Budget Office (CBO), a trusted source for such studies, found that acquisition should rise to at least $94 billion (Selinger, 2002).

Congressman Dicks wrote, "The Defense Department acquisition budget is in crisis" (Selinger, 2002). The congressman said acquisition levels were not only inadequate to sustain the force structure, but were driving up operation and maintenance costs because aging weapon systems were not being replaced quickly enough. As an example, he noted that the aging Navy F-14 Tomcat aircraft experienced a 227% increase in maintenance hours per flying hour from 1992 to 1999. Aging equipment, increased equipment complexity, and quality of life issues have increased O&M spending. Not only have O&M costs grown, but they have grown faster than anticipated.

Former Secretary of Defense (SECDEF) Donald Rumsfeld indicated that the low rate of ship production was not a problem in the near term due to the relatively young age of the fleet, averaging 16 years in 2002 (Wolfe, 2002). However, the expected average ship age is projected to increase as fewer ships are built. Secretary Rumsfeld supported the President's FY 2003 budget and was critical of the DoD trend allowing the aging of Navy combat assets (Rumsfeld, 2002). Congressional efforts to add acquisition dollars into the budget over the President's request for recapitalization have been directed towards reducing fleet average age, which also would reduce some O&M costs. The Navy has had to make difficult choices in its budget to fully fund spare parts, munitions and steaming hours, as well as adding capability through ship conversion.

The Navy is not the only service impacted by aging equipment. The Air Force also faces the consequences of an aging fleet. In its FY 2008 budget presentation the Air Force tied the obsolescent fleet to a decline in readiness, saying "Readiness has declined 17% since 2001, due to operating a smaller, older fleet" (Faykes, 2007, p. 15).

The Future Years Defense Plan (FYDP) in 2003 called for a build rate of 5 ships in FY 2004, 7 in FY 2005, 7 in FY 2006, and 10 in FY 2007. Former SECDEF Rumsfeld indicated that contractor problems and more realistic cost estimates for weapons systems by DoD raised costs and resulted in fewer ships requested in FY 2003 (Defense Daily International, 2002, p. 23). Rumsfeld noted that the military services traditionally have underestimated contract costs. For example, in FY 2003, the Navy paid $600 million for past shipbuilding bills resulting from previously underestimated costs. Rep. Gene Taylor (D-MI) reminded Rumsfeld that the real

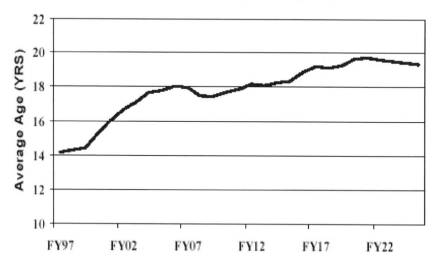

Source: Knox, 2002, p. 27.

Figure 12.2.   Expected average ship age: Navy.

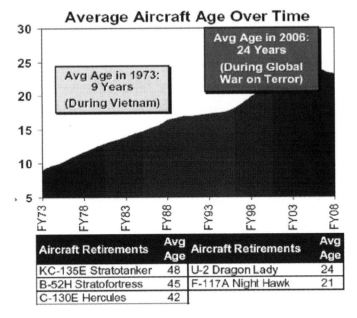

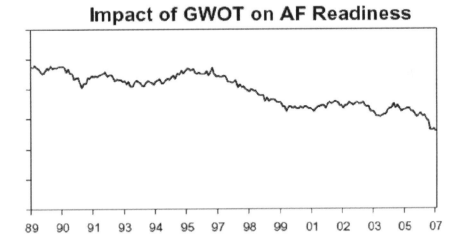

Source: USAF Budget Brief, 2007. President's Budget for FY 2008, Gen. Frank Faykes, Director, Air Force Budget, Feb. 5 2007, p. 14.

Figure 12.3.   Average age of aircraft 2006: Air Force.

Source: Faykes, 2007, p. 15.

Figure 12.4.   Readiness down 17% due to lack of modernization.

problem was not enough contracts and not enough money for shipbuild-ers. Taylor told Rumsfeld that no company would try to build a shipyard, given the poor return on current shipyard operations (Defense Daily International, 2002, p. 23). Later we discuss some of the complications in buying these new weapons systems.

OPERATIONS AND MAINTENANCE VERSUS ACQUISITION INVESTMENT

O&M costs have increased since 1997 and comprised a 39% share of the DoD budget in FY 2002, substantially more than its Cold War share. The decline in procurement from Cold War levels is evident. Moreover, according to the General Accountability Office (GAO), the DoD 2001 FYDP consistently understated cost and overstated savings projections in O&M (GAO, 2000c). The core problem is that planned spending increases for acquisition may be squeezed out to pay for O&M funding shortfalls. For example, there is a gap between Army stated requirements and the DoD planned missile acquisition for the Patriot Advanced Capability-3 missile. According to GAO, analysis of the costs, benefits, and alternatives for defending U.S. forces and assets by DoD is weak and needs to be improved (GAO, 2001a). GAO suggested that better analysis was needed to allow decision makers in DoD and Congress to make decisions on the number of missiles to buy.

Similar issues regarding the vulnerability of surface ships may not be reflected adequately in the budget for ship defense programs. On shipbuilding and maintenance for example, according to Office of Management and Budget (OMB), the Navy ship depot maintenance budget will support 95.5% of O&M requirements and 100% of the shipbuilding and construction, Navy (SCN) account demand for ship overhauls requirement in FY 2003. However, according to DoD and the Navy, the OMB estimate is optimistic and does not accurately reflect the Navy fleet state of readiness. This is an even greater problem after the war in Afghanistan in 2002 and the war in Iraq in 2003 where Navy assets were used heavily. With the decline in the number of battle force ships, the ability to maintain fleet operating tempo is pressed to the limit.

The spiking of ship operating tempos (OPTEMPO) during the war on terrorism has resulted in increased depot maintenance costs (DoN, 2002a). In 1993, the Navy had 108 ships forward deployed; this represented 24% of its 458 ship battle force. In 2003, the Navy projected that 87 of its 308 ships would be deployed—28% of the battle force (DoN, 2002a). In fact, this estimate was low due to the size of the force needed to support the war in Iraq. The high rate of utilization along with the aging of assets inevitably results in depot maintenance for ships and other assets that exceed costs projected in budgets (DoN, 2002a). It would appear that reductions in O&M spending and military personnel appropriations are necessary to generate the savings required to adequately fund ship and other weapons modernization. Supplemental funding for the Iraq war provided by Congress in 2003 helped, but was not enough to offset the difference. Already in the summer of 2003, calls are heard for more personnel for the Army; any move in this direction would further reduce the recapitalization budget and increase the O&M as well as the personnel budgets.

LIMITED FUNDS FORCE ACQUISITION CHANGES

In 2002 Admiral Vern Clark, then the Navy Chief of Naval Operations (CNO), observed that the Navy needed $12 billion more per year than it was receiving to buy aircraft, ships, and other major weapons systems. Clark indicated that the Navy must make a $12 billion a year commitment to shipbuilding to have an adequate Navy in the future. "We can't undo what has happened over the course of years in under-funding acquisition accounts," Clark said in arguing that acquisition accounts must grow over the FYDP. "We must buy more ships and aircraft to meet the needs of tomorrow's Navy" (Aerospace Daily International, 2002, p. 10). Years of underfunded weapons acquisition programs have also contributed to aging aircraft.

The trade-offs forced by limited funding for acquisition produce heated discussion both inside the Pentagon and in Congress. For example, the controversial acquisition of the Joint Strike Fighter (JSF) has competed with and driven out acquisition dollars for other assets for all of the military services. In essence, the JSF acquisition is the initiative of the DoD to attempt to save money by buying a single aircraft type that all services can use for strike capability. Funding for the JSF has been taken from each of the military service budgets, thereby reducing funding to support development and delivery of other systems. Former Secretary of the Navy, Gordon England told the Senate Armed Services Committee in 2002 that the Navy had funded the JSF program adequately. The Navy provided $1.7 billion to the Lockheed Martin JSF when it could have bought other aircraft such as the F/A-18 or other aircraft or ships. The JSF is intended to enhance in Navy strike capability for future war fighting forces (Defense Daily, 2002, p. 4). However, the issue for the Navy and other services is how to sustain acquisition of other assets while paying for this expensive program.

For the Navy, funding the shipbuilding program remains a critical problem. In House defense appropriations hearings, Representative Jo Anne Davis (R-VA) noted that shifting future carrier acquisition by one year and moving the DD-21 acquisition program into research and development program rather than an actual buy would result in a "huge dip" in the future work force at Northrop Grumman Newport News Shipbuilding. "Right now, they're having problems with their work force," she said, adding that it was difficult to hire specialized shipyard workers again once they are let go (Davis, 2002). Congressman Rob Simmons (R-CT) said that laid-off workers at General Dynamics (GD) Electric Boat submarine facility often required at least 2 years to get their security clearances back once they are rehired (Defense Daily International, 2002, p. 23).

Congressman Reed, whose state is home to the GD Electric Boat facility, stated that buying one attack submarine a year would not be adequate to sustain a fleet of 55 attack submarines as planned in the 2001 Quadrennial Defense Review. An attack submarine life is about 30 years. Reed said the acquisition rate of two attack submarines has been proposed for years, but has been continually put off due to budget constraints. He warned that further delays in increasing the rate would create a deeper acquisition shortfall that will be even harder to overcome (Defense Daily International, 2002, p. 23). The shipbuilding association proposed that Congress add $415 million to the Bush administration budget to fund advance acquisition to allow the Navy to reach an acquisition rate of two attack submarines a year by FY 2005.

INADEQUATE FYDP ESTIMATES FOR ACQUISITION

Over the past decade, many proposals to reduce and streamline DoD infrastructure have been debated, and some have been implemented to generate savings to modernize weapon systems (GAO, 2001a). DoD officials have repeatedly emphasized in congressional budget hearings the importance of using resources for the highest priority operational and investment needs. Infrastructure reductions are difficult and painful because achieving significant cost savings requires up-front investment, closure of installations, and the elimination of military and civilian jobs. Further, promised infrastructure savings have not fully materialized. It was expected that the next round of base closure and realignment scheduled for 2005 would address the problem. However, the anticipated Base Reorganization and Closure Committee (BRAC) was not done by DoD and Congress in 2005 and past efforts to cut bases have not saved what was projected. FYDP since this time also have not been realistic or accurate in either force structure projection or cost estimation.

The 1988, 1991, and 1993 base realignment and closure (BRAC) produced decisions to fully or partially close 70 major domestic bases and resulted in a 15% reduction in plant replacement value. Between FY 1996 and FY 2001, no significant savings resulted from infrastructure reforms and the proportion of infrastructure spending in DoD budgets remained constant. The 1995 BRAC was supposed to reduce the overall domestic base structure by a minimum of another 15%, for a total 30% reduction in DoD-wide plant replacement value. However, the 1995 closures and realignments resulted in a total reduction of approximately 21%, 9% short of the DoD goal (GAO, 2001e). BRACs since this time have realized less than anticipated immediate savings. In general, BRAC savings that have been realized have come over a longer period of time than initially projected.

DoD has attempted to resequence its acquisition spending timelines. DoD has reduced planned acquisition in successive FYDPs and has reprogrammed some acquisition to the years beyond the FYDP. Optimistic FYDP planning results in uncertainty with regard to defense priorities. The result of all this is that tough decisions and trade-offs have been avoided and pushed into the future (GAO, 2001a). Figure 12.5 shows that, according to the CBO, a large investment increase will be required to make up for deferred funding. This view conforms to that of the military departments. The challenge for senior DoD officials and program managers is how best to maximize weapons acquisition dollars in an uncertain funding environment where competing demands are numerous. Reflecting on DoD near term cuts in acquisition and procurement rates for force structure by all the services in the FY 2004 budget, Farrell (2003) worried:

> the reality is that the force structure now being cut may never come back, given a projected budgetary environment where deficits are growing and Social Security benefit claims will skyrocket as the baby boomers retire in droves. More than likely, this budget could mark the beginning of a gradual, unannounced force structure decline that is likely to be permanent.

By late 2007 nothing had happened to change the accuracy of this observation.

UNCERTAINTY AND LOW RATES OF INITIAL PRODUCTION

Program managers have been forced to sponsor low rates of initial production (LRIP) of military hardware by industry because of inability to demonstrate that weapon systems will work as designed. However, on

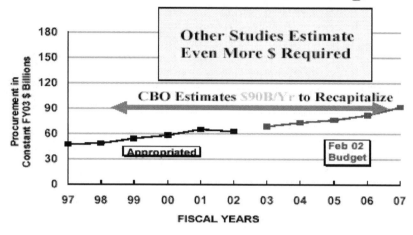

Source: Knox, 2002, p. 32.

Figure 12.5.   Recapitalization needs remain large.

many buys DoD still has begun full-rate production of major and secondary weapons without first ensuring that the systems meet critical performance requirements. For example, the F/A-22 aircraft program involves considerable technical risk because it embodies technological advances that are critical to its operational success. Nevertheless, DoD and the Air Force began production of the F/A-22 aircraft well before beginning initial operational testing, committing to a buy of 70 aircraft at a cost of $14 billion before initial operational testing was complete (GAO, 2003c). The GAO has reported that DoD policy to begin low-rate initial production of weapons with little or no operational testing and evaluation (OT&E) has in some cases resulted in acquisition of substantial quantities of unsatisfactory weapons (GAO, 2003c).

OT&E is the primary means of assessing weapon system performance in a combat-representative environment. It consists of field tests, conducted under simulated "realistic" conditions, to determine the effectiveness and suitability of a fixed asset for use in combat by military users. The options available to DoD and Congress are significantly limited when a system proves deficient. Used effectively, OT&E can be a useful internal control tool to ensure that decision makers have good information about weapon system performance (GAO, 2003b). The decision to proceed with production should be made with OT&E data because, in many cases, the LRIP is also the de facto full-rate of production. The primary problem with low initial rate of production is that unit costs are higher. This puts additional pressure on already underfunded projects (GAO, 2003c).

Too often under current practice, program managers begin production of weapon systems before development and OT&E is complete. When this strategy is used, critical decisions are made without adequate information about demonstrated operational effectiveness, reliability, logistics supportability, and readiness for production. Rushing into production before critical tests have been successfully completed results in purchase of weapon systems that do not perform as intended. Premature purchases have resulted in lower-than-expected availability for operations, higher maintenance, and often have led to expensive modifications. However, there are some advantages to properly managed and executed LRIP. LRIP can shorten the acquisition lead-time when utilized in combination with risk mitigation techniques such as adequate testing and evaluation (Yoder, 2003).

UNDERESTIMATION OF PROGRAM COSTS

The prevailing DoD acquisition culture continually generates and supports acquisition of new weapons. This is the role expected of spending agencies in the budget process (McCaffery & Jones, 2001, pp. 164-165). Also typical is the presence of incentives, augmented by special interest lobbying, to override the need to meet weapon requirements at minimal cost. As a result of this, the desire not to underestimate costs and to obtain some slack in the budget, program managers often exhibit a tendency to overestimate future funding and to underestimate program costs. This results in the creation of more programs than can be executed. When inadequate estimates are discovered, it also tends to undermine trust in the program manager (PM) and program.

In defense of the PM, there are increased costs of doing business where budgets are constrained, as noted above with low initial rates of production. For example, program managers often have to reduce, delay, and stretch out programs, substantially increasing the acquisition and lifecycle cost of systems. In addition to the higher unit costs caused by program stretch-outs, the primary downside to the misestimation problem is inability to address valid requirements when resources have been consumed on lower priority projects that were thought to be affordable due to poor cost estimation (GAO, 2003b).

PRESSURE TO REDUCE ACQUISITION CYCLE TIME AND UNRELIABLE COST ESTIMATION

DoD has a compelling need to accelerate the weapon systems acquisition cycle so that weapons are fielded more quickly. It is not in the interest of the fighting forces to take 10 to 15 years from planning for requirements to fielding weapons. An additional incentive to reduce cycle time is to lower average unit production costs. DoD has a goal to increase acquisition spending through "recapitalizing" about $10 billion dollars per year in the next decade by shifting funds to acquisition from other accounts. The DoD inspector general (IG) expressed doubt that planned actions will free up the amount of funding required. The IG has reported that a significant gap exists between weapon systems modernization requirements plans and planned funding shifts and savings (DoD, 2003b).

To make matters worse, in some cases program sponsors have made unrealistically optimistic assumptions about the pace and magnitude of technical achievement, material costs, production rates and savings. Some savings anticipated as a result of increased industry competition to keep cost low, presenting attractive milestone schedules, appear to be unjustified. Budget problems are inevitable when weapon systems cost more to acquire than estimated, take longer to field, and do not perform as promised. This occurs when careful cost estimation is secondary to the speed of fielding a new system (GAO, 2002b). Here, the PM faces a "Catch 22" situation; either meet the need for faster acquisition or reduce costs—the PM is required to promise both results when this outcome is unlikely.

DoD continues to pursue a number of Major Defense Acquisition Programs on the assumption that savings needed to complete an economical buy will materialize. Far too much weapons acquisition planning is based on overly optimistic assumptions about the maturity and availability of enabling technologies. The result is that DoD program and spending plans generally cannot be executed within the funding available (GAO, 1997b). Numerous problems persist in DoD and congressional budgeting and spending practices for weapon system acquisitions, suggesting that wants and needs are not balanced within affordability limitations.

For example, the availability of several billions of dollars in funding that the Air Force has projected for space system expansion is highly uncertain (GAO, 1997b). Figure 12.3 shows the average age of Air Force aircraft and demonstrates the need for new systems. If new funding is not forthcoming, an alternative to acquisition of new aircraft is to invest in superior space-based weapons systems to replace aircraft mission requirements. However, in late 2002, DoD, the President, and Congress had not agreed on funding an increased allocation overall for DoD or by service for the first 6 years of the 18-year FYDP force projection (FY 2000 to FY

2005). The FYDP projects force structure for 18 years and dollars for 6. Despite this problem, for the last 12 years of the FYDP projection (FY 2006 to FY 2017), the Air Force has planned funding increases for program modernization without identifying funding sources, thus creating additional uncertainty and putting the expansion of space systems in jeopardy. Future acquisition spending for aircraft will very likely have to be reduced to fund planned space operations and also cover additional requirements underestimated in the plan and budget, for example, system maintenance.

OBSOLETE SYSTEM REQUIREMENTS

Some weapon systems are still under development and production rates that were designed to meet war fighter needs of a decade past or more. Even though the Cold War threat upon which they were justified has disappeared, many anti-Soviet designed systems are still purchased by DoD. Obsolete requirements and solutions are wasteful and consume dollars needed elsewhere in the acquisition budget (GAO, 1997b). Continued acquisition of weapons and systems that do not satisfy the most critical current and future weapon requirements, and commission of plans for more acquisition to programs than cannot reasonably be expected to meet future needs (and may not even be available in future) impairs efficient resource allocation in DoD. This situation is intensified as the cost for obsolete weapons systems continue to rise while performance becomes increasingly inadequate to the war fighter. Further, delivery schedules continue to slip on modernization, some of which fail to meet current needs. Figure 12.6 shows the rising average cost of weapon replacement.

DoD continues to generate and support acquisition of existing weapons, supposedly modernized to meet future requirements. However, when system upgrades fail to work or are not made, money is wasted. Inherent in the acquisition culture are powerful incentives that influence and motivate what may be termed dysfunctional self-interest behavior by participants in the process: the military departments, Office of the Secretary of Defense (OSD), Congress, and industry. The result is that acquisition money is wasted.

It is not unusual to discover DoD incentives to resist comprehensive force modernization and replacement of assets that coincide with special interest pressure to override the need to meet the most critical new weapon requirements. For example, the Air Force C-17 aircraft continued production despite analysis which showed that if the C-17 program were halted at 40 aircraft, 64 commercial wide-body aircraft could be added to the existing airlift fleet for an estimated life-cycle savings of $6 billion

Fielded and Replacement
System Cost Comparison

<table>
<tr><td colspan="2" align="center">Fielded System
Unit Cost*</td><td colspan="3" align="center">Replacement System
Unit Cost*</td></tr>
<tr><td></td><td></td><td></td><td align="center">Service Est</td><td align="center">CAIG Est</td></tr>
<tr><td>F-15C</td><td>45.2</td><td>F-22</td><td>99.8</td><td>124.0</td></tr>
<tr><td>SSN 688</td><td>845.0</td><td>SSN 774</td><td>1,615.1</td><td>1,620.0</td></tr>
<tr><td>A/V-8B</td><td>29.9</td><td>JSF</td><td>51.6</td><td>70.0</td></tr>
<tr><td>OH-58D</td><td>9.8</td><td>RAH-66</td><td>23.3</td><td>27.5</td></tr>
</table>

*Average Procurement Unit Cost in Constant Year 2000 $M

Note: CAIG estimate made by DoD Cost Analysis Improvement Group.

Source: Knox, 2002, p. 36.

Figure 12.6.   Fielded and replacement system cost comparison.

when compared with acquisition of a fleet of 120 C-17 aircraft. The Air Force acknowledged that there were considerably cheaper alternatives to meet airlift requirements than full production of the C-17, but delayed making a change, instead launching a new study to determine an optimal mix of aircraft to meet airlift requirements (GAO, 1994a). The C-17 continued in production despite significant schedule delays, performance shortfalls (e.g., problems with wings, flaps, and slats) and cost overruns. Political pressure on DoD from members of Congress helped keep the production line alive and caused the Air Force to incur substantial funding opportunity costs. Cultural resistance to change, service parochialism, and public and congressional concern with the economic effects of reduced or cancelled weapons contribute to reluctance to consider cutting programs that may no longer be effective. Moreover, since promotion goes to managers of successful programs, personal career advancement needs result in some program managers pressing for continued support of their programs whether or not they meet the needs of the forces.

A number of decision points in the acquisition process exist to support the need for research on alternative systems before continuing existing weapons programs. The analysis of alternatives (AOA) is one such tool to determine whether weapon systems are needed. The AOA is an analysis of proposed system operational effectiveness related to its life-cycle costs compared to various other alternatives to meet the mission need. Although the military services conduct extensive analyses to justify major acquisitions, these often are narrowly focused and do not fully consider alternative solutions.

The PM's job is to provide analysis, advice and counsel to DoD acquisition executives, particularly regarding the selection and executability of proposed alternatives (DoD, 2002b). Research and technology efforts are not disassociated from weapon programs until they reach the program definition and risk-reduction phase. Historically, military service analyses do not include joint acquisition of systems with other services (GAO, 1997b). Previous failed attempts at joint weapons development such as the TFX Fighter were due to parochialism, cultural biases, and inaccurate requirement determination. In contrast, programs like the JSF may prove to be a success in joint acquisition.

Because DoD does not routinely develop information on joint mission needs and aggregate capabilities, there is little assurance that decisions to buy, modify or retire systems are based upon comprehensive assessment of all appropriate alternatives. This is an area in which the planning component of the Planning, Programming, Budgeting and Execution System (PPBES) and related processes in the acquisition decision system are badly in need of transformation.

OUTDATED ACQUISITION BUDGET AND FINANCIAL SUPPORT SYSTEMS

U.S. defense outlays have purchased many of the world's most capable weapon systems. However, many acquisition contract administration systems and processes are costly and inefficient. DoD continues to rely on a huge number of poorly coordinated and complex networks of financial, logistics, personnel, acquisition, and other management information systems. Roughly 80% of these systems are not under the control of the DoD comptroller or any other DoD official, including the deputy secretary, the under secretary for acquisition, technology and logistics (USD AT&L). These systems gather and store the data needed to support day-to-day management decision making. The Government Performance and Results Act (GPRA) and the Chief Financial Officer Act (CFOA) have not proven effective in forcing DoD reform of this labyrinth of support systems, although some progress has been made.

Many DoD business operations use old, inefficient processes and outmoded "legacy" information systems, some of which were developed as long ago as the 1950s and 1960s. For example, DoD still relies on the

Mechanization of Contract Administration Services (MOCAS) system—which dates to 1968. The development of this network system has not been by design. Instead, like many DoD information systems, it has evolved into an overly complex and error-prone entity that presents many problems in use. These include the usual suspects, a lack of standardization across DoD components, multiple systems performing the same tasks, duplicate data stored in multiple systems, manual data entry into multiple systems, and a large number of data translations and interfaces that combine to reduce data integrity.

The Standard Procurement System (SPS) was intended to replace the contract administration functions currently performed by MOCAS. GAO reported that DoD has not economically justified its investment in SPS because its analysis of costs and benefits was not credible (GAO, 2001e). Although DoD committed to fully implementing SPS by March 31, 2000, this target date slipped by over 3½ years to September 30, 2003, and then slipped further. Whether SPS will ever perform as intended once fully operational remains to be seen as of 2007.

Another example of inadequate support systems is the DoD financial management information system. DoD financial systems have not been able to adequately track and report whether $1.1 billion in earmarked funds that Congress provided for spare parts and associated logistical support were actually used for intended purposes (GAO, 2002a). The vast majority of this funding, approximately 92%, was transferred to military service operation and maintenance accounts. Once the funds were transferred into O&M accounts, DoD could not separately track the use of funds. As a result, Congress lost confidence in DoD's ability to assure that the funds it received for spare parts purchases were used for that purpose. This is only one example of many that indicate the weakness in DoD financial management systems, as explained in greater detail in the previous chapter. It must also be remembered that problems with DoD financial management operations go far beyond accounting and finance systems and processes. Wasteful contract administration practices in some cases have added billions of dollars to defense acquisition costs.

WEAPONS ACQUISITION COST AND RESOURCE MANAGEMENT TRADE OFFS

Some critics believe the defense weapons acquisition process, while pressured from without, also is fundamentally flawed from within. For example, on March 7, 2001, in testimony before Congress, Comptroller General David Walker testified that, "the acquisitions process is fundamentally broken, the contracts process has got problems, and logistics as well" (McCaffery & Jones, 2004, p. 335). In 2005, another GAO study warned that, as a result of inefficient systems and practices, the DoD invited a series of troubling outcomes: "Weapon systems routinely take much longer to field, cost more to buy, and require more support than provided for in investment plans" (GAO, 2005a, p. 68). GAO observed:

> For example, programs move forward with unrealistic program cost and schedule estimates, lack clearly defined and stable requirements, use immature technologies in launching product development, and fail to solidify design and manufacturing processes at appropriate junctures in development. As a result, wants are not always distinguished from needs, problems often surface late in the development process, and fixes tend to be more costly than if caught earlier. (GAO, 2005a, p. 68)

Defense acquisition has long been beset by problems related to both politics and efficiency, as defense business executive and acquisition expert Norman Augustine (2005) observed:

> On the surface, defense acquisition appears to have little in common with commercial acquisition. For starters, defense acquisition occurs in a monopsony. Further, it is replete with mini-monopolies. (From how many places could one have purchased, say, an additional B-2?) Defense acquisition also operates in a governmental system that intentionally traded optimal efficiency for strong checks and balances—such as those implicit in separating the Legislative and Administrative branches. Nonetheless, there are certain fundamentals of sound management which are applicable virtually everywhere, including in the defense acquisition process. They are just much more difficult to apply in government, where the stakes are higher, authority less hierarchical, and the spotlight much brighter.
>
> The problems in defense acquisition—and there are many—tend to be widely misunderstood. Outright dishonesty, for example, is extraordinarily rare...but when it occurs its impact is particularly devastating. Over the years, toilet seats, coffee pots and screwdrivers have also received an abundance of ink, but they are not the problem either. A number of studies of the defense acquisition process have been conducted since the genre was born with the Hoover study in 1949. There is remarkable agreement as to the problems which need to be addressed. The difficulty resides in having the will to do anything about those problems.
>
> Gil Fitzhugh's study in 1966 observed that a fundamental problem is that everyone is responsible for everything and no one is responsible for anything. Dick DeLauer's study in the 1970's concluded that the problem was "turbulence"—perpetually changing budgets, schedules, requirements and people. Dave Packard's somewhat more recent study pointed to the shortage of experienced managers as the root cause of many problems.... But it is important to note that in spite of such criticisms, the Department of Defense's acquisition process has provided our armed forces with the

equipment that is the envy of the world's military forces. It's just that it could, and should, do even better. (p. 43)

Congressional and DoD transformation initiatives under Defense Secretary Donald Rumsfeld focused on greater reliance on commercial products and processes and more timely infusion of new technology into new and existing systems. Commercial product usage is implemented with an understanding of the complex set of impacts that stem from use of commercial off-the-shelf technology (Oberndorf & Carney, 1998). Procurement solicitation requirements are written to include performance measures. If military specifications are necessary, waivers must first be obtained. Solicitations for new acquisitions that cite military specifications typically encourage bidders to propose alternatives. The DoD has made significant progress in disposing of a portion of its huge inventory of military specifications and standards through cancellation, consolidation, conversion to a guidance handbook, and replacement with performance specifications and nongovernment standards.

Despite all of this change, the primary criticisms of the acquisition process remain—that it is too complex, too slow, and too costly (Barr, 2005). In some cases it also may produce weapons that are "over-qualified" or irrelevant to the task at hand when they are finally put in the field because the threat and war fighting environment have changed since acquisition and procurement decisions were made to contract for weapons platforms, systems and components. Annual budget cycle procedures and politics within the DoD and between the DoD and Congress add complexity, turbulence and some degree of confusion to this mix. This chapter explores these complexities and offers some explanations for their occurrences. In the following two chapters (13 and 14) we propose some recommendations for improvement. However, before we get to reform concepts we explore the fiscal environment for acquisition, and review some basic budgetary terminology important to the acquisition process. Many of these terms are not found in the normal budget process (e.g., milestones, ACAT, Milestone Decision Authority).

COMPLICATIONS IN ACQUISITION FUNDING

To understand current issues in funding for acquisition we need to review how the budget process operates in terms of Budget Authority (BA), Total Obligational Authority (TOA), and outlays. To return briefly to what we have covered in previous chapters, BA is provided to DoD through appropriation by Congress. BA provides DoD permission to spend money to make or buy necessary defense assets. BA is appropriated for 1 year or for multiple years, for example, 3 years for aircraft acquisition, 5 years for ship construction and so forth. BA allows departments and agencies to incur obligations and to spend money on programs. Thus, BA results in immediate FY or future year obligations and outlays. TOA is a budget term that indicates the total of all money available from prior FYs and the current FY for spending on defense program in the current FY. Typically, asset acquisition is paid for using both current and prior year appropriations and extends over a multiple year time horizon. TOA is, in effect, the accumulation of annual Budget Authority.

As with all federal departments and agencies, DoD attempts to spend all the funds appropriated to it for the purposes specified by Congress. By law, unexpended BA for which spending authority expires before obligations are incurred is returned to the Treasury and is no longer available for DoD to spend. New budget legislation from Congress may reappropriate expiring BA for continuation of authorized programs or transfer unobligated BA where the purpose of the original funding has changed. presidential rescissions may cut existing BA. Net Offsetting Receipts—money collected directly from the public—may be applied by Congress to replace (offset) BA.

BA is spent via the obligation process. Hiring personnel, contracting for services, and buying equipment all are ways of incurring obligations against budget authority. Outlays then are the actual expenditures that liquidate government obligations. Before passage of FY 1989 defense authorization and appropriation legislation, prior year unobligated balances were reflected as adjustments against TOA in the applicable program year only. However, since then, both the CBO and OMB have scored (recorded) such balances as reductions to current year BA. Previously, reappropriations were scored as new budget authority in the year of legislation. However, in preparing the amended FY 1989 budget, CBO and OMB directed scoring of reappropriations as BA in the first year of availability (Candreva & Jones, 2005; DoN, 2002a). The change reduced DoD spending flexibility in out-years.

Figure 12.7 illustrates the positions of the four main DoD appropriation accounts. These accounts show spending for more than 96% of DoD appropriations over the time period indicated. O&M, personnel, and procurement clearly show the cost of military operations in Afghanistan and Iraq in 2004-2006. The President's budget predicted a decline in O&M funding in FY 2007 and FY 2008, with a very modest decline in procurement and then a buildup to the end of the period.

In Figure 12.8, it can be seen that Army procurement will be cut, while procurement for Navy and Air Force will continue to grow. This figure also shows the tremendous impact on the military departments of the drawdown of defense from the Reagan peak into the 1990s. Since this is a

DOD TOA by Appropriation

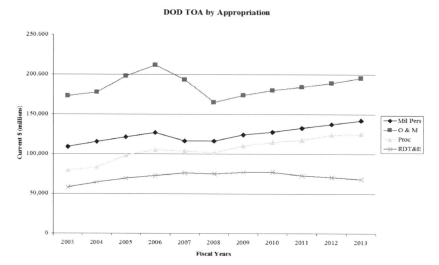

Source: Authors, 2007, derived from DoD Comptroller, Greenbook, 2007: President's FY 2008 Budget. Table 6-1, DoD Major Appropriations by Title (in current dollars, in millions), p. 67.

Figure 12.7.   DoD Budget Authority by appropriation title, FY 2003-2013.

constant dollar display, we can see that Air Force and Navy do not have the purchasing power they had in 1991, to say nothing of at the Reagan peak and even if the funds forecast out to 2013 come true, they still will be short. Meanwhile, as we discuss later, replacement weapons systems cost more, thus the same amount of money as 1985 would allow fewer weapons systems to be purchased. Of course, their capabilities would be greater, but there is no guarantee that these would be the 'right' capabilities.

It is important to realize that the acquisition accounts (and RDT&E or research, development, training and engineering) are fundamentally more complex than the "readiness" accounts like military personnel and O&M. Budget managers in the latter accounts have a 1-year period of obligational availability, while managers in the acquisition accounts have multiple years of obligational availability, depending on what is being purchased. This alone makes for more complexity. The outcomes of this are most apparent not at the lower levels of DoD where participants know which rules to follow or can seek guidance if they do not, but at the upper levels, where top level policymakers pressed by the needs of the next year's budget may want to juggle the investment (acquisition) accounts to help pay for increasing operating expenses in the personnel and O&M

Procurement by Military Department

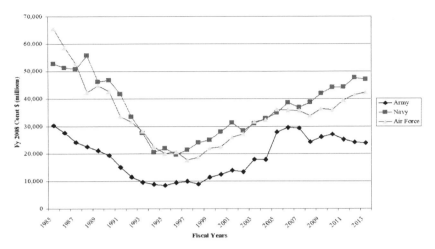

Source: Authors, 2007. Derived from DoD Greenbook, 2007: Army TOA by Title FY 2008 Constant Dollars, Table 6-16 p. 153; Navy TOA by title FY 2008 Constant Dollars, Table 6-17, p. 159; Air Force TOA by title FY 2008 constant dollars, Table 6-18, p. 165. (1985-2006 actual expenditures, 2007 current year appropriation and 2008 to 2013 estimates.)[1]

Figure 12.8. DoD procurement account trends by military department.

accounts. In doing so know there will be multiple opportunities to repair the programs they borrowed from in the future. This pattern is quite evident from the latter part of the 1990s to the present. For example, in 2006 Deputy Commandant and Director of the Marine Corps Programs Division General Raymond Fox, observed that the U.S. Marine Corps (USMC) was facing increasing cost growth in personnel and the O&M accounts and that when offsets (cuts) had to be made to meet cost growth, offsets would come from the investment accounts and not from the budgets of the other military services or from supplemental appropriations. Fox (2006) noted, "Investment accounts pay the offsets, not other services, not supplementals" (Fox 2006, p 10).

One of the most serious problems faced by DoD is how to fund the replacement of used assets that have served far beyond their projected depreciation term. In an ideal world, the President would request and Congress would appropriate new money to pay for new weapon assets as systems reached the end of their useful operational life or were super-seded by superior systems. The world as it exists is much more complex. Sometimes there is no appetite for increased DoD funding and no apparent threat, hence the procurement holiday of the 1990s. Sometimes weapon systems are bought in order to keep the industrial base and certain skill sets alive (thus submarines in the early 1990s). Sometimes weapons are purchased because they keep thousands of people employed in important electoral districts. Sometimes the short run needs of one service preempt the needs of the other services, as seems to be the case with the Army and its need for explosion resistant vehicles in 2006 and 2007; while DoD procurement increased in these years, much of it went to Army procurement. Sometimes the immediate need even when fulfilled has important consequences for that service's long run needs; for example the provision of tank-like vehicles appeared in 2007 to be drawing funds from and endangering the Army's plans for its Future Combat Systems Program. Then, too, sometimes reality intervenes to change the direction of weapons procurement; until their utility is demonstrated; for example, in 2005 tanks and tank-like vehicles were low on the Army's list of priorities, but with the deteriorating situation in Iraq, their value (and priority) increased. For all of these reasons and more, acquisition of military hardware is a lot more complicated than simply buying from a carefully drawn and prioritized list of equipment, regardless of what has been approved in the military department Program Objective Memorandum which is supposed to resolve issues related to system choices so as to set priorities.

Most of the issues with respect to weapons acquisition choice and financing lie within DoD itself. For example, sometimes what seem to be "no-brainer" decisions seem to fall "through the cracks." O'Bryon (2007) relates the story of the Air Force's failure to install automatic ground collision avoidance systems in certain U.S. fighter aircraft when the need for this capability was clearly demonstrated in 1999 and was standard equipment on a fighter produced in Sweden in 2000 (p. 53). Estimates indicated that if the system was installed in the F-16, this would have saved 150 lives and $7.5 billion in airframes over the period studied. Seemingly, the Air Force had made promises to install systems in the F-16, F-22 and F-35, but did not do so due to budget cutting and weight concerns. In August of 2007, the Air Force again promised to install the automatic ground collision avoidance system in fighters, but with the budget stresses it faces, whether and when this will happen is open to question.

As with the Army, notwithstanding budget increases related to the conflicts in Iraq and Afghanistan, all of the military service plans for modernization have come under pressure. Defense analyst Muradian (2007) noted,

> We're beginning to see that transformational programs are ebbing away in the FY2009 budget submissions of the military services. The Air Force is cutting both the Joint Tactical Radio System and its next-generation cruise missile radar. It doesn't have the money for C-17's. It doesn't have the money for the Alternative Infrared Sensor System. We are seeing the fraying of the investment agenda as a result of budget pressures. (p. 4)

That these were real pressures was indicated in a FY 2008 Air Force budget brief that indicated:

- personnel costs were up 59% while the number of personnel had shrunk 8% over the last 22 years.
- investment accounts had decreased 19% over the last 22 years (1986-2006).
- operating costs were up 179% over last 10 years and continued to stress the ability to recapitalize.
- 14% of the (aircraft) fleet was grounded or had mission-limiting restrictions.
- Air Force operating costs were up 179% between 1996 and 2006, while the aircraft inventory had shrunk.
- average cost of the flying hour program increased 10% between FY 2007 and FY 2008 while the budgeted cost of inflation (allowed AF by DoD) was up 2.4% (Faykes, 2007).

In the summer of 2007, cross-pressured by the need to meet current operating obligations and the need to modernize, the Air Force was perplexed as it faced a $100 billion shortfall over the next 6 year FYDP period. The United States Air Force (USAF) had trouble deciding which programs to keep alive. As explained by Muradian (2007),

> while the service can afford individual programs—whether buying more C-17's or F-22's—keeping both in production as well as acquiring new tankers, developing a new bomber, fielding the F-35 JSF and covering the cost for classified and other space programs are the backbreakers. Like the Navy, the Air Force is being asked to make cuts to cover urgent war costs for the Army and Marine Corps. (p. 4)

At the end of the summer in 2007, even though the Air Mobility Command had been asked to develop a rationale for thirty additional C-17's, the USAF had only put two C-17's on its unfunded priority list, and this provided, "little justification to the Hill to spend the money needed to keep the production line (open)" (p. 4) Defense expert Loren Thompson of the Lexington Institute observed, "The reality is that C-17 backers in Congress are running out of patience with the service's (USAF) failure to signal a requirement" (p. 4). If the Air Force, or any service, will not state a requirement, even if it is for only one aircraft, then it is hard for congressional backers, who would like to support the weapon system because of constituency reasons, to mobilize the necessary voting support to do so. Also, one Air Force official noted that FY 2009 was a transitional year and that there would be a new presidential administration's policies and budget to deal with for FY 2010 no matter which party won the 2008 elections. Thus, since they would have to do FY 2010 again, the version of FY 2010 in the FY 2009 budget submission was probably not the time for broad and "revolutionary" strategies.

Particularly in the acquisition accounts, budget and program planners have to "strategize" about the full 6 years of the FYDP. In the recent past this has meant that as each new fiscal year arrived, the FYDP was changed and the numbers of ships, aircraft, and other weapons systems were decreased in the near future or even across all 6 years because there was not enough money in the investment accounts to fund them. For example, the F-22 Raptor was originally planned as a 750 airplane buy to replace the F-15, but cuts over the last fifteen years have reduced it to 183 planes (Muradian, 2007, p. 4).

Figure 12.9 illustrates turbulence in the "near years" in the Navy aircraft procurement budget. The slash marks indicate where aircraft buy quantities have changed downward, primarily due to cost pressures.

Turbulence in the FYDP is a fact of life in DoD. There is no reason to expect that a program cannot 'get well in the out-years' while suffering near year cuts, but as a practical matter since the mid-1990s, when

Aviation Quantities

	FY07	FY08	FY09	FY10	FY11	FY12	FY13	FY08-13
JSF	0	8/6	32/8	36/18	23/19	40	42	133
F/A-18E/F	30/34	24	20	22/24	44/19	21	0	108
EA-18G	12/8	18	22	20/16	10/8	2	0	68
MV-22B	14/13	19/21	31/30	25/30	37/30	30	30	171
AH-1Z/UH-1Y	18/8	19/26	23/28	23/28	23/28	24	24	149
MH-60S	18	20/18	26/18	26/18	26/18	18	18	108
MH-60R	25	25/27	31	32/28	31/28	25	27	166
E-2C	2	0	0	0	0	0	0	0
E-2D AHE	0	3	3	3	4	4	4	21
CH-53K (HLR)	0	0	0	0	0	0	6	6
P-8A (MMA)	0	0	0	6	8	10	13	37
C-40A	0	0	0/1	0	1	1	1	4
T-45C	12	0	0	0	0	0	0	0
T-6A/B(JPATS)	24/20	46/44	46/44	46/44	46/43	43	22	240
KC-130J	4/2	4	4/2	2	2	2	2	14
VH-71	0	0	4	3	4	4	4	19
BAMS UAS	0	0	0	0	4	4	4	12
MQ-8B (VTUAV)	4	3	5	6	6	9	10	39
F-5E	5	0	0	0	0	0	0	0
TOTAL	165/151	199/188	257/213	267/228	253/222	237	207	1,295

Source: Bozin, 2007, p. 10.

Figure 12.9. Turbulence in naval aircraft in the FYDP.

programs have been cut in the out-years they stayed cut, and many other programs also were reduced. There was a time in DoD (e.g., the mid-1980s) when programs that were cut in the near term were later expanded in the out-years and there was some realistic chance that this could happen. Thus, those suffering the cuts were not pessimistic about 'getting well' in the out-years. DoD planners could move more of a buy to the out-years, take a cut in the current year to accommodate budget pressures, but schedule postponed units into the FYDP out-years (usually years 5 and 6). It was assumed that everyone would "get well" in the out-years. However, in the mid and late 2000s the risk has been that the projected out-years increases never arrived. Figure 12.10 illustrates the difference between the FYDP and the actual appropriation.

What this graphic indicates is that there was a severe disjunction between DoD plans and what was provided in the actual budgets, and that DoD tended to stick with linear projections from its base year but reality turns out rarely to be linear. We see that the FYDP is often off target and is particularly poor in predicting "tipping" points. For example, the graphic clearly shows the mis-estimation of the continuation of the early and mid-1980s defense buildup. As far as Pentagon planners were concerned it was going to go on in a straight line into the late 1980s. They did not predict the slowdown after the mid-1980s due to deficit politics. They could

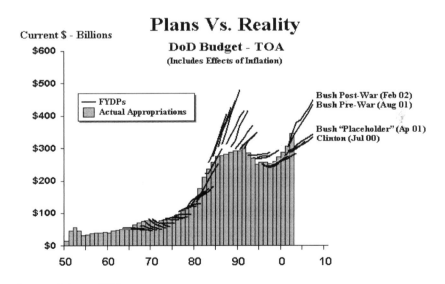

Source: Spinney, 2002.

Figure 12.10. President's Budget FYDP projections vs. actual defense budget.

hardly be faulted for failing to predict the collapse of the USSR, and FYDP planners did get the direction of the Reagan buildup right until the mid-1980s; they also were correct with the direction if not the amplitude of the drawdown of the 1990s.

A critical reaction to this figure might be that FYDP projections have seldom been accurate, and this is generally a correct perception (see chapter 14 recommendations for terminating the FYDP). Still, it must be remembered that the FYDP is a 6 year profile embedded in a 1 year budget system; each year the DoD offers a 6 year FYDP profile and Congress almost always changes the DoD part of the President's budget proposal. Moreover, 6 years covers three congressional elections and one presidential election and priorities always change over this period of time. It is important to note in the exhibit how linear DoD plans are (the FYDP line) and how curvilinear reality (the budget line) turns out to be. To some extent DoD is caught up by its own estimations, for example, military personal expenses increase as pay and benefits grow; most supporting O&M expenses grow at inflation adjusted rates, and the procurement accounts tend to be dominated by multiyear purchases. It is no wonder that defense planners tend to see DoD growth as inevitable and in a linear profile. Additionally, some Presidents (Ronald Reagan and George W. Bush for example) gave DoD topline guidance for growth equal to or above inflation—this justified a linear and increasing FYDP. Also, the FYDP is a DoD plan for what it projects will be needed in terms of capability to meet the threat in the future. However, sometimes the President and Congress cannot provide all the funds DoD would like to have without breaking necessary commitments in other areas such as deficit reduction and entitlement spending, or in other discretionary appropriations. Figure 12.11 reminds us that the federal government funds numerous functions other than defense and that DoD budget requests will be compared either directly or indirectly to needs in other areas of federal spending. In all circumstances, DoD needs for future acquisition will be in direct competition with funding for a variety of public needs including medical care and social security.

Finally, the threat environment rarely changes in a linear fashion. This was true for the end of the Cold War and after the terrorist attack of 9/11/ 2001. The United States has demonstrated it will pay whatever it takes when national security is threatened, but in the long-term quest to recapitalize the military, the stakes are high and the budget battle is fought one year at a time, with an inclination to satisfy short term needs and put off sophisticated and expensive weapons systems that meet the threat of 2030 but seem to meet no current need.

As the "squeeze" on modernization dollars has continued (most experts date its start to 1986), the Air Force and Navy, as capital intensive

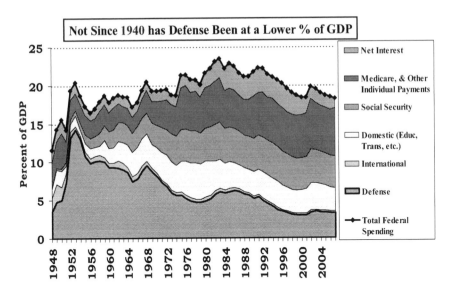

Source: Daly, 2004, slide 30.

Figure 12.11.   Shares of the federal budget.

services, have since the mid-1990s designed programs and budgets to recapitalize by trimming personnel and O&M costs. The DoN embarked on a drawdown of military manpower in 2001 termed Sea Enterprise. By 2006, DoN had reduced the Navy by about 40,000 sailors and the Navy intended to use whatever was freed up to strengthen its shipbuilding and aircraft acquisition programs. Some of the savings did go to these programs, but much of the rest went to operational needs.[2] As can be seen from the FY 2008 budget this drawdown was expected to continue.

Many critics outside the Pentagon do not realize the extent to which "corporate" DoD, led by the SECDEF, the USD AT&L and others can be opposed to the acquisition plans of the military departments. Many observers are well acquainted with former Secretary Rumsfeld's battle with the Army over canceling the Crusader system in 2001. While SECDEF does not always win visible public struggles over high profile weapons systems (e.g., SECDEF Cheney's efforts to terminate the V-22 program in the early 1990s were unsuccessful), SECDEF almost always wins the "big acquisition" battles in annual DoD budget negotiations. In 2007, some experts viewed DoD as "siphoning off" money for other uses that the Air Force and Navy were saving through personnel reduction to put into weapons modernization. All of this took place in a fiscal

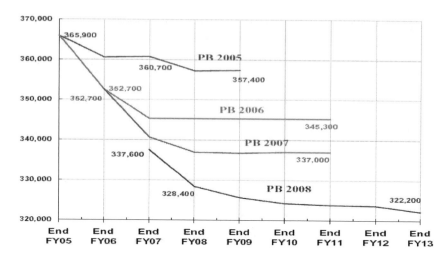

Source: Department of Navy, 2007, FY 2008 budget, February, Figure 29, p. 56.

Figure 12.12.   Trends in naval manpower.

environment in which some critics, including Democrats in Congress, charged that DoD was "lavishly" funded. Muradian (2007) asked, for example, how it was possible that the DoD, funded at more than $680 billion in 2007, could not afford to buy new weapons. The answer, he said, was, "The global operations, particularly the costly wars in Iraq and Afghanistan, the demand to funnel needed gear to soldiers in the field, combined with growing fuel and benefits costs are squeezing strategic modernization accounts" (p. 4). In the process, the Air Force and Navy, he added, "have borne the brunt of cuts to cover war costs" (p. 4). Muradian observed that in 2005 the Air Force,

> launched a plan to retire hundreds of airplanes and cut tens of thousands of airmen to save money for modernization programs. But those savings, and others since, have been effectively repossessed by the Pentagon to pay for the mounting war costs. (p. 4)

Those who might perceive this budget trade-off as solely a dilemma about future weapons systems should remember General Faykes' assertion that 14% of the USAF fleet was unable to fly required missions and that readiness as measured by USAF metrics was down 17% from 2001 as a result of wartime OPTEMPO and flying a smaller, older fleet of aircraft. As we explore throughout this book, the program and budgetary battles about weapons for today impact defense capabilities tomorrow.

THE FIT OF ACQUISITION AND BUDGET SYSTEMS

The DoD resource allocation system (PPBES) includes capital and operational budgeting within a long-term "requirements" setting process based on the FYDP and then separate acquisition and budgeting process, as we describe in the next chapter. This set of systems locks most of the lifecycle costs of a weapons system in before it enters into the budget process. In the case of the F/A22A Raptor airplane acquisition, it was four years from the concept phase to flight testing of the prototypes in 1990 and another year before a design and manufacturer were selected. In those 5 years the USSR dissolved and the Cold War ended, but the U.S. needed a tactical air superiority fighter airplane to meet projected future threats. From 1991 onward, the design would go through a number of changes until 2001 when it was approved for production. At this point the number of airplanes to be bought had been decreased by more than half and the cost per unit had increased considerably. This was a high profile weapons system and congressional cuts had twice limited its development, in 1993 and 1997. In 2002, in an effort to widen the utility of the system, DoD added ground attack capabilities to the requirement, and cost per unit skyrocketed with an accompanying decrease in the number to be bought. All in all, this system took over 20 years to design and deploy.[3] Figure 12.13 tells the story of F/A-22 Raptor development.

What is of more than casual interest is that most of the life cycle cost of any weapons system is designed in long before the system enters the budget process. If the graphic below is applied to the Raptor, approval for production happened in the Raptor program in 2001, 15 years after the concept was first initiated; this would be at point IOC (initiation of concept) in Figure 12.13. The Raptor story changes somewhat with the addition of ground attack capability, which almost doubled its unit price from $187 million to $345 million, but this is also symptomatic of what happens to other advanced weapons acquisitions. As costs rise, DoD and Congress inevitably make a series of program-to-budget adjustments with price increases that are offset with quantity decreases, or by decreases in weapon capability, as perhaps was the case with the automatic ground avoidance radar systems. Another way of avoiding the consequences of increased cost is by justifying that the platform does more so it *should* cost more, or by increasing system versatility, which in the Raptor's case substantially increased cost. These are fact of life issues in the acquisition process which are compounded and intensified when acquisition appropriation accounts are used to pay operating accounts bills, for example, for war fighting.

In some respects, the Raptor case is typical for the acquisition of more sophisticated defense acquisition assets. On the one hand, the

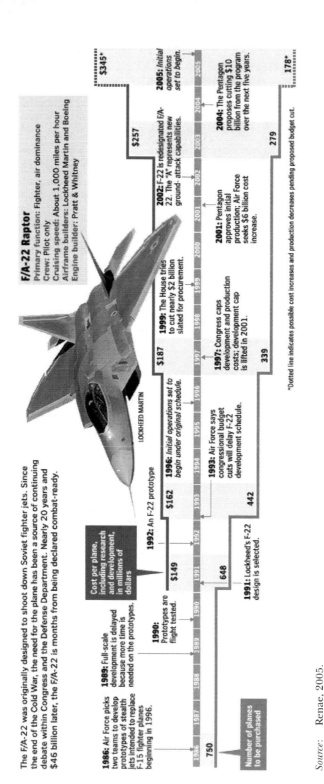

The F/A-22 was originally designed to shoot down Soviet fighter jets. Since the end of the Cold War, the need for the plane has been a source of continuing debate within Congress and the Defense Department. Nearly 20 years and $46 billion later, the F/A-22 is months from being declared combat-ready.

F/A-22 Raptor

Primary function: Fighter, air dominance
Crew: Pilot only
Cruising speed: About 1,000 miles per hour
Airframe builders: Lockheed Martin and Boeing
Engine builder: Pratt & Whitney

LOCKHEED MARTIN

Cost per plane, including research and development, in millions of dollars

1986: Air Force picks two teams to develop prototypes of stealth jets intended to replace F-15 fighter planes beginning in 1996.

1989: Full-scale development is delayed because more time is needed on the prototypes.

1990: Prototypes are flight tested.

$149

1991: Lockheed's F-22 design is selected.

1992: An F-22 prototype

$162

1993: Air Force says congressional budget cuts will delay F-22 development schedule.

1996: *Initial operations set to begin under original schedule.*

$187

1997: Congress caps development and production costs; development cap is lifted in 2001.

1999: The House tries to cut nearly $2 billion slated for procurement.

$257

2001: Pentagon approves initial production; Air Force seeks $6 billion cost increase.

2002: F-22 is redesignated F/A-22. The "A" represents new ground- attack capabilities.

$345*

2004: The Pentagon proposes cutting $10 billion from the program over the next five years.

2005: *Initial operations set to begin.*

178*

Number of planes to be purchased

750   648   442   339   279

*Dotted line indicates possible cost increases and production decreases pending proposed budget cut.

Source: Renae, 2005.

Figure 12.13.   The F/A-22 Raptor program and its changes.

development of the aircraft conceived in 1981 was hopelessly prolonged, complex, and expensive. "The program has gotten bogged
down by competing interests that want their products in the aircraft,"
said a former systems analyst for the DoD and a designer of the F-16
and the A-10 Warthog. In the end, the Air Force wanted the Raptor to
be a technological marvel packed with everything that would fit resulted
in poor planning and the program's exorbitant cost. The former DoD
official observed, "They made it so complicated and hopeless that it
took forever.... It's just disgraceful."

On the other hand, the first F/A-22 Raptor squadron commander testified to the excellence of the product, as follows "Lt. Col. Mike Shower,
squadron commander for the first Elmendorf Raptors, said no enemy aircraft even comes close to the F-22." "Our old stuff is essentially on par,"
said Shower, who has piloted both the Raptor and the F-15. "There is a
significant amount of threat out there, but the F-22 absolutely dominates
when we fly" (Halpin, 2007). In this program we see staggering program
complexity, enormous technical sophistication, a continuous trade off
between cost and number of aircraft procured and a program that took 25
years to field from concept to operational deployment, but a weapon that
absolutely dominates its battle space.

Once a weapons system has entered into production, routine cost
growth can lead to billions of dollars of cost increases, as it has in the
Navy shipbuilding program. It is important to note how early decisions
in the acquisition process set the course for total ownership costs, as
shown in Figure 12.14. A GAO analysis of Navy shipbuilding programs
identified the long construction times for ships and the fact that the
total cost for a ship must be budgeted for in its first year of construction, which leads to uncertainties in estimating costs and resultant cost
growth (see Appendix B based on a Rand Corp. report on reducing
shipbuilding costs). Figure 12.15 illustrates how long it takes to produce selected weapons systems.

In 2005, after an analysis of cost growth for a set of eight Navy ships of
different types under construction, GAO (2005) found,

> Increases in labor hour and material costs together account for 77% of the
> cost growth on the eight ships. Shipbuilders frequently cited design modifi-
> cations, the need for additional and more costly materials, and changes in
> employee pay and benefits as the key causes of this growth. (p. 2)

Figure 12.16 shows components of shipbuilding cost growth.

Altogether GAO estimated a cost growth of $1.95 billion on a FY 2005
budgeted amount of $18.5 billion for the eight ships in the study, a cost
growth of about 10.6% on the ships studied as a result of the factors

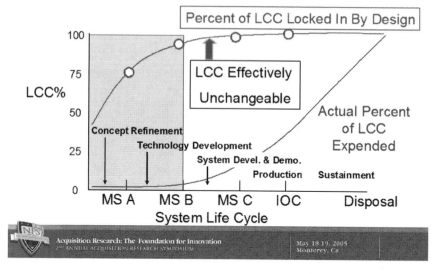

Source: Boudreau, 2005.

Figure 12.14.   Early decisions set the course for total ownership costs.

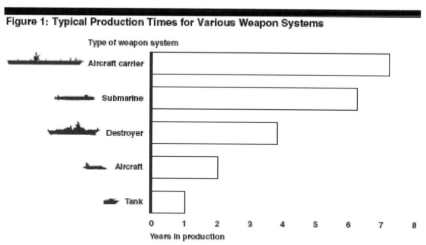

Source: GAO-05-183, February 2005, p. 6.

Figure 12.15.   Production times for weapons systems.

Components of Cost Growth

5%

Overhead rate and labor rate increases

17%

Material increases

38%

40%

Labor hour increases

Percentage of overall cost growth due to shipbuilder construction costs

Percentage of overall cost growth due to cost of Navy-furnished equipment

Source: Shipbuilder and Navy (GAO: 05-183, p. 2) analysis.

Source: GAO: 05-183, p. 2.

Figure 12.16. Cost growth by component in shipbuilding.

illustrated in Figure 12.16. GAO also suggests that costs may grow another billion by the end of these programs (GAO, 2005, p. 8). GAO continued by criticizing some Navy cost estimation and budget practices and offered recommendations for improvement. However, our interest here is longer term: these are large and sophisticated assets and the process of building them is often more a voyage of discovery in new materials, new methods, and new requirements than it is one of routine construction. It is no wonder that costs increase. As GAO noted, "U.S. Navy warships are the most technologically advanced in the world. The United States invests significantly to maintain this advantage" (p. 1).

CONCLUSIONS

Acquiring weapons platforms such as aircraft, ships, tanks, and weapon support systems for military forces is central to accomplishing the mission of the DoD. Each year, the President submits the defense acquisition budget as part of the overall budget for the DoD to Congress for review and appropriation. Threats to national security, perceived or actual, and political priorities drive the amount of defense funding requested and appropriated for weapons acquisition. Congress, representatives from the executive branch and DoD, industry lobbyists, analysts in defense think tanks, media experts and a variety of others debate the merits of spending and programmatic alternatives and maneuver to receive resources. Congress expects DoD to provide quality products that meet war fighter needs while sustaining program stability, and more recently to shorten acquisition program cycle times, to develop more innovative approaches to weapons research, development, design, testing, evaluation, production, support, and use.

DoD program managers promote and attempt to garner funding for their programs in the annual budget process. Following appropriation, Congress and DoD provide directives and guidance to assist the military services in weapons acquisition. The weapons asset investment budget is constrained by a relatively stable defense budget top-line and has been squeezed by increasing operating and support costs for aging weapon systems, and since 2003 by the events in Afghanistan and Iraq where the ongoing costs of war have limited DoD ability to develop, purchase, and deploy new weapon systems. In this chapter we have explored many of the issues that complicate budgeting and acquisition and, as such, challenge the managerial capabilities of DoD decision makers, acquisition program managers, and financial managers.

NOTES

1. With the outcome in Iraq uncertain, it is important to note when reviewing DoD future year estimates that they are estimates and could change substantially with new events, as they did after 1990 with the end of the Cold War, after 2001 with the attack on 9/11 and after 2003 with the invasion in Iraq.

2. For more on this see Jason Miller, "An Analysis of the Sea Enterprise Program." June, 2005. NPS: Monterey, CA. Miller's research concluded that about 60% of the $46 billion in savings taken over the FYDP from FY 2004 to 2009 would be realized, although estimates of total savings differed. The other 40% basically was siphoned off to meet ongoing needs (see Miller, 2005, pp. 69-70.

3. The "first ever" deployment of the F-22A took place in Alaska in Northern Edge 2006 a joint training total force exercise (Faykes, 2007, p. 6). The concept dates to 1981.

CHAPTER 13

IMPROVING LINKAGES BETWEEN RESOURCE MANAGEMENT AND ACQUISITION SYSTEMS

INTRODUCTION

This chapter analyzes the relationships between the defense budgeting system, the capabilities and requirements determination process, and the Defense Acquisition System (DAS). We identify strengths and weaknesses, identify issues of contention, and make some modest suggestions for change. To do this we have to start with a review of some basic concepts in acquisition.

For purposes of clarification for the analysis that follows it is necessary to define some terms. ACAT I programs are Major Defense Acquisition Programs (MDAP). The acronyms MDAP and ACAT I are used interchangeably here. An MDAP is defined as a program estimated by the under secretary of defense for acquisition, technology and logistics (USD AT&L) to require eventual expenditure for research, development, training, and engineering (RDT&E) of more than $365 million (in FY 2000 constant dollars) or acquisition of more than $2.190 billion (in FY 2000 constant dollars), or other programs designated by the USD (AT&L) to be ACAT I. There are three major acquisition categories stipulated by

Budgeting, Financial Management, and Acquisition Reform in
The U.S. Department Of Defense, pp. 565–601
Copyright © 2008 by Information Age Publishing
All rights of reproduction in any form reserved.

regulation in Department of Defense (DoD): ACAT I, ACAT IA, and ACAT II. ACAT ID (defense) and ACAT IC (component or individual service) are analyzed in more detail subsequently.

Responsibility for coordinating the components of the acquisition funding process rests ultimately on individual program managers (PM). As the lead acquisition official for a program, the PM is charged with integrating the administrative demands of the Planning, Programming, Budgeting, and Execution System (PPBES) (e.g., how much to ask for in the budget process) with the management demands of the acquisition process (e.g., meeting milestone requirements in acquiring weapons and weapons support systems). The process for major acquisition programs (ACAT I) is the most complicated, has the highest level of importance and, normally, the largest financial exposure. Consequently, this category receives the greatest oversight attention from various offices within DoD and from Congress. The ACAT I PM has to meet strict qualification requirements, higher than those for non-ACAT I PMs. Training and experience requirements are designed to select PMs who possess a broad base of acquisition experience. PMs must be or become adept at the skills of managing acquisition while also operating effectively as proponents and defenders of their programs in the programming and budget processes. Without proper positioning in the PPBES process, a weapon system can be designed and tested and still not be acquired because it does not survive the gauntlet of the PPBES and congressional review and decision processes.

Within PPBES, once a system has been identified as necessary to meet mission capability requirements (theoretically in parallel with the planning phase of PPBES); it has to be reviewed for inclusion in the Program Objective Memorandum (POM). The POM is constructed separately by each military service and department. For example, the Navy official responsible for preparing the POM is the chief of naval operations (CNO) and the CNO Office of the Chief of Naval Operations (OPNAV) staff. The POM and budget processes used to operate separately. However, as noted in previous chapters, in 2001 reform under former Secretary of Defense (SECDEF) Donald Rumsfeld required that the POM and budget processes be merged and that programming and budget review take place simultaneously.

According to the new process, once the merged POM and budget are approved by the service chiefs, for example, the CNO for the Navy, the POM is sent to the service secretary (secretary of the Navy) and then the program review staff of the Office of the Secretary of Defense (OSD). Military service and department (MILDEP) acquisition executives (assistant secretaries for acquisition and related functions also referred to as component acquisition executives or CAEs) compile and review requirements for defense acquisition programming and budgeting, before the POM and budget are forwarded to SECDEF staff, specifically to the under secretary for defense for acquisition (USDA, or by other titles—these titles change over time—the current title is under secretary of defense for acquisition, technology and logistics.

The under secretary of defense (AT&L) and staff review requests from the military services and departments to establish an acquisition program baseline (APB) and budget for all of DoD. This step requires an agreement between the Milestone Decision Authority (MDA) manager in the OSD for acquisition and the PM on the cost, schedule, and performance objectives and thresholds of all acquisition programs. The APB contains the most important cost, schedule, and performance parameters and is updated as required.

Weapons program officials establish the APB to document the cost, schedule, and performance objectives and thresholds of for their program. The PM prepares the APB at program initiation for acquisition category programs, and at each subsequent major milestone decision, and following a program restructure or an unrecoverable program deviation. APBs contain objectives for cost, schedule, and performance parameters, as noted. The specificity and number of performance parameters evolve as the program is better defined. The schedule parameters include program initiation, major milestone decision points, initial operating capability and any other critical system events. These critical events are proposed by the PM and approved by the MDA for each program.

Maximizing PM flexibility to make cost, performance, and schedule trade-offs without "too much" higher-level review and micromanagement is deemed essential to achieving programmatic objectives from the view of the program office. Therefore, creating an executable agreement and sustaining consistent milestone reporting in conformance with the APB is a critical task for the PM. The level of ACAT designation (e.g., I, II etc.) normally is assigned after approval of the operational requirements document by the MILDEP and USD (AT&L). A proposed ACAT designation is provided in the requirements document.

THE DEFENSE ACQUISITION BOARD AND THE DEFENSE PROGRAM AND BUDGET

What follows is a simplified overview of the decision process for acquisition. However, we note that it is almost impossible to keep this description up-to-date due to the fact that the process seems to be in almost continuous change as new reforms are developed and implemented, one after another. In the acquisition resource decision process, the intended weapon's user (e.g., the Army) identifies an operational capability that cannot be satisfied by anything but a fixed asset (weapon, system, or platform) and produces a mission need statement (MNS) that includes capabilities assessments. Once the MNS is approved by the service chief of staff and validated by the Joint Chiefs of Staff (JCS) Joint Requirements Oversight Council (JROC), the deputy secretary of defense convenes the Defense Acquisition Board (DAB). The DAB is the defense department senior level forum for advising the USD (AT&L) on critical decisions concerning ACAT I programs. Some programs are administered at the DoD level and others by the military departments (MILDEPS). The USD (AT&L) is the MDA for "ACAT ID" programs where the "D" stands for defense. The service component is the MDA for "ACAT IC" programs, where the "C" stands for component, for example, Army, Navy, Air Force. The Navy MDA, for example, is the assistant secretary of the Navy (research, development, and acquisition).

The DAB is comprised of DoD senior acquisition officials. The DAB reviews the MNS and makes recommendations to the MDA, for concept studies of a minimum set of alternatives. This review and MDA approval constitute the milestone 0 decision point. The MDA oversees the "concept studies" and approval process, and directs the initiation of phase 0, concept exploration and definition, with an acquisition decision memorandum.

Milestones are major decision points for weapons systems. The milestone review process is predicated on the principle that systems advance to higher acquisition phases by demonstrating that they have met prescribed technical specifications and performance thresholds. For all ACAT I programs, a life cycle cost estimate is prepared by the PM in support of program initiation and all subsequent milestone reviews. For example, the Navy PM establishes, as a basis for life-cycle cost (LCC) estimates, a description of the salient features of the acquisition program and of the system itself (Department of the Navy, 2002b). The LCC estimate plays a key role in the management of an acquisition program. At each milestone decision point, including the decision to start a program, LCCs, cost, performance, and schedule trade-offs, cost drivers, and affordability constraints are major considerations. Here the primary purposes include providing input to acquisition decisions among competing major system alternatives. LCC help determine requirements. Cost drivers are identified among alternatives. LCC also provide an index of merit for trade-off evaluations in design, logistics, and manufacturing and the basis for overall cost control.

In budget preparation, the components initiate the process and state their spending and execution priorities. For example, the Navy component acquisition executive, assistant secretary of the Navy, research, development and acquisition (ASN RD&A) prepares the budget estimate request for Navy ACAT I programs in support of Milestones II and Milestone III. Once the budget is enacted, the ASN (RD&A) exercises line management over program executive offices (PEOs) and direct reporting PMs (Department of the Navy, 2002b). The PEO generally relies on hardware systems commands for administrative support, including comptroller functions for financial management. Once the budget begins the execution phase, the fund-flow for both PEO and hardware systems commands funds are within single conveyance, via a normal path for appropriations. The PEO exercises control of designated resources within the hardware systems command allocation.

Once the programs and budgets for each military department have been reviewed by the senior leaders group, they are included in the approved SECDEF POM and, subsequently, in the DoD budget. The defense acquisition budget is merged with the budget comprised of all spending accounts by the under secretary of defense, comptroller and once approved by SECDEF, is sent to the President's Office of Management and Budget as a component of the President's budget submitted to Congress annually.

LINKAGE BETWEEN PPBES AND THE ACQUISITION DECISION SYSTEM

The architecture of the PPBES interacts with two other major systems for acquisition planning, decision making, and execution. These two systems are:

1. The Joint Capabilities Integration and Development System (JCIDS) that is employed for determining war fighting requirements and capabilities, and
2. The Defense Acquisition System (DAS), a system used for planning, decision and execution for research and development, test and evaluation and then procurement of capital assets.

Three systems—PPBES, JCIDS, and the DAS—comprise the core of the DoD financial resource and acquisition decision making, allocation, and execution process. Let us examine the JCIDS and the DAS more closely.

JOINT CAPABILITIES INTEGRATION AND DEVELOPMENT SYSTEM

JCIDS has replaced what used to be known as the Requirements Generation System (RGS). Through the JCIDS, defense decision makers apply the prevailing precepts of national and defense strategy to create joint fighting forces capable of performing the military operations required by the nature of the threat faced by U.S. armed forces—something that is constantly changing. The JCIDS process is shown below.

The JCIDS was developed to identify joint war fighting requirements and to emphasize a top-down orientation to decision making. Instead of the former process—in which MILDEPS and services determined mission requirements and identified joint needs to increase program funding attractiveness as they prepared and routed their acquisition program proposals up the chain of command—in JCIDS, the chairman of the Joint Chiefs of Staff (CJCS) first determines if the required capability exists, then pushes it down to the resource sponsor in the MILDEPS and services for acquisition. If jointness in acquisition and procurement is required, then the program is essentially "born joint." In addition, the term "capabilities-based" is a recent refinement of guidance for the entire purpose of the acquisition decision system. In the JCIDS, gaps in war fighting capability, either current or those programmed in the Future Years Defense Plan (FYDP), are identified—and any risks associated with gaps are quantified. JCIDS decision makers then determine future capabilities to address existing gaps. In doing so, it is important that the

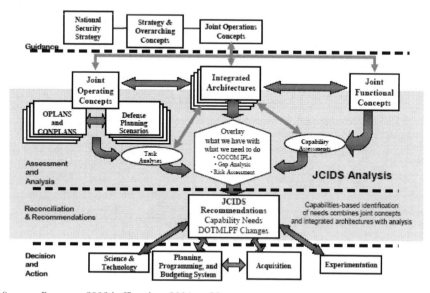

Source: Bowman, 2003 in Fierstine, 2004, p. 22.

Figure 13.1.   The JCIDS process.

decision makers be specific enough about a new capability to include key attributes with appropriate measures of effectiveness, supportability, time, distance, effect (including scale) and obstacles to be overcome. Additionally, the capability needs be general enough not to prejudice decisions in favor of a particular means of implementation.

THE DEFENSE ACQUISITION SYSTEM

Whereas top level DoD decision makers use the JCIDS to identify capability requirements as current and future threat scenarios emerge, DAS evaluates JCIDS-defined capability gaps, and initiates and executes acquisition and procurement programs to field systems to bridge these gaps. In situations where the technology exists to fill a requirement, the DAS exists to acquire a tailored and capable product quickly and in a cost-efficient manner. When new technology is required to fill a capability gap, it is through the DAS that the DoD develops, tests, demonstrates and deploys the new technology in a timely manner and at a fair and reasonable price. In either case, the DAS is forward-looking and tries to ensure that systems fielded support not only today's fighting forces, but also those of the future.

The DAS exists in a highly dynamic and political environment. Since defense acquisition in aggregate involves billions of dollars each year, the process, participants and individual programs are linked to powerful stakeholders. These include the executive branch of the federal government with the DoD acting as its agent, the legislative branch where the Senate and House Armed Services and Appropriations Committees decide what assets will be acquired and funded, private industry where large defense contractors compete for business, market share, and product continuity, in which the subcontractors and small businesses seek a piece of the business, and state and local governments where the defense industrial base is located, where the workforce lives, where dollars are spent and taxes are collected. These stakeholders are both supportive in seeking dollars for defense acquisition and rivals for business. This is true not only in the private sector, but between the MILDEPS and the DoD, the MILDEPS and each other, and within the MILDEPS as potential programs compete for approval and budget.

Since the DoD determines DAS policies and procedures, negotiates each annual budget, makes decisions regarding acquisition programs and the awarding of lucrative contracts to private industry, each major player in the process with authority may attempt to exert influence in the DAS, be it for efficiency reasons, career or organizational ambition or relative to other sources of motivation. Ultimately, Congress holds the power of the purse and must balance defense and nondefense spending. Nonetheless, all these stakeholders compete for some sort of corporate, organizational and professional gain. DoD acquisition is performed in the highly competitive, but only partially transparent, environment of the nation's capitol.

To do their jobs well, those who manage projects within the DoD must understand the political, social, and economic aspects and consequences of the defense acquisition process. From the lowest echelons of program management to the top, the USD, AT&L, all DoD participants must be both knowledgeable and sensitive to the competing forces and attempt to craft each program and project so that, ultimately, war fighters are provided the best assets to support national security policy. The key stages or milestone points of the DAS process move from requirements setting and concept design to determine weapon system needs by the end users—the fighting forces—through technology and systems development to production (procurement) and deployment to war fighters, and, finally, to postdeployment operations and support.

According to the DoD directive 5000.1 (May 12, 2003), defense acquisition is, "the management process by which the DoD provides effective, affordable and timely systems to users" (DoDD, 2003. p. 2). Decision makers use JCIDS to identify capability requirements as the current and future threat dictates. When new technology is required to fill a capability gap, it is through the DAS that DoD develops, tests, demonstrates, and deploys the new technology, "in a timely manner, and at a fair and reasonable price" (p. 2).

ACQUISITION SYSTEM CHANGES

In late 2002, then Deputy Secretary of Defense Paul Wolfowitz canceled the existing set of DoD 5000 series acquisition regulations. In his memorandum, he explained that the acquisition system as defined by these regulations was not flexible, creative, or efficient enough to meet the needs of the DoD. Therefore, he ordered a revision of the acquisition process and a reissue of the directives to, "rapidly deliver affordable, sustainable capability to the war fighter that meets the war fighter's needs" (Wolfowitz, 2003, p. 1).

The DAS process breaks the project lifecycle into three general stages: presystems acquisition, systems acquisition, and sustainment. These three stages are further divided into five distinct subphases: concept refinement (CR), technology development (TD), system development and demonstration (SDD), production and deployment (P&D), and operations and support (O&S). These processes guide a program from initial exploration

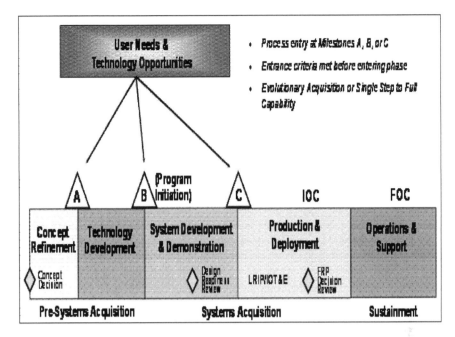

Source: U.S. Department of Defense, 2003. *Operation of the Defense Acquisition System.* DoDI 5000.2. Washington D.C.

Figure 13.2.   The Defense Acquisition System: Major Phases/Milestones.

of required capability (as detailed in an initial capabilities document (ICD), to the P&D of a technologically mature weapons system, including required operational support.

Additionally, each program has a distinct chain of command through which decisions are made. Depending on the size and visibility of a particular program, there may be up to four levels in the chain of command before the ultimate decision is made by the MDA. Complex programs are sometimes divided into smaller elements and assigned groups of acquisition professionals across a range of functional disciplines. These groups are called integrated process teams (IPTs). Some serve as executors of their respective functional program area. Others serve as advisory bodies.

The PM is at the bottom of the chain of command. According to DoDD 5000.1, the PM, a middle-range military or defense civilian (O-5/O-6) is the individual with responsibility for and authority to accomplish program objectives for development, production, and sustainment to include "credible cost, schedule, and performance reporting to the MDA" (Defense Acquisition University (DAU), 2003, p. 2). The PM reports to a

PEO. The PEO, a one- or two-star flag officer or senior executive service equivalent, is responsible for a group of like programs within each military department and service. PEOs report to component acquisition executives (CAEs). Each service has one CAE responsible for the management direction of their respective procurement system. The secretary of the Navy has delegated this position to the assistant secretary of the Navy for research, development, and acquisition (ASN RDA). Finally, the CAE reports to the defense acquisition executive (DAE). The DoD has only one DAE, the USD (AT&L). The USD (AT&L) is authorized under Title 10, U.S. Code to be,

> the Principal Staff Assistant and advisor to the Secretary and Deputy Secretary of Defense for all matters relating to the DoD acquisition system; research and development; advanced technology; developmental test and evaluation; production; logistics; etc.

Also, as the DAE, he presides over the military department and service secretaries and, "is responsible for establishing acquisition policies and procedures for the Department. He also chairs the DAB, and makes milestone decisions on Acquisition Category (ACAT) ID programs" (DAU 2003, p. 31). Programs are categorized by whether they are a DoD-wide asset or an asset for one service and by estimated dollars to be expended, with different rules applying to different-sized programs. The MDA, that is, overall responsibility for all programs, may be delegated to anyone in this chain of command. The MDA for many small programs is the PM, whereas MDA for the large procurement programs and the most politically sensitive programs is usually held at the top by the USD (AT&L).

In the DAS decision process, program movement through the three DAS stages is strictly controlled through a series of six decision points and program reviews. The first stage of the DAS is presystems acquisition. Presystems acquisition activities are focused on refining material solutions to needs as defined in a published ICD. This stage is split into two phases: concept resolution and technological development. As the first phase concludes, and the second major decision point (Milestone A) is reached when the MDA approves both the preferred solution supported by the AoA (analysis of alternatives) and the TDS (technology development stage).

Once Milestone A is achieved, the TD stage begins. With the exception of some high-dollar shipbuilding programs, an official acquisition program has still not considered to have been initiated at this point. Therefore, funding is restricted to work that is done in this phase, the intent of which is to, "reduce technology risk and to determine the appropriate set of technologies to be integrated into a full system" (DAU,

2003, p. 6). This stage is iterative in that the technologies to be refined are continuously developed and processed through close interaction between the S&T community, the users and the developers. As such, the TDS is constantly reviewed and updated with each incremental effort as the technology demonstrations gradually show the proposed solution to be, "affordable, militarily useful, and based on mature technology" (DAU 2003, p. 6).

The TD phase ends when either the MDA decides to terminate the effort, or the third major decision point (Milestone B) is achieved. To be granted Milestone B approval, the second major JCIDS analysis, the capability development document (CDD), must be approved through the JCIDS process, and the MDA must approve both the acquisition strategy and the APB. The MDA must be satisfied that an affordable increment of militarily useful capability has been identified, the technology for that increment has been demonstrated in a relevant environment, and development and production of a system can be achieved within a relatively acceptable time frame, normally less than 5 years. With an ICD providing the context, and an approved CDD describing specific program requirements, Milestone B approval is achieved, signaling the availability of sufficient technology maturity. When funding is approved by Congress and apportioned from the DoD—critical steps—then a formal acquisition program is born and moves forward in the DAS process.

If a program is to be executed in increments or spirals through an evolutionary acquisition process, each increment will be its own program from the development and demonstration phase forward. Each increment or spiral must have its own Milestone B and C approval. Additionally, increment-specific KPPs (key performance parameters) must be delineated in the CDD for each increment or spiral. Finally, before beginning this phase, and with the current increment TDS as a basis, the PM must build and the MDA must approve an acquisition strategy for follow-on increments. Solutions to capability needs can come from a variety of sources, including COTS (commercial off-the shelf-technology) as well as previously discovered mature technologies that heretofore had no obvious DoD application. As such, not all acquisition efforts need start in CR. Some programs can enter the DAS at later stages; the SDD stage marks the first point at which a more mature technology with an approved ICD and CDD may enter the DAS for further refinement without undergoing the scrutiny of CR or TD.

SDD has two main purposes: system integration and system demonstration. Systems integration involves integration of both mature technologies and component subsystems into one complete design that meets the stated requirement. Additionally, at this point, design detail should be achieved as well as trade-offs considered between risk and technology maturity. Risk is defined as how much less capability is allowable while still providing the war fighter with a system that meets the intent of the ICD. Thus, decisions must be made to ascertain what is necessary and what is achievable based on the maturity of the technologies involved. During this stage, such risk decisions must be objectively determined by the program decision makers to limit program costs and the overall time required for systems development.

Systems integration is considered complete when a working prototype has been designed, tested, and documented as functional in an environment appropriate to that in which the user will employ it. Another decision, the design readiness review (DRR), must be successfully negotiated to move to the next part of SDD: systems development. The DRR is a mid-phase assessment of the design to document the complete system in terms of the percentage of drawings completed; planned corrective actions to hardware/software deficiencies; adequate development testing; an assessment of environment, safety and occupational health risks; a completed failure modes and effects analysis; the identification of key system characteristics and critical manufacturing processes; an estimate of system reliability based on demonstrated reliability rates; and so forth (DAU 2003, p. 8). This phase is complete when both the whole system is verified as useful and capable, and the appropriate industrial capability exists to allow the program to move on to the next phase, P&D. Additionally, to gain Milestone C approval, the MDA needs to be satisfied that the program is ready to be committed to production. Otherwise, the MDA must terminate the program. Finally, the CPD must be obtained through the JCIDS process. This step declares that the performance required to exit the SDD phase and the forecasted production capability required to successfully accomplish the P&D phase are in place.

The objective of the fourth phase of acquisition, P&D, is to establish the full operational capability of the program, the ability to produce it in an optimal manner, and to ensure that the final system meets original JCIDS intent as stated in the ICD. P&D begins with Milestone C approval that commits the DoD to production of the program. As such, it authorizes the program to enter either low-rate initial production (LRIP) for large programs that require this approach, full production for smaller programs that do not, or limited deployment and test for information systems that are software intensive.

There are two aspects to P&D. The first is operational test and evaluation (OT&E), including both initial (IOT&E) and follow-on (FOT&E). The test products used come from the production line (either LRIP or otherwise as applicable) and the director, operational test, and evaluation (DOT&E)—for those products requiring DOT&E oversight—or the appropriate operational test agency determines the number of production-line units required for the testing regimen. The other aspect to the P&D phase is the ability of the established production line to handle the job of producing the required units at the rate required by contract. For large-scale production efforts, LRIP is required to ensure adequate and efficient manufacturing capability, to produce the minimum quantity necessary to provide units for IOT&E, to establish an initial production base for the system, and to permit an orderly increase in the production rate for the system (sufficient to lead to full-rate production upon successful testing) (DAU 2003, p. 9).

For programs requiring LRIP, the final decision analysis, provided in the Full Rate Production Decision Review, is required before moving into full-speed production. This decision is made by the MDA after consideration of, "initial operational test and evaluation and live fire test and evaluation results (if applicable); demonstrated interoperability; supportability; cost and manpower estimates; and command, control, communications, computer, and intelligence supportability and certification (if applicable)" (DAU 2003, p. 56).

Finally, as the first production units are delivered to the user, the O&S phase begins. There is an overlap in the last two phases, and the PM must maintain oversight of both. O&S has two distinct parts: sustainment and disposal. Logistics and readiness matters at this point include maintenance, transportation, manpower, personnel, training, safety, survivability, etc.; these matters are a primary focus of the PM during sustainment. There are a number of postdesign and production factors, such as the fleet logistics capability for the Navy for example, that must be addressed and tested during this phase before ascertaining the supportability of the program through established channels, be they military or commercial. Assets also are tested for efficiency to determine system ability to effectively provide support to the user in the most cost-efficient manner to achieve the lowest possible lifecycle cost and, to the extent possible, total ownership cost. Since many programs stay in the field for years, even decades, the PM must work with the user to document the O&S requirements to continuously evaluate the lifecycle costs, making improvements or service life extensions as necessary in attempt to control and contain total ownership costs.

The last phase of the DAS, disposal, is focused on meeting the costs associated with the end of the useful life of an asset. Throughout the design process, the PM must detail hazards that will affect end-of-life costs and must estimate and plan for eventual disposal costs. When the system finally reaches the end of its useful life, the PM is responsible for ushering it through the process of demilitarization and disposal, "in accordance with all legal and regulatory requirements and policy relating to safety (including explosives safety), security, and the environment" (DAU, 2003, p. 11)

In summary, from the description above it is clear that the DAS is a highly complex, protracted decision process and management control system, which explains in part why it takes so long to acquire new defense assets. Could this process be reduced in terms of complexity, number of decision steps, players, and decision cycle-time through process reengineering? To answer this question, we turn our attention to how the DAS systems relates to the needs requirements system.

LINKAGE OF JCIDS WITHIN THE DAS DECISION PROCESS

The JCIDS and the DAS systems are tied to each other in a number of different ways. The primary goal of the DAS is to acquire capabilities for the DoD as directed through the Joint Chiefs. This relationship is carried out formally through the four formal JCIDS documents as well as through the many required DAS program reviews. They are also informally linked through the leaders of each process, some of whom have multiple roles to play in both.

As noted, the JCIDS documents include the initial capabilities documents (ICDs), capability development documents (CDDs), capability production documents (CPDs) and the capstone requirements documents (CRDs). These are directly and formally linked to DAS events. They are governed by policy and regulation and provide critical information to DAS leaders with respect to critical program elements like performance criteria, program size, impacts and constraints. They also help specify the level of administrative oversight required.

Generally, different JCIDS documents are required before each DAS milestone review; also, DAS players have to submit documents to JCIDS players for approval before a program can proceed past a milestone; for example, before milestone B approval, "the CDD must be received from the JCIDS leadership. For the JCIDS decision makers to approve the CDD, they must receive data from the DAS representatives and review the progress of the program" (Fierstine, 2004, p. 55). This represents a formal relationship where documents are passed back and forth between players in these two systems, with one set providing data and the other approving it before the first may give milestone approval. Notice in the schematic how each of the milestone decision points (MS A, MS B, MS C) is accompanied by input from the JCIDS via JROC and DAS via DAB.

Additional critical formal links are created between the two systems when the same players hold important positions in both systems. First among these is the SECDEF and his staff, the deputy secretary of defense,

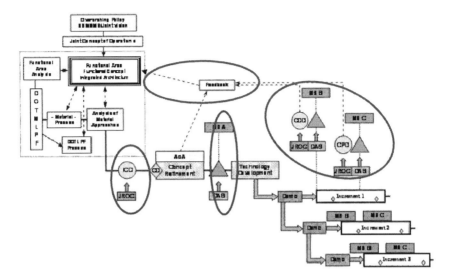

Source: DoDI 5000.2, 2003, p. 3. Circles indicate linkage zones between JCIDS and DAS.

Figure 13.3.   Linkage between JCIDS and DAS.

the under secretary for AT&L, and the under and assistant secretaries including the DoD comptroller and chief financial officer, and the assistant secretary for planning, analysis and evaluation. The USD (AT&L) is central to this process as he chairs the DAB and MDA for all the large procurement programs. He also has the authority to ask the JROC to review a program at any time. This gives him a powerful hand in both the JCIDS and DAS processes. The under secretary of defense chairs the DAB, as noted, where all the important decisions are made which involve both JCIDS and DAS items. Assistant secretaries for acquisition from the MILDEPS serve on the DAB and on the functional capabilities boards. An additional important participant is the assistant secretary of defense for networks and information integration.

On the military side of the house, the most important link at the joint level is the vice chairman of the Joint Chiefs of Staff (VCJCS) who functions as chairman of the JROC and is vice chair of the DAB. The senior military officer in this position also served as vice chair of the Senior Leader Review Group (SLRG) under former Secretary Rumsfeld. The SLRG was chaired by the deputy secretary of defense and convened to address major defense policy issues, some of which included asset acquisition decisions. However, the SLRG was never intended to replace or reduce the importance of the DAB. Representation in the decision process by the JCS also is important. These include the offices of J-8 (the Joint potential designator (JPD) gatekeeper), J-7, (the executive agent for transformation), and J-6 (the agent who ensures IT/NSS interoperability and provides review, coordination and certification functions in support of the JCIDS and DAS) (CJCS, 2004, p. B-4).

Within the MILDEPS, the vice chiefs of each service sit on the JROC, and as noted the service acquisition secretaries sit on the DAB. It should be remembered that individual military personnel form the lion's share of representation on oversight and analysis bodies related to both processes. Also, the military services are the sponsors for every program and research effort, and they staff the program offices. Furthermore, the military services contribute as participants in the JCIDS analysis processes.

Since the JCIDS and the DAS are event-driven systems, they follow similar patterns and are linked through their programs and documentation. In contrast, the PPBES is a calendar-driven sequence of events. JCIDS or DAS events may or may not fit neatly in the POM/budget cycle. DAS events may or may not fit neatly into the off-year or on-year cycle. For example, when a major program gets a "go" signal in an off-year, what this does to the basic concept of off-year is yet to be determined. It hardly seems like the program will be told to wait until next year, but if resources then are committed, does this mean that decision space is preempted from the following on-year? Does this mean the on-year becomes an off-year? What if the "go" signal occurs in the first year of a presidential regime? Will this mean a wait? If it is a major capacity-enhancing acquisition, what will this mean for the Quadrennial Defense Review (QDR) scheduled to arrive some 12 months later? Will strategy and doctrinal changes be preempted? What if a large program appears about to fail a major milestone, but it has been counted on as a part of a presidential legacy in the fourth year of a presidency: will the program be "forced" and the assumption made that it will get well (that its difficulty will be corrected) in the off-years (e.g., the USMC V-22 Osprey aircraft)? These decisions have consequences for each other, just as the battlefield concept in the late 1990s when the decision about armoring Humvees was made; doctrine appears to have envisioned a front-line/rear-area split with little need to armor Humvees because only a few would be used in or near the front line. Iraq did not turn out that way, hence the scramble to up-armor Humvees.

The point is that any procurement effort can span multiple annual PPBES cycles, be under the influence of a series of layered PPBES decisions and feed data back into any number of current and future PPBES phases. The link to PPBES formally comes from the SPC which develops the Strategic Planning Guidance (SPG). The SPC includes the combatant commanders (COCOMS) and virtually all of the senior leadership in the

DoD, civilian and military, including 19 four-star billets, the service secretaries and various OSD-level representatives. While the SPC produces the SPG, technically it is rendered under the authority of SECDEF, and in the official acquisition process and in budgetary terms it belongs to him. The SPG sets the agenda for the programming and budget processes, feeding directly into the POM. It identifies and sets DoD-wide trade-offs and identifies joint needs, excesses and gaps within and between programs. It focuses on capabilities required as threat changes, on war-plan analysis, new concepts, and lessons learned from ongoing war operations when the nation is at war.

For example, one lesson learned might be that U.S. forces have to be prepared and to fight in nontraditional battle environments (e.g., Iraq and Afghanistan). Such lessons have significant consequences for both military doctrine and attributes of war fighting platforms. If Humvees were going to be at risk of taking direct and high-powered fire wherever they went (i.e., virtually everywhere under the Global War On Terrorism [GWOT] scenario), then their armor needed to change, and heavily armored Humvees were procured and fielded. But the lesson went further, leading the Army to be authorized to develop and acquire a new light armored vehicle (Mine Resistant Ambush Protected vehicle or the MRAP).

The POM process also is informed by issues surfaced by the COCOMSs routed through an extended planning process to the joint staff. The result of this input of information is the chairman's program recommendation and the joint planning guidance (JPG), which help integrate joint capabilities into the POM process. The link between the DAS and PPBES is that the JCIDS capabilities analysis model is used to examine current and forecasted capability needs.

At the MILDEP and service level, a number of other interactions exist. In the Department of the Navy, for example, during the POM and budget build/review processes, the Navy requires officers and analysts under N7 and the financial managers and analysts under N8 independently conduct their own campaigns, scenario and program analyses. In doing so, they use the same scenarios, simulations and models as are used in the JCIDS by OSD, the joint staff and the rest of the MILDEPS. Additionally, all the data regarding past, current, and future program cost comes from the program offices who manage the services' acquisition programs.

At the most basic level, the PPBES and DAS are linked through program cost data. Program offices build OSIPs (Operational Safety Improvement Programs); these are used to create the budget line items that detail program cost data and to feed that data through their budget offices for their programs to the Navy Budget Office (FMB); here, it is used during program-cost analysis throughout the year. When

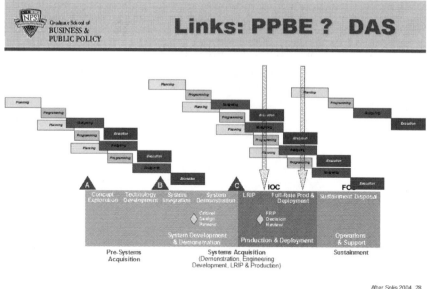

Source: Solis, 2004, p. 28.

Figure 13.4.   Links between PPBES and DAS.

the FMB asks questions about a program or recommends changes, those are answered or completed based on the data provided in these OSIPs. These questions may happen during the budgeting phase, when marks and reclamas (appeals of budget cuts) are made, or during budget execution. The analysts in N7, who represent the warfare requirements community, and the analysts in N8, who are the budgeters and linked to the PPBES, closely monitor the acquisition programs. In the current year, if a program is under-executing, then the program and budget analysts will make adjustments as necessary to ensure that money is diverted to those programs that will spend it by the end of the appropriation period.

The result is that the war fighting-needs system, the acquisition system (DAS), and the PPBE system focus around various points of integration and articulation—from an assessment of the threat in the SPG to a design for joint capabilities in the JPG through the POM building process and into the annual budget preparation and review processes. While formal documents provide for coordination, some coordination happens by forcing decisions on different aspects of defense needs through the same sets of players. Formal documents are required and reviewed by these players before decisions are made initially and at subsequent important check points, be they milestones, POM, or budget decisions. Additionally, staffs of analysts in different organizational locales have responsibilities for data production and review in program creation, implementation, and execution. They tend to be focused on a single-issue—on, for example, the best weapon system, or the most weapon systems for the money available this year. These players assume coordination and integration is done at levels above them or prior to program starts, or whenever the POM is built and reviewed, or whenever the threat changes or when new capabilities are needed or old capabilities may be foregone, or even when a strike in a tin mine in South America may imperil the pace of a program.

There is no doubt but that this is a complicated arrangement. Perhaps the single most confounding factor in these equations is time. Weapon systems take time to develop and build. The V-22 for the Marine Corps has been in development of one sort or another since the late 1980s, the Navy LPD-17 since 1998. The engineering and deploying of the surveillance drone in Afghanistan in 18 months is the exception to the rule. Most weapons acquisition programs take years to develop. The procurement effort can span multiple annual PPBES cycles, be under the influence of a series of layered PPBES decisions and feed data back into a number of current and future PPBES phases.

What this means is that when complicated programs (all weapons programs are complicated) are conceived and developed, they move through the multiphase PPBES process. What this means in practice is that they also are reviewed by different individuals. Turnover in personnel in the DoD is high. This happens by law and practice for military leaders; the effect is that turnover happens every 2 to 3 years. This level of turnover is just as true on the civilian side. Thus, the Marine Corps V-22 program has seen six different secretaries of defense. It was begun Under Secretary of Defense Caspar Weinberger and continued under Secretaries Dick Cheney, Les Aspin, William Perry, William Cohen, Donald Rumsfeld, and Robert Gates. In fact, the average tenure of senior leadership in the DoD is 1.7 years. Thus, coordination by position is riskier than it seems. If the distance between milestones A and B or B and C is more than two years, it is highly likely that most of the participants serving on the DAB will have changed. And, even when they are the same people, they may be sitting in new positions that changed the interests they represent. This is true for both civilian and military leaders. The result is that one should not count on the effectiveness of coordination by position. This leaves coordination by process as the reality.

Fiscal climate is also a complicating factor as we have explained throughout this book. Weapons systems that take years to develop and field will go through varying fiscal climates: for example, the Marine

Corps V-22 aircraft started in a rich procurement environment in the mid-1980s and was kept alive in the procurement holiday in the 1990s. Change also comes from change in the threat situation or battlefield doctrine: Secretary Rumsfeld's goal of transforming the Army to a lighter, agile, and more lethal organization doomed the Crusader artillery system. Another aspect of this happens when a service can not decide on the capabilities it wants and, thus, decides to maximize all capabilities; this is roughly what happened to Navy air plans in the early 1990s. The result was a years-long delay for plans for new aircraft. Thus, the passage of time means that people, resources, and doctrine change. These are all threats to the orderly integration of the war fighting requirements, DAS and PPBES.

PROBLEMS WITH PPBES AND DAS ALIGNMENT

During the years we have researched this topic a great number of interviews were conducted in the Pentagon on the degree of fit between PPBES and Acquisition decision systems. A number of current and past DoD process players in and around the Beltway were interviewed, including some now working in the private sector doing business with the DoD. Those interviewed in this project included representatives of Navy contractors, representatives from Navy air and sea system commands, Washington-based Navy resource management officials, OSD acquisition officials and active and retired JCS officials. Interviews were supplemented by discussions and briefings by high-level military officials in the Office of Program Analysis and Evaluation and the Joint Chiefs' staff (J-8).

We make no claim that our interview findings are definitive, but they provide insight into potential (perceived as real) dysfunctions within and between the PPBES and DAS analysis and decision processes. First, interviewees voiced concern with what we may term political issues: that all levels of the chain of command produce budget estimates that are above guidance, that the political sensitivity of large weapons programs affects requirements analysis and resource decisions, and that many decision makers use political clout to stave off directives from higher authority. Second, they criticized process: that a small number of people in the processes have disproportionate influence, that decisions are adversely affected by time compression—compounded by the lack of sufficient information—and that decisions are adversely impacted by the existence of too many approval levels in the acquisition chain of command. Third, they focused on management and cost issues: that there is excessive duplication within and between the PPBES and DAS processes at all levels; that repetitious calculation of program costs in response to program and budget "drills" has an adverse effect on motivation, and that absence of clarity and consensus on costs causes significant difficulty in execution when budgeted funds are lower than required. Insofar as transformation is concerned, they reported that concurrent program and budget review in the new PPBES process has caused a significant increase in workload without a significant increase in benefits; they felt that transformation has not resolved the issue of communicating appropriate information to decision makers, and transformational change actually has slowed down many stages of the review and decision processes. They identified barriers to change to include: (a) emergent user needs not addressed adequately; (b) an over-reliance on correct verbiage in the OSIPs; (c) ill-defined and cumbersome blanket joint requirements, (d) inequitable distribution of common funds; (e) innovation hindered by the type of rigid control exercised over multiyear procurements constrains program flexibility; (f) with regard to program documentation, required process forms and "semantics" sometimes confuses intent; (g) budgetary constraints that drive changes in schedule and/or performance requirements that, in turn, have an unintended and negative impact on cost control.

Some interview respondents thought that, as budgets "moved up" the organizational hierarchy, there was a tendency to overestimate dollars to get the correct amount of war fighting capability; they believed this resulted in budgets exceeding guidance. Some respondents also felt that the large and expensive weapons systems which were built in several congressional districts or states were, perhaps, not subjected to as searching a warfare analysis scrutiny as they should have been. Respondents were concerned that, "leadership can and does direct funding for programs deemed important, yet not supported by the analysis, given the info available to mid-level experts" (Fierstine, 2004, p. 99). They also said the lack of time and insufficient data or expertise impacted the quality of the budget decisions that were made. Respondents also worried about the degree of overlap and churn in the system. We speculate further on this below.

The MILDEPS and services, the joint staff and OSD all do very similar analyses using the same data, models, and simulations. All of this adds time and manpower effort to the process without necessarily reducing the necessity for guesswork and intuition. With respect to transformation, respondents felt that the PPBES was still a work in progress and had not produced a significant increase in benefits. We would observe the primary difficulty here is that the budgeters begin to work on the budget before a POM package has been completed. Further, in the budget and programming process, people routinely make decisions without a full grasp of all the facts and data. This was evident at all levels, from those in the program and requirements offices who had to route paperwork through people unfamiliar with their platform, to those in FMB making spot judgments due to time constraints. Finally, everyone interviewed complained about the length of time it takes to route paperwork and receive decisions.

Respondents also worried that emergent needs were not identified and integrated into the system soon enough—in effect, that joint needs had priority, and some programs were identified as joint and given priority when the likelihood of their being used in a joint environment was low. They also criticized the cumbersome procedures necessary to gain approval in the JCS review process. Some of those interviewed expressed the view that some program and requirements officer emergent needs for existing programs are not adequately addressed in the current system. Most argued that a big part of the current problem is the fact that the comptrollers are tied to the exact terminology in the OSIPs; therefore, anything not specifically delineated in the OSIPs has to endure the lengthy delay of a new program start-up. They all complained about the difficulty of navigating through the vague joint requirements required of all communication gear; these requirements force them to route all associated programs and upgrades through numerous joint wickets, even though many of the programs would not be used in such a manner as to require the joint standard. Finally, a few interviewees took issue with the equitable distribution of funds in programs that took money from everyone in order to provide commonality to all platforms. They claimed that these funds were effectively an under-the-table system for certain systems to get their capability funded by everyone else.

Those interviewed explained that playing the game carefully is important. One interviewee had a list of the correct words to use when writing justification for dollars in different appropriations. Although a number of terms were virtually synonymous and would appear to mean approximately the same thing, a word that was wrong for the account could lead to a turndown or a do-over. For example, a careful analyst would use the terms "investigate or research" when writing justification for an RDT&E account, but use the terms "analyze or assess" when doing the same activity for an aircraft procurement Navy justification. And an O&M request using these words would be looked upon unfavorably. The word "track" is probably as close as the O&M accounts get to in depth analysis.

Respondents were concerned with innovative adaptations to organizational stress. Here we point out how requirements change (downward) as programs fail to meet requirements; we will illustrate how PMs have found that if they can move their programs to a multiyear profile, they can fend off much of the churn that is driven by the annual budget process, particularly one that takes place in an era of scarce resources. Programmers have begun to increasingly use multiyear procurement strategies in an attempt to fence off programs from the annual churn that is inevitable. For example, breaking a multiyear procurement contract is a tremendously powerful argument to ward off a cut. The programmers also have used BTRs (below threshold reprogramming) to their advantage to protect their accounts from raids during execution. This has the added benefit of cushioning them against the end of the year need to spend their money or lose it by designating a recipient for unspent funds and then possibly getting reciprocation from that recipient after the new budget comes along.

Much of what we found has been reported in reports by General Accountabiliy Office (GAO). In its work comparing best practices in industry and DoD acquisition programs, GAO sent out surveys to 185 category I and II DoD programs managers in April, 2005 (GAO, 2005, see pp. 19-20 for a discussion of methodology). The response rate was 69%. Results from this study indicate that the problem is that the DoD has an inherent flaw in what could be called its capital budget process: it starts too many programs and fails to prioritize programs in process so that resources may be shifted to the most appropriate program when necessary in a distressed fiscal environment (e.g., when costs of raw materials or labor rise). The GAO indicated as follows:

> The primary problem, according to many PMs and verified by GAO's work, is that DoD starts more programs than it can afford and does not prioritize programs for funding. This creates an environment where programs must continually compete for funding. Before programs are even started, advocates are incentivized to underestimate both cost and schedule and over-promise capability. (pp. 8-9)

PM comments tend to blame the OSD for part of the problem. It must be remembered that OSD is the final arbiter in DoD of who will get what money for what systems; when times get tough, it is SECDEF who must decide what can not be funded or how big any across the board cuts may be. As can be seen below, PMs are not sympathetic to the OSD role.

As Figure 13.5 implies, PMs believed that they were operating in an environment where there was unfair competition for funding (GAO, 2005, p. 40). The next two figures indicate some of the dimensions of the problem. For example, in Figure 13.6, most PMs believed that the parameters of their program were reasonable at the start, with about 24% falling in the some (18%) or little or no (6%) categories.

Then, in response to an open-ended question on biggest obstacles (Figure 13.7), 36% of the managers responded that funding instability was the biggest obstacle, almost three times the number who mentioned requirements instability, the next category. What these evidences seem to

- OSD staff has reduced funding without any understanding or appreciation for program impacts. It appears that the staff makes arbitrary cuts.

- OSD has a very near-term execution year focus, resulting in great instability. In reality, it should provide much more strategic vectors for the Department instead of short-term adjustments to fix more tactical-level funding needs.

- My experience is that the [service] and OSD typically cut programs to pay top down bills.

- There is no such thing as funding stability in DOD. Funding reductions and program stretchouts are the norm due to top down fiscal bills that occur during the execution year. The Pentagon must pay the bills, therefore it takes funds from the programs, thereby contributing to program stretchout, cost increases, inefficiencies, etc.

- Unstable funding results in pressure to do aggressive things in order to minimize the impact of budget cuts on schedule and performance. I believe this has been a major factor in recent...program execution problems.

- Our product is considered a support function. When funding gets tight, we have been considered a bill payer for others, even if it has "broken" our program.

Source: GAO, 2005, p. 40.

Figure 13.5.   Program manager comments on competition for funding.

hint is that much of the cause of acquisition turbulence lies in the funding mechanism which produces funding instability with the passage of time.

We also observe that another significant issue for acquisition and resource allocation is that an overwhelming amount of redundancy exists at all levels of the chain of command. This finding is supported by a study by the Center for Strategic and International Studies warning, "that various military bureaucracies 'unnecessarily overlap,' resulting in duplicative and, in some cases, overly large staffs that require wasteful coordination processes and impede necessary innovation" (Schmitt, 2004, p. 48). GAO came from a slightly different perspective but found much the same phenomenon.

In the Figure 13.8 (GAO, 2005, p. 59), PMs reported on what types of authority they thought they needed. The implications are clear: PMs believe they need more authority to execute their programs and efficiently allocate the resources they have been given, without undue and unnecessary oversight, without needlessly complicated reporting requirements. The GAO found that PMs expressed frustration with the time required of them to answer queries of oversight officials, "many of which did not add value. Some PMs, in fact, estimated that they spent more than 50 percent of their time producing and tailoring and

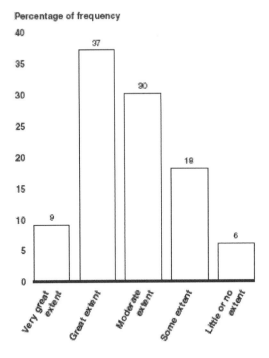

Percentage of frequency

Source: GAO, 2005, p. 43.

Figure 13.6.   Extent that parameters of programs
are reasonable at program start.

explaining status information to others" (p. 46). The GAO also noted,
"Program managers commented that requirements continue to be added
as the program progresses and funding instability continues throughout.
These two factors alone cause the greatest disruption to programs,
according to program managers" (p. 45).

Another way of looking at this is to understand that almost every
secretary, under secretary, assistant secretary and high ranking military
officer with a required signature anywhere in the JCIDS-DAS-PPBES
decision-making processes has their own group of analysts to recheck,
reverify and recertify the data provided them from others (all of which are
in or near the Pentagon). An example with regard to aviation would be
how the individual programs, system command Budget financial
managers, the offices within N7, N8 (under the chief of naval operations)
and the OSD all have cost-analysis experts on staff looking at the same
data, yet coming up with different conclusions. Although risk reduction is

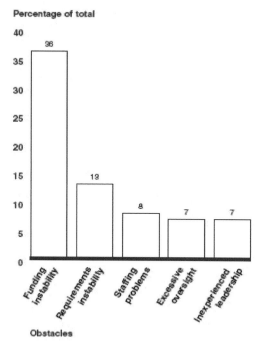

Percentage of total

Source: GAO, 2005, p. 44.

Figure 13.7.   How program managers respond to questions on obstacles faced.

important, it seems that DoD analysis capability has grown (in aggregate) past the point of diminishing returns.

In our view the results of this research project call for efforts to reduce such redundancies through business process analysis and reengineering to eliminate duplication of effort, thereby increasing the quality of analysis provided to DoD decision makers, and to streamline decision making overall so that the time lapse between problem recognition and resolution is shortened (Hammer, 1990; Jones & Thompson, 1999, pp. 47-106). Reengineering requires a careful, step by step analysis of work processes to assess the value of work performed at each step, eliminating duplication of effort, unnecessary work and excessive procedural complexity. In addition we would like to see improved integration and more systematic communication within and between the PPBES and acquisition decision processes. This in effect may be achieved only through improved alignment of the PPBES and DAS so that they work together rather than in tandem as separate systems. Realignment in

- Program managers need to have more ability to control their funding in order to make more efficient system and production trade-offs. Program managers also need more ability to work with the warfighter to pursue moderate or even high risk strategies when the payoff for the warfighter warrants such a change. Program managers also need the ability to directly interface with OSD and with Congress and should not be restricted through service staffs in order to facilitate communications.

- Program managers should be able to select and award most contracts versus going to the PEO or service acquisition executive for a decision.

- I believe program managers should be allowed to spend small amounts of underrun as they see fit for their program. Too often, any underrun is taken to pay for other programs.

- [We need] more authority to budget for and manage management reserves. The [planning and budgeting] process is too slow to react to new funding requirements to mitigate program risks.

- In the current environment, we do not control the numbers of military, civilian, or contractor personnel that work in the program office. We do not have the authority to hire and fire personnel, or to seat personnel in our office space. We do not have the authority to get adequate tools for our people to do their work, such as computers, printers, copiers, telephones, etc.

- Once appropriated by Congress, program managers should have more flexibility to transfer between program elements and budget accounts, and also the service and major commands should have less ability to remove funds that are being properly executed in order to transfer them to other programs.

- Program managers should be given authority to move funds between colors of money. Colors of money greatly reduce the flexibility that program mangers often need to make tradeoffs within their programs.

- [We need authority] to be able to fire or replace people immediately or affect their bonus.

- [We need authority] to give monetary awards to support professionals.

- The key is not more authority; it is allowing program managers to fully exercise the authority they already have. No program manager minds reasonable oversight, but the current level of oversight is unreasonable.

Source: GAO, 2005r.

Figure 13.8.   Program manager comments on types of authority needed.

essence required matching the two systems step by step in their analysis and decision stages to insure that timing and scheduling fit together so that mutual dependencies are addressed properly, with changes made when phases of decision do not coordinate effectively. The overall goal of realignment is to match the processes of decision and to make the necessary changes in organizational structure to fit both the mission and the product demand functions necessary to meet market changes, for example, changes in military mission and field asset requirements as is taking place presently in fighting the war on terrorism. When organizational mission forces changes in strategy and asset requirements, realignment typically is needed in complex organizations to match structure to strategy. This principle is well understood in the private sector but not so in government (Jones & Thompson, 1999, pp. 127-178). In this regard, we would go so far to say that effective realignment of PPBES and the DAS may only be realized through merger of the two separate systems into a single decision process.

As a step toward achieving better linkage of the type, we suggest is necessary without radical decision system redesign and realignment we support the creation of a new information system to record and communicate real-time, highly detailed, accurate and useful programmatic cost, schedule, and performance information to decision makers (Fierstin, 2004, p. 110). Included in this system should be highly detailed prioritization lists so that when decisions have to be made at subsequent levels of the budget an acquisition processes, those having to make decisions are better able to determine what should be implemented and what must be cut when necessary, and what should be bought when additional funding is available.

We suggest that such a system would increase decision efficiency and shorten decision time since top leadership officials would be able to make decisions based on readily available data without having to "drill" back down into the military/service budget and program offices to get data to satisfy their needs. Our study indicates a need for simplifying the entire acquisition document-and-review process through business process reengineering of the type suggested by Jones and Thompson (1999). We find that current operators are reducing the risk of making the wrong decision by increasing the time to make the decision. We also worry that currently there is no satisfactory way to address ideas or concerns that "bubble up" from the field that would add small increases in capability in the near-term. Currently such decisions are divided between existing programs that require attention and emergent ideas that require immediate funding and could be fielded quickly and at low cost.

An example of a less urgent nature includes F-14 adaptation of the Air Force LANTERN pod. This upgrade was on the United States Air Force air community "top-ten upgrade list" for years, but was only able to secure funding after a monumental demonstration of innovation. Had the acquisition decision pipeline been able to rapidly and cost-effectively address this need, then neither war fighters nor DAS staff would have had to contend with the protracted process of test, evaluation and demonstration. Since changes such as this are relatively small and tend to be focused on the short term versus the JCIDS horizon of decades, war fighter operators typically are unable to enter the funding debate without great difficulty. We would argue that this issue appears small but in fact is important to field users. This example and our overall observations suggest that a better system needs to be established to allow adequate prioritization and swift communication of field level needs up the chain of command and into and through both the PPBES and DAS processes. It is also useful to observe that these recommendations call for reduction in staffs to eliminate redundancies, and also for the installation of a comprehensive real-time information system that would serve the same information to all participants; additionally, we support creation of a failure-analysis unit and system. The risk here is that adding a new and complex information system and a new organizational entity to systems already rife with information systems and complexity is problematic.

ABSENCE OF LINKAGE BETWEEN ACQUISITION AND PPBES PROCESSES

Our findings relative to integration of the PPBES and DAS decision cycles indicate there are numerous points at which substantial and reinforcing linkages exist, and others where the systems operate separately. The question is to what degree should parts that are not integrated presently be better integrated in the future? We conclude that significant business process reengineering and organizational realignment are needed to improve the overall integration and the consequent effectiveness of DoD resource management and acquisition. We believe that improved integration and more systematic communication within and between the PPBES and acquisition decision processes is necessary. This in effect may be achieved only through a long-term commitment to process reengineering and initiatives to improve alignment and linkage within and between the PPBES and DAS so that they work together rather than in tandem as separate systems (Thompson & Jones, 1994; Thompson & Jones, 1999). We conclude that effective realignment probably can be realized only through DAS reengineering and significant reconfiguration of PPBES, as we describe in chapter 14.

As part of marginal or more sweeping reform of DoD acquisition and PPBES, we recommend that a new and better information system be created with the expressed purpose of providing data extracted from both systems to DoD decision makers in as close to real time as possible. We suggest that an enterprise data management approach would be most appropriate to implement this improvement effectively. However, we caution that the costs of enterprise system development and implementation must be controlled carefully through rigorous system design examination and close management control when systems are proposed and developed by contractors. Poor system design and weak, ineffective oversight control over development of new management information system design, along with ineffective contract management continue to be bugaboos that plague DoD.

We find that the war fighting needs definition system, the DAS, and PPBES perform and focus on different tasks and their decision processes are not well integrated at many points—from the assessment of the threat in the SPG to a design for joint capabilities in the JPG through the POM building process and into the annual budget preparation and review processes. Most, if not all, of the top leaders in the DoD hold multiple responsibilities in these systems. While formal processes intend to provide coordination, much co-ordination only happens by forcing decisions on different aspects of defense needs through the same sets of players—and by going around much of the procedural steps and complexity of the individual decision systems. Excessively complicated documentation and overly formal procedural requirements are required to be incorporated and adhered to before decisions can be made even initially—and at subsequent important check points, be they intermediate milestones, POM formulation, or budget preparation and decision making stages. Typically, staffs of analysts in different organizational locales in and around the Pentagon often have duplicative responsibilities for data production, analysis and review in the great number of steps moving from program creation through decision, implementation, and execution.

In addition to the complexity inherent in these systems, we conclude that the passage of time itself has important consequences for defense acquisition. Weapon systems take time to develop and build. The procurement effort can span multiple annual PPBES cycles, be under the influence of a series of layered PPBES decisions and feed data back into any number of current and future PPBES phases. The passage of time means that people, resources, and doctrine change. These are all threats to the orderly integration of war fighting requirements between the DAS and PPBES.

Operators in the process express concern and frustrations with the outcomes produced by the DAS and PPBES. These range from process duplications, repetitive calculations of program costs by different staffs, inflated budget estimates for programs and concerns about the efficiency of the concurrent program- and budget-review processes. We suggest that one way to improve the acquisition process is to change the budget process to a multiyear format. We believe this could reduce end-of-year turbulence and churn and allow for greater rationalization of DoD decision-making systems. We reiterate that reengineering and realignment of the relationships between PPBES and the DAS is needed to achieve efficiency gains both within and between these systems.

With respect to the Rumsfeld era PPBES transformation initiatives, some measures to improve decision integration were designed to be successful in combining the programming phase or "endgame"—the last part of the programming phase—with budget review, but significant problems in implementing this reform have come up over the past 5 years, as discussed under budget execution, and in the final chapter of this book. The endgame is where the DAB (and sometimes the SLRG when it functioned under Rumsfeld) reviews, approves and sometimes is forced to cut major acquisition programs. For example, since 2004 the defense secretary, deputy secretary, the USD AT&L and, consequently, the DAB have had to consider both significant increases in acquisition along with reductions forced by the tight fiscal constraints of future POMs and defense budgets. The DAB review, forced by the need to reduce spending projections due to the costs of the GWOT and other budgetary costs including those for recapitalization, and for personnel and personnel entitlements programs, resulted in some major acquisition program shifts and reductions. These included approval of the Navy decision to retire an aircraft carrier early (the Kennedy), cancellation of the C-130J buy for the Air Force, and reductions in the size of buys in submarines and surface vessels for the Navy, modularization for the Army (the acquisition portion of this initiative), and cuts in the Joint Strike Fighter and the F/A-22 aircraft program for the Air Force.

THE IMPORTANCE OF EXECUTION

Despite all the time and effort devoted to "getting things right" and balanced in the DAS and financial management resource process, in the end many problems of matching resources to program needs have to be addressed during program and budget execution. In this regard, many of the problems we have identified as inherent in execution in chapter 8 and elsewhere will persist as far as we can ascertain, for example., where budgets drive programs when it should be the other way around. Some of this is inevitable—as a result of congressional politics that produce changes in defense budgets and acquisition programs beyond the ability of the DoD to resist, and from changes in the war fighting environment. When such reordering of priorities from outside of DoD occurs, it causes significant disruption in both programs and budget—especially in preparation of future budgets and in the execution of current appropriations. It also forces changes in both the structure and content of the POM and QDR, causing the programming process to have to move in reverse to accommodate budget changes in a way that almost always causes discontinuity and drives up costs in acquisition program management and execution.

Begun under former Secretary Rumsfeld and continued Under Secretary Robert Gates and his successor a number of changes were initiated intending to improve the manner in which the PPBES serves as a decision system for DoD to better integrate financial decisions with acquisition decision making. This began, in part, as a result of former Secretary Rumsfeld's demand for better information upon which to base decisions and his willingness to listen carefully and to question vigorously the data and options provided to him from his staff. In addition, it was an intended result of the changes made in the PPBES cycle to better connect the process to SECDEF procedural preferences. However, the jury is still out on whether these changes in PPBES have achieved their objective. We believe they have not, up to this point at least.

We also conclude that the acquisition system has been strengthened in terms of program review (e.g., for interoperability) in a number of ways including improvement in the role and performance of JCS staff (J8) where not just defense-wide acquisition programs (as was the case before Rumsfeld's transformation), but all DoD acquisition programs now are supposed to be reviewed for jointness, interoperability, capability, and feasibility. We note, however, that this rule has not been inculcated thoroughly throughout DoD as non-ACAT I programs have been afforded options to circumvent joint review (described to us as "doors and windows" to avoid joint review). And, in fact, experts on the DAS insist that joint review is not required for many service specific programs. Be that as it may, both the PPBES and DAS continue to reveal excessive procedural complexity that should be eliminated through rigorous business process reengineering at minimum, and the absence of sufficient integration that only may be addressed through structural realignment or even replacement of these two systems with better business processes.

A key problem we have noted for both PPBES and acquisition is the relationship between programming and budgeting. This may be further understood by pointing out differences in perspectives between staff working in different parts of the two decision systems. For example, the DoD comptroller staff who have the chief responsibility for budgeting within PPBES contend that budgeting always integrates acquisition programming. However, many military department/service officials, DAS PMs (and even some OSD level programmers) do not share this view, contending that too many budget decisions drive the POM, rather than the other way around as the system is supposed to operate. This may have changed to some extent over the past four years with the goal of programming and budget analysis and decision coupling, but there is insufficient evidence available about resolution of this problem to convince us that there is demonstrable improvement in how DoD budgeting and acquisition systems operate with regard to both integration and reduction of duplication of effort.

What we have found is that DoD budgeting and acquisition systems and staff have had to work harder to respond to changes in the threat and war fighting environment in the years since September 11, 2001 due to the demands of war. However, there is a difference between working harder and working smarter through the use of what may be termed better practice. Our findings suggest strongly the need for reengineering and realignment of both the PPBES and acquisition decisions systems and linkages to reduce duplication of effort, to achieve better integration between systems, and to eliminate unnecessary procedural complexity in DoD resource decision making.

CONCLUSIONS

The major challenge facing the DoD in the period 2008-2012 and beyond is how to continue to modernize the fighting forces and increase the pace of business transformation while paying a high price in waging a war on terrorism. In essence, what the DoD must fund and support in the short-term must be traded-off against longer-term investment intended to improve both business-management efficiency and force readiness. Given this dilemma, it is clear that DoD leadership faces critical resource management challenges in the next decade. In this cost constrained environment it is essential for DoD to attempt to reengineer and realign separate budgeting/financial management and asset acquisition systems to improve integration of overall defense management decision making. Any changes such as those suggested in this book would come in an environment already highly impacted and, frankly, confused by too much change.

Given the analysis and conclusions reported in this chapter and other reservations expressed in chapter 14 and in our chapter on PPBES, we cannot at present paint a rosy picture of our results of past and the most recent "transformation" of resourcing for acquisition. We observe that at the program and project-management level (within budget execution from the financial management perspective), there remains a high level of uncertainty regarding financial stability and management control. While macro changes at the DoD level may make participants in the OSD believe that the system has been changed (and they probably are right with respect to their position perspective), the larger question remains whether macro system changes have improved the cost, performance, speed of delivery of weapons and weapons systems in reality. This improvement will only result from better management and management control at the point of relationship of the buyer (DoD) and the supplier (the private-sector contractors). It is evident from preliminary analysis (and from the experience-based knowledge of serving and retired program and project managers) that there still is much to be improved in the nature of contracting, contract management, and enforcement of DoD and government controls through a properly designed and enforced management control system (Jones & Thompson, 1994).

The dilemma is, in part, a result of management failure on the part of government in assuming that private-sector contractors will obey DoD and federal acquisition rules and guidelines and the restrictions built into contracts, without sufficient DoD leadership, oversight, and enforcement of law and contracts. Is the blame for project-cost overruns the fault of greedy contractors who attempt to take advantage of government incompetence or lax enforcement? Is the blame due to this absence of control on the part of the DoD? It appears that both are causes of the problems of costs exceeding estimates, the extended time taken to develop and deliver new and increasingly more technologically complex weapons systems, late delivery, system failures (despite higher-than-projected costs), inadequate documentation provided for training of end-users, installation deficiencies and many other problems with the quality and performance of systems delivered to the fighting forces.

Our point is that it is unwise and incorrect to gloat about or claim victory in the battle to make acquisition and its funding more efficient at the top levels of the Pentagon when, at the level at which programs and projects must be managed, so little has changed to achieve the improved efficiency and effectiveness goals of business transformation (Dillard, 2005, 2006). No amount of change at the top levels of the Pentagon will achieve these goals. To bring meaningful reform, change must reach down to the level at which spending occurs and programs are executed, where the government and contractor interface and relationship is so crucial to improving performance and results.

How can business process reform reach down to the program and project level? Some may argue that a great deal of effort has been exerted toward deregulating and contracting out, much to the benefit of the DoD generally and acquisition specifically. Undeniably, deregulation of the FAR (federal acquisition regulations) and DAR (defense acquisition regulations), the DoD 5000 series, and other DoD administrative rules, along with passage of acquisition reform legislation by Congress, has shown promise and the intent to engage in meaningful change. However, a significant number of initiatives to improve management of acquisition programs at the government/contractor interface have concentrated on auditing.

The problem with this approach is that of closing the barn door after the horses have escaped. It is fine to discover contractor overcharging ex ante and to extract penalty payments from contractors as a result. However, this is merely a financial transaction that does little or nothing to improve the services to and benefits for the end-user—the war fighter. When unworkable products are manufactured and delivered, no matter what the cost to government, the result for the end-user ranges from frustration in the best of circumstances to casualties and death under the worst of circumstances.

It may be argued that what is needed is not more deregulation, but adequate level of effort in enforcing the rules that are in place, which can only happen through high-quality, knowledgeable and skilled leadership. In addition, improving the application of the correct tools in contracting is essential to any meaningful progress. All of this, in turn, implies investment in the education of leaders and decision makers (current and future), increased continuity of leadership, and the ability to manage looking forward—rather than backward in the manner that characterizes the "reform by audit" mentality. Whoever advances the proposition that auditors are the best source of the management knowledge and expertise needed to improve business practice should be better informed. Even the audit community itself would not advance this proposition.

So, where do we go from here? We believe the knowledge about how to improve acquisition management at the ground level resides, to a great extent, with those who have done the job, that is, experienced (and often retired) program and project managers. If this were not the case, then why would the private sector hire and pay these people so well to represent them in dealing with the DoD? The question of leadership in ground-level reform, where it will make the most difference for the end-user, thus becomes how to retain this expertise rather than force it into retirement to engage in profit generation for contractors?

Further, as noted, improvement in the design and skill in managing contracting instrumentation is vital—and much effort has gone into this initiative in the past several decades. As a colleague remarked, "What kind of cost-plus contract haven't we tried to create the right incentives to perform and deliver the results? We have tried them all!" We would suggest that it is one thing to write a good and enforceable contract and another to actually enforce it. Learning how to do this is one obstacle; getting the attention of a revolving crew of leaders to either do it or permit it to be done is another. Our hope is that pointing out that improved management and control is needed is a start to moving in the right direction, to be realized through adopting the appropriate control system design and execution strategy, and should be a prime target for reform equally worthy to the reformulation of the DAS and PPBES (Jones & McCaffery, 2005; Thompson & Jones, 1994).

With respect to the continuing pace of reform throughout the DoD, no SECDEF can alone manage an enterprise as complex as the DoD. And in fact, it is important to point out that in the past and presently, input to program and budget decisions in the DoD is provided by the deputy secretary of defense and staff, the under secretary for acquisition, technology and logistics, the position in the DoD that bears a large part of the responsibility for actually attempting to manage DoD acquisition. In addition, the under secretary comptroller, and under and assistant secretaries for other OSD and military department and service functional areas including program analysis and evaluation, policy, force management and personnel, legislative affairs, health, reserve affairs, and others, all provide views and analyses to guide program and budget decision making.

In conclusion we want to point out that numerous business process change initiatives beyond improved financial management, PPBES and acquisition process reforms are in progress in DoD. In the areas of both acquisition and logistics, change toward increasing spiral (continuous and simultaneous) development, and to "sense and respond logistics" processes is underway. Improving information technology (IT) for management of inventory systems in real time to permit managers to know how much and where material is located on a worldwide basis also has been addressed and is fully operational in the Air Force. In the area of information technology, network-centric combat information systems are under development and fielding in all of the military services. Such systems coordinate various types of data to a single command point in real time to improve the ability to see and manage military operations (Jones & Thompson, 2007, pp. 231-259).

Applications of network-centric IT in the area of business management may be the next step, although this will be costly. However, such applications are one approach to coordination of decision making in flatter, network-types of organization (i.e., hyperarchies), rather than through traditional bureaucratic forms of organizing to solve complex and sometimes "wicked" problems (Jones & Thompson, 2007; Roberts, 2000). Given the vital importance of information technology, it is essential for the DoD to address the knowledge, skills and abilities of its workforce to fully leverage the potential of IT and other business process methods.

These and the other initiatives identified in this chapter are only a sample of the many reform measures currently under some degree of implementation and experimentation in DoD. Given the progression from the industrial age to the age of technology in an increasingly global commercial marketplace, capitalization on new technologies is a key part of transformation to create "knowledge warriors" for significant battlefield advantage. These initiatives are not under implementation independent of budgets constraints and cost accountability requirements—virtually all are expected to reduce costs while cutting cycle time with either improvement of quality or, at least, no diminution of quality of service to customers. The business models and plans developed for these initiatives often are based on business processes used in the private sector.

Business process reform also has stressed continuous learning and the creation of self-learning organizations that can observe and orient themselves more quickly to new threat environments; they must then make decisions and take action to learn more quickly by trial-and-error in a cycle of restructuring, reengineering, realignment, and rethinking of both means and objectives (Jones & Thompson, 2007). Further, critical issues related to transition management, organizational change, organizational design and appropriate institutional arrangements are raised whenever DoD institutional reform is attempted on a large scale. We now turn to give additional attention to these aspects and variables related to more comprehensive reform in the last chapter of this book.

CHAPTER 14

BUDGETING AND ACQUISITION BUSINESS PROCESS REFORM

INTRODUCTION

In this book we have introduced the basics of the federal budget process, provided an historical background on the foundation and development of the budget process, indicated how defense spending may be measured and how it impacts the economy, described and analyzed how Planning, Programming, Budgeting and Execution System (PPBES) operates and should function to produce the annual defense budget proposal to Congress, analyzed the role of Congress in debating and deciding on defense appropriations and the politics of the budgetary process including the use of supplemental appropriations to fund national defense, analyzed budget execution dynamics, identified the principal participants in the defense budget process in the Pentagon and military commands, assessed federal and Department of Defense (DoD) financial management and business process challenges and issues, and described the processes used to resource acquisition of defense war fighting assets, including reforms in acquisition and linkages between PPBES and the defense acquisition process.

This concluding chapter serves three purposes, based upon the material provided in the previous chapters of the book. Our first purpose is to assess

---

Budgeting, Financial Management, and Acquisition Reform in
The U.S. Department Of Defense, pp. 603–661
Copyright © 2008 by Information Age Publishing

the future of PPBES and related budget reform and to suggest that it may take more that a marginal adjustment to the current PPBES process to plan and budget most effectively for national defense. In this regard we recommend that DoD, and the federal government as a whole, adopt a capital budgeting process. The second purpose is to review and assess previous acquisition reforms in DoD, many of which continue into the present. The third purpose is to assess modification of the current acquisition process to achieve improvements in business processes imbedded within this system as well as to make the overall process operate more efficiently.

REFORM OF PPBES AND DEFENSE BUDGETING: WHERE TO NEXT?

As explained in detail in chapter 5, PPBES changes have created a combined 2-year program and budget-review decision cycle (but not a biennial budget), with a complete review in year one, followed by limited incremental review in year 2. This change in cycle from a full-program review and a full-budget review each year to a combined review with a comprehensive review happening every other year was meant to reduce the inefficiencies of unnecessary remaking of program decisions; the program should drive the budget rather than the opposite. With the programming and budgeting cycles operating contemporaneously, decisions are intended to be arrived at more effectively, whether they are made in the off- or on-year. Changes made in each off-year cycle are intended to have quicker effect by compressing the programming and budgeting cycles while still preserving the decisions made in the on-year cycle through the off-year by limiting reconsideration of decisions to only the most necessary updates. In essence, decisions flow from the Quadrennial Defense Review (QDR) and other studies; then, a structure is erected in the Strategic and Joint Planning Guidances that provides direction for the remaining years of a presidential term.

The processes summarized above will remain in place, in theory to best assimilate and adapt DoD financial management and budgeting to dramatic changes in worldwide threat and, correspondingly, defense capability requirements. Year-to-year changes in the program structure and budget then are made only to adjust to incremental fact-of-life changes. Also, this new process put the secretary of defense (SECDEF) into the decision environment at an earlier stage than in the old PPBS process; it put him "in the driver's seat," in the words of one DoD official. Decisions under the reformed PPBES are intended to reach the secretary while options are still open, and while important and large-scale changes still can be proposed—before the final decision has become a foregone conclusion at the military department and service level. When the defense secretary's input came at the end of the stream of decisions, some changes that could have been made were preempted because they would have caused too much "breakage" in other programs. This problem persists and, as indicated in this chapter, this is only one of many business practice problems that need to be reassessed and changed to improve resourcing for national defense. The purpose of this section is to assess whether significant changes in DoD financial processes are warranted, and if so, what future change options should be considered.

Up to this point in time, under former Secretary Rumsfeld and continuing under Defense Secretary Gates and his successor presumably, a number of changes have been made and implemented to varying degrees with the intention of improving the manner in which the PPBES serves as a decision system for DoD to better integrate financial decisions with acquisition decision making. We also conclude that this linkage has been strengthened somewhat, although not enough, through program review by the Joint Chiefs of Staff (JCS J8) where all DoD acquisition programs now are reviewed for jointness, capability, and feasibility.

With respect to budget formulation as opposed to execution we might wonder what would happen to DoD resource decision making if the POM were eliminated and replaced by a process of longer-term budgeting. In traditional budgeting, budget submitting offices have to answer several important questions as they ascertain what they need in the budget and as they justify their requests to funding sources. These questions include "what," "why," "when," "where," and "how." The answer to "how much" flows from the answers to the prior questions. All of these questions are important, but possibly the two most important questions in this set are the "what" and "why" questions. They set the stage for the fact-finding that causes answers to the how, where, and when questions to surface.

For example, if there is no need for a ship or a tank, then there is no need to define when you might need it, where you might need it, or how it might be configured or delivered. This interrogative pattern is the whole cloth upon which budget decisions are based. Much academic research has focused on the concept of incrementalism, that is, that budgets change only by small amounts on the margin and not much as a percentage of the total from one year to the next. This is a tested analytic finding, but not one that is useful for the PPBES decision makers because they do not build budgets by focusing on percent of change. Rather, they first determine what it is they need (capability and requirement). They do this by analyzing the world around them and its impact on the organization and its systems. They then establish what is needed to improve or operate more efficiently or effectively than in the previous planning period or fiscal year (FY). Finally, they evaluate in detail what this will cost and what can be executed in the annual budget.

With the implementation of the PPBS in 1964 under Robert McNamara, the defense budget system split the focus of these questions into three parts. The planning and programming functions (in which the Strategic Planning Guidance [SPG] and Program Objective Memorandum [POM] are built) deal with the "what" and "why" questions, and to some extent "where" and "when." Most of what is left for the budget process is the task of answering the question, "how much this year?" Still, budget formulators do have to present their fully justified budget to reviewers in the DoD, the Office of Management and Budget (OMB), and Congress. This means that they have to convey the part of the POM that answers the "how" and "what" questions along with the request for "how much." To do this, budget offices have to put back together the pieces of the program that is built in different places for different purposes by different sponsors. Asking what the best profile for the ingredients for an aircraft carrier battle group over the next 10 years (a planning and programming question) is different from asking how much is needed to operate the battle group for the next year. However, in PPBES, to decide "how much," the budgeters have to know what the total program will look like in practice.

As long as there is clear articulation and separation of these processes and one feeds carefully into the other, this system can work—as long as the POM feeds information into the budget process. For the most part, budgeters may have been happy to have many of the big resource questions decided for them, leaving them to focus on pricing out next year's needs. For their part, programmers have developed rules that allowed them to develop a good POM for each cycle. Usually, this means everyone gets something, but no one gets everything they want.

With the passage of time, dysfunctions appeared in this scenario. First, the military departments (MILDEPS) created POMs that were more conducive to their needs than to joint war fighting needs. The Goldwater-Nichols Act reforms (1986) were intended to rectify this situation. Then, with the drawdown after the fall of the Soviet Union, budget offices were placed in the awkward position of having to make decisions because the calendar said it was time to do so—even when the POM had not been completed—because those who built the POM could not decide which was the best way to downsize while maintaining the capacity to deter or fight future wars. Military department and DoD budget offices were, by and large, unhappy at having to make programmatic drawdown decisions under this circumstance. However, now in the past few years, the program decision-making process has not been completed in time to meet the needs of the budget part of the process.

Most recently, this is allegedly due to the combined program- and budget-review process under the PPBES. Also, various changes have been made to the processes of planning and programming for weapons acquisition, but none has been fully successful. Part of the problem is the overly complicated programming and budgeting process. Former SECDEF Rumsfeld and others have characterized the PPBES process as too slow and too complicated. As part of his transformation effort, Rumsfeld and DoD staff changed PPBES so that the programming and budgeting analysis and decision phases could be roughly concurrent. The POM process begins first, but both the budget and the POM process are supposed to end at the same time. In effect, the failure of the programming system to reach decisions may be viewed as having broken the budget process.

In reality, the budget process can only reach the "how much" question by answering the "what" and "why" questions. If the answers to these questions all appear at the same time, or when they are not answered at all, then the budget process has to, in effect, duplicate what is supposed to be done in the POM process to produce a budget on time. Indeed, under the new PPBES process, some parts of the budget process have had to operate as if there was no POM process.

This leads to the question: is there a genuine need to prepare a POM, especially if budgeting were done on a longer-term basis of 2 to 5 years? Perhaps it would be useful to take the transformational PPBES reform one step further and discard the separate POM process by simply incorporating the POM questions and POM process outputs into the budget process? This is a more sizeable task than it appears due to the existence of a bureaucracy which produces the POM. A first response is that participants in this bureaucracy might resist fearing their loss of jobs. However, this is perhaps a less sizeable task than it seems because the military staff involved in the POM process have other career lines and can perform functions as war fighters, and/or players in the defense-acquisition process or the warfare-requirements-setting system. There would be some civilian positions, mainly those in the Pentagon, that would disappear in this new integrated POM/budget cycle—a cycle that could perhaps be called the Planning, Budgeting and Execution System (PBES). Despite this problem, replacement of the entire PPBES with longer range budgeting is the option we prefer, primarily because it would restore an orderly and complete analytical process while decreasing some of the repetitiveness and needless rework of the annual budget process.

RELATED BUDGET REFORMS

While creating a two-phase planning and budgeting system of the type outlined above would rationalize the operation of PBE within the DoD, a useful further step would be to create a longer-term appropriation period. DoD fiscal execution patterns are needlessly complicated by the rush to spend one-year appropriations before the close of the FY. And the mixing of different appropriation periods for different appropriations needlessly complicates administration for those who execute budgets.

Most of the DoD budget functions on a multiyear pattern—longer for military construction and procurement of long-lived assets such as ships and aircraft, and shorter for personnel and supporting expenses (operations and maintenance or O&M). However, even if personnel is legally an annual appropriation, in reality the force size and composition is relatively fixed and will remain so until some external crisis event forces review and change. Personnel could as well be a two, three, of even five-year appropriation. We suggest that the DoD budget is, in effect, a multiple-year budget now. It would make sense to recognize it as such and to appropriate for multiyear periods for all accounts, and to extend the obligation period for short-term accounts beyond 1 year at minimum.

A 2-year appropriation (or obligation period) for personnel and O&M accounts would be a useful starting point for Congress, as we have noted. Critics of such an approach often point to Congress's need to exercise oversight through the budget. However, Congress can exercise whatever oversight it cares to in various ways, for example by focusing on execution review in off-budget years in a 2-year cycle. A 2-year budget also would reduce the opportunity for Congress and the President to insert what all recognize as "pork" into defense appropriations. The suggestions we make here would reduce opportunities for pork, but would also allow for meaningful oversight by Congress, and would reduce the size of the Pentagon bureaucracy while releasing additional military officers from administrative jobs for return to duty in their warfare specialties.

It must be noted that the task of defense resource planning and budgeting is part managerial and part political. Thus, from our perspective, no amount of budget process, PPBES or business process reform will reconcile the different value systems and funding priorities for national defense and security represented by opposing political parties, nor will it eliminate the budgetary influence of special-interest politics. Value conflict was evident in the early 1980s when public support, combined with strong presidential will and successful budget strategy, produced unprecedented peacetime growth in the defense budget, in particular in the investment accounts. Constituent and special-interest pressures make it difficult for Congress and the DoD to realign the defense budget. While we applaud the spirit of many of the changes made in DoD during the period 2001-2005, reform of the defense budgeting process does not mean that producing a budget for national defense politically will be much easier in the future than it has been in the past. Threat perception, capabilities assessment and politics drive the defense budget, not the budget process itself (McCaffery & Jones, 2004). Additionally, the size of the deficit and rate of increase in mandatory expenditures make top-line financial relief for the DoD unlikely.

We also may observe that a sequence of annual budget increases for national defense in the early and mid-2000s has not brought relief to many accounts within the DoD budget. At the same time, requirements of fighting the war on terrorism have intensified the use of DoD assets and the costs of military operations. Because the need for major asset renewal has been postponed for too long, new appropriations have gone and will go in the future largely to pay for new weapons system acquisition, and for war fighting against terrorism. What this means is that accounts such as those for O&M for all branches of the armed services will continue to be under pressure and budget instability; restraint will remain a way of life for much of the DoD. This places a heavy burden on DoD leadership, analysts and resource-process participants to achieve balance in all phases of defense budgeting and resource management.

Ending what we know as programming and the POM would be a major change to PPBES. In our view, programming is only effective, if at all, at the end-game anyway but preparing and processing the POM wastes huge amounts of valuable DoD staff time and energy that can be put to better use. Also, ideally, the period for obligation of *all* accounts in the new DoD budget process would permit obligation over a period of 2 or 3 years for all accounts—including fast spend accounts including O&M, military personnel, and so forth. The reason for multiyear obligation for all accounts is to enable more effective budget execution and end the highly wasteful and inefficient end-of-year "spend it or lose it" incentive syndrome. (Some will argue that there would still be a rush to spend at the end of whatever the appropriation period is; for starters, we will gladly accept a 50% improvement if it happens every 2 years rather than every year.) This change would, of course, require the approval of Congress. However, the DoD could implement long-range budgeting (including capital budgeting) as a part of the overall reform—while Congress continues to operate on the annual budget cycle it prefers (for a number of reasons related to serving constituent and member interests). No change in the federal budget process can be made unless it permits Congress to continue to do its business according to the incentives faced by members. To think otherwise is naive. Still, as noted above, the only part of the reform advocated here that would require explicit congressional action is lengthening the obligation period for all accounts to two or three years, as has been done internationally, in the United Kingdom and other countries, for example. And, in fact, it occurred in the United States in a small way before the elimination of what was termed the "M" account in the early 1990s due to illegal use of this account by the Air Force in financing the B-1 bomber and other programs. DoD had substantially greater flexibility in managing money for which the obligation period had expired. Under the M account process expired funding was allowed to be retained and reallocated by DoD for a period of 3 years.

The change to extend the obligation period for 1-year appropriations to 2 years would require Congress only to modify certain provisions of appropriation law. Otherwise, the DoD could implement a long-range accrual based budgeting system on its own, subject to gaining approval of and support for it from Congress—but this would not require change in law. In essence, it is incumbent on the DoD to persuade Congress to support such change, and this will only occur if the DoD is able to show members how they, the DoD and the American taxpayer will be better off as a result of the reform.

REFORMS LEADING TO CAPITAL BUDGETING IN DoD

If budget reforms are going to be made, management reforms must be made simultaneously to ensure that change is properly implemented and all persons involved are aware of and are willing and able to make the appropriate organizational and process adjustments. This is especially true if one of the reforms is decentralizing part of the decision-making process. Decentralizing the decision-making process should in our view involve the use of capital budgeting, where additional authority for capital asset purchases could be further shifted down to the military department level to program managers. Even though former SECDEF Rumsfeld's requests for "broadened discretionary powers" in the Defense Transformation Act and in other appeals were generally denied by Congress, with the exception of giving DoD authority to develop a new personnel system, many of these ideas had considerable merit (McCaffery & Jones, 2004).

Since federal agencies have much tighter constraints than businesses in the private sector, it is difficult to provide incentives for agencies to better manage their capital assets. However, Congress could adopt policies similar to those in the United Kingdom, Australia, and New Zealand and allow departments and agencies, including DoD, to raise and keep revenues from selling or renting out existing assets (President's Commission to Study Capital Budgeting, 1999). Further, as suggested by the Defense Acquisition Performance Assessment (DAPA) report (2006) Congress and DoD should establish a capital reserve account to improve financial stability for acquisition. If good capital budgeting processes were established in the budget process and agencies are allowed to keep revenues from the sale of assets, there are at least two incentives for agencies to manage their assets well.

If capital budgeting was implemented, the strategic plans of the departments could be more easily and efficiently integrated into both resource management and acquisition decision and execution processes within DoD. Although the Government Performance and Results Act (GPRA) requires agencies to submit 5-year strategic plans, the plans are currently not used directly in considering appropriation requests for capital assets and spending. Additionally, it would be useful for planning purposes if the strategic plans and budgets were tied to the life cycles of the capital assets. Although the capital programming guide directs agencies to consider life-cycle costs and compare them to expected benefits, the life-cycle costs are not directly linked to the agency's strategic plans. If capital asset life-cycle costs were tied to strategic plans, funding for the maintenance and replacement of assets could be better planned. In our opinion, capital budgeting should be done on an accrual basis so that program and budgetary plans would include all future outlays for capital asset acquisition, especially for new weapons systems. If life cycles are estimated for assets, then the department would determine and commit more explicitly to replacement of obsolete equipment and systems (President's Commission to Study Capital Budgeting, 1999).

In an effort to assist agencies in making decisions on capital asset investments, the agencies should continue to prepare annual financial statements as required by the Chief Financial Officers Act (CFOA). It should be noted, however, that preparation of financial statements simply for CFOA compliance should not be the goal. The goal should be preparation of financial statements that are used to aid in better decision making. In addition, departments and agencies would prepare and use the detailed inventories of existing capital assets required by the CFOA. The information in these reports would be consolidated by DoD and used to guide DoD and the MILDEPS in preparing long-term capital plans, similar to and replacing the Future Years Defense Program (FYDP). This would assist Congress in reviewing and assessing these plans.

Most states have separate capital budgets. Analysis of case studies of state capital budgets adds fuel to the debate over whether there should be a separate capital budget at the federal level. While there are many critics of a separate capital budget in the federal government, proposals for instituting separate capital acquisition funds (CAF) at the agency level have been advanced and analyzed by President's Commission to Study Capital Budgeting, as noted above.

In implementation, the process would require all federal departments and agencies to prepare and submit to OMB (or in the case of DoD, to submit directly from Office of the Secretary of Defense [OSD] to Congress) a separate capital budget. Following this, once capital budgets were negotiated between agencies, analyzed and approved by Congress as part of the annual budget process, a segment of the department's appropriations enacted by Congress would be placed in the department's capital acquisition fund and could only be used for acquiring long-lived capital assets. This is the application of capital budgeting that would fit most comfortably into the existing federal budget process.

A more comprehensive change approach would be to establish a single capital acquisition fund for the entire federal government as a separate fund account entity. Under this approach, or the agency based option, a CAF would borrow from the Treasury to buy capital assets and Treasury would charge operating units a debt service amount based on an "equitable" rate of interest (e.g., at the federal prime rate, or possibly discounted for internal government borrowing). Additionally, the CAF would inherit all of the agency's existing capital assets in an effort to capture all agency costs of capital.

The argument in support of the CAF approach is that a single fund or multiple separate funds for capital acquisition would help agencies better plan and budget for capital assets. In addition, agencies would be better held accountable for planning and budgeting and, presumably, would be more likely to use their resources efficiently. These funds would also smooth out the BA required by agencies and would help to reduce potential spikes in the budget associated with full funding requirements. An important aspect of introducing separate CAF, however, is the definition of capital assets. OMB would have to issue guidance on what constitutes a capital asset to ensure implementation is consistent throughout the agencies (President's Commission to Study Capital Budgeting, 1999).

While the Government Accountability Office (GAO) originally agreed with and supported the President's Commission to Study Capital Budgeting recommendation to implement CAF, GAO then published a study concluding that the proposed benefits of CAFs could be achieved through simpler means (GAO, 2005). GAO asserted that CAFs, as a financing mechanism for federal capital assets, would ultimately increase management and oversight responsibilities for the Treasury Department, the OMB, the Congressional Budget Office (CBO), and the departments and agencies that would utilize CAFs.

While recognizing that CAFs might improve decision making and remove many of the spikes and troughs in BA associated with large dollar capital assets, GAO noted that some federal agencies now use different approaches to address capital investment planning and decision making.

GAO research on capital-intensive federal agencies, coupled with interviews they undertook with officials from Congress, Treasury, and OMB, led to their conclusion that CAFs, as proposed by the President's Commission to Study Capital Budgeting, would be too complicated for implementation because of the additional budget complexities that they would create. Interviews with executive and congressional officials led GAO to believe that a proposal to institute CAFs, even on a pilot basis, would have few, if any, proponents. Because of these reasons, GAO recommended that the focus should be placed on improvement and widespread implementation of improved asset management and cost accounting systems to address the problems for which CAFs were proposed as a solution (GAO, 2005).

We regard the GAO criticism of CAF as correct in that it would be a significant departure from how budgeting for long-lived assets in done presently in the federal government, and that it is not entirely compatible with the current congressional budget process. However, to reject this proposal for this reason is to miss the point about the need for and advantages of capital budgeting. Thus, we believe the GAO analysis, while accurate, misses the point. What we recommend for DoD concurs generally with the assessment by President's Commission to Study Capital Budgeting on what is needed. We assert that the benefits of capital budgets include following:

- Improved assessment of the condition of existing capital assets,
- Better estimates of the funding needed for maintaining assets,
- More clearly and directly assigning priorities for capital asset investment in a separate capital budget (or budget component).
- Performing and applying better cost information from DoD accounting systems to assist budgeting decisions.
- Investing to achieve necessary improvements in basic DoD transactional and cost accounting systems so they are capable of fully informing capital planning and budgeting decisions in real time and in discounted present value terms.

Our recommendations represent a mix of the methods used by the private sector and similar to approaches practiced by most U.S. state governments.

To conclude this section of the chapter, as we have explained, part of the basic business model for asset acquisition to be applied under a reengineered system is a private sector-oriented capital budgeting process in which asset and financial resource planning are completely integrated into the budget and resource management processes rather than separated as is the case with existing DoD acquisition and resource management systems (i.e., PPBES). The new business model would employ a single, fully integrated Enterprise Resource Planning Information Technology system and database rather the multiple systems and databases that characterize existing DoD systems.

From a managerial perspective, leading the DoD capital budget process and redesigned acquisition process would still be the task of the under secretary of defense for acquisition, technology and logistics (USD AT&L) and the small acquisition staffs of the MILDEPS, along with input from combatant commanders, to determine the capabilities desired for war fighting. Capital budgeting would not change or reduce this set of responsibilities.

Under DoD capital budgeting, a prioritized list of desired capabilities would be established under the sole authority delegated to the USD AT&L, under the advice of a small JCS staff, but without the JCIDS process because, in our view, this process has only added unnecessary complexity to a review and analysis process that was already overcomplicated. We acknowledge that the JCS should perform analysis of interoperability, jointness of asset use, and system compatibility, but this should be done with much less procedural complexity that is present in the Joint Capabilities Integration and Development System (JCIDS) process as it has operated since its inception. We do not believe JCIDS represents a better way of doing business than the admittedly inefficient process it replaced—or more aptly put—only augmented.

The SECDEF, except symbolically, would not be a player in the reengineered capital budget process and system based on the fact that except in extraordinary instances he is not a player in the system as it presently functions, and also based on modern business management theory and principles of delegation of authority and matching responsibility/accountability to "let managers manage." Once the prioritized capabilities list was set, the estimated costs (assuming a high degree of uncertainty in many cases, e.g., research, development, training, and engineering or RDT&E) of acquiring capabilities would be matched up with estimates of the availability of resources with data drawn from the single long-range budgeting system. And, as capital budgeting is performed in the private sector and in many U.S. state governments, a line would be drawn, determined on affordability, at someplace on the list. All assets to be acquired that fell above the line would be contracted for development and RDT&E by the private sector. All assets that fell below the line of affordability would not be started. In terms of how the current acquisition milestone process is organized and operates, we advocate simplifying the process by reducing it to fewer basic stages.

A REVIEW OF PROCESS CHANGES IN
WEAPONS ACQUISITION AND RESOURCE MANAGEMENT

Numerous reforms since the 1950s have attempted to improve the defense acquisition process, and almost all of these have included some form of resource management changes, large and small, intended to improve how DoD buys weapons, weapons platforms, and equipment. Recent reforms including more open competition, streamlined acquisition procedures, elimination of obsolete regulations, and more effective program management are some of the substantial changes made in DoD in the last 15 years to improve acquisition budgeting and management. Establishing open competition also is a significant part of recent acquisition transformation initiatives. Changes in acquisition information technology (IT) resulting from the passage of the Clinger-Cohen Act and other legislation by Congress, the use of cost as an independent variable (CAIV) as a means of reducing acquisition costs, plus spiral acquisition are other changes that have been intended to yield positive results.

This chapter reviews a number of the more important procedural, regulatory, and legislative reforms to the defense acquisition process initiated and implemented over roughly the past 15 years. Some of the reforms noted have been completed, but have implications for current processes. Even though these changes are no longer under implementation, understanding their intent helps to paint a picture of how the system evolved to where it is today. For example, the Federal Acquisition Reform Act (FARA) and the Federal Acquisition Streamlining Act (FASA) have been incorporated into other DoD acquisition administrative law, referred to as instructions by number in DoD, for example, DoD 5000.2R. In each example of reform, policy decisions and legislation have been intended to address significant acquisition reform problems.

ASSESSING PAST AND
CONTINUING ACQUISITION REFORM INITIATIVES

Pervasive problems persist in the process for acquiring defense assets. These problems include affordability, cost control, keeping to schedules, and performance estimating errors. Accurate estimates of program affordability, and for acquiring weapons often are based on optimistic assumptions about the maturity and availability of enabling technologies (GAO, 1997b). The use of outdated information systems makes the ability to accurately track and measure acquisition costs even more difficult. Thus, weapons acquisition reform is driven by a myriad of factors and borne out of the desire to acquire the best weaponry at the least cost. Beyond technical issues, the politics of acquisition are complex and present additional challenges to be overcome. In summary, continual tension persists between top-level policy and budget process players including Congress, and defense acquisition executives, and midlevel DoD officials including program managers and comptrollers confronted with limited resources and a complex set of constraints in the form of laws, rules, regulations and guidance.

In assessing acquisition reforms past and present it must be emphasized that the DoD budget is reviewed and appropriated in competition with other priorities. In that respect the world has changed significantly in the last two decades, as the DAPA report commissioned by Acting Deputy SECDEF Gordon England in 2005 concluded:

> The fundamental nature of defense acquisition and the defense industry has changed substantially and irreversibly over the past twenty years.... In 1985, defense programs were conducted in a robust market environment where over 20 fully competent prime contractors competed for multiple new programs each year. The industrial base was supported by huge annual production runs of aircraft (585), combat vehicles (2,031), ships (24) and missiles (32,714). Most important, there were well-known, well-defined threats and stable strategic planning by the Department. Today, the Department relies on six prime contractors that compete for fewer and fewer programs each year. In 2005 reductions in plant capacity have failed to keep pace with reduction in demand for defense systems, (188 aircraft, 190 combat vehicles, 8 ships, and 5,072 missiles). (DAPA, 2005, p. 6)

The panel's key findings as summarized in the graphic below focus on process stability, increased trust, decreased oversight and continued accountability. We offer this only to remind readers that this is a highly complex area where problems are many and seemingly easy solutions are often ruled out by the necessity for checks and balances between branches of government, the continued need for oversight between government and the private sector, and continuing demands for vigilance in the use of public money. As a result, solutions sometimes are easy to prescribe, but hard to bring about.

THE FEDERAL ACQUISITION STREAMLINING ACT OF 1994

DoD issued an update to its regulations governing the acquisition of major weapon systems on 13 October 1994. Among other things, the update incorporated new laws and policies, including FASA, separated mandatory policies and procedures from discretionary practices, and

MAJOR FINDINGS

• *Strategic technology exploitation - key US advantage*
 • *The world has changed*
  - *Goldwater-Nichols era (post 1986)*
    • *20+ primes,*
    • *multiple new starts*
    • *huge annual production runs (585 aircraft, 2,031 vehicles, 24 ships, 32,714 missiles)*
  - *Today*
    • *Six primes DoD can't live without*
    • *Few new starts*
    • *Low rates of production (188 aircraft, 190 combat vehicles, 8 ships/subs, 5,702 missiles)*
    • *Plus a need to be agile*
	• *The acquisition system must deal with external instability, a changing security environment and challenging national issues*
	 • *DoD management model based on lack of trust*
	  - *oversight is preferred to accountability*
	 • *Oversight is complex, it is program focused - not process focused*
	 • *Complex acquisition processes do not promote success - they increase cost and schedule*
	 • *DoD elects short term savings and flexibility at the expense of long term cost increases*

For incremental improvement (applied solely to the acquisition process) to achieve success, DoD processes must be stable – they are not

Source: Defense Acquisition Performance Assessment, 2005, p. 5.

Figure 14.1.   Major findings on acquisition reform from the DAPA.

reduced the volume and complexity of the regulations. FASA required the SECDEF to define cost, schedule and performance goals for all of the Major Defense Acquisition Programs (MDAP) and for each phase of their acquisition cycle. Highlights included streamlined proposal information or page count, shortened proposal submission time, reduced evaluation team size or evaluation time, and limited source selection factors pertaining to cost, past experience, performance, or quality of content. FASA called for full and open competition, to be obtained when, "all responsible sources are permitted to submit sealed bids for competitive proposals" (Federal Acquisition Regulations [FAR], 2000). Full and open competition is achieved through open specifications (U.S. Code 253a (1) (A)).

FASA establishes a clear preference for acquisition of commercial items in the federal government. It requires agencies to reduce impediments to buying commercial products and to train appropriate personnel in the acquisition of such products. One such impediment is the use of design specifications, which restrict competition and make acquisition of commercial products difficult. Design specifications typically tell a vendor how a product is to be made or how a service is to be performed. A commercial vendor, whose product has been developed for public use, seldom conforms to government design specifications. FASA instilled flexibility, and timeliness into the acquisition process.

THE FEDERAL ACQUISITION REFORM ACT—
THE CLINGER-COHEN ACT

The major pieces of legislation affecting acquisition and IT were the FARA and the Information Technology Management Reform Act. While originally passed as two separate initiatives, their impact on each other made it impossible to consider each separately. The two acts were later combined and renamed the Clinger-Cohen Act (1996). The major impact on IT was the repeal of the Brooks Act and its associated restriction on acquisition of resources. The Clinger-Cohen Act encouraged the acquisition of commercial off the shelf (COTS) IT products and allows the Office for Federal Acquisition Policy to conduct pilot programs in federal agencies to test alternative approaches for acquisition of IT resources. The Clinger-Cohen Act directs agencies to use "modular contracting" based on successive acquisitions of "interoperable increments." (Federal Register, 1996, p. 27). The Clinger-Cohen Act created the position of chief information officer for the DoD, and combined life cycle approvals for weapon systems and IT systems into a single instruction, the DoD 5000.1 series.

FARA and FASA have been overtaken or superseded by other DoD reform initiatives applicable to MDAPs and weapons acquisition. Still, both FARA and FASA are valid and enforceable. FARA, among many other things, expanded the definition of "commercial items" to include those things not only sold to the general public, but also those *offered* to the general public. These initiatives were pushed by industry, primarily because under the two Acts, firms participating in government acquisitions with qualified "commercial" products are exempted from over 100 statutory and regulatory requirements. For example, firms may be exempted from the Truth in Negotiations Act that requires firms to certify cost and pricing data on negotiated actions greater than $550,000 (Yoder, 2003).

Additional reforms have involved fostering the development of measurable cost, schedule, and performance goals and incentives for acquisition personnel to reach those goals. Among other things, program managers, as well as senior DoD and military department officials, now must establish cost, schedule, and performance goals for acquisition programs and annually report on their progress in meeting those goals. They must establish personnel performance incentives linked to the achievement of goals. Program executive offices also must submit recommendations for legislation to facilitate the management of acquisition programs and the acquisition workforce.

In this respect, it should be noted that each service has an acquisition executive responsible for acquisition and contracting workforce education and training among other things. For example, in the Navy the director of acquisition management is responsible for all Navy acquisition career management issues, both military and civilian, including, but not limited to:

- Promotion parity analysis
- Reservist policies
- Congressional and legislative education/training issues
- Defense acquisition university mandatory training
- Acquisition workforce tuition assistance
- Business and financial management

Contracting out services has been a major initiative since 2000, under the guidance of the OMB. In 2000, federal agencies procured more than $235 billion in goods and services. Overall, contracting for goods and services accounted for about 24% of federal government FY 2001 discretionary resources (OMB, 2003a). About 38% of acquisition personnel government-wide are either already eligible to retire or will be eligible by September 30, 2007 (OMB, 2003a). At DoD and Department of Energy— the two largest contracting agencies—39% of the acquisition workforce will be eligible to retire by FY 2008 (GAO 2003d). What this means is that the human capital skill mix will change dramatically as retirements proceed and new personnel are hired. In the meantime, new requirements, tasks, and skills are demanded of both old and new acquisition managers as a result of federal and acquisition regulatory reform efforts.

COMMERCIAL OFF-THE-SHELF ACQUISITION

FAR applies to all contracting regulations. The pertinent part of the FAR with regard to COTS reforms is part 12, which indicates in essence that federal government organizations should perform market research to maximize the use of commercial products. DoD enforcement of Part 12 of the FAR over the past 5 years has caused weapon program managers to evaluate and, where appropriate, purchase commercial or nondevelopmental items (CNDI), when they are available from industry, when they meet the organization needs. Defense contractors are required to incorporate CNDI to the maximum extent possible.

Initial feedback on the success of this initiative is highly positive. It appears that the change has permitted commercial firms to produce the kinds of outcomes from their development of new products that meet DoD needs. Specifically, firms that developed sophisticated products in significantly less time and at lower cost than their predecessors have been rewarded with contracts. However, to some extent the quality and credibility of commercial firm cost information available to DoD acquisition decision makers remains a problem. The long-term life cycle support costs associated with utilizing potentially rapidly obsolete commercial items has yet to be fully documented (Yoder, 2003).

COST AS AN INDEPENDENT VARIABLE

DoD directive 5000.1 directed a new development in cost analysis termed "Cost as An Independent Variable" or CAIV. System performance and target costs are to be analyzed on a cost-performance trade-off basis. The CAIV process is intended to make cost a more significant constraint as a variable in analysis of effectiveness and suitability of systems. CAIV is intended to reduce acquisition costs. After Desert Storm and before the war on terrorism began on September 11, 2001, threats were not increasing in perceived capability at as fast a rate. The DoD acquisition budget decreased accordingly. Under these circumstances, it was more appropriate to make cost a stronger driver in system design due to decreased budgets. Such an approach also was consistent with commercial practices in new system developments, where market forces drive the price of new systems.

CAIV helps the program manager (PM) recognize that the majority of costs are determined early in a program life cycle. Consequently, the best time to reduce life-cycle costs is early in the acquisition process. Cost reductions are accomplished through cost and performance trade-off analysis, which is conducted before an acquisition approach is finalized. Incentives are applied to both government and industry to achieve the objectives of CAIV. Awards programs and "shared savings" programs are used creatively to encourage generation of cost-saving ideas for all phases of life-cycle costs. Incentive programs target individuals and government and industry teams. The PM works closely with the user to achieve proper balance among cost, schedule, and performance while ensuring that systems are both affordable and cost-effective. The PM, together with the user, propose cost objectives and thresholds for MDAP approval, which will then be controlled through the APB process (lifecycle costs). The PM searches continually for innovative practices to reduce life-cycle environmental costs and liability.

Research by Coopers and Lybrand identified over 120 regulatory and statutory "cost drivers" that, according to contractors surveyed, increased the price DoD pays for goods and services by 18% (Lorell & Graser, 1994). Some of the more egregious cost drivers included government imposed accounting and reporting standards and systems such as cost accounting standards (CAS) and complex contract requirements and statements of work (Lorell & Graser, 1994). The basic goal of this study was to develop a more "commercial-type" defense acquisition process. This included reducing regulator burden, transferring more program cost, design and technology control authority and responsibility to the contractor, exploiting commercially developed parts, components, technologies and processes, and making cost/price a key requirement. This study was compatible with the goals of the Revolution in Business Affairs under the Clinton administration and Transformation of Business Affairs under the administration of President George W. Bush.

THE SINGLE PROCESS INITIATIVE

In 2002, former SECDEF Donald Rumsfeld directed DoD to change the management and manufacturing requirements of existing contracts to unify them within one facility, where appropriate (Rumsfeld, 2002). This initiative is called the block change or single process initiative (SPI). Program managers are tasked with ensuring SPI reduces weapon acquisition costs. Allowing defense contractors to use a single process in their facilities is a natural progression from the contract-by-contract process of removing military-unique specifications and standards initiated in FASA. Contractors will incur transition costs that equal or exceed savings in the near term. Moving to common, facility-wide requirements is intended to reduce government and contractor costs in the long term.

DoD 5000.2R Transformation From Regulatory to Policy Guidance

In 2002, Secretary Rumsfeld directed that DoD 5000.2R be converted from a regulatory tool to a more functional and flexible policy guidance document. The 5000 series has in the past been regarded as administrative law. It demanded user requirements including the preparation operational requirements documents and estimation of initial operational capability. The 5000.2R acquisition requirements had been firm and not subject to modification without specific waivers (Rieg, 2000). However, SECDEF, the services, and program managers recognized the need for greater flexibility to manage acquisition.

The revised DoD 5000.2-R document promised to piggy-back on other acquisition reforms, allowing greater flexibility and control for acquisition leadership. DoD 5000.2-R was revised to recommend that integrated process teams be used during program definition, to aid the definition of requirements and system supportability. In addition, program structure changes are directed to include an acquisition strategy of open systems. To maximize program effectiveness, the PM is directed to use commercial sources, risk management, and CAIV. The PM should use program design incorporating integrated product and process development and place system engineering emphasis on production capability, quality, acquisition logistics, and open system design (Oberndorf & Carney, 1998).

DIRECTOR OF ACQUISITION PROGRAM INITIATIVE

In past practice, annually the director of acquisition program integration determined if each MDAP had reached 90 percent or more of cost, schedule, and performance parameters when compared to APB thresholds. The appropriate decision authority must make a similar determination for nonmajor acquisition programs. If 10% or more of program parameters are missed, a timely review is required. The review addresses any breaches in cost, schedule, and performance and recommends suitable action, including termination.

Major acquisition defense program baselines must be coordinated with the DoD comptroller before approval. Cost parameters are limited to RDT&E, acquisition, the costs of acquisition of items procured with operations and maintenance funds, total quantity, and average unit acquisition cost. As the program progresses through later acquisition phases, acquisition costs are refined based on contractor actual costs from program definition and risk reduction, engineering, manufacturing and development, or from initial production lots. Cost, schedule, and performance objectives are used as described above in the CAIV process to set the APBs. Cost, schedule, and performance may be traded-off by the PM, within the range between the objective and the threshold without obtaining MDAP approval. This initiative intends to improve executive level oversight and program management reporting. In addition, it may enhance executive and PM flexibility in the best use of available funding.

A REVISED CAPITAL ACCOUNT PROCESS: FURTHER SUPPORT FOR CAPITAL BUDGETING

The 2006 QDR has recommended that DoD establish a capital account for major acquisition programs. This would be a major change for the acquisition process. The recommendation mirrors the outcome of the DAPA study directed by Deputy SECDEF Gordon England. In its findings in December, 2005, this study recommended:

The SECDEF should establish a separate Acquisition Stabilization account to mitigate the tendency to stretch programs due to shortfalls in the DoD non-acquisition accounts that ultimately increases the total cost of programs. This will substantially reduce the incidence of "breaking" programs to solve budget year shortfalls and significantly enhance program funding stability. (DAPA, 2005, p. 10)

In effect, the panel recognized that acquisition account leaders could not protect the acquisition accounts from acting as a bank for the operating accounts during budget execution—thus the recommendation that DoD procurement, research and development budget be separated from the overall defense budget. This separation would help prevent the kind of financial whiplash that causes cost overruns according to retired Air Force Lt. Gen. Ronald Kadish, panel director and a vice president at Booz Allen Hamilton, a prominent defense consulting firm. The panel found that every dollar taken from a program induces $4 of cost increases in later years. "Though many in Washington blame the uncertainty of the annual budget approval process on Congress, most of the damage was self-inflicted by the Pentagon. It is largely a 'government-induced' instability" (Ratnam, 2005).

In Secretary England's confirmation hearings, both the Senate and House Armed Services Committees expressed an interest in improving acquisition practices, an interest that was specified in the reports on the DoD authorization bill. For example, the Senate report accompanying S1042, the Senate version of the Defense Authorization Bill, notes that after nearly 20 years of reform since the Packard Commission Report and Goldwater-Nichols, "major weapons systems still cost too much and take too long to field." The committee added, "Funding and requirements instability continue to drive up costs and delay the eventual fielding of new systems. Constant changes in funding and requirements lead to continuous changes in acquisition approaches" (House Conference Report, 2005, pp. 354-356; Senate Conference Report, 2005, p. 345). This culminated in the recommendations and findings made in the QDR in language that went beyond the establishment of a capital account, to include a capital budgeting process:

To manage the budget allocation process with accountability, an acquisition reform study initiated by the Deputy SECDEF recommended the Department work with the Congress to establish "Capital Accounts" for Major Acquisition Programs. The purpose of capital budgeting is to provide stability in the budgeting system and to establish accountability for acquisition programs throughout the hierarchy of program responsibility from the PM, through the Service Acquisition Executive, the Secretaries of the Military Departments and the OSD. Together, these improvements should enable senior leaders to implement a risk-informed investment strategy reflecting joint war fighting priorities. (QDR, 2006, pp. 67-68)

This process would be supported by a procedure that would rest on joint collaboration among the war fighter, acquisition and resource communities, with the war fighters assessing needs and time frame and the acquisition community contributing technological judgments on technological feasibility and "cost-per-increment" of capability improvement. The budget community's contribution would be an assessment of affordability. These inputs would be provided early in the process, before significant amounts of resources are committed. The QDR also recommended that the DoD, "begin to break out its budget according to joint capability areas. Using such a joint capability view—in place of a military department or traditional budget category display—should improve the department's understanding of the balancing of strategic risks and required capability trade-offs associated with particular decisions" (QDR, 2006, pp. 67-68). The DoD promised to explore this approach further with Congress. History indicates that Congress clings tenaciously to the appropriation structure currently in place because it serves Congress's purposes, but it is good to remember that all that is now familiar was once new.

It is clear that the defense acquisition process has long been beset by problems related to both politics and efficiency. Numerous reforms since the 1950s have attempted to improve the acquisition process. Recent reforms including more open competition, streamlined acquisition procedures, elimination of obsolete regulations and more effective program management are some of the substantial changes made in DoD in the last ten years to improve acquisition budgeting and management. Establishing open competition also is a significant part of recent acquisition transformation initiatives. Changes in acquisition IT resulting from the passage of the Clinger-Cohen Act and using CAIV as a means of reducing acquisition costs is another change expected to yield positive results.

Congressional and DoD reform initiatives have focused on greater reliance on commercial products and processes and more timely infusion of new technology into new or existing systems. Commercial product usage is implemented with an understanding of the complex set of impacts that stem from use of commercial products (Oberndorf & Carney, 1998). Solicitation requirements are written to include performance measures. If military specifications are necessary, waivers must first be obtained. Solicitations for new acquisitions that cite military specifications typically encourage bidders to propose alternatives (SECDEF, 2002a). DoD has made significant progress in disposing of the huge inventory of military specifications and standards through cancellation, consolidation, conversion to a guidance handbook, or replacement with a performance specification or nongovernment standard.

Some reforms already have had unanticipated consequences. For example, FARA and FASA eliminate, with minor exceptions, the requirement for "certified cost and pricing data" under the Truth in Negotiations Act (TINA). This has been heralded as a blessing for industry, but has caused problems for contracting officers who are mandated to determine "fair and reasonable" cost and price prior to award of contract. Specifically, there are instances where firms have claimed "commercial item exemptions" from TINA, when not one single item has ever been sold to the general public, and hence, there is little or no standard for determining the reasonableness of the price. Without TINA and cost analysis, the contracting officer may be awarding without solid factual benchmarks, standards, or measures of what is "fair and reasonable" (Yoder, 2003).

The Defense Acquisition Corps has increased education and training requirements for key positions such as for the critical acquisition position (CAP). CAPs are the most senior positions in the defense acquisition workforce, including program executive offices, program managers, deputy program managers of MDAP ACAT I defense acquisition programs and the program managers of significant non-MDAP ACAT programs. Maximizing program manager and contractor flexibility to make cost/performance trade-offs without (unnecessary) higher-level permission is essential to achieving cost objectives. Therefore, the number of threshold items in program requirements documents and APBs has been reduced. All of these changes add up to significant, albeit incremental, transformation of the DoD acquisition system.

The primary criticism of the acquisition process is that it is too complex, too slow, and too costly. It may also produce weapons that are "overqualified" to the task or irrelevant to the task at hand when they are finally put in the field as the threat has changed. Annual budget cycle politics adds to this mix; the continual purchase of weapons because they are good for congressional electoral districts irrespective of defense needs is wasteful. In addition, there is the fact of life adjustment of the 1990s; there was a procurement holiday and it has resulted in increased maintenance costs for older weapons systems. The outcome is increased O&M budgets and a gap in the procurement budget that reaches into the tens of billions of dollars, a gap that will not be closed in the near future. Add to this mix the fact that almost 40% of the federal and defense acquisition community will be eligible to retire in 2008. This would seem to leave a problem of immense magnitude. However, as we have documented above, these are not new problems.

The defense acquisition process has almost always appeared to be broken, but the irony of this is that the products it produces are among the best in the world. That is why Marines went into battle in their fathers' helicopters and some pilots flew their grandfathers' bombers over Iraq, why the main U.S. battle tank has been superior to anything on the field for over a decade. Moreover, this broken process engineered and deployed missile firing drone aircraft while the war in Afghanistan was in progress. The system can and has reacted quickly. America, the society of disposables, fast food, and microwave cuisine has also produced weaponry that is excellent and durable. The process is cumbersome, overly expensive, complicated—and highly political, but it does work.

In the best of worlds, DoD would acquire weapons assets in an environment of stable funding and management. Acquisition process reform over the past ten years has sought to provide a more stable environment in which to acquire better, more efficient weapons. However, the era following the end of the Cold War and the advent of the war on terrorism has made acquisition more difficult. Further, reform of acquisition and PPBES processes have created their own turbulence as change has been continuous as we explain further in the next section. At times, it is difficult for program managers and others involved in the DoD acquisition process to stay up-to-date on the status of change because one wave of reform spills over into the next. Continuous improvement of weapons acquisition budget estimation, execution and management has and will continue to present a challenge to all participants in the process. The pattern of continuous reform of acquisition and budgeting for weapons systems over the past several decades is a fact of life. Why should anyone expect the future to be different? We attempt to answer this question in the next section of this chapter.

AQUISITION PROCESS AND RESOURCE MANAGEMENT REFORM

Reform of the entirety of DoD budget, financial management, and acquisition decision making systems and business processes is a huge and ambitious topic to analyze, much less to accomplish. Our intent here is to advances our views on the practical underpinning for reform of defense resource, acquisition and business process management. We argue the necessity for relying on capital and longer term budgeting and resource management methods, more stringent application of business process reengineering, and increased use of markets and the private sector in moving from bureaucratic approaches toward smarter systems of organization and operation.

In any dialogue on the topic we acknowledge that reform of DoD acquisition is not an easy task. Part of the problem is that so much reform has been attempted since the 1980s and the results of these efforts have been mixed, much as we have explained in chapters 13 and 14. To some extent, the dynamics of constant reform are part of the problem, and many recent changes have not been as successful as anticipated. As Dillard concluded:

> In the last three years, there has been a great deal of turbulence in U.S. defense acquisition policy. This has contributed to confusion within the acquisition workforce in terminology, major policy thrusts, and unclear implications of the changes. The new acquisition framework has added complexity, with more phases and delineations of activity, and both the number and level of decision reviews have been increased. Decision reviews are used as top management level project control gates, and are also a feature of centralized control within a bureaucracy. Although the current stated policy is to foster an environment supporting flexibility and innovation, the result is a continuous cycle of decision reviews. Program Managers may now have fewer resources to manage their programs as they spend much of their time, and budgets, managing the bureaucracy. Moreover, the implicit aspects of the still new model have not been fully realized, and may result in policy that actually lengthens programs -- counter to goals of rapid transformation. The framework, and its associated requirements for senior level reviews, are opposed to the rapid and evolutionary policy espoused, and are counter to appropriate management strategies for a transformational era. (Dillard, 2005, p. 72; see also Dillard, 2004)

Another prominent acquisition policy expert summarized the challenge of reform as follows:

> The DoD [is in] a transformative period—leveraging emerging technologies to develop a net-centric warfare capability—while actively conducting military operations, throughout the spectrum of conflict, in support of the global war on terror. As a result, DoD is struggling to meet these competing requirements and reconcile ... spending between traditional and new programs. Therefore, creating a more efficient acquisition system is a top priority. High-quality research in the area of acquisitions is necessary to ... improve performance, reduce acquisition cycle times, and reduce the costs of DoD acquisitions, even as the Department confronts rapidly changing external and internal environments. (Gansler & Lucyshyn, 2005, p. 1)

In this section of the chapter we outline and articulate our proposals for fundamental reform of defense budgeting, resource and acquisition management systems and decision processes, based on and integrated with the many of the principles of enterprise organization and management developed largely in the private sector, along with capabilities-based analysis, decision making and implementation. First, however, let us summarize why we and many others believe significant acquisition business process reform should be undertaken in DoD.

In our view, there is much that is wrong with DoD resource decision processes and their relationship to the defense acquisition system (DAS), as has been explained in previous chapters. Too often PPBES and budgeting get in the way of efficient acquisition management. On its own, we believe that DoD resource management and acquisition decision processes are flawed to the extent that that they continuously propagate analytical and decision errors. They are excessively bureaucratic to the extent that they should be significantly redesigned, reengineered, and de-bureaucratized. Many existing work processes should be replaced completely by new processes to enable improved capital asset investment analysis of alternatives, decision making and execution in a much shorter period of time, involving far fewer participants, and in synchronicity with long-range planning and accrual budgeting principles that place emphasis on measurement of performance and results rather than input and process variables. These two systems (PPBES and the DAS), as they operate presently, are an incredible and wasteful triumph of process over substance. In short, we believe that if we really want to run DoD like a business (i.e., using smart business practices) the best way to accomplish this goal is to adapt smart systems into DoD and federal government organizations, with careful attention to the differences in purpose between government and the private sector, and in part to further move much of what is in our view nongovernmental work to private business—through increased devolution and redirection of essentially nongovernmental functions into the private sector.

With respect to the need for reform of DoD acquisition, budgeting, and related processes, we are not alone in rendering the conclusion that such action is needed. Then acting Deputy SECDEF Gordon England explained to the Senate Armed Services Committee (2005), "the entire acquisition structure within the DoD needs to be reexamined and in great detail ... there is growing and deep concern about the acquisition process within the DoD and in the Committee." On its part the Committee reported,

> The committee is concerned that the current Defense Acquisition Management Framework is not appropriately developing realistic and achievable requirements within integrated architectures for major weapons systems based on current technology, forecasted schedules and available funding. (House Conference Report # 109-89 -HR-1815 -Title VIII—Acquisition Policy, Acquisition Management, and Related Matters, p. 355)

A detailed study of the DAS by a select panel of experts (DAPA, 2006) tasked by Deputy Defense Secretary England in June 2005 came up with similar conclusions. In his tasking letter England wrote, "Simplicity is desirable....Restructuring acquisition is critical and essential" (England, 2005). The DAPA panel reviewed over 1,500 documents to establish a baseline of previous recommendations, held open meetings and maintained a public Web site to obtain public input, heard from 107 experts, received over 170 hours of briefings, and surveyed over 130 government and industry acquisition professionals (DAPA, 2006, p. 7). In December, 2006 the DAPA panel reported that the primary problem faced by acquisition executives and managers was program and funding instability, which is caused by the forces we have identified in this book. The panel reached the following conclusions:

- The acquisition system must deal with external instability, a changing security environment and challenging national issues.
- The DoD management model is based on lack of trust —oversight is preferred to accountability.
- Oversight is complex, it is program-focused – not process-focused.
- Complex acquisition processes do not promote success—they increase cost and schedule.
- DoD elects short term savings and flexibility at the expense of long term cost increases.
- Because ... major processes are not well integrated:

  We have an unrecognized, government-induced and long-standing cycle of instability which causes unpredictability in costs, schedule, and performance that ultimately results in development programs that span 15-20 years with substantial unit cost increases leading to loss of confidence in DoD acquisition systems. (DAPA Executive Summary, 2006, pp. 9, 12)

With respect to improving the performance of the system the DAPA recommended the following, organized into seven categories: organization, workforce, budget, requirements process, requirements management and operational test, acquisition strategy, and industry:

Organization
- Realign authority, accountability and responsibility at the appropriate level and streamline the acquisition oversight process.

Workforce
- Rebuild and value the acquisition workforce and incentivize leadership.

Budget
* Transform the budgeting process and establish a distinct acquisition stabilization account to add oversight throughout the process.

Requirements – Process
* Replace JCIDS with combatant commanders-led (COCOM) requirements procedures in services, and DoD agencies must compete to provide solutions.

Requirements – Management and Operational Test
* Add an "operationally acceptable" test evaluation category. Give program managers explicit authority to defer requirements

Acquisition –Strategy
* Shift to time-certain development procedures.
* Adopt a risk-based source selection process

Industry
* Overcome the consequences of reduced demand by sharing long range plans and restructuring competitions for new programs with the goal of motivating industry investments in future technology and performance on current programs (DAPA Executive Summary, 2006, p. 14).

Specifically, related to budgeting for acquisition the Panel recommended the following:

* Enhance the budget process by establishing a distinct acquisition stabilization account for all post-Milestone B programs. Add practical management reserve at the service level.
* Establish a separate acquisition stabilization account to mitigate the tendency to stretch programs due to shortfalls in DoD accounts that ultimately increase the total cost of programs.
* Create a management reserve in this account by holding termination liability at the service level.
* Adjust program estimates to high confidence when programs are base-lined in this account (DAPA Executive Summary, 2006, p. 17).

The distinct acquisition stabilization account and management Reserve recommended by the DAPA panel constitutes in our view a step towards establishing both a capital budget and a capita reserve account within the CAF for DoD, as we recommend. It is important to recognize that while

R&D (research and development), design and prototyping, production, and other contracted work would be paid for from the CAF under our proposed reform, the CAF would provide such funding from separate internal accounts based upon the legal requirements imbedded in statutory law for separation of appropriations by type. However, we would suggest that DoD make the case to Congress to fund a capital reserve account within the CAF to accommodate change more quickly that does the annual budget process and to provide additional stability to DoD acquisition and contractor defense firms.

The conclusions developed by the DAPA Panel as rendered in its December 7, 2006 report was carried forward subsequently in July, 2007 when, in response to a reporting requirement from the 2007 Defense Authorization Act sponsored by Senator John Warner (R -VA), Kenneth Krieg, USD AT&L at the time, submitted the SECDEF's "Defense Acquisition Transformation Report to Congress." This report formally asked Congress to enact a number of the recommendations indicated above from the 2006 DAPA panel report into law as part of the FY 2008 National Defense Authorization Act (Defense Acquisition Transformation Report to Congress, 2007).

Beyond capital budgeting, as is clear from the conclusion reached by the DAPA Panel and recommendations to Congress made by the USD AT&L, business process redesign and reengineering are key to successful acquisition process and related resource management reform.

BUSINESS PROCESS REENGINEERING: THE BASICS

A major component of any DoD acquisition reform strategy will require very stringent application of business process reengineering that results in implementation of new and more efficient organizational processes based on organizational redesign of roles and responsibilities and how work is performed. However, before we indicate specifically how business redesign and process reengineering would be applied in DoD, let us briefly review what this technique entails and how it is applied.

Business process reengineering is an attractive initiative to public management reformers because reducing costs, cutting service production cycle time and improving quality and productivity so often depends on moving beyond the constraints imposed by traditional highly bureaucratic ways of performing work. Business process redesign and reengineering endeavors to establish efficient work processes. At the most fundamental level, reengineering concentrates on "starting over" rather than on trying to "fix" existing process problems with marginal or incremental "band-aid" solutions. Barzelay (1992) has characterized traditional types of marginal organizational reform as, "paving the cow paths." In contrast, business process reengineering requires thinking about processes and not functions and positions in organizational hierarchies. The goals of reengineering are increased customer satisfaction and improvement in service quality combined with greater efficiency as measured primarily by reduced cycle time and cost. Reengineering takes advantage to the greatest extent of computer and other information technologies. It requires repeated pilot testing of alternatives proposed to replace existing work processes prior to implementation of new systems and processes.

Only a brief attempt is made here to define reengineering as much has been written about it, most notably by Hammer (1990), Hammer and Champy (1993), and Hammer and Stanton (1995). Reengineering is a top-down process wherein the organization, typically driven by resource constraints and competitive market pressures, attempts to serve its customers better by reducing work process cycle time which, in turn, can reduce costs either in the short or long-term.

Reengineering does not attempt to modify existing processes. Rather, it replaces existing processes with more efficient ways of doing business. Critical to accomplishing the goals of reengineering is increased use of computer and other information technologies to allow fewer employees to do the work formerly performed by more people. Reengineering alters work flow and sequential or reciprocal task dependent relationships, short-cutting older processes in part through substituting computer assisted data gathering, analysis, decision, and management for manual human labor. However, the key is not so much replacing people with technology as much as it is working smarter, eliminating unnecessary, duplicative, paper-heavy work methods.

Not surprisingly, reengineering can result in organizational redesign, for example, flattening or "delayering" as fewer lower and midmanagement employees are needed to do the same or better work after processes have been reengineered. This enables redeployment of some personnel to direct customer service, depending on demand, ability, aptitude, and training. Essential to reengineering is investment in education and training of staff to operate new processes effectively. Reengineering success examples are numerous (Hammer, 1996, pp. 174-190; Hammer & Champy, 1993, pp. 150-199; Hammer & Stanton, 1995, pp. 204-227, 254-273) and often show reduction of work process steps of 70 to 90%, cuts in cycle time of 60 to 80% and reduction of costs from 20 to 80%. In other words, reengineering is intended to make quantum rather than marginal performance improvements.

The process of reengineering involves a commitment by executives to fully support the initiative, the selection and prioritization of processes to be reengineered, assignment of project responsibility to work teams, selection of work team members representing older processes and many or all of the stakeholders in the process outcomes, assignment of team leadership and chaptering/liaison responsibilities, analysis of existing processes, development of alternatives to the status quo, pilot testing, and evaluation of alternatives tested, integration of trial and error lessons in redevelopment of alternatives, refinement of the best alternative and, finally, implementation of the new process and discontinuance of that which it replaces.

Some simultaneous operation of old and new processes may be necessary temporarily. Selection and tasking of work teams is critical to achieving desired results. Continuity of executive support for testing and insulation for failure is essential. Some or many errors should be expected in attempting to define new processes. Full commitment of resources to see the reengineering initiative through also is essential. Staff time, technological support and funding must be provided as needed by process action teams. Furthermore, support for the effort must be virtually open-ended in terms of time schedule, that is, teams must be free to work on alternatives until they have succeeded. Setting artificial end-dates by which process must be reengineered is not productive. Instead, teams should be asked to work until they "get it right."

The bottom line for evaluating the success of reengineering is improved customer satisfaction (i.e., results). Cycle time and cost reduction are not ends in themselves. Rather, they are the results of better work processes. Metrics are critical to determining whether reengineering is successful and, consequently, methods for evaluating results and comparing them to those achieved under previously used processes have to be built into the reengineering effort. Without a means for measuring quantitatively and qualitatively the improvement in service reengineering is virtually pointless. There are simpler ways to cut costs if this is the only objective. This means that results indices must be identified, data bases and collection procedures designed and constructed, data must be gathered, analyzed and compared. Accounting data must be related to results measures to permit cost analysis as well as consumer response to process alternatives whose costs differ. Typically, different parts of the customer base will prefer different mixes of service quality and cost. Reengineering must attempt to accommodate such preferences, which is also the objective of change.

Proponents of reengineering recognize that many organizational work flows, job designs, control mechanisms, and structures are either superfluous or obsolete. Reengineering processes, accompanied by restructuring and downsizing, intends to improve administrative performance and, by slimming the organizational bureaucracy, save money. As Michael Hammer (1990) explains,

It is time to stop paving the cow paths. Instead of embedding outdated processes in silicon and software, we should obliterate them and start over. We
should reengineer our [organizations]; use the power of modern information processing technology to radically redesign our ... processes in order
achieve dramatic improvements in their performance.... We cannot achieve
breakthroughs in performance merely by cutting fat or automating existing
processes. Rather we must challenge the old assumptions and shed old
rules. (Hammer, 1990, pp. 104, 107)

APPLICATION OF SYSTEM REDESIGN AND BUSINESS PROCESS REENGINEERING TO DoD ACQUISITION

Rigorous business process reengineering could be applied in DoD to the
extent that much of the work and many of the decision steps in the current acquisition decision process would be eliminated, along with the
need for the staffs, both civilian and military, that perform this work. This
approach assumes than much of the work performed in the DoD acquisition process may be replaced by the application of IT or can be eliminated because this work adds no value relative to planning, decision
making or program execution.

This is a somewhat harsh indictment of the current process; however
we believe our assumption about the need to eliminate many of work
steps can and should be accomplished. Further, we assert that as a result
of installation of smart IT systems and elimination of duplicative and
unnecessary work, the reduction of cycle time for decision making and
execution will result in substantial increases in productivity and output
and reduction of cycle time from initial proposal to fielding systems, while
at the same time increasing the quality of decisions and products and, as a
result, reduce acquisition costs dramatically.

Such an outcome is easy to prescribe but not so easy to implement. It is
easier to define in general terms what work should remain and what a
redesigned and reengineered process would look like than it is to list what
would disappear as a result of radical process reengineering. Essentially,
what should remain is the role of the central decision makers with whom
the responsibility for acquisition capabilities and requirements determination, analysis and decision making rests, for example, the USD AT&L,
the acquisition chiefs in each of the MILDEPS, the MILDEPS combatant
commanders, and the JCS. Most importantly, the responsibility to manage programs assigned to program managers should be matched by
authority, with fewer accountability reviews and less oversight, to manage
the programs for which they are responsible from a total systems
approach including full integration of life cycle analytical methods.

The challenge to the overall DAS is that the short-term needs of the war fighter commanders have to be balanced against the medium and longer-term demands of MILDEPS and services for recapitalization. To accomplish this war fighter requirements for capability have to be articulated by the COCOMs and integrated quickly into programs and budgets. In doing so the issue of interoperability must be addressed, whether it is required or not. Interoperability is needed in the medium and long-term due to the necessity both for satisfactory joint war fighting operations and staying within budget constraints. Thus, the DAS has to allow multiple lines of acquisition and procurement to operate simultaneously to meet short, medium and longer term needs (Dillard, 2007). For example, as Humvee vehicles in Iraq and Afghanistan have been "armored up" for the Army, Marines, special forces, and other users, simultaneously the Army initiated buying a new and better armored vehicle (the Mine Resistant Ambush Protected vehicle or MRAP), and is in the process of designing and buying a new light armored vehicle from Textron corporation (the Peacekeeper II) to deploy in the battlefield of the future.

While it is axiomatic to say that the war fighters' short-term needs must be met and, therefore, the combatant commanders have to play a potent role in the capabilities/requirements proposal; longer term recapitalizations cannot ever be ignored. Thus, in setting requirements and responding to contingencies as they emerge both shorter and longer term capabilities have to be balanced against each other. The input from the COCOMs has to come up from the MILDEPS and services, as do all proposals for new acquisition programs. On the other hand, the role of the MILDEP acquisition executives and, ultimately, the USD AT&L is to assess whether longer range needs are balanced with what the COCOMs want. And the role of JCS is to insure interoperability to the extent possible.

This is essentially how the DoD acquisition system works presently, and the reform we suggest would not alter the basic structure of this overall program proposal and decision making process. However, we believe strongly that the overall process can and should be simplified and streamlined significantly. In a redesigned acquisition process decision makers would be assisted by smaller staffs to perform analysis. When we say smaller we mean on the order of perhaps a dozen to twenty total staff persons in each office. Using the best and brightest minds, and IT and other tools of modern technology, these staffs would perform virtually all of the analysis of system requirements, planning, performance specifications, presentation of options to decision makers and the other tasks leading to the actual contracting for RDT&E and acquisition. Gone would be the many offices and staffs that now perform such analysis, for example, for preparation of the POM.

In the Navy this would result in the complete elimination of N81 for example. Staffs that presently perform program and project planning in and around the Pentagon and in Navy systems commands that are not involved in program execution would be reduced. The only duplication of effort in analysis of capability requirements would be between the small staffs of the USD AT&L and each of the staffs of the individual Secretariats and of the MILDEPS. In turn, as is the case presently, the MILDEPS would be responsible for input from the war fighting commands, although such input also would continue to flow to the JCS staff. In this regard, the JCIDS process, as it operates presently (or is supposed to operate) would be eliminated. This however would not relieve the JCS staff from conducting interoperability and jointness review for ACAT I programs. This function should remain a responsibility of the JCS staff.

Under such reform, what would happen to the requirements to build the FYDP and the POM? Under this reform approach there would be no FYDP because it is unneeded, always out of date and virtually useless for the purposes it was designed to meet in the 1960s under SECDEF Robert McNamara. However, there would be a capital budget schedule to structure capital asset planning as we indicate in this chapter. Additionally, the POM drill that repeatedly rebuilds the defense program assets would disappear as unnecessary—because it is unnecessary to constantly rebuild a known base of assets to be acquired. As with zero-base budgeting, the POM "build it from the bottom all the way to the top" exercise is a complete waste of time and effort. All that really matters in the POM build are the decisions about new starts. The base will take care of itself on autopilot at the insistence of the MILDEPS, at least until it reaches Congress. As noted earlier, the PPBES process as it operates now would be discarded entirely, replaced by a process of long-range budgeting, and program and budget execution.

As for the acquisition planning and decision process, all work that is not involved in program execution would be performed by the staffs of the USD AT&L, Joint staff and the MILDEP secretariats and military side of the departments, but by far fewer staff with far fewer reviews by succession of committees. As one former senior PM told us, "If you want to get a decision in the Pentagon, don't try to do it by committee. Someone has to be responsible for decisions and held accountable accordingly."

One of the ways to streamline the DAS process is to eliminate duplicative reviews of program proposals by successive committees that tend to ask the same questions but cannot resist the proclivity to add to program complexity by requesting new and previously nonexistent requirements to weapons platforms and systems. Often such add-ons appear to be motivated by the desire of military officers to enhance their careers through recommendation of additional requirements as a "career accomplishment" rather than based on evidence that add-ons are essential to mission performance. The cost of successive add-ons is increased program and budget turbulence and instability, plus a lot of additional work to accommodate or reject the proposed change by program sponsors.

Another factor that inevitably slows down system acquisition analysis and decision making is the competition and sometimes strong disagreement between different parts of the MILDEPS (e.g., between the Office of the Chief of Naval Operations [OPNAV] and systems commands in the Navy), and within organizations including systems commands. One military PM we interviewed said,

> I knew politics would be a major part of this job but I thought the source of problems would be Congress ... but I spend much more time 'politicking' to keep my program alive within my own [systems] command and with OPNAV than with Congress by far.

He stressed that he had to obtain multiple approvals even for "minor decisions I should be able to make myself" from multiple levels within his systems command and in OPNAV which slowed down the progress of his program, which made it more difficult to keep it on schedule and, where modifications were requested, within cost.

Another part of the problem as we see it is that in the current system DoD asks too much of contractors relative to their incentive to take work in the first place, and perform well on contracts once they are awarded. We address this issue at the conclusion of this section.

AN EXAMPLE OF A SIMPLIFIED ACQUISITION PROCESS

To gain perspective on proposals for simplification, redesign and reengineering opportunities we may observe first that the acquisition process may be divided into four basic stages: concept and technology development, system development and demonstration, production and deployment, and operations and support (sustainment).

Second, we observe that the questions that have to be answered to acquire a weapons platform or system are relatively simple in the abstract: (1) What does the entity responsible for acquiring an asset want, and why? (2) How does the intended user of the asset want to use it? (3) What does the asset need to do in terms of performance (Dillard, 2007)? (4) How much money do we have to acquire the asset? Answering these basic questions is not nearly as easy as stating them.

Third, the participant roles in the process and functions to be performed have remained relatively the same (but have become much more complex) since the beginning of the nation. Generally speaking, these roles and functions are performed in nine sequential steps: (1) some entity identifies a capability request, (2) the capability identified has to be validated initially as a legitimate requirement, (3) a weapons platform or weapons system (e.g., equipment) has to be designed to meet the validated capability requirement, (4) DoD contracts for development, test and evaluation, which is intended to and often does lead to design improvements; this work is performed by firms that want to compete for the right to produce and sell the asset to the government, (5) the acquisition of the asset has to be planned (programmed in DoD terms) and then proposed in the defense budget sent to Congress by the President, and then Congress has to appropriate money to buy the asset, (6) DoD performs the role of buyer from the private sector using a myriad management tools for soliciting initial proposals (RFPs or request for proposals), evaluating bids and eventually selecting of the supplier(s), and then contracts for the R&D, prototypes and other work required to develop the prototype (7) DoD evaluates the asset and decides whether to move forward in to full-scale production, (8) the builder/producer must determine how best to manufacture the asset and supply it to DoD within a highly comprehensive and typically tight set of constraints over design, cost, schedule, and so forth (9) assets are delivered to DoD and provided (deployed) to the user, that is, the war fighters.

The components of the acquisition process that we point to as candidates for redesign and reengineering cut across all of these functional stages identified above, although we give less attention to the user phase. Still, we do suggest several new proposals with respect to fielding of weapons systems as we indicate subsequently. We envision a significantly reengineered and simplified acquisition decision and execution process. However, as experienced observers will note, some of what we advocate already is done by DoD, but perhaps not quite in the way we envision it in the model that follows. To illustrate what we advocate we provide as an example a simplified version of what a redesigned and reengineered acquisition process would consist of that is organized into seven main phases.

The Jones-McCaffery Model for Acquisition System Redesign and Reengineering

1. The initial phase is proposal of a desired capability by the MILDEPS and services. This proposal could come from a war fighter command or more centrally from the military chiefs (e.g., from OPNAV or elsewhere in the Navy). The proposal would undergo one comprehensive review and analysis by the staff of the

MILDEP acquisition secretariat and then a decision on whether to proceed development (advanced development latter phase) upon by the service assistant secretary for acquisition. Analysis of proposed systems would be assisted by information on requirements from the COCOMs where available so that, ideally, it is assured to the extent possible that the systems proposed meet a real war fighter need.

This first phase assumes implicitly that the military services have a really good idea what they want, even at the operational capabilities level. However, we note that a number of experienced acquisition practitioners have identified the requirements process as one of the weaknesses of defense acquisition. As one seasoned former PM put it, "In my opinion, this [inability to define requirements adequately] is due primarily to (a) a chronic deficiency of human capital, (b) a dysfunctional and complicated bureaucratic structure, and (c) a perpetual desire to mix needs with prescribed solutions." Another highly experienced critic put it more bluntly, "Do you assume that the war fighter or the military departments and services really know what they want?"

We acknowledge the potential and real weaknesses that exist presently in defining what assets should be acquired for the war fighter. Our first response is that the shift to identifying and specifying the capability desired rather than the specifics of asset performance requirements that has taken place in DoD as a result of transformation over the past five years or so is a step in the right direction to improve the requirements setting process. Secondly, we propose in the second phase of our model of the redesigned process a check to weed out poorly defined capability requests and requirements proposals. This would be (and is now) part of the responsibility initially of the MILDEP acquisition professionals and then of the USD AT&L and JCS. This is not a significant departure from how business is performed presently. However, we wish to point to the statement of USD AT&L John Young, Jr. included below indicating that improvements are needed to state capabilities more clearly, to define requirement more carefully, and to kill off bad proposals earlier in the process (Dillard, 2006).

2. The MILDEP request for the capability and a specific system to meet the capability requirement would be analyzed simultaneously and together by a combination of the staffs of the USD AT&L and the JCS, with a single recommendation issued together to USD AT&L for decision. The USD AT&L would decide on a "go or no go" basis to approve or disapprove the "capabilities and system

request" and this decision would represent the choice of the SEC-DEF, as is the case presently. No separate review by SECDEF would be made except where the secretary took the initiative to do so. It is presumed that some necessarily approximate design requirement and some specifications would be determined by this stage in the process. Still, many issues with respect to feasibility of design, engineering, technological feasibility and cost would inevitably remain to be resolved subsequently. However, we agree with Under Secretary Young in stressing that the culture of "just move it along" in initially approving the capability and requirement has to be changed. Too many asset proposals are approved for development by the MILDEPS and this absence of discipline is as much a cultural phenomenon as it is a failure to perform work diligently. If the MILDEP culture endorses the "let's fly it up the flag pole and see who salutes" approach, then insufficient screening results in wasting time and energy assessing less desirable systems that, in turn, takes time away from analysis and development of systems that are really needed. As Mr. Young put it, "troubled programs share common traits … programs were initiated with inadequate technology maturity and [without] an elementary understanding of the critical program development path." This type of error has to be eliminated, and such discipline will become increasingly necessary as money for DoD weapons acquisition declines, as it will inevitably based on historical analysis of the peaks and valleys for defense funding that we identified in chapter 4.

3. Once a "capabilities and systems request" was approved by USD AT&L, the MILDEPS would request the private sector to prepare and submit design and R&D proposals. The responses from private firms would include bids for their designs, including costs for meeting the required capability and system requirements. Again, this is not much different than what is done in acquisition and contracting presently, with the exception that no R&D would be assigned to government labs. All R&D would be done in the private sector. Notably different from current practice is that competition for the right to produce would be open to United States as well as non-U.S. firms from selected foreign nations.

4. Then, first the MILDEP program office, and second a committee or board representing the combined staffs of the USD AT&L, the JCS and the MILDEPS, would review private sector proposals, each of which would contain the design specifications determined by the private firms and the costs estimated to meet the requirement with a specific platform, system or equipment asset. The second step, the combined review which would include the JCS analysis of inopera-

bility, would result in the recommendation to USD AT&L of one or more contractors for prototype production and related R&D, or that more bids be solicited if none of the bids are deemed satisfactory. Notably, this recommendation would be made by the MILDEP acquisition executive. While the analysis performed during this phase would involve participants representing a number of stakeholders, the primary agent responsible for analysis would be the MILDEP staff. Still, the final *decision authority* to move forward on a system must rest *solely* with the USD AT&L. The USD AT&L and staff would assess the recommendation from the *single* (not multiple) combined committee and staff review of proposals and decide on which to accept and which to reject. Ultimately, in any organization, final decision authority and accountability for asset acquisition decision making has to be assigned to one official. This principle is firmly imbedded in the lessons derived from effective corporate management in the private sector. Management by committee is not management at all. Rather, it is a recipe for error just as is excessive and duplicative reviews of systems leading up to the point of decision. In this respect DoD systems, structures and work processes are weak and wasteful. Too many duplicative reviews by too many entities are performed with the result that it takes longer to reach the point of decision.

5. The next step in the process would be preparing and issuing the contracts for prototype production, and for additional RDT&E where needed. The types of contracts used for prototyping, and later for full-scale production, would be determined as we explain elsewhere in this chapter, based on what is appropriate relative to the capability and system characteristics we identify. Both fixed and flexible price contracts would be used and, as is the case now, the tendency would be to use flexible and incentive based contracts (with strict penalties for failure to perform within cost and time constraints written into the contract) for programs where uncertainty is higher at the front-end of development, of complex systems for example, and then moving to more fixed price contracts where uncertainty was reduced as system designs and characteristics became known and they moved toward and into production. As is the case presently, after bid and award of contracts, most of the technical and financial risks involved in design all the way to production are assumed by the private sector.

   With respect to funding, RDT&E and the latter phases of design and then production would be paid for using appropriations made by Congress—nothing new here as this is a Constitutional requirement. However, because in this model we assume adoption of a capital budget by DoD, financing for acquisition would be provided from the capital investment fund and the term of financing would depend on the needs of the government and the contractor. The primary objective of the CAF would be to stabilize the funding and budget process for weapons and system development, acquisition and deployment. Money to fund the DoD CAF would be appropriated by Congress as is the case presently, for example, by different types of appropriation (i.e., different colors of money) through the regular appropriation process. However, money thus appropriated would be deposited into different accounts within the CAF according to color of money requirements and restrictions provided by Congress. Again, this would not cause Congress to have to make any change in the way it appropriates money for DoD acquisition.

With regard to political considerations, it is highly evident to us that CAF would work best for DoD if Congress would provide funding with maximum flexibility, e.g., with "no-year" end dates, extended time for obligation, higher thresholds for DoD reprogramming without approval from Congress, and, ideally, delegating some between appropriation transfer authority to DoD, subject to reporting but not approval by Congress. We might argue the advantages to Congress of adopting accrual budgeting which would provide multiple year and forward funding for acquisition to replace the annual appropriations budget that Congress prefers, but we acknowledge that Congress is unlikely to ever accept this approach to budgeting, although it is commonplace in the private sector. This reluctance stems from the fact, regrettable at times, that Congress tends to be more concerned with where money is spent and who gets DoD contracts for production of warfare assets than it is with the efficiency of the DoD acquisition process, the performance of program management, or the productivity of the private sector. However, if Congress genuinely wants DoD to provide stable financing for acquisition then members must realize that DoD needs help from them to do so (see more on this area under the politics of reform section of this chapter).

The CAF approach would require a very different system of financing and accounting for appropriation by DoD. As we have recommended, DoD would use a longer term budget and resource management system in which financial obligations for acquiring assets would be managed and accounted for on a full accrual basis, using a separate capital budget to support the financing of systems acquisition. To do so, it would be desirable for all money deposited into the CAF to have the period for obligation extended as we have proposed, and under the most desirable circumstances DoD would request Congress to appropriate what constitute capital outlay appropriations (to buy long-lived assets) on a "no year" basis as explained above, that is, with no end year specified. This would enable DoD CAF managers to provide much more stability to program managers for system development, acquisition and deployment than is the case at present.

6. The private sector would be required to perform additional design work if necessary to produce the final asset prototype with all of the technical and performance attributes intended for the asset once put into full scale production. Further R&D would be done by private firms with government oversight both in terms of performance and cost in the contractor's production facilities similar to the current process often employed, but with more emphasis on product performance and schedule in addition to cost to meet the required program capability requirement. The private sector would supply DoD with prototype models ready, in basic form at least, to test and evaluate realistically for fielding. This final prototype would be jointly and simultaneously tested by the contractor and the PM team on behalf of DoD. Under the conditions of the contract, DoD would have the option to accept or reject the asset. To reemphasize the point, the primary responsibility for satisfying DoD's role and responsibility in test and evaluation would be performed by the MILDEPS program management staff (herein referred to as the contract team), as is much the case now, but with oversight by a single representative from the combined USD AT&L and Joint staff review committee. The purpose of this oversight is to provide another "back-up" check to balance the system in evaluating the prototype. Thus, at this point a combined member evaluation contract team consisting of the PM and staff, the USD AT&L, JCS and the contractor would work together in one place at one time to evaluate the asset.

We propose, in addition, contracting for multiple competing prototypes along with evaluation through collaboration of government and industry teams. This is a component of reform that has received strong support at the DoD executive level. In a memorandum to the Secretaries of the MILDEPS, the Chairman of the JCS, Commander of U.S. Special Forces Command and Directors of Defense Agencies dated September 19, 2007 Acting USD AT&L John Young, Jr. wrote:

Many troubled programs share common traits - the programs were initiated with inadequate technology maturity and an elementary understanding of the critical program development path. Specifically, program decisions were based largely on paper proposals that provided inadequate knowledge of technical risk and a weak foundation for estimating development and procurement cost. The Department must rectify these situations. Lessons of the past, and the recommendations of multiple reviews, including the Packard Commission report, emphasize the need for, and benefits of, quality prototyping. The Department needs to discover issues before the costly System Design and Development (SDD) phase. During SDD, large teams should be producing detailed manufacturing designs—not solving myriad technical issues. Government and industry teams must work together to demonstrate the key knowledge elements that can inform future development and budget decisions. To implement this approach, the Military Services and Defense Agencies will formulate all pending and future programs with acquisition strategies and funding that provide for two or more competing teams producing prototypes through Milestone (MS) B. Competing teams producing prototypes of key system elements will reduce technical risk, validate designs, validate cost estimates, evaluate manufacturing processes, and refine requirements. In total, this approach will also reduce time to fielding. Beyond these key merits, program strategies defined with multiple, competing prototypes provide a number of secondary benefits. First, these efforts exercise and develop government and industry management teams. Second, the prototyping efforts provide and opportunity to develop and enhance system engineering skills. Third, the programs provide a method to exercise and retain certain critical core engineering skills in the government and our industrial base.... Based on these considerations, all acquisition strategies requiring USD(AT&L) approval must be formulated to include competitive, technically mature prototyping through MS B. (Young, 2007)

We presume that, under our proposal, some bids and, consequently, prototypes would come from non-U.S. firms (see our recommendations on the "Buy America Act" and similar laws that have been passed by Congress which would have to be either repealed or modified to permit non-U.S. firms to participate as we recommend).

At this point in the process, the PM led contract team and all other DoD test and evaluation participants would be constrained to requesting only very minimal changes to the asset prototype produced by the private firm. Significantly, changes would be held to a strict cost constraint based on a specified percentage of the projected per unit cost of the asset once it entered full scale production (best guess estimate of "should cost") and would only be approved if the contractor could complete minor modifications within 90 to 120 days. The MILDEP PM would have responsibility, assisted by the acquisition team, for testing along with the contractor any modifications allowed under the contract once the modified prototype was available for further test and evaluation. Once such T&E (test and evaluation) was completed the PM would have sole authority to recommend to the MILDEP acquisition decision authority and the USD AT&L, whether to move to contracting for full-scale production. In this respect we want to empower the PM beyond what is authorized within the existing DAS.

7. Once accepted by PM, in consultation with his/her contract team, the contract to move into full-scale production would be awarded and the purchase funded. For this to happen quickly requires simultaneous alignment of the DAS and the financing process as we recommend under capital budgeting. With respect to final DoD decision authority, the USD AT&L would be required to approve the proposal for contracting for full scale production. In addition, as Gansler has suggested we would place the Assistant Secretary for Networks and Information Integration (N&II) under the USD AT&L (which would change the USD's title to IAT&L) "to emphasize the importance of information-centric systems, both for warfare and for infrastructure" (Gansler, 2007, p. 15). What we intend at this point is essentially direct contact between the PM and the USD AT&L or someone fully authorized and designated on his staff to give final approval. The role of the MILDEP acquisition decision authority at this point would be to step in only to terminate a program. As one former senior PM put it, "It is never too late to kill a program." And this dictum should apply to the full scale production phase that follows final DoD approval for movement to full scale production. Thus, stopping programs such as the infamous Navy A-12 aircraft or the Army Crusader should be regarded as the norm rather than the exception. When program failure is imminent, allowing the decision to terminate it to drag out for years simply wastes money and work effort that should be applied to programs for which the need is highly apparent.

The CAF would supply a stable base for funding production of a specified quantity of the asset. No changes in the design, engineering or technology of the asset would be permitted during the initial production run. While this to some extent deviates from the principles of continuous improvement and spiral acquisition, such control is necessary to protect the contractor from constant changes that while they might be attractive to some DoD entities, in practice cause programs production schedules to slip, increase costs beyond "should cost" government estimates, contractor estimates upon which bids were tendered, and the amount of funding made available for procurement, that is, as evidenced by the perennial cost over-runs that Congress, GAO and even DoD deride (on assessing the risks of spiral development see Dillard & Ford, 2007).

8. The final phase of the redesigned acquisition process would be acceptance of the asset by the war fighter commands as meeting a required capability—through official certification by the appropriate COCOM (more than one COCOM could be involved in this certification). If the war fighter rejected an asset as not meeting capability requirements, or for reasons of poor or nonperformance, DoD would have the authority, prescribed previously within full production contracts, to require a repayment (i.e., a penalty) to the Treasury of a portion of the production contract funding received by the contractor. This innovation to the overall acquisition system would require passage of new legislation by Congress authorizing such action by DoD. Also, it is clear that contractors would not support such legislation giving DoD so much leverage to reject deployed assets combined with a repayment penalty. However, if the goal of the acquisition and financing process is to field systems that meet war fighter needs, this type of legislation is needed to assure complete accountability for assets performance by contractors.

In addition, because contractor expertise typically is required in training and supervision of the use of the asset by war fighters in some instances, a separate contract would be entered into under circumstances on an as needed basis, to finance all or part of the cost of fielding and training with the clear requirement that all assets be fully supported by user manuals, other documentation and required software where applicable *prior to* the point of installation on existing platforms in the case of system replacement or augmentation, or fielding of new platforms or systems. Under the current financing and fielding process RDT&E money cannot be used for anything beyond installation. Technically, training of the type that is often needed cannot be funded or provided by contractors out of RDT&E or production appropriations. However, the advantage of the CAF is that funding stability for training and installation would be paid for from an existing pool of money for this purpose. This aspect of the CAF would require Congress to appropriate funds specifically for such types of contracts and contractor work.

How would the more rapid progress of weapons system RDT&E, development, and the rest of the process be tracked by USD AT&L, the Joint Staff and especially the MILDEP program management team? As

we have explained in chapter 13, this should be done using a single integrated computer system for US AT&L and each of the MILDEPS.
RDT&E would be performed by the private contractors as is, to a significant extent, the case presently.

We acknowledge fully that the redesigned and simplified process example outlined above is just that, an outline of one approach to a reengineered process. Additional analysis is needed to determine how the process
would be implemented beyond what we have stipulated and what parts of
the existing acquisition process would be molded to fit with the reengineered process. Without passage of new legislation by Congress the last element of the reengineered process could not be implemented by DoD.

In evaluating the process outlined here an obvious question is when
there is so little incentive for private sector contractors to bid and perform work for DoD, would not many of the elements of what we suggest
be done further reduce this incentive? What would stimulate the five
major U.S. defense contractors to continue to want to maintain their
defense lines of products and this part of their highly diversified businesses? Part of the answer to this question is greater reliance on competition in a global marketplace and more off-the-shelf buying by DoD. If
large domestic contractor firms decided to abandon their defense business, we presume this would create opportunity for non-U.S. contractors.
Further, our proposal could create incentives for U.S. defense firms to
seek joint ventures with non-U.S. firms. We address this topic in the last
section of the chapter.

Contracts, Risk, and Accommodation of Uncertainty

We are aware that there is a whole layer of contracting and contract
management that must take place to cause private firms to bid to meet
DoD RDT&E and asset acquisition needs.

In addressing the topic of contracting and instrumentation, we
recognize the complexity of contracting has to take into account that risk
and uncertainty are related to the types of contracts used to acquire
services and assets from the private sector (Thompson & Jones, 1986). We
understand the principle that where risk and uncertainty are high,
flexible price and incentive-type contracts are the best tools for getting
the performance desired from contractors. In contrast, where risks are
lower, because of less uncertainty, fixed price contracts may be employed
usefully.

As is the case under current practice, DoD has to be careful to apply
the type of contract tools that are matched to the nature of the performance required under a contract. But this is only part of the equation. As
explained by Thompson and Jones (1986), the choice of management

control system used by DoD has to be matched to the nature of the market (competitive versus noncompetitive), the nature of the asset to be acquired (homogeneous or heterogeneous; known versus unknown product characteristics), and the level of uncertainty and risk (low versus high) involved from R&D to production and eventual fielding of the assets. The advantages of fixed price contracts are in most cases obvious, for example, where COTS is applied to contract for purchase of an asset that already has been produced and is available for purchase without modification. Further, where flexible price contracts are used to appropriately accommodate uncertainty and risk, it is highly advantageous to build in incentives to stimulate contractor performance, for example, incentive bonuses. However, when this and similar approaches are used, and this approach has been employed very successfully by DoD, it is necessary to make sure that the incentive to perform on one part of the contract (e.g., producing an improved radar system on time) does not draw energy and attention away from achieving performance standards on the overall contract, for example, for building a ship.

We also acknowledge neglect to some extent of logistics reform in our model, although we have made reference to the significant advances in spiral logistics and adoption of new systems at the end of chapter 13. Such advances contribute to the task of meeting war fighter needs quickly and efficiently and should be applauded. We note in this regard how DoD has applied private sector methods along with smart practice employed within DoD in reforming logistics processes and practices. Such advances reinforce the supposition that acquisition and related financial management reforms also can be modeled to an extent on private business practices, and that reform initiatives can succeed given appropriate design and implementation, sufficient executive support, and time to mature.

Additional Consequences of Redesign and Reengineering

In a significantly redesigned and reengineered acquisition process several additional and major changes should be made as a consequence of reform. For example, as indicated, all government R&D laboratories that perform defense work would be eliminated because the work they perform can be obtained from private labs at lower cost. Likewise, some relatively unused or under used government production facilities would be closed and terminated, for example, Navy shipyards that never build ships and could not out-perform private yards in terms of price and schedule if required to operate on a nonsubsidized basis and level playing field. In this regard, moving shipyards to mission funding and away from working capital funding, as has been done to some extent in the Navy, removes any incentives to increase productivity that might have been present before this change was made.

Additionally, some of the work performed by the MILDEP system commands would be redefined if more systems are bought off-the-shelf rather than made (contracted for) by these commands. This is not to conclude that all work performed by systems command is unnecessary. In fact, just the opposite is the case. We advocate that increased authority to match responsibility be provided to MILDEP program managers. Still, if DoD moves further towards a "buy" rather than "make" and a "pull" versus "push" acquisition and procurement strategy as we suggest in the last section of this chapter, then less work related to the "make" approach to acquisition would be available to be performed by system commands. As Gansler (2007) has put it, "The DoD must shift from a 'supply push' system to a 'demand pull' system based on 'sense and respond' and secure IT (for 'total asset visibility')" (Gansler, 2007, p. 26).

Smart Practice Examples

One approach to determining how to reform acquisition not explored in this book to any great extent is to review carefully what has worked with successful acquisition programs. One example is the Navy DDG-51 Arleigh Burke class AEGIS guided missile destroyer program. Originally designed to defend against Soviet aircraft, cruise missiles, and nuclear attack submarines, this higher capability ship is used in high-threat areas to conduct anti-air, anti-submarine, anti-surface, and strike operations. The mission of the Arleigh Burke-class DDG-51 is to conduct sustained combat operations at sea, providing primary protection for Navy aircraft carriers and battle groups, as well as essential escort for the U.S. Navy and Marine Corps amphibious forces and auxiliary ships, and can perform independent operations as necessary. These ships contain a myriad of offensive and defensive weapons designed to support maritime defense needs well into the 21st century. The DDG 51 was the first Navy ship designed to incorporate shaping techniques to reduce radar cross-section to reduce detectability and likelihood of being targeted by enemy weapons and sensors. DDG 51s were constructed in flights, allowing technological advances during construction (DDG-51, 2007).

The DDG-51 acquisition program is a "smart practice" example of successful acquisition in that it was (a) managed within cost, (b) came in on schedule, and (c) met war fighter requirements. Causal factors included (a) experienced and consistent PM leadership, (b) good program management teamwork, (c) clear identification of requirements and what the platform was supposed to do, (d) good relations between the PM office and contractors, (e) a highly competent contractor, (f) realistic contracts, (h) relatively stable funding due to justification and defense of the program by the PM and Navy to DoD and on the Hill. As one high level Navy official who headed the PM office said about the program,

> The DDG-51 acquisition was managed by a highly motivated and dedicated government-industry team with extremely clear lines of communication. The PM was charged with "cradle to grave" management of entire system (ship and all the weapons on it). The DDG-51 was the first ship built from the keel up as an entirely integrated weapons system. In terms of cost and schedule, costs were well-contained from the beginning but the schedule for first ship was unrealistic and a contract modification was needed to deal with this problem. We made that happen with full involvement of the Secretariat, the Navy uniformed leadership and the Hill. (Greene, 2007)

For another view on improving the system, see Appendix B that summarizes recommendations from a Rand Corporation study on reducing the costs of Navy shipbuilding.

In conclusion, we maintain that the DAS as it functions presently is not broken as much as it is abused by too much process, too many work steps and too many participants that force too many changes that drive up costs and time to production and fielding. Steps in the process that do not add value more than cost need to be eliminated. Participation in the decision process purely for the sake of participation is wasteful and results in a myriad of negative consequences. When Deputy SECDEF England called for simplification (England, 2007), for us this meant, pure and simple, that some procedural steps and the philosophy of review, re-review and then re-review again had to be stopped. Some stakeholders who participate in the acquisition review and decision process need to be removed, and there is no reason to expect they will like this change.

As Wildavsky (1964) observed long ago, change in political and managerial decision processes inevitably produces winners and losers. Continuous adding of new requirements to systems ultimately causes schedules to slip and costs to rise inordinately. In execution, PMs need greater stability and this means they need fewer changes in the programs they are managing, and they need to be able, on their own, to say no when late system add-ons are proposed, or where production problems emerge that cannot be corrected without incurring greater costs than benefits, for example, as with the A-12 aircraft program. And while virtually all observers continue to applaud the value of continuous improvement through spiral acquisition, several questions always need to be addressed. First, how much will the proposed additional change add to cost and time to delivery? Second, is the integration of new and better technology (e.g., software for example) worth adding 2 years and $10 million to system costs? Third, how much change is too much for program managers and contractors to accommodate within cost and schedule constraints? Addressing such questions has a lot to do with establishing greater stability in the acquisition system that all seem to favor, at least in principle.

These observations are not new, but now they need to be heeded; this is our primary point. As we have noted, the acquisition system isn't broken but it is abused for careerist, bureaucratic, and private purposes horribly, and the result is that weapons and equipment take far too long to field, cost too much and, too often, the result is that fewer units are procured than are needed or products that have consumed considerable financial resources are never delivered to the war fighter.

We fully concede that the type of business process redesign and reengineering reform we advocate is unlikely on its own to correct much of what is deficient in the performance of work within the DoD acquisition, contracting and financial management bureaucracy. We believe major changes are needed to what is done now internally within DoD and in concert with private sector defense contractors. We assert that much work now done within government could and should be performed almost entirely outside of government. If one accepts the viability of this assertion, the questions then become, how would these different approaches to reform be put into practice and what are the implications of each in terms of changing existing DoD organization and business processes? We have attempted to address a variety of issues in our analysis but we accept criticism to the effect that implementing the type of change we advocate is more complicated and faces more hurdles than we have identified. Some of what we have recommended is under implementation, at least in part, as we write. Other suggestions are beyond the range of political or organizational acceptability at present. Some of our proposals simply may be ill-advised. In defense, we assert that what we have tried to accomplish in analysis of acquisition process redesign, reengineering and simplification is provided to stimulate more thinking and dialogue on reform within and outside of DoD in the broader acquisition community of practice.

Finally, in recommending acquisition system redesign, process restructuring and reengineering we want to go on record in stating that increased use of contractors to perform what are essentially government functions has gone too far and needs to be reduced dramatically. We advocate continued outsourcing of only what we and others deem to be essentially nongovernmental work. Whether this means that government employment should increase correspondingly depends entirely on the continued need for the types of work that have been outsourced over the past decade, and the politics of the budgetary process. Finally, with respect to acquisition reform, we recommend beyond redesign and reengineering of business processes an increased use of commercial off-the-shelf acquisition and procurement, relying more extensively than what is done presently on an international marketplace instead of buying almost exclusively from domestic producers. To this topic we now turn our attention.

GLOBALIZATION OF DEFENSE ACQUISITION

The DoD should take greater advantage of the competitive dynamics of an international defense capital asset market in the same way that large firms in the private sector operate presently. As Jacques Gansler (2007) has explained,

> The Security world has changed dramatically—especially since 9/11/01 (geopolitically, technologically, threats, missions, war fighting, commercially, etc.).... However, the Defense Industrial Structure, the controlling policies, practices, laws, and the Services' budgets and "requirements" priorities have not been transformed to match the needs of this new world. (Gansler, 2007, p. 3)

We see the need for transition to a system in which, as noted, the product is the exclusive focus of decision effort. If one accepts the potential viability of this approach, the question then becomes, how would this be done? How would such a system operate and what are the most important issues to be resolved in privatizing DoD weapons systems acquisition? In our analysis we take into account how contemporary business corporations operate, compete and, at times, cooperate presently in a global marketplace. We argue that to operate defense acquisition in a more business-like manner it is necessary to understand the forces and market dynamics that have caused the corporate sector all over the world to adopt new forms of structure, behavior and performance. The DoD needs to take advantage of competition in the emerging global marketplace. As Gansler (2007) has noted, there is now, "A 'globalized defense market' [to enable] technology transfer with allies and buying from the best—with proper risk-based concern regarding security" (p. 12). What is needed in terms of the characteristics of the most desirable defense industrial base in the midtwenty-first century is, among other things, an acquisition strategy that, "draws fully on commercial and global technologies" (p.11).

We assert that the key advantage of the global acquisition reform approach is to use the leverage inherent in the competitive dynamics of an international defense capital asset market in the same way that large firms in the private sector operate rather than relying on the system and process DoD uses now which is, in essence, a gigantic, disconnected and inherently ineffective government bureaucracy. This structure resembles in form the Cold War era Soviet-style long-range planning hierarchy in which *the process becomes the product*. We argue for a transition to a system in which the product is the focus of decision effort. If one accepts the potential viability of this assertion, the question then becomes, how would this be done? How would such a system operate and what are the most important issues to be resolved in privatizing DoD weapons systems acquisition?

For DoD the basic argument we advance is movement towards a buy rather than make acquisition strategy in most cases, and for DoD to try to buy COTS weaponry, systems and equipment not just from U. S. firms but from the international marketplace. If most war fighting assets were bought in this way, this would allow vast reduction in DoD tasks to be performed in planning, building, contracting and execution, and would permit significant leaning of the acquisition part of the organization. Proper execution of this approach to eliminate nonvalue added work so as to increase time devoted to high value added tasks is the key. Further, we advocate outsourcing to the private sector or eliminating all work that is not core governmental in kind. And, we have indicated in this chapter why and how business process should be applied and roughly how much the DoD acquisition bureaucracy should be cut. These conclusions and recommendations also apply to those we have made relative to the abandonment of PPBES and adoption of long-range capital and performance-based budgeting and resource management. Where steps and stages of the work process are eliminated from the existing, highly cumbersome, DoD resource planning, and budgeting processes, the workforce should be reduced accordingly.

We believe that DoD should consider buy versus make to a much more extensive degree than is the case presently. There are some examples where DoD is in fact taking this approach now. The Army is buying a helicopter from Australia, the Marine Corps acquired a fast moving marine troop carrier vessel based on an existing Australian boat design, and members of Congress, including Senator John McCain (R-AZ) have suggested that the Air Force consider competition for refueling aircraft acquisition from Airbus in addition to Boeing. The Marine Corps bought the Harrier aircraft from the United Kingdom long ago. Many other examples abound where the U.S. military is buying equipment from foreign nations.

We would advocate that DoD further consider acquiring major war fighting assets such as strategic and tactical aircraft, missiles, ships, submarines, tanks, armored personnel carriers, trucks, and the rest from overseas producers. As we have explained, DoD should take advantage of competition and even create such competition for supply in the international marketplace, much as it has done in the past in the U.S. defense industry. And just as international corporations have moved production offshore, the U.S. defense industry can move offshore (some already have done so) to take advantage of lower labor costs so as to compete for business from DoD. Further, our proposal would create an incentive for U.S. defense firms to consider joint ventures with foreign firms.

If DoD can buy an existing platform or system that supplies the capability needed from abroad at a lower cost, why should it continue to support what has become essentially monopolistic supply from U.S. firms? Economic theory teaches us that monopolists eventually will set prices too high and will under-produce to exploit their monopoly position. Over the past 15 or so years the U.S. defense industry has consolidated through merger and acquisition to the effect that three large firms dominate the market. They have argued that such strategy was and is necessary for them to survive and make a profit. We do not dispute these claims. However, we do dispute that DoD is better off buying weaponry and supporting systems and equipment from an oligopolistic market when we know from economics that such market structure results in overpricing and underproduction.

The reform to have DoD acquire weapons systems from the international marketplace is not advanced in ignorance of the very real concerns related to the security risks associated with buying from foreign firms. Espionage is a concern both domestically and abroad and the standard assumption is that the risks are higher abroad than at home. We think that achieving security anywhere in the world, given some obvious constraints in some nations, is a matter of how much is invested to achieve it and how it is managed so that all security risks are addressed. In our view if the same security precautions are taken with all firms, foreign and domestic, then we do not see the differences between risks overall. This assumes that the U.S. buys assets from allies who have a mutual stake in cooperative security arrangements in their regions of the world. For example, we do not expect that the U.S. would buy medium or long-range missiles from China, although it could. But could the U.S. buy submarines or ships from South Korea? Most critics would answer that this is not possible, but is that necessarily the case if an asset produced by a foreign firm most cost effectively met the capability requirements of the U.S. military? The longer term nature of the security relationships between nations will always govern who does business with whom in international markets.

A similar concern with our globalization approach is related to the consistent and long-term availability of spare parts and customer support, for example, for software. Our concerns for software are mitigated by the view that all software for war fighting platforms and systems would have to be developed or supported by U.S. firms, partly out of security requirements and partly so that competition for software development and upgrading would remain relatively open. Further, in our vision of how capital budgeting would operate on a longer term and accrual basis, part of the way in which supply of spares is ensured is to buy what is needed up front with the purchase of the major weapons asset as part of the same contract. Would this not build in intentional obsolescence sooner than needed if upgrades and new systems are developed by the supplying firm and as the system and equipment needs of the U.S. military change? Would international buying create a situation in which needed upgrades could not be purchased at all? Our answer is that this situation exists for DoD and the military presently and we do not see how the risk of buying under conditions where there is more competition to provoke innovation to meet U.S. defense needs is greater than at present. If markets are allowed to work as they should, where demand exists, supply emerges to meet the demand. Will this always be the case? Not necessarily, nor is there any guarantee that requirements and capability will remain stable. In fact, the virtual assurance of change in the threat environment and, consequently, in capabilities required argues for the advantage of markets as adaptive mechanisms to lead technology development and availability in ways far better that any comprehensive, planned bureaucratic system can achieve.

THE POLITICS OF REFORM

We recall the statement by former Defense Secretary Dick Cheney who presided over one of the largest cutbacks in weapons system programs, stopping more programs under development than had ever been done before—even after World War II and the Vietnam War (Jones & Bixler, 1992, pp. 129-171). When asked about the effects of such sweeping cuts Cheney replied that it was not the responsibility of the SECDEF to maintain the health and stability of the U.S. defense industry. What was implicit in Cheney's observation is that the responsibility for advocating the cause of U.S. contractors belongs to the contractors themselves, to their lobbyists and, ultimately, those who represent their interests in Congress.

Critics of the view we advocate point out that Congress would not permit DoD to engage in wide-scale international shopping and buying, and they are right—if current law is any indication. For example, the "Buy America Act" prohibits much of the type of business with foreign firms that we indicated is needed. Further, as Gansler (2007) put it,

> significant changes must be made in the ITAR, Export Controls, the Berry Amendment, [in rules governing] specialty metals, etc. to recognize the

[need to operate in a] global defense market (with appropriate risk-based consideration of security and vulnerability concerns) ... [and also to] remove barriers to commercial firms (e.g., CAS) and encourage their participation (via OTA, FAR Part 12, etc.). (pp. 24, 27)

Thus, for DoD to implement our recommendations, some provisions of these and other laws and rules would have to be repealed or modified. Such change is no small order of business and we acknowledge this fact. Congress is interested in keeping defense production at home to protect U.S. labor interests and to supply jobs for their constituents in part because this behavior is what gets members elected and reelected. Further, members of Congress do not shirk from adding assets produced in their states or districts into defense appropriations whether DoD and the military have asked and budgeted for these assets. Pork barreling in support of special interests is endemic in Congress to the extent that it is simply business as usual and DoD is forced to go along with this practice, trading off what is needed badly for what is needed less or not at all, so as to obtain support for its other budget priorities. Further, once a program has been forced into the defense budget in Congress, DoD and the military services are coopted into supporting the program in the future. Thus, pork barreling and earmarking of funds for special purposes by Congress is something that DoD often supports, for example, the V-22 aircraft. However, at the same time that Congress creates and protects American jobs and industry it rails (assisted by GAO and other audit agents) against DoD for asset production cost over-runs, inefficiently low rates of production, failure to set priorities, long cycle time for moving from requirement specification to production and fielding of war fighting assets, and general mismanagement and inefficiency. Our point in this regard is, as the cartoon character Pogo observed, "We have met the enemy and he is us."

The acknowledged excesses of democratic decision making notwithstanding, how long can or should the DAS, the U.S. military and the U.S. taxpayer, have to suffer the consequences of what, at best, is congressional and DoD waste of money and time in coercion of the process of buying war fighting assets, or at worst, behavior that probably is (or should be) criminal—literally—in violation of statutory and administrative law? The answer to this question, based on historical precedent is that such practices have been normal in Congress from the eighteenth century and the beginning of the union (McCaffery & Jones, 2001). Why then should we demand a change now? Our answer is that Congress, DoD as well as the rest of the federal government need to put their money and support where their mouths are ... in support of the incorporation of better business practices in DoD and elsewhere.

Members of Congress and the executive branch speak loudly and often about the need for better business practices in DoD and government. This trend is not new and did not originate under the initiative of former SECDEF Donald Rumsfeld (all Rumsfeld did was try to implement the advice he and other SECDEFs had received). Antecedents may be found across the twentieth century in the recommendations of various Hoover and other commissions and special "Blue Ribbon" studies (e.g., from the Grace Commission and the Packard Commission in the 1980s). Congress has passed innumerable DoD acquisition reform bills into law, in theory, to improve DoD efficiency and effectiveness. Congress has approved GPRA and GMRA (Government Management Reform Act of 1995) and much similar legislation over the past 20 years, much of it aimed at improving government and DoD efficiency and cost consciousness and performance. GAO auditing is used by Congress with the goal of improving efficiency.

Our point is that elected and appointed officials appear to want to be perceived as desirous of stimulating efficiency, higher performance, and productivity. They often speak of the need to "support our fighting forces in the field," particularly in time of war. However, these same officials then perform an about face when it comes to authorization of defense programs and appropriation of defense spending authority. Apparently, to paraphrase the famous dictum of President Harry Truman, "the buck doesn't stop here."Apparently, in terms of real accountability for matching word to deed, the buck does not stop anywhere in the federal government. As we have noted elsewhere, simultaneously, the federal budget and process is overcontrolled and out of control.

Why should we expect Congress to begin to better discipline itself? One reason is that Congress has, in fact, adopted self-denying legislation in the recent past, for example, by creating and living with the consequences of base realignment and closure (BRAC) law where at the end of a deliberate process of analysis, Congress must accept or reject a list of bases to be closed as an up or down vote, as it did in 1988, 1991, 1993, 1995, and 2005. Might we expect similar behavior with respect to congressional review and voting in approval of defense acquisition programs and spending?

What is the likelihood that Congress would, for example, agree to vote without any changes, either for or against a capital budget proposal sent to it by the President as part of the DoD budget? This is precisely what we recommend be done. We suggest that if Congress is faced with an "all or nothing" choice it will make the correct decision just as it has with BRAC. We challenge Congress to pass legislation that creates authority for DoD to prepare and submit a capital budget. and to approve accompanying legislation that requires a congressional vote for or against the acquisition capital budget package submitted to it by DoD without changes, exactly as is the case with BRAC.

Whether Congress is willing to do what we argue for is an open question. First, members would have to perceive that doing this would somehow provide them advantage in the political process. But, it has worked for BRAC and in this process all members have had to give up something to get what is desired for the whole. Could the same be true for defense acquisition proposals?

A second area of resistance to the ideas for increased and open market competition for DoD business that must be anticipated is that which would emerge inevitably from American defense industry and organized labor. We mention this but will not explore it to any extent. Suffice it to state that in a democratic political system all parties have the right of access to the political process to defend their interests, even if those interests advocate in favor of less or no competition, oligopoly and higher versus lower labor costs. However, if the market were to dictate the answer to how war fighting assets are acquired by DoD, we may draw some conclusions by comparing the U.S. defense industry to the U.S. auto industry, that is, there may be a need to compete with and in some cases merge with international competitors to survive. And for organized labor, some jobs are better than no jobs.

Furthermore, DoD does not have to wait for Congress to change the annularity that drives how it authorizes, appropriates, and performs oversight of its program approval and spending roles to begin to change and operate its reformed acquisition process, nor its multiyear budget processes. Congress is unlikely to change its ways that are based on sustaining options and the ability to assert priorities in resource allocation due to the incentives of the political system (Jones & Bixler, 1992). However, we argue that DoD can restructure and reengineer itself and adopt different business models and processes without any change in the congressional budget and oversight processes. Some minor adjustments from Congress would help, for example, extending the obligation period for 1 year appropriations to permit more realistic and efficient defense spending but, overall, DoD can operate a long-range resource management system of its own design on its own as long as it still translates the outputs into formats acceptable to Congress as it does with annual appropriation legislation. DoD does this now with is existing acquisition, procurement and PPBES processes, for example, in use of the milestone authority decision process and cross-walking from program to program elements to appropriations formats. We argue that it is incumbent on DoD leadership to demonstrate to Congress how DoD can operate more effectively and efficiently rather than to depend on congressional and GAO oversight to determine what smart systems and practices should be adopted and how they should be implemented.

CONCLUSIONS

DoD could operate more efficiently similar to multinational corporations but it is hamstrung by bureaucratic inefficiency and multiple layers of overlapping managerial and political control. From our perspective, the question of reform is one of how to structure and operate the organization so as to better match capability with mission.

In our view, the defense acquisition decision process is so excessively bureaucratic that, as with the PPBES process, it should be replaced completely by a new process that would enable capital asset investment analysis of alternatives, decision making, and execution in a much shorter period of time, involving far fewer participants, and in synchronicity with a long-range planning and accrual budgeting process that places emphasis on performance rather than input and process variables. Both the DAS and PPBES processes, as they operate presently, are an incredible and wasteful triumph of process over substance. We believe that if we really intend to run DoD as a business (i.e., using smart business practices) the best way to accomplish this goal is, literally, to make it a business—through privatization of what we perceive as essentially nongovernmental functions performed in the DoD acquisition process to the private sector. In our view, much of what the DoD acquisition and contracting bureaucracy does presently, sometimes well but sometimes very badly, could and should be performed entirely outside of government.

Part of the reform problem is alleged to be "politics" that is, having to operate under the constraints of a democratic political system. But, in fact, free and democratic political systems force compromise under conditions of a high degree of transparency. The result is the same as for competitive firms as opposed to those protected by highly rigid economic regulation and political systems such as the former Soviet Union that force solutions regardless of the nature of the problem. Democracy is in fact ugly and slow at times but it beats the competition provided by other political systems in the long run in terms of mission and financing choice—but not production of the assets needed for national defense, for example, China.

Here is where DoD has the obligation to lead political leaders in the right direction. But, what do we do instead? We organize and operate under the constraints of a highly inflexible, slow, torpid bureaucracy and blame the design and constraints on the political system when in fact the problem lies far more with DoD structure and resistance to operating in markets as a free buyer and seller. And in light of the purpose of this study, we fail in essence to take much or any advantage of the worldwide market in defense assets. To be sure, the problem of moving to more open buying of defense assets and to buy rather than make is both political and organizational. However, we argue that politics follows rather than leads in the definition of better structural/organizational fit to mission and market dynamics. The critical question is whether DoD leadership is willing to take the risks associated with leadership of competitive market oriented reform and privatization of noncore functions that requires adoption of a radically different business model.

We also observe that where the production of privately consumed goods and services is concerned, private organizations are usually more efficient than state-owned enterprises. We assert that the same is true, for reasons explained by economics, for production of assets needed by DoD. Consequently, DoD should increase its reliance on the private sector worldwide in acquisition of war fighting capital assets. Also, we noted that reducing the cost of information should increase the efficacy of markets relative to organizations and of nongovernmental organizations relative to government. Improved communications technology, logistics, and IT all have reduced the cost of information, and have thus increased efficiency in the private sector. Value chain analysis is needed to make significant improvement in DoD acquisition and resource management, and implementation of the results of such analysis will require adoption of more rigorous business process reengineering and reduction in the workforce size and scope of work demanded of the existing DoD acquisition and resource management bureaucracy. DoD has the same opportunity to take advantage of the methods used by private industry to increase the efficiency and efficiency or acquisition, procurement, contracting and resource management, and we argue that part of this opportunity potentially could be provided by using a more competitive market strategy of buying from the international marketplace to the greatest extent possible.

We also have asserted that there is little reason to question the pace of change and contingency in the cultures and environments within which DoD must operate in today's world, nor the fact that DoD must respond to such change. We believe that not all such change will involve evolution towards organizational netcentricity and replacement of bureaucracy with hyperarchy (where appropriate and feasible; see Jones & Thompson, 2007). More moderate adjustments to change are far more likely to be made before such organizations consider more radical reformulation of their design, structure and modes of operating internally and in conjunction with other organizational entities. However, we have provided support for the argument that as a result of threat and other environmental changes and increased contingency, some movement towards hyperachic design and netcentric operation is inevitable if DoD is to become more responsive and better able to accomplish its primary mission in the twenty-first century. As threats change so must the national defense organizations that develop the capabilities to meet the demands that result from new environmental threat and international security circumstance.

We accept that comprehensive reform for both resource management systems including PPBES, and the defense acquisition process may not be politically feasible presently, and therefore we advance a marginal adjustment strategy using capital budgeting and radical reengineering of DoD acquisition, procurement, contracting and resource management as the more feasible option until the political climate is ready for more comprehensive change. And, in fact, both capital budgeting and reengineering may be undertaken at the same time as DoD continues to experiment with global and open market acquisition of COTS platforms, systems and equipment. In this regard the internal DoD business process reforms we advance are complementary with what we advise for DoD externally to take greater advantage of the global marketplace in acquisition of military war fighting assets.

EPILOGUE

Problem: If a cop in Anytown, USA, pulls over a suspect, [ideally] he checks the person's ID remotely from the squad car. He's linked to databases filled with Who's Who in the world of crime, killing, and mayhem. In Iraq, there is nothing like that. When our troops and the Iraqi army enter a town, village, or street, what they know about the local bad guys is pretty much in their heads, at best. Solution: Give our troops what [some of] our cops have. The Pentagon knows this. For reasons you can imagine, it hasn't happened.... This is a story of can-do in a no-can-do world, a story of how a Marine officer in Iraq, a small network-design company in California, a nonprofit troop-support group, a blogger and other undeterrable folk designed a handheld insurgent-identification device, built it, shipped it and deployed it in Anbar province. They did this in 30 days, from December 15 to January 15. Compared to standard operating procedure for Iraq, this is a nanosecond.... Before fastening our seatbelts, let's check the status quo. As a high defense department official told the journal's editorial page, "*We're trying to fight a major war with peacetime procurement rules.*" The department knows this is awful. Indeed, a program exists, the Automated Biometric Identification System: retina scans, facial matching, and the like. The reality: This war is in year 4, and the troops don't have it. Beyond Baghdad, the U.S. role has become less about killing insurgents than arresting the worst and isolating them from the population. Obviously it would help to have an electronic database of who the bad guys are, their friends, where they live,

Budgeting, Financial Management, and Acquisition Reform in
The U.S. Department Of Defense, pp. 663–664
Copyright © 2008 by Information Age Publishing
All rights of reproduction in any form reserved.

tribal affiliation—in short the insurgency's networks.... The Marine and Army officers who patrol Iraq's dangerous places know they need an identification system similar to cops back home. The troops now write down suspects' names and addresses. Some, like Marine Major Owen West in Anbar, have created their own spreadsheets and PowerPoint programs, or use digital cameras to input the details of suspected insurgents. But no Iraq-wide software architecture exists.... On the night of January 20, Major West, his Marine squad and the *jundi* (Iraq army soldiers) took the MV 100 and laptop on patrol. Their term of endearment for the insurgents is "snakes." So of course the MV 100 became the Snake Eater. The next day Major West e-mailed the U.S. team digital photos of Iraqi soldiers fingerprinting suspects with the Snake Eater. "It's one night old and the town is abuzz," he said. "I think we have a chance to tip this city over now." A rumor quickly spread that the Iraqi army was implanting GPS chips in insurgents' thumbs.... Over the past 10 days, Major West has had chance encounters with two marine superiors—Major General Richard Zilmer, who commands the 30,000 joint forces in Anbar, and Brigadier General Robert Neller, deputy commanding general of operations in Iraq. He showed them the mobile ID database device.... I asked General Neller by e-mail on Tuesday what the status of these technologies is now. He replied that they're receiving advanced biometric equipment, "like the device being employed by Major West." He said "in the near future" they will begin to network such devices to share databases more broadly. Bottom line: The requirement for networking our biometric capability is a priority of this organization. As he departs, Major West reflected on winning at street level:

> We're fixated on the enemy, but the enemy is fixated on the people. They know which families are apostates, which houses are safe for the night, which boys are vulnerable to corruption or kidnapping. The enemy's population collection effort far outstrips ours. The Snake Eater will change that, and fast.

You have to believe he's got this right. *It will only happen, though, if someone above his pay grade blows away the killing habits of peacetime procurement.* [comments in brackets, italics, and bold added by report authors] (Henninger, 2007, p. A14).

APPENDIX A

Definitions of Abbreviations and Acronyms

AACAT – Acquisition Category

ADM – Acquisition Decision Memorandum

AOR – Area of Responsibility

APB – Acquisition Program Baseline

APN – Aircraft Procurement Navy

ASD – (SO/LIC) Assistant Secretary of Defense for Special Operations and Low Intensity Conflict

ASN (RD&A) – Assistant Secretary of the Navy for Research, Development and Acquisition (CAE for the Navy)

ATN – Alliance Test Network

BA – Budget Authority

BAH – Basic Allowance for Housing

BAM – Baseline Assessment Memorandum

BCP – Budget Change Proposal

BES – Budget Estimate Submission

BGM – Budget Guidance Memorandum

BOR – Budget OPTAR Report

BOS – Base Operating Support

BR – Concurrent Resolution on the budget

BRAC – Base Reorganization and Closure Committee

BSO– Budget Submitting Office

C4I – Command, control, communications, computers and intelligence

CAD – Computer Aided Design

CAE – Component Acquisition Executive

CAIG – Cost Assessment Improvement Group

CAIV – Cost as an Independent Variable

CAP – Critical Acquisition Position Description

CNDI – Commercial or Non–Developmental Items

CBO – Congressional Budget Office

CCR – Concurrent Resolution on the Budget; also CBR, BR.

CEB – CNO Executive Review Board

CENTCOM – Commander in Chief, Central Command

CFE – Commercial Furnished Equipment

CFFC – Commander Fleet Forces Command

CINC – Commander in Chief

CIVPERS – Civilian Personnel

CLF – Commander, U.S. Atlantic Fleet

CLINGER–COHEN ACT of 1996 – Information Technology Reform Act of 1996

CNDI – Commercial or Non–Developmental Items

CNO – Chief of Naval Operations

COMNAVAIRPAC – Commander Naval Air Forces Pacific

COMNAVSUBPAC – Commander Naval Submarine Forces Pacific

COMNAVSURFPAC – Commander Naval Surface Forces Pacific

COMOPTEVFOR – Commander, Operational Test and Evaluation Force

COTS – Commercial off the Shelf

CP – Capability Plan

CPAM – CNO Program Assessment Memorandum

CPF – Commander, U.S. Pacific Fleet

CPA – Chairman's (of the Joint Chiefs) Program Assessment

CPR – Chairman's (of the Joint Chiefs) Program Recommendation

CRA – Continuing Resolution Appropriation

CVN 68 – NIMITZ Class Nuclear Powered Aircraft Carrier

DAB – Defense Acquisition Board

DDG 51 – Arleigh Burke Class Aegis Destroyer

DFAS – Defense Finance and Accounting Service

DIT – Design Integration Test

DOD – Department of Defense

DoDD –Department of Defense Directive

DODIG – Department of Defense Inspector General

DON – Department of the Navy

DPG – Defense Planning Guidance

DUSD – Deputy Under Secretary of Defense

DW – Defense–wide

EA – Executive Agent

EMD – Engineering, Manufacturing and Development Phase

ESPC – Energy Savings Performance Contracts

EUSA – Eighth United States Army

FAD – Funding Authorization Document

FARA – Federal Acquisition Reform Act of 1996

FASA – Federal Acquisition Streamlining Act of 1994

FASAB Federal Accounting Standards Advisory Board

FFMIA Federal Financial Management Improvement Act

FHCR – Flying Hour Cost Report

FHMP Family Housing Master Plan

FHP – Flying Hour Program

FHPS – Flying Hour Projection System

FMB – Navy Budget Office

FMB Director, Navy Office of Budget

FO – Flying Hours Other

FP – Force Protection

FY – Fiscal Year

FYDP – Future Years Defense Plan

GAO – General Accounting Office

GFE – Government Furnished Equipment

GPRA – Government Performance and Results Act:

HAC – House Appropriations Committee

HASC – House Armed Services Committee

HQ – Headquarters

H.R. – House Resolution

IA&I – Industrial Affairs and Installations

IDTC – Interdeployment Training Cycle

IG –Inspector General

IMD – International Institute for Management Development

IPDE – Integrated Product Data Environment

IPPD – Integrated Product and Process Development

IPT – Integrated Process Team

ISPP – Integrated Sponsor Program Proposal

IT – Information Technology

IWAR – Integrated Warfare Architecture

JCS – Joint Chiefs of Staff

JP – Joint Publication

JROC – Joint Requirements (of JCS) Oversight Committee

JSF – Joint Strike Fighter

LAN –Local Area Network

LBTE – Design Integration in a Land Based Test Environment

LPD-17 – Marine Amphibious ship used for embarking, transporting and supporting troops

LRIP – Low Rate Initial Production

MDA – Milestone Decision Authority

MDAP – Major Defense Acquisition Program

MEB – Marine Expeditionary Brigade

MFP – Major Force Program

MHPI – Military Housing Privatization Initiative

MILCON –Military Construction

MIPR – Military Interdepartmental Purchase Request

MOCAS – Mechanization of Contract Administration Services system

MOU – Memorandum of Understanding

MS – Milestone

MSA – Master Settlement Agreement

MUHIF – Military Unaccompanied Housing Improvement Fund

MWR – Morale, Welfare, and Recreation

NAVAIR – Naval Aviation Systems Command

NAVSEA – Naval Sea Systems Command

NDI – Nondevelopmental Item

NHBS – Navy Headquarters Budgeting System

NMCI – Navy and Marine Corps Intranet

NMSD – National Military Strategy Document

No. – Number

NOR – Net Offsetting Receipt – Collections from the public

NSS – National Security Strategy

OAC – Operating Agency Code

O&M – Operations and Maintenance

O&MN – Operations and Maintenance, Navy

O&MMC – Operations and Maintenance, Marine Corps

O&S – Operation and Support Costs

OBAD– Operating Budget Activity Document

Obligation – legal set aside funds for a future payment, as in letting a contract

ODC (P/B) – Office of the Deputy Comptroller (Program/Budget)

OFPP – Office for Federal Procurement Policy

OGA – Other Government Activity

OFPP – Office for Federal Acquisition Policy

OMB – Office of Management and Budget

OMN – Operation and Maintenance, Navy

OPNAV – Office of the Chief of Naval Operations

OPTEMPO – Operational Tempo

ORD – Operational Requirements Document

OSD – Office of the Secretary of Defense

Outlay – An Expenditure or Liquidation of Obligations

PACFLT – U.S. Pacific Fleet

PACNORWEST – Pacific Northwest

PACOM – U.S. Pacific Command

PBAS – Program Budget and Accounting System

PBCG – Program Budget Coordination Group

PBD – Program Budget Decision

PCP – Program Change Proposal

PDRR – Program Definition and Risk Reduction

PE – Program Element

P.L. – Public Law

PM – Program Manager

POM – Program Objective Memorandum

PPBS – Planning, Programming and Budgeting System

PPBES – Planning, Programming, Budgeting and Execution System

PR – Program Review

PRESBUD – President's Budget

QDR – Quadrennial Defense Review

R3B – Resource Requirements Review Board

RBA – Revolution in Business Affairs

Reappropriations – Extending previously appropriated funds

RDT&E – Research, Development, Training and Engineering

Rescissions – Canceling new Budget Authority or Unobligated Balances

RMA – Revolution in Military Affairs

ROE – Return-on-equity

SAC – Senate Appropriations Committee

SASC – Senate Armed Services Committee

SECDEF – Secretary of the Defense

SECNAV – Secretary of the Navy

SGL – Standard General Ledger

SLAN – Secure Local Area Network

SO – Special Operations

SOC – Special Operations Command

SOF – Special Operations Forces

SPI – Single Process Initiative

SPP – Sponsor Program Proposal

SPS – Standard Acquisition System

SRM – Sustainment, Restoration, and Maintenance

SSBN – Strategic Ballistic Missile Submarine

SSGN – Tomahawk Launch Capable Converted Strategic Ballistic Missile Submarine

Subunified – Subordinate Unified

SV – Service

TOA – Total Obligational Authority—Value of the direct defense program in a given year from this year and previous years.

TOC – Total Ownership Costs

T-POM – Tentative Program Objective Memorandum

TSOC –Theater Special Operations Command

TSP – Thrift Savings Plan

TYCOM – Type Commander

UFR – Unfunded Requirement

UMD – Unmatched Disbursement: payment that can not be matched to an existing obligation

U.S. – United States

USARSO – United States Army South

USC – United States Code

USCENTCOM – United States Central Command

USCINCSOC – Commander in Chief, United States Special Operations Command

USD (AT&L) – Under Secretary of Defense for Acquisition, Technology and Logistics

USD(C) – Under Secretary of Defense (Comptroller)

USEUCOM – United States European Command

USFK – United States Forces Korea

USJFCOM – United States Joint Forces Command

USNORTHCOM – U.S. Northern Command

USPACFLT – United States Pacific Fleet

USPACOM – United States Pacific Command

USSOCOM – United States Special Operations Command

USSOUTHCOM – United States Southern Command

APPENDIX B

REDUCING SHIPBUILDING COSTS

The following excerpt from a Rand Corporation National Defense Research Institute (2006, p. 59) study of increases in shipbuilding costs is provided to support in part our argument for acquisition and related financing and business process reform in DoD.

What can the Navy do to reduce its ship costs? While this study did not conduct an exhaustive search of ways that these costs could be reduced, our interviews with shipbuilders and other knowledgeable sources elicited a dozen preliminary ideas related to the issues we found:

- Increase investments in shipbuilding infrastructure aimed at improving producibility
- Increase shipbuilding procurement stability
- Fund shipbuilding technology and efficiency improvements
- Improve management stability
- Change GFE-program management controls
- Employ batch production scheduling
- Consolidate the industrial base
- Encourage international competition and participation
- Build ships as a vehicle
- Change the design life of ships
- Buy a mix of mission-focused and multi-role ships
- Build commercial-like ships.

Some of these ideas are highly speculative and, given the current fiscal and legislative environment, have dubious prospects for implementation. Nonetheless, we present all of them for completeness of discussion...

Arena, M. V., Blickstein, I., Younossi, O., & Grammich, C. A. (2006). *Why has the cost of navy ships risen? A macroscopic examination of the trends In U.S. naval ship costs over the past several decades.* Santa Monica, CA: Rand Corporation, National Defense Research Institute.

BIBLIOGRAPHY

Abramowitz, A., & Saunders, K. (2005). *Why can't we all just get along? The reality of a polarized America*. The Forum, *3*(2), 1-22.

Adelman, K., & Augustine, N. (1990). *The defense revolution: Strategy for the brave new world*. San Francisco: Institute for Contemporary Studies Press.

Aerospace Daily. (2001). Washington, DC, AD 54: 4.

Aerospace Daily. (2002). Washington, DC, AD 55: 10.

Ahearn, D. (2002). Lawmakers seek more funds for ships: CBO outlines cuts. *Navy News Week, 23*(13), 31.

Alberts, D. S., & Hayes, P. (2003). *Power to the edge: Command and control in the Information Age*. New York: Harper & Row.

Allan, H. (2002, September 26). *Interview: N801 programming issues*. Makalapa, HI: COMPACFLT.

Allen, J. (2005, July 4). House passes fiscal 2006 bills in record time. *CQ Weekly*.

American Enterprise Institute. (1982). *Congressional Budget Process*. Washington, DC: Author.

Anthony, R. N. (2002). Federal accounting standards have failed. *International Public Management Journal, 5*(3), 297-312.

Appleby, P. (1957). The role of the budget division. *Public Administration Review, 17*(Summer), 156-158.

Armed Forces Management. (1969, October). Mel Laird: Coach, quarterback, or both? *Armed Forces Management*, 34.

Arrow, K. (1969). *The organization of economic activity: Issues pertinent to the choice of market versus non-market allocation. The analysis and evaluation of public expenditure: The PPB system*. Washington, DC: U.S. Government Printing Office, U.S. Congress, Joint Economic Committe.

Art, R. (1985). Congress and the defense budget: Enhancing policy oversight. *Political Science Quarterly, 100*(2), 227-248.

Ashdown, K. (2004). Defense spending pork fest. The waste basket. *Taxpayers for Common Sense, IX*(45), 10.

Ashdown, K. (2005). Emergency spending bill pays for political pork (News release). *Taxpayers for Common Sense, X*(31), 6.

Augustine, N. R. (1982). *Augustine's Laws*. New York: American Institute of Aeronautics and Astronautics.

Augustine, N. R. (2005). *Forward: Defense acquisition performance assessment, executive summary: A report by the assessment panel of the defense acquisition performance assessment project for the acting deputy secretary of defense*. Washington, DC: DoD.

Barnett, T. P. (2005, July). Donald Rumsfeld: Old man in a hurry. *Esquire Magazine, XVII*(7), 34-36.

Barr, S. (2005, February 21). Congress growing impatient with longtime "high risk" areas of financial waste. *The Washington Post*, B02.

Barzelay, M. (1992). *Breaking through bureaucracy*. Berkeley, CA: University of California Press.

Barzelay, M. (2003). Introduction: The process dynamics of public management policymaking. *International Public Management Journal, 6*(3), 251-282.

Barzelay, M., & Campbell, C. (2003). *Preparing for the future: Strategic planning in the U.S. Air Force*. Washington, DC: Brookings Institution Press.

Barzelay, M., & Gallego, R. (2005). From "New Institutionalism" to "Institutional Processualism": Advancing knowledge about public management policy change. *Governance, 12*(3), 31-45.

Barzelay, M., & Thompson, F. (2005). Case teaching and intellectual performances in public management. In I Geva-May (Ed.), *Thinking like a policy analyst* (pp. 83-108). New York: Palgrave-Macmillan.

Bath Iron Works. (2001). *An overview: DDG 51 1998-2001*. Retrieved from www.gdbiw.com/company_overview/shipbuilding/lpd17/default.htm

Bath Iron Works. (2001). *Overview: LPD17 1998-2001*. Retrieved from www.gdbiw.com/company_overview/shipbuilding/lpd17/default.htm

Behn, R. D. (2003). Why measure performance? Different purposes require different measures. *Public Administration Review, 63*(5), 586-606.

Bellamy, C., & Taylor, J. (1998). *Governing in the Information Age*. Buckingham, England: Open University Press.

Bendorf, C. (2002). *Can the current acquisition process meet operational needs?* Maxwell Air Force Base, AL: Air War College, Air University.

Bennett, J. T., & Muradian, V. (2007, March 26). Interview Kenneth Krieg, Pentagon Acquisition Chief. *Defense News*, 30.

Bennet, J. T. (2007, July 30). CSBA: Black spending doubled since 1995. *Defense News*, 22.

Berger, P. L., & Luckman, T. (1980). *The social construction of reality: A Treatise in the sociology of knowledge*. New York: Insington.

Berman, L. (1979). *The office of management and the budget and the presidency, 1921-1979*. Princeton, NJ: Princeton University Press.

Blechman, M. (1990). *The politics of national security: Congress and U.S. defense policy*. New York: Oxford University Press.

Bloom, T. R. (2003, March 31). *Statement by the Director, Defense Finance and Accounting Service to the Subcommittee on National Security, Emerging Threats, and Inter-*

national Relations of the House Government Reform Committee, U.S. House of Representatives. Washington, DC: U.S. Congress.

Bolles, A. S. (1969). *The financial history of the United States from 1774 to 1789* (2nd ed.). New York: D. Appleton. (Original work published 1896)

Borcherding, T. E. (1988). Some revisionist thoughts on the theory of public bureaucracy. *European Journal of Political Economy, 4,* 47-64.

Bouckaert, G., & Pollitt, C. (2000). *Public management reform: A comparative analysis.* Oxford, England: Oxford University Press.

Boudreau, M. (2005, May). *Acquisition reform.* Presentation at the Acquisition Research Conference. Monterey, CA: NPS.

Boutelle, J. (2002. June 25). *Testimony by Director of Commercial Pay Services, Defense Finance and Accounting Service, to the Subcommittee on National Security, Veteran's Affairs and International Relations of the House Government Reform Committee, U.S. House of Representatives.* Washington, DC: U.S. House of Representatives.

Boutelle, J. (2003, March 31). *Testimony by the Director of Commercial Pay Services, Defense Finance and Accounting Service, to the Subcommittee on National Security, Emerging Threats, and International Relations of the House Government Reform Committee.* Washington, DC: U.S. House of Representatives.

Bowman, K. (2003, March 21). *Introduction to the Joint Capabilities Integration and Development System: JCIDS* (Power Point Presentation). Washington, DC: DAU.

Bowsher, C. (1994, April 12). *Testimony by the U.S. Comptroller General to the Senate Governmental Affairs Committee.* Wahsinton, DC: U.S. Senate.

Boyne, W. (2003). *Operation Iraqi Freedom: What went right, what went wrong, and why.* New York: Forge Books.

Bozin, S. (2007, February). Director, Office of Budget, Department of the Navy FY 2008. *PB08 Press Briefing,* 10.

Bradley, S., Hausman, J., & Noland, R. (1993). *Globalization, Technology, and competition: The fusion of computers and telecommunications in the 1990s.* New York: Harvard Business School.

Brooks, D. (2002, November 14). *Interview. CPF-N00F2 Issues.* Makalapa, HI: COMPACFLT.

Browne, V. J. (1949). *The control of the public budget.* Washington, DC: Public Affairs Press.

Brunsson, N. (1989). *The organization of hypocrisy: Talk, decisions and actions in organizations.* New York: Wiley.

Bryce, D. (2002, June 25.). *Joint Service Lightweight Technology Suits—JLIST. Testimony by the Program Manager, Nuclear, Biological and Chemical Defense Systems, Marine Corps System Command to the Subcommittee on National Security, Veterans Affairs, and International Relations of the House Government Reform Committee.* Washington, DC: U.S. House of Representatives.

Brynjolfsson, E., & Hitt, L. M. (2000). Beyond computation: Information technology, organizational transformation and business performance. *Journal of Economic Perspectives, 14*(4), 23-48.

Budget of the United States for Fiscal Year 2006. (2005). *Historical Tables.* Retrieved from www.whitehouse.gov/omb/budget/fy2006/pdf/hist.pdf

Budget of the United States Fiscal Year 2004. (2003). *Historical Tables.* Washington, DC: Government Printing Office.

Buell, R. C. (2002). *An analysis of improvisational budgeting from calendar year 1990 to 1999.* Master's thesis. Monterey, CA: Naval Postgraduate School.

Burk, J. (1993). Morris Janowitz and the origins of sociological research on armed forces and society. *Armed Forces and Society, 19*(2), 167-185.

Burkhead, J. (1959). *Government budgeting.* New York: Wiley.

Burlingham, D. M. (2001, January). Resource allocation: A practical example. *Marine Corps Gazette, 85,* 60-64.

Bush, G. W. (2007, May 2). *Veto message from President of the United States.* House Document 110-31.

Business Management Modernization Program. (2003). *DoD Response to P.L. 107-314, National Defense Authorization Act for FY2003, Transition Plan.* Washington, DC: Department of Defense. Retrieved from www.dod.mil/comptroller/bmmp

Business Management Modernization Program. (2003, May). *Organizational Readiness Assessment, Table 2.1, and "Organizational Readiness Summary of Raw Data.* Retrieved from www.dod.mil/comptroller/bmmp

Buzzanco, R. (1996). *Masters of war: Military dissent and politics in the Vietnam Era.* Cambridge, England: Cambridge University Press.

Citizens Against Government Waste. (2002). *Pig Book 2002.* Washington, DC: CAGW. Retrieved from www.cagw.org

Citizens Against Government Waste. (2005, April 6). *2005 Pig Book Exposes Record $27.3 Billion in Pork.* Washington, DC: CAGW.

Caiden, N., & Wildavsky, A. (1974). *Planning and budgeting in poor countries.* New York: Wiley.

Caldwell, L. K. (1944). Alexander Hamilton: Advocate of executive leadership. *Public Administration Review, 4* (Spring), 145-161.

Candreva, P. J. (Ed.). (2004). *Practical financial management: A handbook of practical financial management topics for the DoD financial manager* (5th ed.). Monterey, CA: Naval Postgraduate School.

Candreva, P. J., & Jones, L. R. (2005). Congressional delegation of spending power to the defense department in the post-9-11 period. *Public Budgeting and Finance, 25*(4), 1-19.

Carlin, T., & Guthrie, J. (2001). Lessons from Australian and New Zealand experiences with accrual and output-based budgeting. In L. R. Jones, J. Guthrie, & P. Steane (Eds.), *Learning from international public management reform* (pp. 89-100). New York: Elsevier Press.

Carter, L. B., & Coipuram, T., Jr. (2005, May 23). Defense authorization and appropriations bills: A chronology, FY 1970-FY 2006. *CRS Report for Congress,* 98-756C.

Catton, V. (2002, September 25). *Interview.* Makalapa, HI: COMPACFLT.

Caudle, S. (2003). Implications in establishing the Department of Homeland Security. *PA Times, 26*(5), 22.

Chairman of the Joint Chiefs of Staff. (2004). *Joint capabilities integration and development system. Instruction 3170.01D.* Washington, DC: CJCS, B-4.

Cheney, E. D. (2002). *Analysis of the Anti-Deficiency Act in the Department of the Navy.* Master's thesis. Monterey, CA: Naval Postgraduate School.

Chiarelli, P. W. (1993, Autumn). Beyond Goldwater-Nichols. *Joint Forces Quarterly,* 71-81. Retrieved from http://www.dtic.mil/doctrine/jel/jfq_pubs/index.htm

Chief Financial Officer CFO Act of 1990 (1990). *Public Law,* 101-576. Retrieved from www.gao.gov/policy/12_19_4.pdf.

Chief of Naval Operations. (2002a, October). *Chairman's program recommendation.* Retrieved from http://cno-n6.hq.navy.mil/n6e/ppbs/ppbsprocess/planning/cpr.htm

Chief of Naval Operation. (2002b). *N6E Analysis of Alternatives.* Washington, DC: CNO.

Chief of Naval Operations (2002c). *Summer Review and POM/PR Issue Papers.* Retrieved from http://cno-n6.hq.navy.mil/n6e/ppbs/ppbsprocess/programming/sumrevw&pom-prispprs.htm

Chief of Naval Operations. (2002d). *OPNAV Notice 5400: Standard Naval Distribution List.* Washington, DC: Department of the Navy.

Chief of Naval Operations. (2002e). *Chairman's program assessment.* Retrieved from http://cnon6.hq.navy.mil/N6E/PPBS/ppbsprocess/Programming/ChrmnsPrgrmAssmnt.htm

Chief of Naval Operations. (2002f). *Joint planning document.* Retrieved, from http://cno-n6.hq.navy.mil/N6E/PPBS/ppbsprocess/planning/jpd.htm

Clemens, A. (2005). *Defense pork reaches record high. Taxpayers for common sense.* Washington, DC: Defense Appropriations Database.

Clinger-Cohen Act. (1996, February 10). *Information Technology Management Reform Act. Public Law 104-106.* Washington, DC: U.S. Congress.

Coase, R. (1937). The nature of the firm. *Economica, 4,* 386-405.

Cohen, S. (2007, April 2). Hurricane Katrina cuts a wide swath of fraud—600 charged. *Monterey County Herald,* A3.

Congressional Budget Office. (1997). *Reducing the deficit: Spending and revenue options.* Retrieved from http://www.fas.org/man/congress/1997/cbo_deficit/def07.htm

Congressional Budget Office. (1999). *Emergency funding under the Budget Enforcement Act: An update.* Washington, DC: CBO.

Congressional Budget Office. (2001, March). *Supplemental appropriations in the 1990s.* Washington, DC: CBO. Retrieved from www.cbo.gov

Congressional Budget Office (2003). *The long term implications of defense plans.* Washington, DC: Author.

Congressional Quarterly Almanac. (1975-2001). Washington, DC: CQA.

Congressional Quarterly Almanac. (1990-1999). *Lifecycle costs.* Washington, DC: CQA, XLVI-LV.

Congressional Quarterly Weekly. (1998). Washington, DC: CQA.

Congressional Quarterly Weekly. (2001). 337.

Congressional Record. (1997, November 6). U.S. Senate Bill S11817-8. Washington, DC: U.S. Senate.

Coram, R. B. (2002). *The Fighter pilot who changed the art of war.* Boston: Little, Brown.

Cordesman, A. H. (2003). *The Iraq War: Strategy, Tactics, and military lessons.* Washington DC: Center for Strategic and International Studies.

Cottey, A., Edmunds, T., & Forster, A. (2002). The second generation problematic: Rethinking Democracy and civil-military relations. *Armed Forces and Society, 29*(1), 31-56.

Coyle, J. (2002, June 25). *Testimony by the representative from the Center for Supply Chain Research, Department of Business Logistics, Pennsylvania State University, to the Subcommittee on National Security, Veterans Affairs, and International Relations of the House Government Reform Committee.* Washington, DC: U.S. House of Representatives.

Cox, M. (2007, June 4). Too Late, XM8. *Defense News,* 38-42.

Daggett, S. (2006, June 13). Military operations: Precedents for funding contingency operations in regular or in supplemental appropriations bills. *Congressional Research Service Report for Congress,* 2.

Daggett, S., & Belasco, A. (2002). *The defense budget for FY 2003: Data summary.* Washington, DC: Congressional Research Service.

Daly, P. J. (2004). *Joint Perspective on defense budget issues. Brief to Naval Postgraduate School.* Monterey, CA: Naval Postgraduate School.

Daniels, M. (2001). *A-11 Transmittal letter from the director of OMB to agencies.* Washington, DC: Office of Management and Budget.

Daniels, M. (2002). *Testimony to the Senate Armed Services Committee.* Washington DC: U.S. Senate.

Defense Acquisition University. (2003). *Introduction to defense acquisition management* (6th ed., K. E. Soundheimer, Ed.). Fort Belvoir, WA: Defense Acquisition University Press.

Davenport, T. H. (1993). *Process innovation: Reorganizing work through information technology.* Boston: Harvard University Press.

Davis J. W., & Ripley, R. B. (1967, November). The Bureau of the Budget and Executive Branch Agencies: Notes on their Interaction. *Journal of Politics, 29,* 749-769.

Davis, J. A. (2002). *Statement to the House Defense Appropriations Subcommittee.* Washington, DC: U.S. House of Representatives

DDG-51 (2007). *Global Security.org.* Retrieved from http://www.globalsecurity.org/military/systems/ship/ddg-51.htm

Defense Acquisition Performance Assessment. (2006). *Executive summary.* Washington, DC: Department of Defense.

Defense Daily International. (2002). *Untitled, 3*(14) 23.

Defense Daily International. (2002). *Untitled, 3*(15), 12.

Defense Daily International. (2002). *Untitled, 2*(13), 53.

Defense Finance and Accounting Service (n.d.). "Overview," www.dfas.mil

Defense Link. (2002, September). *Navy announces DDG 51 multiyear contract.* Retrieved from http://www.defenselink.mil/news/Sep2002/b09132002_bt470-02.html

Department of the Air Force. (2002a). *Budget process. Deputy assistant secretary—budget.* Retrieved from http://www.saffm.hq.af.mil

Department of the Air Force. (2002b, November). *The planning, programming, and budgeting system and the air force corporate structure primer.* Retrieved from http://www.saffm.hq.af.mil

Department of the Army. (2002). *Budget process. Deputy assistant secretary of the army, budget office.* Retrieved from http://www.asafm.army.mil/budget/budget.asp

Department of Defense. (1998). *The road ahead: Accelerating the transformation of department of defense acquisition and logistics processes and practices for FY 1998.* Washington, DC: DoD.

Department of Defense. (1999). *Thirty second annual department of defense cost analysis symposium.* Retrieved from http://www.ra.pae.osd.mil/adodcas/slides/lpd_17.pdf

Department of Defense. (2000). *Defense Acquisition Regulations: DoD 5002.2R,* Washington, DC: DoD.

Department of Defense. (2000, February 7). *Defense Link News, Department of Defense Budget for FY2001* (News release). Washington, DC: DoD.

Department of Defense. (2002a). *Defense Acquisition Regulations: Part 217.* Washington, DC: DoD.

Department of Defense. (2002b). *Critical acquisition position description and DAC qualification requirements.* Washington, DC: DoD.

Department of Defense. (2003a). *Transformation planning guidance.* Washington, DC: DoD.

Department of Defense. (2003b). *News release. Office of the Secretary of Defense, 353-03.* Washington, DC: DoD.

Department of Defense. (2003c). *Greenbook: 2003: National Defense Budget Estimates for FY 2004. Office of the Principal Deputy Under Secretary of Defense Comptroller.* Washington, DC: DoD. Retrieved from www.dod.mil/comptroller /defbudget/fy2004

Department of Defense. (2003d). *Financial Management Modernization Program: Transition Plan Strategy Version 2.1. Program Management Office.* Washington, DC: DoD.

Department of Defense. (2003e, February). *Procurement Programs P-1: DoD Component Summary.* Washington, DC: DoD. Retrieved from http://www.dtic.mil/comptroller

Department of Defense. (2006). *Quadrennial Defense Review.* Washington DC: DoD.

Department of Defense. (2007, March). *Greenbook, FY2008. National defense budget estimates for FY2008.* Washington, DC: DoD.

Department of Defense. (2007). *FY 2007 Supplemental request for the Global War On Terror.* Washington, DC: Washington DC: DoD.

Department of Defense. (2007). *FY 2008 National Defense Authorization Act, Defense Acquisition Transformation Report to Congress.* Washington DC: DoD.

Department of Energy. (1996). *Budget execution manual: Legal bases for budget execution (DOEM 135.1-1).* Washington, DC: Office of the Chief Financial Officer, Budget Execution Branch, CR-131.

Department of the Navy. (1998). *Financial Guidebook for Commanding Officers* (NAVSO, P-3582: IV-1). Washington, DC: DoN.

Department of the Navy. (2001, December). *Budget Highlight Book, Office of Budget.* Washington, DC: DoN.

Department of the Navy. (2002a). *Operation and maintenance, Navy BSS4 base support, FY 2003 budget estimate submission. Exhibit OP-5.* Retrieved from http://navweb.secnav.navy.mil/pubbud/03pres/db_u.htm

Department of the Navy. (2002b). *Special Acquisition Considerations, Office of Budget.* Retrieved from http:/www.navweb.secnav.navy.mil

Department of the Navy. (2002c). *Budget Highlights Book, Office of Budget.* Washington, DC: DoN.

Department of the Navy. (2002d). *Department of the Navy Budget Guidance Manual.* Retrieved http://dbweb.secnav.navy.mil/guidance/bgm/bgm_frame_u.html

Department of the Navy Budget Office. (2003a). *FY03 Program Budget Decision 130.* Washington, DC: DoN.

Department of the Navy Budget Office. (2003b). *FY03 Program Budget Decision 721—Reduction in Total Ownership Costs Initiatives.* Washington, DC: DoN.

Department of the Navy. (2003c). *Exhibit: From National Security Strategy to Budget Execution* (PPBES Briefing Materials). Washington, DC: DoN.

Department of Navy. (2007, February). *FY2008 budget.* Washington, DC: DoN.

Department of the Navy. (n.d.). *Budget Manual.* Washington, DC: DoN.

Department of the Treasury. (2002). *Financial report of the U.S. Government 2001.* Washington, DC: DoT.

Dewey, D. R. (1968). *Financial history of the United States.* New York: Longmans, Green.

Dillard, J. T. (2003). *Toward centralized control of defense acquisition programs: A comparative review of the decision framework from 1987 to 2003* (Technical Report). Monterey, CA: Naval Postgraduate School.

Dillard, J. T. (2004, September). *Centralized control of defense acquisition programs: A comparative review of the framework from 1987-2003. Acquisition research program.* Monterey, CA: Naval Postgraduate School. Retrieved from http://www.acquisitionresearch.org/_files/FY2004/NPS-PM-04-021.pdf

Dillard, J. T. (2005). Controlling risk in defense acquisition programs: The evolving decision review framework. *International Public Management Review, 6*(2), 72-86. Retrieved from www.ipmr.net

Dillard, J. T. (2006). *When should you terminate your own program? Bad Business: The JASORS debacle. Acquisition research program.* Monterey, CA: Naval Postgraduate School. Retrieved April DAY?, YEAR?, from http://www.acquisitionresearch.org/_files/FY2006/NPS-PM-06-083.pdf

Dillard, J. T. (2007). Personal interview with authors, September 27.

Dillard, J. T., & Ford, D. N. (2007, May 16-17). *Too little too soon? Modeling the risks of spiral development.* Paper presented at the 4th annual Acquisition Research Symposium of the Naval Postgraduate School, Monterey, CA.

Dixit, A. (2002). Incentives and organizations in the public sector: An interpretative review. *Journal of Human Resources, 37*(4), 696-727.

Douglas, J. (2002). *Statement on defensive acquisition spending.* Washington, DC: Aerospace Industries Association.

Doyle, R., & McCaffery, J. L. (1991). The Budget Enforcement Act of 1990—The path to no-fault budgeting. *Public Budgeting and Finance, 20*(2), 25-41.

Doyle, R., & McCaffery, J. L. (1992). The Budget Enforcement Act after one year. *Public Budgeting and Finance, 21*(2), 3-22.

Doyle, R. (2003). *Budgeting for Defense in Democracies. International Defense Acquisition Resource Management Program.* Unpublished manuscript Monterey, CA: Naval Postgraduate School.

Dreher, R. (2007, April 2). War wrecking National Guard. *Monterey County Herald,* A-8.

Duma, D. (2001). *A cost estimation model for CNAP TACAIR aviation depot level repair costs.* Master's thesis. Monterey, CA: Naval Postgraduate School.

Eisler, P. (2007, September 4). House staffer sped up Humvee steel. *USA Today,* 2.

England, G. (2005, October 19). *Deputy Secretary Of Defense, Memorandum For Service Secretaries, CJCS, VCJS, and Service Chiefs, Subject: FY07 Budget.* Washington, DC: DoD.

England, G. (2005, June 7). *Deputy Secretary of Defense. Testimony to the Senate Armed Services Committee.* Washington, DC: U.S. Senate.

Evans, A. (2006). *Long-term military contingency operations: Identifying the factors affecting budgeting in annual or supplemental appropriations.* Master's thesis. Monterey, CA: Naval Postgraduate School.

Evans, P. B., & Wurster, T. S. (1997, September-October). Strategy and the New Economics of Information. *Harvard Business Review,* 71-82.

Farrell, L. D. (2002). *Statement by the President of the National Defense Industries Association.* Washington, DC: NDIA.

Faykes, J. A. (2007, February 5). *Director USAF Budget. FY08 President's Budget: We Are America's Airmen* (Press Brief). Washington, DC: USAF.

Feaver, P. (1996). The civil-military problematique: Huntington, Janowitz and the question of civilian control. *Armed Forces and Society, 23*(2), 42-61.

Federal Acquisition Regulations. (2000). Washington, DC: U.S. Government Printing Office.

Federal Register. (1996). September, 27. U.S. Government Printing Office. Washington, DC: U.S. GPO.

Feinstein, D. (2003, September). *Senator Feinstein's votes: Defense. news from Senator Diane Feinstein of California.* Retrieved from http://feinstein.senate.gov/votes/defense.htm

Feltes, L. A. (1976, January-February). Planning, programming, and budgeting: A search for a management philosopher's stone. *Air University Review.* www.airpower.maxwell.af.mil/airchronicles/ aureview/1976/jan-feb/feltes.html

Fenno, R. (1966). *The power of the purse.* Boston: Little, Brown.

Fesler, J. W. (1982). *American public administration: Patterns of the past.* Washington, DC: American Society for Public Administration.

Fierstine, K. (2004). *Investigating incompatibilities among the PPBE: Defense acquisitions and the defense requirements setting process.* Master's thesis. Monterey, CA: Naval Postgraduate School, Graduate School of Business and Public Policy.

Fisher, L. (1975). *Presidential spending power.* Princeton, NJ: Princeton University Press.

Fountain, J. (2001a). *Building the virtual state.* Washington, DC: Brookings.

Fountain, J. (2001b). Public sector: Early stage of a deep transformation. In R. Litanand & A. Rivlin (Eds.), *The economic payoff from the Internet revolution* (pp. 235-268). Washington, DC: Brookings.

Fox, R. (1988). *The defense management challenge.* New York: Harper & Row.

Fox, R. (2006, April 14). *General, Deputy Commandant and Director of the Marine Corps Programs Division: Investment & Resourcing Information and Strategies; Past, Present & Future* (Press brief).

Fox, R., with Field, J. L. (1988). *The defense management challenge: Weapons acquisition.* Boston: Harvard Business School Press.

Frenzel, B. (2005). *Budgeting in Congress: How the budget process functions. Testimony before the U.S. House of Representatives, Budget Committee.* Washington, DC: U.S House of Representatives.

Friedman, S. (2001). *Transforming department of defense financial management—A strategy for change. Report to the secretary of defense.* Retrieved from www.dod.mil/ news/jul2001/d20010710finmngt.pdf

Friedman, S. (2002, June 4). *Testimony to the Subcommittee on National Security, Veterans Affairs, and International Relations of the House Government Reform Committee.* Washington, DC: U.S. House of Representatives.

Friedman, T. L. (2007, March 8). U.S. soldiers shortchanged. *Monterey County Herald*, A-8.

Gaebler, T., & Osborne, D. (1993). *Reinventing government.* New York: Penguin Books.

Gailey, C., Reig, R., & Weber, W. (1995). *A study of the relationship between initial production test articles used in a system development program and the success of the program* (Technical Report TR2-95). Fort Belvoir, VA: Defense System Management College Press.

Gansler, J. S. (1989). *Affording defense.* Cambridge, MA: MIT Press.

Gansler, J. S. (2007). *Desired characteristics of the 21st century defense industrial base. Acquisition research program.* Monterey, CA: Naval Postgraduate School. Retrieved May 16, 2007, from http://www.acquisitionresearch.org/_files/ FY2007/NPS-AM-07-056.pdf

Gansler, J. S., & Lucyshyn, W. (2005). *A strategy for defense acquisition research acquisition research program.* Monterey, CA: Naval Postgraduate School Retrieved from http://www.acquisitionresearch.org/_files/FY2005/UMD-AM-05-021.pdf

Garfield, J. (2001). *Statement by the Chairman of the House Appropriations Committee in 1879, in Congressional Budget Office, Supplemental Appropriations in the 1990s.* Washington, DC: Congressional Budget Office.

Garvin, D. (1993, July-August). Building a learning organization. *Harvard Business Review*, 5-17.

General Accounting Office. (1984). *Improper use of industrial funds by defense extended the life of appropriations which otherwise would have expired* (AFMD-84-34). Washington, DC: GAO.

General Accounting Office. (1991). *Principles of federal appropriations law* (2nd ed.). Washington, DC: GAO.

General Accounting Office. (1994a). *The C-17 Program Update and Proposed Settlement: Military Update* (GAO/T-NSIAD-94-166). Washington, DC: GAO.

General Accounting Office. (1994b). *Acquisition reform: Implementation of Title V of the Federal Acquisition Streamlining Act of 1994* (GAO/NSIAD-97-22BR). Washington, DC: GAO

General Accounting Office. (1994c). *Weapons acquisition: Low rate initial production used to buy weapon systems prematurely* (GAO/NSIAD-95-18). Washington, DC: GAO.

General Accounting Office. (1996a). Managing for results: Achieving GPRA's objectives requires strong congressional role, testimony (GAO/T-GGD-96-79). Washington, DC: GAO.

General Accounting Office. (1996b). *Defense infrastructure: Budget estimates for 1996-2001 offer little savings for modernization* (GAO/NSIAD-96-131). Washington, DC: GAO.

General Accounting Office. (1997a). *Managing for results: Analytic challenges in measuring performance* (HEHS/GGD-97-138). Washington, DC: GAO.

General Accounting Office. (1997b). *Defense weapon system acquisition problems persist* (GAO/HR-97-6) Washington, DC: GAO.

General Accounting Office. (1997c). *DoD budget: Budgeting for operation and maintenance activities* (GAO/T-NSIAD-97-222). Washington, DC: GAO.

General Accounting Office. (1997d). *Defense weapon systems acquisition* (HR-97-6). Washington, DC: GAO.

General Accounting Office. (1997e). High risk areas: Eliminating underlying causes will avoid billions of dollars in waste (GAO/T-NSIAD/AIMD-97-143). Washington, DC: GAO.

General Accounting Office. (1997f). *The Government Performance and Results Act: 1997 Government-wide implementation will be uneven.* Washington, DC: Author.

General Accounting Office. (1998). *Managing for results: Measuring program results under limited federal control* (GGD-99-16). Washington, DC: GAO.

General Accounting Office. (1999a). *Managing for results: Opportunities for continued improvements in agencies' performance plans* (GGD/AIMD-99-215). Washington, DC: GAO.

General Accounting Office. (1999b). *Increased accuracy of budget outlay estimates* (GAO/AIMD-99-235R). Washington, DC: GAO.

General Accounting Office. (2000a). *Managing for results: Views on ensuring the usefulness of agency performance information to Congress* (GGD-00-35). Washington, DC: GAO.

General Accounting Office. (2000b). *Managing for results: Challenges agencies face in producing credible performance information* (GGD-00-52). Washington, DC: GAO.

General Accounting Office. (2000c). *Future years defense program: Risks in operation and maintenance acquisition programs* (GAO-01-33). Washington, DC: GAO.

General Accounting Office. (2000d). *FY 2000 contingency operations costs and funding audit* (GAO/NSIAD-00-168). Washington, DC: GAO.

General Accounting Office. (2001a). *Defense infrastructure: Budget estimates for 1996-2001 offer little savings for modernization.* Washington, DC: GAO.

General Accounting Office. (2001b). *Defense inventory: Information on the use of spare parts funding* (GAO-01-472). Washington, DC: GAO.

General Accounting Office. (2001c). *Navy inventory: Parts shortages are impacting operations and maintenance effectiveness* (GAO-01-771). Washington, DC: GAO.

General Accounting Office. (2001d). *Defense systems acquisitions.* Washington, DC: GAO.

General Accounting Office. (2001e). *Major management challenges and program risks: Department of defense* (GAO/01-244). Washington, DC: GAO.

General Accounting Office. (2001f). *High risk update* (GAO-01-263). Washington, DC: GAO.

General Accounting Office. (2002a). *Weapons systems support* (GAO-02-306). Washington, DC: GAO.

General Accounting Office. (2002b). *DoD high risk areas: Eliminating underlying causes will avoid billions of dollars in waste.* Washington, DC: GAO.

General Accounting Office. (2002c). *DOD financial management: Important steps underway but reform will require a long term commitment* (GAO-02-784-T). Washington, DC: GAO.

General Accounting Office. (2002d). *Defense acquisitions navy needs a plan to address rising prices in aviation parts* (GAO-02-565). Washington, DC: GAO.

General Accounting Office. (2002e). *DoD financial management: Integrated approach, accountability, transparency, and incentives are keys to effective reform.* Washington, DC: GAO.

General Accounting Office. (2002f). *Canceled DOD appropriations: Improvements made but more corrective actions are needed.* Washington, DC: GAO.

General Accounting Office. (2002g). *DoD financial management: Important steps underway but reform will require a long term commitment* (GAO-02-784-T). Washington, DC: GAO.

General Accounting Office. (2002h). *OMB leadership critical to making needed enterprise architecture and e-government progress* (GAO-02-389T). Washington, DC: GAO.

General Accounting Office. (2003a). *Defense inventory* (GAO-03-18). Washington, DC: GAO.

General Accounting Office. (2003b). Defense Systems Acquisitions (GAO-03-150). Washington, DC: GAO.

General Accounting Office. (2003c). *Low rate initial production used to buy weapon systems prematurely.* Washington, DC: GAO.

General Accounting Office. (2003d). *Federal procurement: Spending and workforce trends* (GAO-03-443). Washington, DC: GAO.

General Dynamics Corporation. (n.d). *General Dynamics awarded $3.2 billion contract to construct six new DDG 51-Class Destroyers* (Press Release). Washington, DC: Author.

Gibbons, R. (2003). Team theory, garbage cans and real organizations: Some history and prospects of economic research on decision-making in organizations. *Industrial and Corporate Change, 12*(4), 753-787.

Giddens, A. (1984). *Structuration theory.* Cambridge, MA: Harvard University Press.

Global Security. (2002). *AAAV specifications: Lifecycle costs.* Alexandria VA: Author.

Godek, P. (2000). *Emergency supplemental appropriations: A department of defense perspective.* Master's thesis. Monterey, CA: Naval Postgraduate School.

Government Accountability Office. (2005a). *High risk program update* (GAO-05-207). Washington, DC: GAO.

Government Accountability Office. (2005b). *Improved management practices could help minimize cost growth in navy shipbuilding programs* (GAO-05-183). Washington, DC: GAO.

Government Accountability Office. (2005c). *Capital financing: Potential benefits of capital acquisition funds can be achieved through simpler means.* Retrieved from http://www.gao.gov/new.items/d05249.pdf

Government Accountability Office. (2007). *Defense business transformation: Achieving success requires a chief management officer to provide focus and sustained leadership* (GAO-07-1072). Washington, DC: GAO.

Government Management Reform Act of 1994. (2003). Retrieved from www.npr.gov/npr/library/misc/s2170.html

Government Performance and Results Act of 1993 (2003). Retrieved from www.doi.gov/gpra

Granetto, P. J. (2003). *Statement to the Subcommittee On Government Efficiency And Financial Management of the House Government Reform Committee*. Washington, DC: Defense Financial Auditing Service, Office of the Inspector General of the Department of Defense.

Greene, J. B. (2007, October 9). DDG-51, e-mail to authors.

Greenspan, A. (2002, February 5). *Testimony to the Senate Budget Committee*. Washington, DC: U.S. Senate.

Grossman, E. M. (2002). *House Panel irate at Pentagon sluggishness in cleaning up accounting*. Washington, DC: Inside The Pentagon.

Gulati, R. (1998). Alliances and networks. *Strategic Management Journal, 19*(4), 293–317.

Gullo, T. A. (1998). *How states budget and plan for emergencies: Testimony to the House Budget Committee*. Washington, DC: Congressional Budget Office.

Hadley, R. T. (1988). Military coup: The reform that worked: The Goldwater-Nichols Military Reform Act of 1986. *The New Republic, 198*(3), 17-18.

Hagar, G. (1991a, April 20). Relevance of process: A tricky question on the hill. *Congressional Quarterly Weekly,* 962

Hagar, G. (1991b, May 31). The budget resolution: Appropriators come close to matching guidelines. *Congressional Quarterly Weekly,* 1360-1362.

Halpin, J. (2007, August 8). On Raptor's arrival, critics call it impractical. *Associated Pres.* Retrieved from www.f-16.net

Halperin, M. H., & Lomasney, K. (1999). Playing the add-on game in congress: the increasing importance of constituent interests and budget constraints in determining defense policy. In L. V. Sigal (Ed.), *The changing dynamics of U.S. defense spending* (pp. 85-106). Westport, CT: Praeger Press.

Hammer, M. (1990, July-August). Reengineering work: Don't automate, obliterate. *Harvard Business Review,* 104-112.

Hammer, M. (1996). *Beyond reengineering: How the process-centered organization is changing our work and our lives.* New York: Harper Business.

Hammer, M., & Champy, J. (1993). *Reengineering the corporation: A manifesto for business revolution.* New York: Harper Business.

Hammer, M., & Stanton, S. A., 1995. *The reengineering revolution: A handbook.* New York: Harper Business.

Hamilton, A. (1961). The Federalist Papers. In J. E. Cooke (Ed.), *The Federalist.* Middletown, CT: Wesleyan University Press.

Hamre, J. J. (1994, April 14). *Testimony to the House Armed Services Committee.* Washington, DC: U.S. House of Representatives.

Hardin, G. (1968). The tragedy of the commons. *Science, 16*(2,), 1243-1248.

Harrison, J. S., & St. John, C. H., (2002.). *Foundations in strategic management* (2nd ed.), Denver, CO: South-Western Press.

Hartung, W. D. (1999). The shrinking military pork barrel: The changing distribution of Pentagon spending, 1986-1996. In L. V. Sigal (Ed.), *The changing dynamics of U.S. defense spending* (pp. 29-84). Westport: Praeger Press.

Heclo, H. (1975). OMB and the Presidency: The problem of neutral competence. *Public Interest, 38*(Winter), 80-98.

Heifetz, R. A. (1993). *Leadership without easy answers.* Cambridge, MA: Harvard University Press.

Helicopter History Site. (2000). *Senate panel approves defense bill.* Retrieved from http://www.helis.com

Heniff, B., Jr., & Keith, R. (2004). Federal Budget Process Reform: A Brief Overview. CRS Report for Congress. CRS RS21752, Washington, DC. Updated 8 July, www.opencrs.com on June 29.

Heniff, B. (2005, January 25). *Congressional Budget Resolutions: Selected Statistics and Information Guide. CRS Report for Congress.* Washington, DC: CRS.

Henninger, D. (2007, February 8). The Snake Eater. Dow Jones Reprints from *The Wall Street Journal*, A14.

Hinricks, H., & Taylor, G. (Eds.). (1969). *Program budgeting and benefit-cost analysis.* Pacific Palisades, CA: Goodyear.

Hleba, T. (Ed.). (1999). *Practical financial management: A handbook of practical financial management topics for the DoD financial manager.* Monterey, CA: Naval Postgraduate School.

Hleba, T. (Ed.). (2002). *Practical financial management: A handbook of practical financial management topics for the DoD financial manager* (3rd ed.). Monterey, CA: Naval Postgraduate School.

Holcombe, R. (1998). Rainy day fund. In J. Shafritz (Ed.), *International Encyclopedia of Public Policy and Administration* (pp. 71-93). Boulder, CO: Westview Press.

Holmes, E. (2007, September 24). USAF head: Drawdown saving less than expected. *Defense News*, 8.

Hormats, R. (2007). *The price of liberty.* New York: Henry Holt.

House Government Reform Committee. (2001a, May 22). *Federal Government Acquisition. Subcommittee on Technology and Procurement of on Federal Government Acquisition.* Washington, DC: U.S. House of Representatives.

House Government Reform Committee. (2001b, March 7). *DoD Financial Management Reform, Hearing of the Subcommittee on National Security, Veterans Affairs, and International Relations.* Washington, DC: U.S. House of Representatives.

Hughes, T. (2003, July 15). Interview with former N-82. Monterey, CA: NPS.

Hughes, T. P. (1998). *Rescuing Prometheus: Four monumental projects that changed the world.* New York: Pantheon Books.

Huntington, S. P. (1957). *The soldier and the state: The theory and politics of civil-military relations.* Cambridge, MA: Harvard University Press.

Hyde, A. C. (1978). A review of the theory of budget reform. In A. Hyde & J. Shafritz (Eds.), *Government budgeting* (pp. 71-77). Oak Park, IL: Moore.

Intergraph Corporation. (1998). *Avondale alliance uses intergraph systems to deploy production integrated product data environment for LPD-17.* Retrieved from http://www.intergraph.com/press98/f_avon.htm

Intergraph Corporation. (2003). *Solutions for integrated data environments.* Retrieved from http://www.intergraph.com/solutions/profiles/documents/lpd17.pdf

Janowits, M. (1960). *The professional solider: A social and political portrait.* Glencoe, IL: Homewood.

Joint DoD/GAO Working Group on PPBS. (1983). *The Department of Defense Planning, Programming and Budgeting System.* Washington DC: DoD/GAO.

Jonas, T. (2002, June 4). *Defense Department Financial Management: Testimony by the Deputy Under Secretary of Defense for Financial Management to the House Government Reform Committee, Subcommittee on National Security, Veterans Affairs, and International Relations.* Washington, DC: U.S. House of Representatives.

Jones, D. C. (1982, March). Why the Joint Chiefs must change. *Armed Forces Journal International,* 64.

Jones, D. C. (1996, Autumn). Past organizational problems. *Joint Forces Quarterly,* 23-28. Retrieved from http://www.dtic.mil/doctrine/jel/jfq_pubs

Jones, L. R. (2001a, December 12). *UK Treasury use of performance measures.* Interview with Jeremy Jones, UK Treasury official, Rome, Italy.

Jones, L. R. (2001b, November 17). Management control origins. Interview with Robert Anthony, North Conway, New Hampshire.

Jones. L. R. (2002a, February 7). An update on budget reform in the U.S. *IPMN Newsletter, 2,* 1.

Jones, L. R. (2002b). IPMN Symposium on Performance Budgeting and the Politics of Reform: Analysis of Bush reforms in the U.S. *International Public Management Review, 3*(2), 25-41.

Jones, L. R. (2003). IPMN Symposium on Performance Budgeting and the Politics of Reform. *International Public Management Journal, 6*(2,) 219-235.

Jones. L. R., & Euske, K. (1991). Strategic misrepresentation in budgeting. *Journal of Public Administration Research and Theory, 3*(3), 37-52.

Jones, L. R., & Bixler, G. C. (1992). *Mission budgeting to realign national defense.* Greenwich, CT: JAI.

Jones, L. R., & McCaffery, J. L. (2005). Reform of PPBS and implications for budget theory. *Public Budgeting and Finance, 25*(3), 1-19.

Jones, L. R., & Thompson, F. (1999). *Public management: Institutional renewal for the 21st century.* New York: Elsevier Science.

Jones, L. R., Thompson, F., & Zumeta, W. (2001). Developing Relevant and Integrated Curricula in Public Management. *International Public Management Review, 2*(2), 23. Retrieved from http://www.ipmr.net

Jones, L. R., & Thompson, F. (2007). *From bureaucracy to hyperarchy in Netcentric and quick learning organizations.* Charlotte, NC: Information Age.

Jones, L. R., & Wildavsky, A. (1995). Budgetary control in a decentralized system: Meeting the criteria for fiscal stability in the European Union. *Public Budgeting and Finance, 14*(4), 7-21.

Jones, W. (1996). *Congressional involvement and relations: A guide for DoD acquisition managers.* Fort Belvoir, VA: Defense Systems Management College Press.

Joyce, P. G. (2002, October 10). *Federal budgeting after September 11th: A whole new ballgame or deja vu all over again?* Paper presented at the conference of the Association for Budgeting and Financial Management, Kansas City, MO.

Kanter, A. (1983). *Defense politics: A budgetary perspective.* Chicago: University of Chicago Press.

Kelly, M. (2007, September 28). How Alaska ferry project floated. *USA Today,* p. 10a.

Kickert, W., Klijn, J., & Koppenjan, J. (Eds.). (1997). *Managing complex networks: Strategies for the public sector.* London: Sage.

Klamper, A. (2005, March). Subcommittee voices dismay over defense supplemental spending. *Congress Daily,* 4.

Knox, B. (2002). *Ten years worth of procurement reforms with specific attention to selected DoN programs.* Master's thesis. Monterey, CA: Naval Postgraduate School, Department of Systems Management.

Korb, L. (1977). Department of defense budget process: 1947-1977. *Public Administration Review, 37*(4), 247-264.

Korb, L. (1979). *The rise and fall of the Pentagon.* Westport, CT: Greenwood Press.

Kosiak, S. (2007). *Historical and projected funding for defense: Presentation of the FY 2007 request in tables and charts.* Washington, DC: Center for Strategic and Budgetary Assessments. Retrieved from www.csbaonline.org

Kovacic, W. E. (1990). The sorcerer's apprentice: Public Regulation of the acquisitions process. In R. Higgs (Ed.), *Arms, politics, and the economy* (pp. 104-131). New York: Holmes & Meier.

Kozar, M. J. (1993). *An analysis of obligation patterns for the department of defense operations and maintenance appropriations.* Master's thesis, Monterey, CA: Naval Postgraduate School.

Kozar, M. J., & McCaffery, J. (1994). DoD O&M obligation patterns: some reflections and issues. *Navy Comptroller, 5*(1), 2-13.

Kucinich, D. J. (2001, March 7). *Hearing of the Subcommittee of the National Security, Veterans Affairs, and International Relations Committee of the House Government Reform Committee.* Washington, DC: U.S. House of Representatives.

Kucinich, D. J. (2002a, June 4). *Opening statement: Subcommittee on National Security, Veterans Affairs, and International Relations of the House Government Reform Committee.* Washington, DC: U.S. House of Representatives.

Kucinich, Dennis J., (2002b, June 25). *Hearing of The Subcommittee on National Security, Veterans Affairs and International Relations of the House Government Reform Committee.* Washington, DC: U.S. House of Representatives.

Kutz, G. (2002a, June 4). *Defense Department Financial Management, Director, GAO Financial Management and Assurance Team, Hearing of the National Security, Veterans Affairs, and International Relations Subcommittee of the House Government Reform Committee.* Washington, DC: U.S. House of Representatives.

Kutz, G. (2002b, June 25). *Defense Department Financial Management, Director, GAO Financial Management and Assurance Team, Hearing of the National Security, Veterans Affairs, and International Relations Subcommittee of the House Government Reform Committee.* Washington, DC: U.S. House of Representatives.

Kutz, G. (2003, June 25). *Testimony to the Subcommittee on Government Efficiency, and Financial Management of the House Government Reform Committee, Director, GAO Financial Management and Assurance Team.* Washington, DC: U.S. House of Representatives.

Labaree, L. W. (1958). *Royal government in America: A study of the British Colonial System before 1783.* New Haven, CT: Yale University Press.

Laird, M. R. (2003). *Melvin R. Laird, Secretary of Defense, January 22, 1969-January 29, 1973.* Retrieved from www.dod.mil/specials/secdef_histories/bios/laird.htm

Lanzillotta, L. (2002, June 4.). *Defense Department Financial Management, Testimony by the Principal Deputy Under Secretary of Defense, Comptroller, and Deputy Under Secretary of Defense for Financial Management Reform to the Subcommittee on National Security, Veterans Affairs, and International Relations Subcommittee of the House Government Reform Committee.* Washington, DC: U.S. House of Representatives.

Lanzillotta, L. (2003, June 25). *Statement to the Subcommittee on Government Efficiency and Financial Management of the House Government Reform Committee.* Washington, DC: U.S. House of Representatives.

LeLoup, L., & Moreland, W. (1978). Agency strategies and executive review: The Hidden politics of budgeting. *Public Administration Review, 38*, 232-239.

Lee R. D., & Johnson, R. (1983). *Public budgeting systems.* Baltimore: University Park Press.

Lewis, J. (2005, February 15). *Chairman, House Appropriations Committee, Report of the Oversight Plans of the House Committee on Appropriations.* Washington, DC: U.S. House of Representatives.

Library of Congress. (1997, November 6). *Congressional Record S11817-8.* Washington, DC: Congressional Research Service.

Library of Congress. (2000). *Exhibit: Defense Appropriation Bill in 2000—Passed on Time.* Retrieved from www.thomas.gov

Lieberman, R. (2002, June 4). *Testimony by the Deputy Inspector General, Department of Defense to the Subcommittee on National Security, Veterans Affairs, and International Relations of the House Government Reform Committee.* Washington, DC: U.S. House of Representatives.

Lindsay, J. M. (1987). Congress and defense policy: 1961-1986. *Armed Forces and Society 13*(3), 371-401.

Lindsay, J. M. (1990). Congressional oversight of the Department of Defense: Reconsidering the conventional wisdom. *Armed Forces and Society, 17*(Fall), 7-33.

Locher, J. R. (1996). Taking stock of Goldwater-Nichols. *Joint Forces Quarterly* (Autumn), 10-17. Retrieved from http://www.dtic.mil/doctrine/jel/jfq_pubs/index.htm

Locher, J. R. (2002). *Victory on the Potomac: The Goldwater-Nichols Act unifies the Pentagon.* College Station, TX: Texas A&M University Press.

Lorell, Mark A., & Graser, J. C. (1994). *An overview of acquisition reform cost savings estimates.* Washington, DC: Coopers & Lybrand.

LPD-17 News. (2002). Retrieved from www.lpd17.navsea.navy.mil/news/main.asp

Luttwak, E. (1982, February). Why we need more waste, fraud and mismanagement in the Pentagon. *Commentary, 73*, 17-20.

March, J. G. (1999). Understanding how decisions happen in organizations. In J. G. March (Ed.), *The pursuit of organizational intelligence.* Malden, MA: Blackwell.

Marks, R. A. (1989). *Program budgeting within the Department of Navy.* Master's thesis. Monterey, CA: Naval Postgraduate School.

Masten, S. E. (1984). The organization of production. *The Journal of Law and Economics, 27*(3), 403-417.

Masten, S. E., Meehan, J. W., & Snyder, E. A. (1991). The costs of organization. *The Journal of Law, Economics, and Organization, 7*(1), 1-25.

Mayer, K. R. (1993). Policy disputes as a source of administrative controls: Congressional micromanagement of the Department of Defense. *Public Administration Review, 53*(4), 293-302.

Mathews, W. (2007, March 26). U.S. House passes $124.3B supplemental. *Defense News, 4*

McCaffrey, J., & Godek P. (2003). Defense supplementals and the budget process. *Public Budgeting & Finance, 23*(3), 53-72.

McCaffery, J. L. (1994). Confidence, competence, and clientele: Norm maintenance in budget preparation. In A. Khan & B. Hildreth (Eds.), *Public budgeting and financial management casebook.* Boston: Kendall, Hunt.

McCaffery, J. L., & Jones, L. R. (2004). *Budgeting and financial management for national defense.* Greenwich, CT: Information Age.

McCaffery, J. L., & Mutty, J. (1999, Summer). The Hidden Process in Budgeting: Budget Execution. *Public Budgeting, Accounting, and Financial Management,* 233-258.

McCaffery, J. L., & Jones, L. R. (2001). *Budgeting and financial management in the federal government.* Greenwich, CT: Information Age.

McGill, M., & Slocum, J., Jr. (1994). *The smarter organization: How to build a company that learns, unlearns and adapts to capitalize on marketplace needs.* New York: Wiley.

McGrady, E. D. (1999). *Peacemaking, complex emergencies and disaster response: What happens, how do you respond?* Alexandria, VA: Center for Naval Analysis.

Mechling, J. (1999). Information Age governance: Just the start of something big? In E. Kamarck & J. Nye (Eds.), *Democracy.com? Governance in a Networked World* (pp. 169-191). Hollis, NH: Hollis.

Merewitz, L., & Sosnick, S. H. (1972). *The budget's new clothes.* Chicago: Markham.

Merle, R. (2005, April 19). *The F-22 Raptor. Interview, Washingtonpost.com.* Retrieved from www.washingtonpost.com/wp-dyn/articles/A37922-2005Apr8.html

Meyer, C. (1993). *Fast cycle time: How to align purpose, strategy, and structure for speed.* New York: The Free Press.

Meyers, R. (1997). Late appropriations and government shutdowns. *Public Budgeting and Finance, 17*(3), 23-42.

Meyers, R. (1994). *Strategic budgeting.* Ann Arbor: University of Michigan Press.

Meyers, R. (2002, October 10.). *Comments on the Federal Budget 2002.* Presentation at the Conference of the Association for Budgeting and Financial Management, Kansas City, MO.

Mihm, C. J. (2002a). *Testimony to the House Committee on Government Reform, Subcommittee on Government Management, Information and Technology, U.S. House of Representatives.* Washington, DC: GAO.

Michael, D. (1992). Governing By learning in an information society. In S. A. Rosell (Ed.), *Governing in an information society.* Montreal, Canada: Institute for Research on Public Policy.

Mihm, C. J. (2000). *Testimony to the House Committee on Government Reform, Subcommittee on Government Management, Information and Technology, U.S. House of Representatives.* Washington, DC: GAO.

Miller, D. (2006). *The fiscal blank check policy and its impact on Operation Iraqi Freedom*. Masters thesis. Monterey CA: Naval Postgraduate School.

Miller, J. (2005). *An analysis of the Sea Enterprise Program*. Masters thesis. Monterey CA: Naval Postgraduate School.

Miller, J. C. (1959). *Alexander Hamilton: Portrait in paradox*. New York: Harper.

Moe, T. M. (1984). The new economics of organization. *American Journal of Political Science, 28*(4), 739-777.

Moore, M. H. (1995). *Creating public value: Strategic management in government.* Cambridge, MA: Harvard University Press.

Morrison, B., Vanden B. T., & Eisler, P. (2007, September 4). When the Pentagon failed to buy enough body armor, electronic jammers and hardened vehicles to protect U.S. troops from roadside bombs In Iraq, Congress stepped in. *USA Today*, A11.

Mosher, F. C. (1954). *Program budgeting: Theory and practice.* New York: Public Administration Service.

Mosley, E. L. (2001). *Statement to the House Committee on Appropriations, Subcommittee on foreign operations, U.S. House of Representatives.* Washington, DC: USAID.

Morris, J. (2002, September 27). *Interview, CPF Comptroller.* Makalapa, HI: COMPACFLT.

Muradian, V. (2007, August 20). USAF struggles with budget shortfall. *Defense News*, 4.

Nadler, D., & Tushman, M. (1988). *Strategic organization design: Concepts, tools, and processes.* New York: Scott, Foresman.

National Defense Authorization Act for FY 2003. (2002). P.L.107-314, U.S. Congress.

Naval Fleet Requirements/Shipbuilding Policy. (2003). Retrieved from http://www.americanshipbuilding.com/init-NavyReqShip.html

Naval Postgraduate School. (2002). *MN-4159 PPBS Brief.* Monterey, CA: NPS.

NAVSEA News. (2002). Retrieved from www.lpd17.navsea.navy.mil/news/main.asp

Newberry, S. (2003). New Zealand's responsibility budgeting and accounting system and its strategic objective: A comment on Jones and Thompson 2002. *International Public Management Journal, 6*(1), 75-82.

Nissen, M. E. (2006). *Harnessing knowledge dynamics: Principled organizational knowing and learning* (Press release). New York: Idea Group.

Northrup Grumman Ship Systems. (1996). Retrieved from http://www.ss.northrupgumman.com/pressrelease/press.cfm

Novick, D. (Ed.). (1969). *Program budgeting.* New York: Holt.

Oberndorf, P., & Carney, D. (1998). *A summary of DoD COTS-related policies. SEI monographs on the use of commercial software in government systems.* Washington, DC: SEI

Obey, D., & Spratt, J. (2003, May 13). *Letter to the Speaker of the House of Representatives.* Washington, DC: U.S. House of Representatives.

O'Bryon, J. (2007, August 20). A promise that needs to be kept. *Defense News*, p. 53.

Office of Force Transformation. (2004). *Office of the Secretary of Defense.* Retrieved from http://www.oft.osd.mil

Office of Management and Budget. (2001a, November 28). *Statement of administration policy, Department of Defense Appropriations Bill for FY 2002.* Washington, DC: OMB.

Office of Management and Budget. (2001b, December 6). *Statement of Administration Policy, Department of Defense Appropriations Bill for FY 2002.* Washington, DC: OMB.

Office of Management and Budget. (2002a, February). *FY03 President's budget 2002, Exhibit P-27 LPD-17, ship production schedule.* Washington, DC: OMB.

Office of Management and Budget. (2002b). *Citizen's guide to the federal budget, FY 2003.* Washington, DC: OMB.

Office of Management and Budget. (2003a). *Budget of the United States Government FY 2004, analytical perspectives.* Washington, DC: Author.

Office of Management and Budget (2003b). *Budget of the United States Government FY 2004, Balances of Budget Authority.* Washington, DC: OMB.

Office of Management and Budget. (2003c). Mid-session review, summary, July 15. Retrieved from www.whitehouse.gov/omb/budget/fy2004 /summary.html#table1

Office of Management and Budget. (n.d.). *Circular A-34: Federal Budget Execution.* Washington, DC: OMB.

Office of Management and Budget. (2007). *Budget of the United States, FY 2008, Analytical perspectives.* Washington, DC: Government Printing Office.

Office of the Secretary of Defense. (2003, February). *Justification for FY 2004 Component Contingency Operations and the OCOTF.* Washington, DC: OSD.

Office of the Chief of Naval Operations. (1999). *N6 PPBS online tutorial.* Retrieved from http://cnon6.hq.navy.mil/N6E/PPBS/ppbs_process.htm

Office of the Chief of Naval Operations. (2002a, March 8). *N801E Memorandum: Mid-year review of the FY 2002 O&MN appropriation.* Washington, DC: Author.

Office of the Chief of Naval Operations. (2002b, October 29). *N801E memorandum: Program review support capability procedures.* Washington, DC: DoN.

O'Toole, L., & Meier, K. (2004). Public management in intergovernmental networks: Matching structural networks and managerial networking. *Journal of Public Administration Research and Theory, 14*(4), 469-494.

Overby, B. (2002, September 26). *Interview: N80 Programming and IWARs.* Makalapa, HI: COMPACFLT.

Owens, M. T. (1990). Micromanaging the defense budget. *The Public Interest, 100*(Summer), 131-146.

Pacific Command. (2002, October). *Area of responsibility,* p. 1. Retrieved from http:/ /www.pacom.mil/pages/siteindex.htm

Pacific Fleet. (2000, June 26). *Instruction 5400.3q Staff Regulations.* Makalapa, HI: DoN.

Pacific Fleet. (2001, July). *Mission statement.* Retrieved from www.cpf.navy.mil/facts/ mission.

Pacific Fleet. (2002a, November). *BSS4 base support FY2003 budget estimate submission: Exhibit OP-5.* Makalapa, HI: COMPACFLT.

Pacific Fleet. (2002b, October 3). *COMPACFLT Comptroller Brief: N00F1.* Makalapa, HI: COMPACFLT.

Pacific Fleet. (2002c, September 17). *COMPACFLT Program and Budget Overview: N001F1*. Makalapa, HI: COMPACFLT.

Pacific Fleet. (2002d, June 26). *Issue 65079*. Makalapa, HI: DoN.

Pallot, J. (1998). The New Zealand Revolution. In O. Olson, J. Guthrie, & C. Humphrey (Eds.), *Global warning: Debating international developments in new public financial management* (pp. 156-184). Bergen: Cappelen Akademisk Forlag.

Pearson, N. M. (1943). The budget bureau: From routine business to general staff. *Public Administration Review, 3,* 126.

Peckman, J. A. (1983). *Setting National Priorities: The 1984 budget*. Washington, DC: The Brookings Institution.

Pedlar, M., Burgoyne, J., & Boydell, T. (1987). *The learning company*. San Francisco: Jossey-Bass.

Perry, W. (1996, March). *Annual report to the President and Congress by William J. Perry, Secretary of Defense*. Washington, DC: USGPO.

Peters, T. (1986, April 28). What gets measured gets done. *Tribune Media Services,* 1.

Pfeffer, J. (1998). *The human equation: Building profits by putting people first*. Boston: Harvard Business School Press.

Philips, W. E. (2001). *Flying hour program cash management at Commander Naval Air Forces Pacific*. Master's thesis. Monterey, CA: Naval Postgraduate School.

Pitsvada, B. (1983). Federal budget execution. *Public Budgeting and Finance, 3*(2), 83-101.

Platts, T. R. (2003, June 25). *Statement to the Subcommittee on Government Efficiency, and Financial Management of the House Government Reform Committee, U.S. House of Representatives*. Washington, DC: U.S. Congress.

Posner, P. (2002, October 10.). *Performance-Based Budgeting: Current Developments and New Prospects*. Paper presented at the conference of the Association for Budgeting and Financial Management, Kansas City, MO.

Powell, F. W. (1939). *Control of federal expenditures*. Washington, DC: The Brookings Institution.

Puritano, V. (1981, August). Streamlining PPBS. *Defense*, 20-28.

Quadrennial Defense Review Report. (2001). Retrieved from www.defenselink.mil/pubs/qdr2001.pdf

Quadrennial Defense Review. (2006, February). Washington, DC: DoD.

Quinn, J. B. (1992). *Intelligent enterprise: A knowledge and service based paradigm for industry*. New York: Free Press.

Ramnath, R., & Landsbergen, D. (2005). IT-enabled sense-and-respond strategies in complex public organizations. *Communications of the ACM, 48*(5), 58-64.

Ratnam, G. (2005, December 7). *DoD should split procurement, R&D From Other Defense Spending*. Retrieved from DefenseNews.com

Reed, J. E. (2002). *Budget preparation, execution and methods at The Major Claimant/ Budget Submitting Office Level*. Master's thesis. Monterey CA: Naval Postgraduate School.

Renae, M. (2005, April 19). *Interview at Washington.com*. Retrieved September, 7, 2007, from www.washingtonpost.com/wp-dyn/articles/A37922-2005Apr8.html

Republican Party Committee. (2005, April 12). *Congress should fund the war with "emergency" spending.* Washington, DC: United States Senate.

Reynolds, G. K. (2000, November 21). *Defense Authorization and Appropriation Bills: A chronology.* Washington, DC: CRS.

Riedl, B. M. (2003). *Ten guidelines for reducing wasteful government spending. Heritage Foundation Backgrounder.* Washington, DC: Heritage Foundation.

Rieg, R. W. (2000, Winter). Baseline acquisition reform. *Acquisition Review Quarterly,* 23-26.

Roberts, L. (1996, December). Shalikashvili Grades Goldwater-Nichols Progress. *Armed Forces Press Service News,* 1-3. Retrieved from www.dod.mil/news/Dec1996/n12181996_9612182.html

Roberts, N. C. (2001). Coping with wicked problems: The case of Afghanistan. In L. R. Jones et al. (Eds.), *Learning from international public management reform* (Vol. 2, pp. 353-376). London: Elsevier Science.

Robinson, D. (2006, February 15). *Democrats want more accountability for Iraq spending.* Washington, DC: Capitol Hill.

Rodriquez, J. (1996). Connecting resources with results. *Budget and Finance, 16*(4), 2-4.

Rogers, D. (2007, March 3). Democrats bill to fund surge, within limits. *Wall Street Journal,* A2.

Rose, J. H. (1911a). *William Pitt and national revival.* London: G. Bell.

Rose, J. H. (1911b). *William Pitt and the Great War.* London: G. Bell and Sons.

Rose, J. H. (1912). *Pitt and Napoleon.* London: G. Bell.

Rumsfeld, D. H. (2001). *Concurrent defense program and budget review. Memorandum to secretaries of the military departments.* Washington, DC: DoN.

Rumsfeld, D. H. (2003). *In elements of defense transformation 2004.* Washington, DC: Department of Defense, Office of Defense Transformation.

Rumsfeld, D. H. (2002, September 11). Interview by Thelma LeBrecht. Washington, DC: Associated Press Wire Service.

Rumsfeld, D. H. (2003, May 22). Taking exception: Defense for the 21st century. *The Washington Post,* 35.

Savas, E. S. (2000). *Privatization and public-private partnerships.* New York: Chatham House.

Scardaville, M. (2002). *Congress must reform its committee structure to meet Homeland Security needs: Executive memorandum #823.* Washington, DC: Heritage Foundation. Retrieved from www.heritage.org/Research/HomelandDefense/EM823.cfm

Schakowsky, J. D. (2002, June 25). *Statement to the Hearing of the Subcommittee on National Security, Veterans Affairs, and International Relations of the House Government Reform Committee.* Washington, DC: U.S. House of Representatives.

Schatz, J. J. (2005, February 28). Urgency driving army line item. *CQ Weekly,* 510.

Schick, A. (1966). The road to PPB: The stages of budget reform. *Public Administration Review, 26,* 243-258.

Schick, A. (1970). The budget bureau that was: Thoughts on the rise, decline and future of a Presidential agency. *Law and Contemporary Problems, 35*(Summer), 519-539.

Schick, A. (1973). A death in the bureaucracy: The demise of federal PPB. *Public Administration Review, 33,* 146-156.

Schick, A. (1980). *Congress and money: Budgeting, spending, and taxing.* Washington, DC: Urban Institute.

Schick, A. (1982). *Reconciliation and the congressional budget process.* Washington, DC: American Enterprise Institute.

Schick, A. (1990). *The capacity to budget.* Washington, DC: The Urban Institute Press.

Schick, A. (2005). *Testimony before the U.S. House of Representatives, Budget Committee.* Washington, DC: U.S. House of Representatives.

Schlecht, E. V. (2002, November 6). Dodging pork missiles. *National Review Online,* 2.

Schlesinger, R. (2003, September 1). Rumsfeld, army leaders in discord. *Boston Globe,* 1.

Schmitt, E. (2004, March 18). Study urges reorganization to streamline the Pentagon. *The New York Times,* A30.

Scott, R. (2002, September 27). *Interview: CPF Aviation Budget.* Makalapa, HI: COMPACFLT

Secretary of Defense. (1990, January 17-26.). *Interviews by authors with DoD comptroller officials.* Washington, DC: DoD.

Secretary of Defense. (2002a). *Specifications and standards: A new way of doing business.* Washington, DC: DoD.

Secretary of Defense. (2002b, April). *Defense Planning Guidance Study #20.* Washington, DC: Dod.

Secretary of Defense. (2002c). *Acquisition program baselines.* Washington, DC: DoD.

Secretary of Defense. (2003a, May 22). *Management Initiative Decision 913.* Washington, DC: Dod.

Secretary of Defense. (2003b). *National defense budget estimates. DoD office of the comptroller.* Washington, DC: Dod.

Secretary of Defense. (2007, July). *Defense Acquisition Transformation. Report to Congress: John Warner National Defense Authorization Act, Fiscal Year 2007, Section 804.* Washington, DC: Dod.

Secretary of the Navy. (1990, January 18). *Interview by authors with navy comptroller official.* Washington, DC: DoN.

Secretary of the Navy. (2001-2003). *Interviews with navy FMB officials, office of the assistant secretary of the navy, financial management and comptroller.* Washington, DC: DoN.

Secretary of the Navy. (2002a). *Department of the Navy Budget Guidance Manual.* Retrieved from http://dbweb.secnav.navy.mil/guidance/bgm/bgm_frame_u.html

Secretary of the Navy. (2002b, April). *Instruction 7000.27.* Washington, DC: DoN.

Secretary of the Navy. (2002c, April 8). *SECNAV Instruction 7000.27: Comptroller organizations.* Washington, DC: DoN.

Secretary of the Navy. (2002d, April 4.). *FY 2004/2005 DoN budget review process. Office of the Assistant Secretary of the Navy, Financial Management and Comptroller.* Washington, DC: DoN.

Secretary of the Navy. (2002e, March 13). Interview with navy FMB official, Office of the Assistant Secretary of the Navy, Financial Management and Comptroller. Office of Budget. Washington, DC: DoN.

Secretary of the Navy. (2002f). *Navy Budget Manual Instruction 7102.2a.* Washington, DC: Department of Navy.

Secretary of the Navy. (2003, January). *Interviews with navy FMB officials.* Washington, DC: DoN.

Seiko, D. T. (1940). *The federal financial system.* Washington, DC: The Brookings Institution.

Selinger, M. (2002, February 15). F/A-18E/F, C-130J, *Helicopters Could Get Increasesin Congress. Aerospace Daily,* 201-232.

Senge, P. (1990). *The fifth discipline: The art & practice of the learning organization.* New York: Doubleday/Currency.

Shalal-Esa, A. (2005, February 9). *Pentagon plays games with war funding requests.* Washington, DC: Capitol Hill Blue.

Sharkansky, I. (1969). *The politics of taxing and spending.* New York: Bobbs Merril.

Shays, C. (2001, March 7). *Statement to the House Committee on National Security, Veterans Affairs, and International Relations of the House Government Reform Committee.* Washington, DC: U.S. House of Representatives.

Shays, C. (2002, June 25). *Statement to the Hearing of the Subcommittee on National Security, Veterans Affairs and International Relations of the House Government Reform Committee.* Washington, DC: U.S. House of Representatives.

Shishido, G. (2002, September 27). *Interview.* Makalapa, HI, COMPACFLT.

Skarin, J. W. (2002, December). *The horizon of financial management for the department of defense.* Master's thesis. Monterey CA: Naval Postgraduate School.

Smith, H. (1988). *The power game: How Washington works.* New York: Random House.

Smithies, A. (1955). *The budgetary process in the United States.* New York: McGraw-Hill.

Snook, J. S. (1999, December). *An analysis of the Planning, Programming, and Budgeting System: PPBS processes of the military services within the Department of Defense.* Master's thesis, Monterey, CA: Naval Postgraduate School.

Solis, T. (2004). Acquisition for the Warfighter. Power Point Presentation. Washington, DC. In K. Fierstine & L. R. Jones (Ed.), *Sources of discontinuity in the PPBES and the Defense Acquisitions Decision Processes* (Working Paper). Monterey CA: Naval Postgraduate School.

Spinney, F. (2002, June 4). *Defense Department Financial Management. Testimony by Tactical Air Analyst, Department of Defense to the Subcommittee on National Security, Veterans Affairs, and International Relations Subcommittee of the House Government Reform Committee.* Washington, DC: DoD.

Steinhoff, J. (2001, March 7). *Testimony before the Subcommittee on National Security, Veterans Affairs, and International Relations of the House Government Reform Committee.* Washington, DC: U. S. House of Representatives.

Stevenson, R. W. (2002, February 2). Bush budget links dollars to deeds with new ratings. *New York Times,* A12.

Stockman, D. (1986). *The triumph of politics.* New York: Avon Books.

Streeter, S. (1999, January 11). Earmarks and limitations in appropriations bills. *CRS Report for Congress*, 98-518.

Streeter, S. (2005). The congressional appropriations process: An introduction. *CRS Report for Congress*, 97-684.

Sullivan, B. (2002, June 25). *Joint purchase card program management, testimony to the Subcommittee on National Security, Veterans Affairs, and International Relations of the House Government Reform Committee*. Washington, DC: U.S. House of Representatives.

Sullivan, R. (Ed.). (2002). *Resource allocation: The formal process* (8th ed.). Newport, RI: Naval War College.

Taft, W. H. (1912, January 27). President of the United States, message on economy and efficiency in the government service, H. Doc. 458 62-2.

Taylor, B. (2002). *An analysis of the departments of the air force, army, and navy budget offices and budget processes*. Master's thesis, Naval Postgraduate School, Monterey, CA.

Taylor, L. B. (1974). *Financial management of the Vietnam conflict*. Washington, DC: Department of the Army.

Tirpak, J. A. (2003, February). Legacy of the air blockades. *Air Force Magazine*, pp. 46-52.

Thompson, F., & Jones, L. R. (1994). *Reinventing the Pentagon*. San Francisco: Jossey-Bass.

Thompson, J. D. (1967). *Organizations in action*. New York: McGraw-Hill.

Thomson, M. A. (1938). *A constitutional history of England* (Vol. IV). London: Methuen.

Tierney, J. F. (2002, June 4). *Statement to the hearing of the Subcommittee on National Security, Veterans Affairs, and International Relations of the House Government Reform Committee*. Washington, DC: U.S. House of Representatives.

A triumph for Pelosi: Review and outlook. (2007, March 24). *Wall Street Journal*, A-10.

Tyszkiewicz, M., & Daggett, S. (1998). *A defense budget primer*. Washington, DC: CRS.

U.S. Air Force. (2002). *The planning, programming, and budgeting system and the Air Force corporate structure primer*. Retrieved from http://www.saffm.hq.af.mil/

U.S. Code 253a 1 A. USC.

U.S. Marine Corps. (2002, November). *Program objective memorandum 2004 guide*. Retrieved from http://www.usmcmccs.org/Director/POM/home.asp

U.S. Senate, Budget Committee. (1998, August). Overview of appropriations and the budget process. *Senate Budget Committee*, 5.

U.S. Senate, Committee on Appropriations. (2003, July 19). FY 2004 Section 302b Allocations (Press Release).

U.S. Senate Armed Services Committee. (2005a, February 17). Hearing on the Defense Authorization Request for Fiscal Year 2006 and the Future Years Defense Program.

U.S. Senate Budget Committee Bulletin. (2005b, July 28). Appropriations update. Republican committee staff to Chairman Judd Gregg.

Utt, R. (1999, April 2). *How congressional earmarks and pork-barrel spending undermine state and local decision making* (Heritage Foundation Backgrounder). Washington, DC: Heritage Foundation.

Utt, R. D., & Summers, C. B. (2002, March 15). *Can Congress be embarrassed into ending wasteful pork-barrel spending?* (Heritage Foundation Backgrounder). Washington, DC: Heritage Foundation.

Vanden Brook, T. (2007, October 3). Spy technology caught in military turf battle: Marines feared army would kill TiVo-like device. *USA Today*, 1.

Walker, David (2001, March 7). *Testimony by the comptroller general to the subcommittee of the National Security, Veterans Affairs, and International Relations Committee of the House Government Reform Committee*. Washington. DC: U.S. House of Representatives.

Walker, D. (2002, February 7). *Testimony by the Comptroller General to the House Committee on Government Reform, Subcommittee on Government Management, Information and Technology*. Washington, DC: U.S. House of Representatives.

Warren, David (2002, June 25). *Testimony by the director, Defense Capabilities and Management Team, GAO to the Subcommittee on National Security, Veterans Affairs, and International Relations of the House Government Reform Committee*. Washington, DC: U.S. House of Representatives.

Weick, K. E., & Sutcliffe, K. M. (2001). *Managing the unexpected: Assuring high performance in an age of Complexity*. San Francisco: Jossey-Bass.

Weisman, J. (2005, February 15). President requests more war funding. *The Washington Post*, A2.

Wheeler, W. (2002, December 9). *Mr. Smith is dead: No one stands in the way as Congress laces post-September 11 defense bills with pork*. Retrieved from http://www .d-n-i.net/fcs/spartacus_mr_smith.htm

White, J. (1985). Much ado about everything: Making sense of federal budgeting. *Public Administration Review, 45*, 623-630.

White, J. (2007, October 16). Pentagon submits budget, and services ask for more. *The Washington Post*, A1.

White, L. D. (1948). *The Federalists*. New York: Macmillan.

White, L. D. (1951). *The Jeffersonians*. New York: Macmillan.

White, L. D. (1954). *The Jacksonians*. New York: Macmillan.

White, L. D. (1958). *The Republican era*. New York: Macmillan.

Wildavsky, A. (1961). Political implications of budget reform. *Public Administration Review, 21*, 183-190.

Wildavsky, A. (1964). *The politics of the budgetary process*. Boston: Little, Brown

Wildavsky, A. (1975). *Budgeting: A comparative theory of budgetary processes*. Boston: Little, Brown.

Wildavsky, A. (1979). *The politics of the budgetary process*. Boston: Little, Brown.

Wildavsky, A. (1984). *The politics of the budgetary process*. Boston: Little, Brown.

Wildavsky, A. (1988). *The new politics of the budgetary process*. Glenview, IL: Scott, Foresman.

Wildavsky, A., & Caiden, N. (1997). *The new politics of the budgetary process*. New York: Addison-Wesley Longman.

Wildavsky, A., & Caiden, N. (2001). *The new politics of the budgetary process* (4th ed.). New York: Longman Press.

Wildavsky, A., & Hammann, A. (1956). Comprehensive versus incremental budgeting in the Department of Agriculture. *Administrative Sciences Quarterly, 10,* 321-346.

Wilmerding, L., Jr. (1943). *The spending power: A history of the efforts of Congress to control expenditures.* New Haven, CT: Yale University Press.

Wlezien, C. (1996, January). The President, Congress, and appropriations. *American Politics Quarterly, 24*(1), 62.

Williams, M. J. (2000, February). Resource allocation: A primer. *Marine Corps Gazette, 84,* 14-15.

Wood, G. S. (2006). *Revolutionary characters: What made the founding fathers different.* New York: Penguin Press.

Wolfowitz, P. (2003, May 6). The Defense Transformation Act for the 21st Century, statement prepared for the House Government Reform Committee. Washington, DC: DoD.

Wolfe, F. (2002, February 14). Seven ship build rate needed in FY03. *Defense Daily International,* 8.

Yoder, C. (2003). *Comments on acquisition reform.* Monterey, CA: Naval Postgraduate School.

Zakheim, D. (2003). *Revised PPBES process.* Washington, DC: Department of Defense, Office of the Comptroller.

ABOUT THE AUTHORS

L. R. Jones is George F. A. Wagner Professor of public management in the Graduate School of Business and Public Policy, NPS, Monterey, California. Professor Jones teaches and conducts research on a variety of public sector budgeting and financial and management issues. He received his BA degree from Stanford University and MA and PhD from the University of California, Berkeley. He has authored more than 100 journal articles and book chapters on topics including national and state budgeting and policy, management and budget control, public financial management, and government reform. Dr. Jones has published 15 books including *Mission Financing to Realign National Defense* (1992), *Reinventing the Pentagon* (1994), *Budgeting and Financial Management in the Federal Government* (2001), *Budgeting and Financial Management for National Defense* (2004), and *From Bureaucracy to Hyperarchy in Netcentric and Quick Learning Organizations* (2007). He was honored by the American Association for Budgeting and Finance in 2005 as recipient of the Aaron Wildavsky Award for Lifetime Scholarly Achievement in the Public Budgeting and Financial Management. Dr. Jones has consulted for a wide variety of national and international organizations and has served as president of the International Public Management Network.

Jerry L. McCaffery is emeritus professor of Public Budgeting in the Graduate School of Business and Public Policy at the Naval Postgraduate School where he taught for over 20 years. He has also taught at Indiana University and the University and worked as a budget analyst for the state of Wisconsin. He has a continuing interest in defense budgeting and financial management and budgeting at all levels of government. He is a past president of the American Society for Public Administration's section on budgeting and financial management.

INDEX